Nucleic Acids and Molecular Biology

For further volumes:
http://www.springer.com/series/881

John F. Atkins · Raymond F. Gesteland
Editors

Recoding: Expansion of Decoding Rules Enriches Gene Expression

Springer

Editors
John F. Atkins
BioSciences Institute
University College Cork
Ireland
and
Department of Human Genetics
University of Utah
and Genetics Department
Trinity College Dublin, Ireland
atkins@genetics.utah.edu

Raymond F. Gesteland
Department of Human Genetics
University of Utah
15N. 2030E.
Salt Late City
UT 84112-5330
USA
ray.gesteland@genetics.utah.edu

ISBN 978-0-387-89381-5 e-ISBN 978-0-387-89382-2
DOI 10.1007/978-0-387-89382-2
Springer New York Dordrecht Heidelberg London

Library of Congress Control Number: 2009938958

Printed on acid-free paper

Springer is part of Springer Science+Business Media (www.springer.com)

Foreword

The literature on recoding is scattered, so this superb book fills a need by providing up-to-date, comprehensive, authoritative reviews of the many kinds of recoding phenomena.

Between 1961 and 1966 my colleagues and I deciphered the genetic code in *Escherichia coli* and showed that the genetic code is the same in *E. coli*, *Xenopus laevis,* and guinea pig tissues. These results showed that the code has been conserved during evolution and strongly suggested that the code appeared very early during biological evolution, that all forms of life on earth descended from a common ancestor, and thus that all forms of life on this planet are related to one another. The problem of biological time was solved by encoding information in DNA and retrieving the information for each new generation, for it is easier to make a new organism than it is to repair an aging, malfunctioning one.

Subsequently, small modifications of the standard genetic code were found in certain organisms and in mitochondria. Mitochondrial DNA only encodes about 10–13 proteins, so some modifications of the genetic code are tolerated that probably would be lethal if applied to the thousands of kinds of proteins encoded by genomic DNA.

In 1986 the 21st amino acid, selenocysteine, which responds to the terminator codon, UGA, when a stem-loop structure in mRNA is downstream of the UGA codon and is recognized by a protein was discovered. In 2002 the 22nd amino acid, pyrrolysine, which responds to the terminator codon, UAG, was discovered. Pyrrolysine is found only in a few species of bacteria.

During the last 40 years a great deal of information has been obtained that shows that some mRNA molecules contain signals in addition to the 64 kinds of RNA codons that modify the translation of codons. These signals may involve intramolecular hydrogen bonding between nucleotides in mRNA such as the formation of hairpin-like stem-loop structures or pseudoknots, certain nucleotide sequences followed by mRNA secondary structure that delay codon translation, or hydrogen bonding between mRNA and ribosomal RNA of the translating ribosomes. These signals add considerable complexity to the translation of mRNA. For example, these signals can alter the reading frame of specific species of mRNA at specific sites within the partially translated mRNA. The signals can specify whether reading frame 1 should be changed to reading frame 2 or to reading frame 3 at specific codons during the translation of the mRNAs. The reading frame can be altered by skipping one nucleotide in the 3' direction, or by going back one nucleotide or two nucleotides in the 5' direction. There is also a mechanism that enables the ribosome to skip 50 bases. Another mechanism evolved that allows ribosomes to translate a specific species of mRNA to a certain point and then continue translation of different molecules of RNA.

Some remarkable and quite beautiful recoding mechanisms have been discovered that function as regulators of gene expression. For example, *E. coli* release factor 2 (RF2) mRNA contains near the beginning of the mRNA a slippery nucleotide sequence before the terminator codon, UGA, followed by a pseudoknot in the mRNA. When the concentration of RF2 protein is high, RF2 protein recognizes the UGA codon and terminates, i.e., aborts the synthesis of RF2 protein. However, when the concentration of RF2 protein is low, one base is skipped in the 3' direction resulting in a shift to reading frame 2 thus enabling the synthesis of full-length RF2 protein. Thus, a frameshift in mRNA translation is used to regulate the translation of RF2 mRNA. Programmed frameshifts are required for the translation of many species of viral RNA, including HIV. Programmed frameshifts also are involved in the translation of some species of mRNA derived from genomic DNA.

Many human genetic diseases have been found that result from mutations that convert a codon for an amino acid to a terminator codon that prematurely terminates the synthesis of the protein. One approach that has been explored is to treat these patients with small molecules such as the aminoglycoside, gentamicin, or other molecules that result in some misreading of codons. This enables premature terminator codons to be translated sometimes as amino acid codons thereby resulting in the synthesis of some full-length proteins.

Another approach that currently is being explored is the use of oligonucleotides that base pair with newly synthesized RNA and prevent defective regions of mRNA from being incorporated into mRNA via alternative splicing. If either approach is successful, many genetic diseases would be alleviated.

Many additional recoding phenomena are described in this book. The book will be useful to investigators in many fields, ranging from molecular biologists to clinical researchers who are interested in the genetic code, regulation of gene expression, or mechanisms of protein synthesis and codon translation.

Marshall Nirenberg

Preface

By 1966 the general nature of readout of the genetic code and codon identity had been established. What was not appreciated then was that decoding is dynamic. Decoding can be altered in an mRNA-specific manner and in a remarkable variety of ways.

The specific meaning of individual codons can be redefined in response to signals in an mRNA. Or a proportion of translating ribosomes can be diverted to a different reading frame at a specific site. And ribosomes can be directed to bypass a block of nucleotides or even to resume on a different mRNA. This book chronicles and analyzes these "recoding" phenomena both to understand the contribution they make to the complexity of gene expression and to understand the mechanisms involved, illuminating the features of ribosomes and mRNA.

These unusual genetic decoding events tell us that the readout of the code itself has been subject to the wiliness of selection, increasing the repertoire of ways to utilize the richness of information encoded in DNA or RNA. A coding sequence in mRNA can specify additional protein products not predicted from standard readout of the classical open reading frame. In some cases the recoding event is a control point for a regulatory circuit. In certain other cases, the key feature is specification of the "special" amino acids selenocysteine and pyrrolysine. Not surprisingly the world of viruses and small mobile chromosomal elements is rich with examples of recoding since their genomes are compact and every mechanism is used to maximize gene density. But, with one viral exception, the cases known so far of specification of the "special" amino acids are for cellular gene decoding.

Deciphering recoding has led to the realization that there is an extra layer of information in messenger RNA that can change the program for its own individual readout. These instructions include a site where the nonstandard decoding event occurs and an assortment of types of signals that greatly stimulate the proportion of ribosomes that perform the recoding event. These stimulatory signals can be 3' or 5' of the recoding site or both. The recoding signals located 3' can be nearby, or distant from the recoding site, and are often in the form of intra-mRNA structures (e.g., single stem-loops or pseudoknots) that somehow influence the ribosome. There are even translation factors that are specialized to specifically interact with some of these signals. Another set of signals involves mRNA pairing with the rRNA of translating ribosomes; in the established cases, the mRNA segment involved is 5' of the recoding site. Yet another signal can be a particular sequence of amino acids in the growing nascent peptide acting within the peptide exit tunnel of the translating ribosome. How the ribosome senses and responds to this variety of signals is still quite unclear but is now becoming amenable to study due to the major advances in knowledge of ribosome structure and an emerging understanding of ribosome conformational changes during the translation cycle.

Redefinition. Carboxy terminal extensions of proteins can be programmed when the meaning of a UAG or UGA stop codon is redefined so that a proportion of ribosomes accepts a near-cognate aminoacyl-tRNA, such as that charged with glutamine (for UAG) or tryptophan (for UGA) instead of a release factor. Translation then continues in the zero frame to synthesize a "readthrough" protein which often contains an additional domain or two. UGA within an open reading frame can also be redefined in a different way, to specify the non-universal, 21st amino acid, selenocysteine, often located at the crucial active site of the enzyme product. Dramatically, multiple UGAs are redefined in selenoprotein P mRNA (10 in human and apparently 28 in sea urchin) for the purpose of transporting selenium. Redefinition of the UGAs in these mRNAs is clearly programmed because it is messenger specific; other UGAs in the same cell specify termination. However, in methanogens when UAG specifies the 22nd amino acid, pyrrolysine, there may be an ambiguous reassignment of the meaning of UAG. But, the specific context of an mRNA may enhance the specification of pyrrolysine.

In the inverse of stop codon redefinition, a sense codon in a specific context can mediate termination. In the case of the StopGo (also called "Stop-Carry on") phenomenon the specific sense codon specifies an amino acid, the protein chain is terminated, and translation continues on to make a second protein from the single ORF.

So far there is no known case of a simple programmed change in the meaning of a standard sense codon – switching one amino acid for another (though there is dynamic redefinition of an exceptional codon for tryptophan at some, but not other, positions in a particular mRNA in the ciliate *Euplotes*).

Redirection of linear readout. Ribosomal frameshifting links two overlapping ORFs, with a variety of mechanisms, a mix of functional results, and with a variety of mRNA-specific signals.

Most programmed frameshifting involves single nucleotide, −1 or +1 shifts (some −2 shifts are known). At least most of these cases involve a dissociation of anticodon:codon pairing, followed by tRNA:mRNA realignment and anticodon re-pairing to mRNA in a new frame (but the situation of Ty3 frameshifting in yeast appears different and in several cases of +1 frameshifting the initial pairing of the tRNA involved is not as stringent as generally occurs).

The known cases of programmed +1 frameshifting involve a slow-to-decode codon in the ribosomal A-site, either a stop codon or a sense codon for which the relevant aminoacyl-tRNA is limiting (a "hungry" codon). There is competition between the peptidyl-tRNA realigning forward and the tRNA or release factor for the zero frame A-site codon. Thus the first nucleotide of the A-site codon can be pivotal for frameshifting-mediated regulatory circuits.

Programmed −1 frameshifting generally yields a fixed ratio of shift to non-shift products: the product whose synthesis involved a frameshift event and the product of standard decoding. The most common type of −1 frameshifting involves tandem dissociation of the anticodon:mRNA pairing of tRNAs in both the P- and A-sites, followed by realignment and re-pairing of both mRNAs in the −1 frame, although re-pairing of only the A-site tRNA is likely to be involved in some cases.

A greatly exaggerated version of dissociation and re-pairing occurs when re-pairing of peptidyl-tRNA to mRNA occurs not at an overlapping codon but at a downstream triplet on the same mRNA, thus bypassing the mRNA sequence in-between. In the best characterized case, 50 nucleotides are bypassed by about half the ribosomes reading the message apparently due to the formation of mRNA structure within the bypassing ribosomes.

In an even more extreme case of redirection, coding resumption occurs on a specific, unique "mRNA," tmRNA. In this case a protein, SmpB, is crucial for resume site selection. tmRNA function was initially thought to be just an elegant mechanism for rescuing ribosomes stuck at the 3' end of aberrant mRNAs that lacked a terminator and for facilitating the destruction of the associated incomplete proteins. However, it is now apparent that tmRNA's role is more extensive as in some cases it is involved in regulation. Also there is emerging evidence of distant 5' nucleotide sequence in several mRNAs that influence tmRNA action.

Examples of Function. Many of the viruses that utilize recoding are of great medical or economic importance, and their mobile chromosomal gene counterparts have had a significant evolutionary impact. The panoply of decoding versatility and sophistication by compact genomes is common and accomplishes diverse goals. For instance, in some plant RNA viruses, frameshifting may be part of the strategy for preventing a logjam of opposing ribosomes and RNA dependent, RNA polymerase acting on the same RNA. In another example, recoding generates the retroviral GagPol polyprotein that results in the precursor form of reverse transcriptase being included in the virion by virtue of its linkage to a small proportion of Gag. This crucial linkage of Gag and Pol could also be accomplished by RNA splicing. But, this would be deleterious because the location of the RNA packaging site would result in virion packaging of subgenomic RNA yielding defective viruses.

Interestingly, the type of recoding utilized by murine leukemia virus for this purpose is programmed readthrough whereas that utilized by HIV is programmed frameshifting – two recoding solutions to the same problem.

Another case of using different types of nonstandard mechanisms to accomplish the same result is the expression of two DNA polymerase subunits from a single bacterial chromosomal *dnaX* gene. In *Escherichia coli*, decoding the standard ORF yields a product containing two carboxy terminal domains that are lacking in the product resulting from a ribosomal frameshift event two-thirds of the way through the ORF. This foreshortened protein likely has a role in translesion polymerase that helps deal with transition through lesions or obstacles on template DNA. Its synthesis is mediated by 50% efficient ribosomal frameshifting with ribosomes in the new frame quickly encountering a stop codon. In contrast, in *Thermus thermophilus*, foreshortened products are derived from translation of the transcripts that result from transcriptional slippage at a run of A residues in the DNA. The population of mRNAs with varying numbers of extra nucleotides at the slippage site result in ribosome termination at now in-frame stop codons.

Evolution of recoding involves selection for both the position and the nature of the recoding site with its requisite stimulatory signals. In the absence of stimulatory signals, sites at which frameshifting or readthrough occur at low levels are, of course present. The current evidence suggests that, at least in bacteria, the most shift-prone sites that are not utilized for recoding are largely confined to poorly expressed mRNAs. For the sites whose "shifty" nature is dependent on scarcity of a particular tRNA, overexpression of an mRNA can lead to an increase in frameshifting raising a cautionary note for expression of high levels of proteins, often in nonhomologous systems, for biotechnological applications.

Scarcity of charged tRNAs can also be caused by amino acid starvation, a not uncommon state for bacteria. Starvation-induced frameshifting might be utilized to retune metabolism in response to the new growth state, so far this has not been shown.

Another consequence of recoding that needs further investigation is a possible under-appreciated role for frameshift-, bypassing-, and readthrough-derived events that do not exist to produce functional products. Ribosomes entering a region of mRNA not accessible by standard translation could have significant consequences on mRNA structure perhaps altering mRNA half-life. Alternatively, frameshifting within a coding sequence that yields early termination in a new frame could also affect mRNA half-life.

Recoding and Human Disease. Much remains unknown about the possible role of nonstandard translation in aging, viral infection, and certain autoimmune diseases. But the beginnings are there.

The stability of some of the proteins derived from ORFs not accessed by standard decoding is of particular interest from an immunological perspective. Preferential display on MHC class I molecules of peptides derived from short-lived proteins for activation of CD8+ T lymphocytes, this is important for the rapid CD8+ T-cell response to viral infection. Though the exact pathway for creating the array of peptides for display is not clear, models invoke rapidly degraded translation products. Some of these could be created by release of short nascent peptides due to ribosomal frameshifting.

Also, frameshifting may influence the severity of some of the triplet repeat diseases. The expanded string of repeats induces frameshifting leading to some product with poly-alanine in place of poly-glutamine.

Other genetic diseases involve frameshift mutations or substitutions that generate premature stop codons. If these new in-frame stop codons happen to be in a favorable context, small molecule drugs that alter translational fidelity can be used to phenotypically partially correct the mutations by stimulating synthesis of even a small portion of full-length product. This could alleviate the symptoms. Clinical trials in cystic fibrosis and Duchenne's muscular dystrophy are in an advanced stage.

It may also be possible to phenotypically correct certain frameshift mutants. Compensatory frameshifting can be stimulated by supplying a small RNA molecule to create a stimulatory signal in the mutant mRNA. Additions to tissue culture cells of such an RNA to create a signal just downstream of a frameshift mutant have yielded some positive results in optimal circumstances, but delivery problems remain.

Recoding events themselves may be targets for beneficial intervention. Since the ratio of Gag to GagPol is critical for HIV propagation, the efficiency of the frameshift event required for GagPol synthesis is a target for drug development. However, success depends on the host not having crucial similar targets. This is just one of the reasons for curiosity about the number of chromosomal genes that utilize the different types of frameshifting.

Foot and mouth disease virus appears to be a case in hand where it appears that the host cell does not use the unique StopGo recoding mechanism that the virus needs for propagation. This StopGo mechanism could be a target for antiviral development.

The path to recoding studies. The origin of knowledge about recoding has several different threads. In the mid-1960s, it was thought that decoding was so rigidly triplet that deviations from it would not be found, i.e., compensatory leakiness of frameshift mutations would not be detectable. And it was thought that mutants of translation components which would violate triplet decoding could not be found, i.e., external suppressors for frameshift mutants would not be isolatable. By 1972, both propositions were known to be incorrect.

Later that decade, an RNA phage-encoded product whose synthesis involved a frameshift event was detected. Also the balance of WT tRNAs was shown to be important for one type of frameshifting, and the relevance of noncognate codon:anticodon interaction was recognized. Nevertheless, the impact of these studies and of the discovery of a DNA phage frameshift product in 1983 was limited.

It was not until 1985–1987 that there were big breakthroughs in the detection of the utilization of specific frameshifting for gene expression. These cases are described in this book.

Redefinition of the meaning of one of the stop codons, UGA, was first discovered in the decoding of the coat protein gene of the RNA phage Qβ in the early 1970s. A proportion of translating ribosomes read through the stop codon by inserting an amino acid at the corresponding position in the protein. Not long afterward, essential readthrough was also shown for some plant viruses to make their RNA polymerase and for murine leukemia virus to make the GagPol precursor protein. This was accepted only slowly since the discovery of RNA splicing in 1977 provided a convenient explanation for accessing alternate open reading frames.

That selenocysteine was directly encoded by specific UGA stop codons, was discovered in 1986 at approximately the same time as the discovery of the initial cases of programmed frameshifting. The common features of reprogramming led to coining of the term “recoding” in 1992.

Recoding versus Reassignment. There seems to be a clear distinction between mRNA, site-specific, reassignment of codon meaning, and the complete reassignment, as for example in certain mitochondria. However, it is usual in biology for boundaries not to be sharp. Ambiguity arises where reassignment has not been fully refined as suggested above in the case of encoding pyrrolysine by UAG codons. For instance, a codon may be especially slow-to-decode, as with AGU and AGA in certain mitochondria. Perhaps surprisingly, the effects of such a codon in a fortuitous context may make a shift-prone site. Such a case may be evident in the common ancestor of the mitochondria of birds and turtles some 200 million years ago. It is thought that an extra nucleotide was present at an internal site in the coding sequence with frameshifting at a fortuitous "shifty" site restoring essential in-frame decoding. The extra nucleotide, and its associated compensatory frameshifting, is inferred to have been lost in many of the descendents of this common ancestor except in the mitochondrial decoding of the majority of extant birds and tortoises.

A parallel situation with an extra nucleotide occurs in a proportion of tracts of nine or more as in certain AT-rich endosymbionts such as *Buchnera aphidicola* which is associated with Aphids. However, in this case, the reading frame is restored by compensatory transcriptional slippage.

In the ciliate, *Euplotes*, UGA is reassigned so that it does not specify termination. It has been proposed that coincident changes in the release factor cause UAA, especially with a 3'A, to become unusually slow-to-decode. There is efficient frameshifting at AAA UAA A in *Euplotes* and required frameshifting occurs at this "terminator" sequence in a remarkable proportion of identified genes. Together with the mitochondrial frameshifting, *Euplotes* decoding illustrates more overlap between recoding and reassignment than encountered in other organisms.

Ancient decoding. Are there any cases of redefined meaning of a codon that are actually ancestral in an evolutionary sense? Consider UGA. Since special signals are required to change the meaning of UGA to specify selenocysteine, it is easiest to consider the standard termination meaning as ancestral. However, in early decoding there may not have been discrimination between cysteine and selenocysteine and perhaps at a stage before divergence of the common ancestor of bacteria, archaea, and eukaryotes, both amino acids were specified by UGN codons. In one version of this scenario, a next step was limitation of cysteine decoding to UGU and UGC, with UGA encoding selenocysteine. As the original anaerobic atmosphere changed to an aerobic one with the advent of an oxygen-rich atmosphere some 2.4 billion years ago, there could have been selection against oxygen-labile selenocysteine except where it was especially advantageous. Perhaps this "restriction stage" is when selenocysteine-recoding signals started to arise, and non-tagged UGA codons later acquired the termination meaning. Such a model is in marked contrast to the obvious one in which the termination meaning was ancestral.

In modern bacteria UGA specifies selenocysteine only if it is followed by a specific stem-loop structure in the mRNA. It is a reasonable supposition, although no more than that, that a 3' nearby stem-loop structure became important for selenocysteine specification in the common ancestor of bacteria, archaea, and eukaryotes.

In modern eukaryotes a specific structure in the 3' untranslated region is required. However, some eukaryotic mRNAs that encode selenocysteine-containing proteins also have some "remnant" of a stimulatory structure just 3' adjacent to the UGA. This element likely preceded the emergence of specific structures in the 3' UTR.

At a much earlier time than selenocysteine specification, during the evolution of decoding itself, it seems likely that primitive readout was incapable of being anything other than slipshod. At this time polyamines may have been playing a protein-like role in primitive ribosomes. The result likely was a plethora of products serving as food for selection. As triplet decoding and codon assignment became locked in, was there a parallel refinement of alternative decoding? Or did the currently observed alternative decoding evolve later as a sophisticated refinement after a period of tediously standard decoding?

Frameshifting for expression of bacterial release factor 2 decoding also has an ancient origin. Its hallmark is stimulation of the frameshift event by pairing between mRNA and rRNA during translation. We can wonder whether this interaction between mRNA and rRNA in ribosomes in the act of translating might not itself have an ancient origin. Could interactions of this type have helped to grip the message?

In modern day ribosomes, it is anticodon pairing that holds the mRNA in place. Detachment and realignment lead to frameshifting, at least in most cases. There is an appealing if somewhat controversial suggestion that standard frame maintenance is maintained by pairing two tRNAs at all times. In this scenario, anticodon pairing by E-site tRNA does not dissociate until A-site aminoacyl-tRNA pairing is established. So strong ribosomal gripping of tRNA would lead to the in-frame grip of the mRNA. However, the E-site appears to be a late addition in ribosome evolutionary history since it is protein-rich. Therefore, before it existed, what served to clasp mRNA? One candidate is the rRNA:mRNA Shine–Dalgarno pairing which was discovered because of its role in initiation of protein synthesis in bacteria. Programmed frameshifting studies have revealed that this interaction is not unique to initiation in that the anti-Shine–Dalgarno sequence of translating ribosomes can scan the mRNA being decoded for potential complimentarity. After such a rRNA:mRNA hybrid forms, the ribosome continues translation for up to 10 nucleotides before the hybrid ruptures. Whether interactions of this type played a role in primordial protein synthesis is of course unknown. But, if so, rather than the primordial coding sequences having been G-rich, perhaps there could have been blocks of coding sequences spanned by G-rich noncoding "anchors" that decoding could bypass. Setting aside such speculative "excesses," recoding studies are clearly contributing to our knowledge of standard decoding and scanning by the anti-Shine–Dalgarno sequences of translating ribosomes is one of several cases in point.

Transcription slippage (also called pseudo-templated transcription or stuttering)

Realignment during transcription parallels translational realignment. A few examples are mentioned above where transcription slippage substitutes for cases of programmed frameshifting. In these cases there has been selection for high-level transcription slippage at specific sites. Such slippage yields mRNAs with inserts of one or more nucleotides – in a bacterial case a diminishing series of mRNAs with up to 15 additional nucleotides and a small minority with deletions of one or a few nucleotides. Standard translation of these mRNAs yields unique products. Instead of the detachment of triplet anticodon pairing, dissociation of the nascent RNA hybrid with template DNA in the transcription bubble is involved. The identity of flanking sequence can delimit the number of extra nucleotides inserted to 1. But whether the flanking sequence can also enhance the frequency, possibly even by the ability of the nascent RNA chain to form a short stem, remains to be seen.

Editing of preformed transcripts can also have consequences similar to several types of recoding. For instance, mRNA editing that changes a stop codon to a sense codon can give the equivalent of stop codon readthrough. Similarities even extend to variable efficiencies of the process and to the importance of mRNA structure. Editing to change the identity of one sense codon to another in a proportion of the mRNAs, constitutes a type of diversity for which there is only one specialized recoding counterpart. It will be fascinating to discover to what extent nonstandard transcription and RNA editing parallel and substitute for their translational counterparts.

Future. As this book attests, our knowledge of recoding has a firm basis but much remains to be done. Together with studies of mutants of ribosomal components, advances in structural information about translation components now are offering the prospect of an understanding of how ribosomes sense and respond to recoding signals. The deluge of sequence information is providing exciting bioinformatic opportunities for comparative analyses to reveal the extent of recoding and transcription slippage. And a dramatic recent advance in determining ribosome location en masse at sub-codon resolution by sequencing vast numbers of mRNA segments protected within ribosomes at a specific time, has great potential in this regard.

Knowledge of the "dark matter" of the genome, those transcribed regions that do not encode mRNA, tRNA, or rRNA, is rapidly showing the complex roles of small RNAs in gene expression. Are some cases of recoding influenced by them?

We look forward to discovering the answers to these and questions not yet asked.

Acknowledgment: We thank Ken Keiler for instigating the American Society of Microbiologists' session that inspired Andrea Macaluso (Springer) to propose this book, our past colleagues, especially Bob (R.B.) Weiss and Alan Herr, and our current colleagues. We also thank Marshal Nirenberg, the pioneer of codon identification, for his generous contribution.

John Atkins and Ray Gesteland

Contents

Contributors

Michael Anikin Department of Cell Biology, School of Osteopathic Medicine, University of Medicine and Dentistry of New Jersey, Stratford, NJ 08084, USA.

Pavel V. Baranov Biochemistry Department, University College Cork, Cork, Ireland.

David M. Bedwell Department of Microbiology, University of Alabama at Birmingham, Birmingham, AL 35294-2170, USA.

Michaël Bekaert School of Biology and Environmental Science, University College Dublin, Ireland.

Marla J. Berry Department of Cell and Molecular Biology, John A. Burns School of Medicine, University of Hawaii at Manoa, Honolulu, HI 96813, USA.

Léa Brakier-Gingras Département de Biochimie, Université de Montréal, Montréal, Québec, H3T 1J4, Canada.

Ian Brierley Division of Virology, Department of Pathology, University of Cambridge, Tennis Court Road, Cambridge CB2 1QP, UK.

Jeremy D. Brown Institute for Cell & Molecular Biosciences, The Medical School, Newcastle University, Newcastle upon Tyne NE2 4HH, UK.

Jonathan D. Dinman Department of Cell Biology and Molecular Genetics, University of Maryland, College Park, MD 20742, USA.

Dominic Dulude Département de Biochimie, Université de Montréal, Montréal, Québec, H3T 1J4, Canada; Centre de Recherche, Hôpital Sainte-Justine, Montréal, Québec, H3T 1C5, Canada.

Philip J. Farabaugh Department of Biological Sciences and Program in Molecular and Cell Biology, University of Maryland Baltimore County, Baltimore, MD 21250, USA.

Olivier Fayet Centre National de la Recherche Scientifique, Laboratoire de Microbiologie et Génétique Moléculaires, Université de Toulouse, F-31000 Toulouse, France.

Andrew E. Firth BioSciences Institute, University College Cork, Cork, Ireland.

David P. Giedroc Department of Chemistry, Indiana University, Bloomington, IN 47405-7102, USA.

Robert J. C. Gilbert Division of Structural Biology, Henry Wellcome Building for Genomic Medicine, University of Oxford, Oxford OX3 7BN, UK.

Vadin N. Gladyshev Department of Biochemistry and Redox Biology Center, University of Nebraska, Lincoln, NE 68588, USA.

Olga Gurvich Cork Cancer Centre, BioSciences Institute, University College Cork, Cork, Ireland.

Dolph L. Hatfield Molecular Biology of Selenium Section, Laboratory of Cancer Prevention, Center for Cancer Research, National Cancer Institute, National Institutes of Health, Bethesda, MD 20892, USA.

Roger W. Hendrix Pittsburgh Bacteriophage Institute & Department of Biological Sciences, University of Pittsburgh, Pittsburgh, PA 15260, USA.

Michael T. Howard Department of Human Genetics, University of Utah, Salt Lake City, UT 84112-5330, USA.

Ivaylo P. Ivanov BioSciences Institute, University College Cork, Cork, Ireland; Department of Human Genetics, University of Utah, Salt Lake City, UT 84112-5330, USA.

Kim M. Keeling Department of Microbiology and Gregory Fleming James Cystic Fibrosis Research Center, University of Alabama at Birmingham, Birmingham, AL 35294-2170, USA.

Kenneth C. Keiler Department of Biochemistry and Molecular Biology, Penn State University, 401 Althouse Laboratory, University Park, PA 16802, USA.

Joseph A. Krzycki Department of Microbiology, Ohio State University, Columbus, Ohio 43210, USA.

Dennis M. Lee Department of Biochemistry and Molecular Biology, Penn State University, 401 Althouse Laboratory, University Park, PA 16802, USA.

Senya Matsufuji Department of Molecular Biology, The Jikei University School of Medicine, 3-25-8 Nishi-shinbashi, Minato-ku, Tokyo 105-8461, Japan.

William T. McAllister Department of Cell Biology, School of Osteopathic Medicine, University of Medicine and Dentistry of New Jersey, Stratford, NJ 08084, USA.

W. Allen Miller Plant Pathology Department, and Biochemistry, Biophysics & Molecular Biology Departments, Iowa State University, Ames, IA 50011, USA.

Vadim Molodtsov Department of Cell Biology, School of Osteopathic Medicine, University of Medicine and Dentistry of New Jersey, Stratford, NJ 08084, USA.

Olivier Namy IGM, CNRS, UMR 8621, F 91405 Orsay, France and Université Paris-Sud, F 91405 Orsay, France.

Knud H. Nierhaus Max-Planck-Institut für Molekulare Genetik, D-14195 Berlin, Germany.

Michael O'Connor School of Biological Sciences, University of Missouri-Kansas City, Kansas City, MO 64110, USA.

Markus Pech Max-Planck-Institut für Molekulare Genetik, D-14195 Berlin, Germany.

Simon Pennell Division of Molecular Structure, MRC National Institute for Medical Research, The Ridgeway, Mill Hill, London NW7 1AA, UK.

Marie-Françoise Prère Centre National de la Recherche Scientifique, UMR5100, Laboratoire de Microbiologie et Génétique Moléculaires, Université de Toulouse, F-31000 Toulouse, France.

Jean-Pierre Rousset IGM, CNRS, UMR 8621, Orsay, F 91405 France, Université Paris-Sud, Orsay, France.

Martin D. Ryan Centre for Biomolecular Sciences, Biomolecular Sciences Building, North Haugh, University of St. Andrews, St. Andrews KY16 9ST, UK.

Dmitry Temiakov Department of Cell Biology, School of Osteopathic Medicine, University of Medicine and Dentistry of New Jersey, Stratford, NJ 08084, USA.

Oliver Vesper Max-Planck-Institut für Molekulare Genetik, D-14195 Berlin, Germany.

Norma M. Wills Department of Human Genetics, University of Utah, Salt Lake City, UT 84112-5330, USA.

Daniel N. Wilson Gene Center, Ludwig-Maximilians-Universität München, D-81377 München, Germany.

Hiroshi Yamamoto Max-Planck-Institut für Molekulare Genetik, D-14195 Berlin, Germany.

Part I
Redefinition

Chapter 1
Selenocysteine Biosynthesis, Selenoproteins, and Selenoproteomes

Vadim N. Gladyshev and Dolph L. Hatfield

Abstract Selenocysteine (Sec), the 21st amino acid in the genetic code, is encoded by UGA. The pathway of Sec biosynthesis in eukaryotes has only recently been discovered. Sec is constructed on its tRNA that is initially aminoacylated with serine and modified to a phosphoseryl-tRNA intermediate with the help of several dedicated enzymes. More than 50 selenoprotein families are now known with most selenoproteins being oxidoreductases. Development of bioinformatics tools led to the identification of entire sets of selenoproteins in organisms, selenoproteomes, which in turn helped explain biological and biomedical effects of dietary selenium and identify new functions of selenium in biology. Roles of selenium and selenoproteins in health have also been addressed through sophisticated transgenic/knockout models that targeted removal or modulation of Sec tRNA expression.

Contents

V.N. Gladyshev (✉)
Department of Biochemistry and Redox Biology Center, University of Nebraska, Lincoln, NE 68588 USA
e-mail: vgladyshev@rics.bwh.harvard.edu

J.F. Atkins, R.F. Gesteland (eds.), *Recoding: Expansion of Decoding Rules Enriches Gene Expression*, Nucleic Acids and Molecular Biology 24,
DOI 10.1007/978-0-387-89382-2_1,

1.1 UGA is Recoded for Sec

1.1.1 Variations in the Genetic Code

The genetic code was deciphered and shown to be universal by the mid-1960s (see Nirenberg et al. 1966 and references therein). All 64 code words in the code were assigned to amino acids or a specialized function. One code word, AUG, was recognized to have a dual function serving to dictate the initiation of protein synthesis and to code for the insertion of methionine at internal protein positions. Three code words, UAG, UAA, and UGA, were assigned specialized roles of dictating the cessation of protein synthesis. It was assumed at that time that there was no more room in the code for another (or other) amino acid(s) and the possibility that code words other than AUG might have dual functions was not considered.

There have been several major variations reported in the genetic code, however, since the mid-1960s. It was initially recognized that not all organelles use the same genetic language, and subsequently, that some organisms use a different genetic language. For example, variations in the universal genetic code were observed in mitochondria and chloroplasts (reviewed in Jukes and Osawa 1990; Yokobori et al. 2001) and in organisms such as mycoplasma that use UGA to code for tryptophan instead of termination (Yamao et al. 1985), *Euplotes* that use UGA to code for cysteine instead of termination (Meyer et al. 1991), and several species of *Candida* that use CUG to code for serine instead of leucine (reviewed in Pesole et al. 1995). Furthermore, some bacteria and archaea use GUG and/or UUG as start codons instead of the universal codon, AUG (Bell and Jackson 1988). Interestingly, evidence in the mid-1980s suggested that the termination codon, UGA, likely had a dual function. The gene sequences of the selenium-containing proteins, glutathione peroxidase 1 (GPx1) in mammals (Chambers et al. 1986) and formate dehydrogenase in *Escherichia coli* (Zinoni et al. 1986), showed that both genes had an in-frame TGA codon in their open reading frames that aligned with Sec in the corresponding proteins. These correlations suggested that UGA coded for Sec, but this assignment could not be made without further experimental evidence as the available data at that time had shown that the serine moiety (Sunde and Evenson, 1987) initially attached to a minor Sec tRNA that decoded UGA was converted to phosphoseryl-tRNA by a kinase (Hatfield, Diamond and Dudock 1982). Thus, it was possible that phosphoserine was incorporated into protein and then posttranslationally modified to Sec making phosphoserine the 21st amino acid in the genetic code. This point was clarified when Sec was indeed shown to be biosynthesized on its tRNA in both bacterial (Leinfelder et al. 1989) and mammalian cells (Lee et al. 1989a). These two studies provided the first direct evidence that Sec was the 21st amino acid and that UGA was therefore recoded for Sec in those organisms that synthesize selenoproteins. The expanded genetic code that includes Sec is shown in Fig. 1.1

	U	C	A	G	
U	UUU Phe	UCU Ser	UAU Tyr	UGU Cys	U
	UUC Phe	UCC Ser	UAC Tyr	UGC Cys	C
	UUA Leu	UCA Ser	UAA Stop	UGA Sec/Stop	A
	UUG Leu	UCG Ser	UAG Pyl/Stop	UGG Trp	G
C	CUU Leu	CCU Pro	CAU His	CGU Arg	U
	CUC Leu	CCC Pro	CAC His	CGC Arg	C
	CUA Leu	CCA Pro	CAA Gln	CGA Arg	A
	CUG Leu	CCG Pro	CAG Gln	CGG Arg	G
A	AUU Ile	ACU Thr	AAU Asn	AGU Ser	U
	AUC Ile	ACC Thr	AAC Asn	AGC Ser	C
	AUA Ile	ACA Thr	AAA Lys	AGA Arg	A
	AUG Met	ACG Thr	AAG Lys	AGG Arg	G
G	GUU Val	GCU Ala	GAU Asp	GGU Gly	U
	GUC Val	GCC Ala	GAC Asp	GGC Gly	C
	GUA Val	GCA Ala	GAA Glu	GGA Gly	A
	GUG Val	GCG Ala	GAG Glu	GGG Gly	G

Fig. 1.1 The genetic code. Sec, encoded by UGA, is highlighted to show that it is the 21st amino acid in the genetic code. A 22nd amino acid, pyrrolysine (Pyl), encoded by UAG, is also shown

It should also be noted that pyrrolysine was recently added to the genetic code as the 22nd amino acid (see Fig. 1.1) (Srinivasan et al. 2002; Hao et al. 2002) which is described in Chapter 3 by Krzycki. The possibility that a 23rd amino acid may also occur in the code has been considered, and although not likely to occur, has not been completely ruled out (Lobanov et al. 2006a). If a 23rd amino acid exists in the code, it would be much less widespread than Sec and may be limited to only a few organisms. Another variation in the genetic code was recently found wherein a single code word can code for two different amino acids, not only in the same organism but also within the same gene (Turanov et al. 2009). UGA was shown to specify the incorporation of Cys and Sec in a single mRNA in the *Euplotes* genus and the structural arrangements of the mRNA preserve the location-dependent dual function of the UGA codon.

1.2 Biosynthesis of Sec

A number of factors had been identified in higher vertebrates over the years that play a role in the biosynthesis of Sec and its insertion into protein. The components involved in the biosynthesis of Sec are discussed below, while the chapter by Berry and Howard (Chapter 2) focuses on those components involved with the incorporation of this amino acid into protein. The principle factors that have been associated with Sec biosynthesis in eukaryotes are Sec tRNA, seryl-tRNA synthetase (SerS), phosphoseryl-tRNA kinase (PSTK), Sec synthase (SecS), and selenophosphate synthetases 1 and 2 (SPS1 and 2). They are described in greater detail below.

1.2.1 Unique Features of Sec tRNA

Sec tRNA is undoubtedly the most unique tRNA identified to date. For example, its transcription begins, unlike any known tRNA, at the first nucleotide within the coding region of its gene (Lee et al. 1987), while all other tRNAs are transcribed with a leader sequence that must be processed. The upstream regulatory sites that govern the transcription of Sec tRNA are unique for tRNA (reviewed in detail elsewhere (Hatfield et al. 1999)). The mature form of the tRNA has a triphosphate on its 5′-end (Lee et al. 1987). It is the longest tRNA sequenced, ranging in length from 90 to 93 nucleotides in some lower eukaryotes (Mourier et al. 2005; Lobanov et al. 2006b) to 95 in *E. coli* and more than a 100 nucleotides in various other prokaryotes (Heider and Bock, 1993). Sec tRNAs in higher vertebrates contain only five modified nucleosides, whereas up to 15–17 modified nucleosides have been identified in other tRNAs. The fact that Sec tRNA is initially aminoacylated with serine, but is the tRNA for Sec, has resulted in it being designated as Sec $tRNA^{[Ser]Sec}$ (Hatfield et al. 1994). The secondary structure of $tRNA^{[Ser]Sec}$ found in mammals and *Plasmodium falciparum* is shown as a cloverleaf model in Fig. 1.2.

The modified nucleosides in $tRNA^{[Ser]Sec}$ are 1-methyladenosine (m^1A) at position 58, pseudouridine (ψU) at position 55, N^6-isopentenyladenosine (i^6A) at position 37, and either 5-methoxycarbonylmethyluridine (mcm^5U) or methoxycarbonylmethyl-2′-*O*-methyluridine (mcm^5Um) at position 34, which is the wobble position of tRNA (Hatfield et al. 2006). The synthesis of the methyl group at position 34 is the last step in the maturation of Sec $tRNA^{[Ser]Sec}$ and this 2′-*O* methyluridine is designated Um34. Interestingly, the synthesis of Um34 is stringently dependent on primary structure and on intact secondary and tertiary structures of $tRNA^{[Ser]Sec}$; i.e., the addition of Um34 cannot occur without the prior synthesis of m^1A, ψU, i^6A, and mcm^5U, and disruption of the secondary or tertiary structure of the tRNA inhibits its attachment (Kim et al. 2000). Furthermore, Um34 formation is dependent on selenium status (reviewed in Hatfield et al. 2006). Under conditions of selenium deficiency, the ratio of mcm^5U/mcm^5Um shifts dramatically in mammalian organs, tissues, and cells from the latter to the former isoforms, and vice versa under conditions of selenium sufficiency (Chittum et al. 1997). Finally, the addition of Um34 to Sec $tRNA^{[Ser]Sec}$ results in striking changes in secondary and tertiary structures.

The above observations relating to the synthesis of Um34 led us to propose that this maturation step was a highly specialized event yielding mcm^5Um, an isoform with a different function in selenoprotein synthesis than its precursor, mcm^5U (Moustafa et al. 2001). This hypothesis was later confirmed as discussed below in the section on **Sec.**

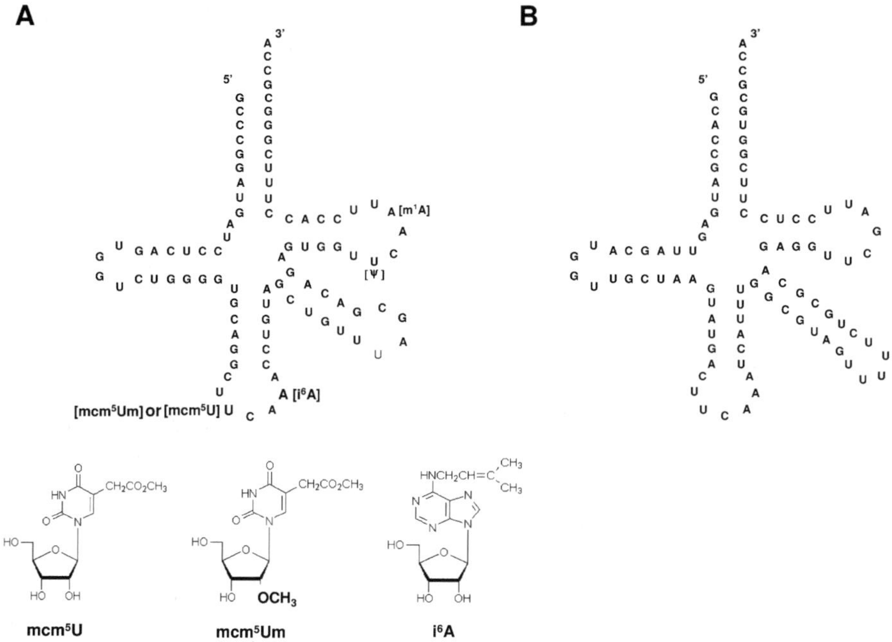

Fig. 1.2 Mammalian and *P. falciparum* Sec tRNA$^{[Ser]Sec}$. (**A**) Mammalian tRNA$^{[Ser]Sec}$. The structures of the modified bases in the anticodon loop of mammalian tRNA$^{[Ser]Sec}$, i^{6}A, mcm^{5}U, and mcm^{5}Um, are also shown. (**B**) *P. falciparum* tRNA$^{[Ser]Sec}$ is 93 nucleotides long and mammalian tRNA$^{[Ser]Sec}$ is 90 nucleotides long and the extra bases occur in the long extra arm (see text). The mammalian tRNA$^{[Ser]Sec}$ structure was determined by sequencing the tRNA (Hatfield et al. 2006), while the *P. falciparum* tRNA$^{[Ser]Sec}$ structure is based on sequencing its gene (Lobanov et al. 2006b), wherein the CCA 3′-terminus, which is added posttranscriptionally, is shown in the figure

1.2.2 tRNA Knockout and Transgenic Mouse Models

Another novel feature of tRNA$^{[Ser]Sec}$ is that it has nine paired bases in the acceptor stem and four in the TψC stem, i.e., it exists in a 9/4 cloverleaf form (Böck et al. 1991; Hubert et al. 1998). Other tRNAs have seven paired bases in the acceptor stem and five paired bases in the TψC stem, i.e., they exist in a 7/5 cloverleaf model. An additional novel feature of tRNA$^{[Ser]Sec}$ is that the D-stem may contain six base pairs while other tRNAs have three to four base pairs in this stem. There are numerous other characteristics of tRNA$^{[Ser]Sec}$ that distinguish it from other tRNAs and these have been reviewed in detail elsewhere (Hatfield et al. 1999).

1.2.3 Aminoacylation of Sec tRNA$^{[Ser]Sec}$

Sec tRNA$^{[Ser]Sec}$ is aminoacylated with serine by SerS which is the initial step in the biosynthetic pathway of Sec (Lee et al. 1989a; Leinfelder et al. 1989). The identity elements in Sec tRNA$^{[Ser]Sec}$ for its aminoacylation therefore must correspond to those in SerS. The identity elements in mammalian tRNA$^{[Ser]Sec}$ have been identified and the major areas are the discriminator base and the long extra arm which have essential roles in aminoacylation (Wu et al. 1993; Ohama et al. 1994). Other regions of tRNA$^{[Ser]Sec}$ that have identity roles are located in the acceptor, TψC, and D-stems (Amberg et al. 1996). Once the tRNA is aminoacylated with serine, the serine moiety serves as the backbone for the synthesis of Sec in prokaryotes and eukaryotes (reviewed in Hatfield and Gladyshev 2002).

1.2.4 Phosphoseryl-tRNA$^{[Ser]Sec}$ kinase

A kinase activity that phosphorylates a minor seryl-tRNA to form phosphoseryl-tRNA was identified many years ago in rooster liver by Maenpaa and Bernfield (1970). About the same time, a minor seryl-tRNA in bovine, rabbit, and chicken livers that recognized specifically the nonsense codon, UGA, was reported (Hatfield and Portugal, 1970). Subsequently, the phosphoseryl-tRNA identified in rooster liver and the UGA decoding seryl-tRNA were later shown to be selenocysteyl-tRNA$^{[Ser]Sec}$ (Lee et al. 1989a, b). The significance of the phosphoseryl-tRNA$^{[Ser]Sec}$ kinase (PSTK) that phosphorylated seryl-tRNA$^{[Ser]Sec}$ to form phosphoseryl-tRNA$^{[Ser]Sec}$ was not assessed until PSTK was isolated and characterized. The kinase activity remained elusive for many years, but was finally identified by combining bioinformatics and biochemistry approaches (Carlson et al. 2004a). That is, we examined completely sequenced genomes for kinase genes occurring in archaea that synthesized selenoproteins, but absent in archaea that lacked selenoproteins, and identified four candidates. The completely sequenced genomes of *Caenorhabditis elegans* and *Drosophila* that were known to synthesize selenoproteins were then searched for homologous sequences to these four kinase genes that were in turn not present in the genome of *Saccharomyces cerevisiae* which did not make selenoproteins. A single candidate kinase was detected using this strategy. Since a gene was present in the mouse genome with homology to the candidate *pstk* gene, it was cloned, its product expressed, characterized, and identified as PSTK (Carlson et al. 2004a). PSTK used seryl-tRNA$^{[Ser]Sec}$ and ATP as substrates and Mg^{++} as a cofactor to yield *O*-phosphoseryl-tRNA$^{[Ser]Sec}$ and ADP. At the time this work was reported, the role of PSTK and its product, phosphoseryl-tRNA$^{[Ser]Sec}$, had not been determined.

1.2.5 Sec Synthase (SecS) and Selenophosphate Synthetase (SPS)

SecS, which was designated SelA in prokaryotes, was initially identified and characterized in *E. coli* by Bock and collaborators (Böck et al. 1991). *E. coli* seryl-tRNA$^{[Ser]Sec}$ served as a substrate for SelA and was converted to an intermediate, which is most likely dehydroalanyl-tRNA$^{[Ser]Sec}$ (reviewed in Böck et al. 1991). The active selenium donor, monoselenophosphate (SeP), is synthesized from selenide and ATP by SPS (SelD) in prokaryotes (Glass et al. 1993). The intermediate, dehydroalanyl-tRNA$^{[Ser]Sec}$, while still bound to SelA, accepts SeP to generate selenocysteyl-tRNA$^{[Ser]Sec}$ which is now ready to incorporate Sec into protein (Böck et al. 1991).

A gene with homology to *selA* was not found in archaea or eukaryotes. However, a candidate SecS was subsequently identified in eukaryotes and archaea by comparative genomic analysis of completely sequenced eukaryotic and archaeal genomes as was carried out in detecting *pstk* (Xu et al. 2006). The survey searching for a eukaryotic and archaeal SecS resulted in the identification of genes co-occurring with known components in the Sec insertion machinery and, in addition, a candidate SecS was detected in mammals. This protein had previously been found in cell extracts from patients with an autoimmune chronic hepatitis as an autoimmune factor that co-precipitated with tRNA$^{[Ser]Sec}$ (Gelpi et al. 1992). This factor was designated as the soluble liver antigen (SLA). SLA was found to be a PLP-dependent transferase (Kernebeck et al. 2001) and also to bind other components involved in Sec metabolism (Xu et al. 2005; Small-Howard et al. 2006). SLA occurred in all eukaryotic and archaeal selenoprotein synthesizing organisms that were examined by comparative genomic analysis, but not in those organisms not synthesizing selenoproteins, nor in any prokaryotic organism whether it did or did not make selenoproteins (Xu et al. 2006).

The mouse gene for SLA (*SecS*) was cloned, the protein expressed, and the function of SLA established by experimental analysis (Xu et al. 2006). *O*-phosphoseryl-tRNA$^{[Ser]Sec}$ was dephosphorylated by SLA to yield Pi and a product that bound to the enzyme. The product that remained bound to SLA, which was an intermediate in the biosynthesis of Sec, was likely not seryl-tRNA$^{[Ser]Sec}$ as seryl-tRNA$^{[Ser]Sec}$ did not itself bind to SLA. Dehydroalanine is likely the intermediate generated by mammalian SecS (Xu et al. 2006), which was the same intermediate identified in *E. coli* (Böck et al. 1991).

selD has two homologous genes in mammals, designated *sps1* and *sps2* (Kim and Stadtman 1995; Low, Harney and Berry 1995; Guimaraes et al. 1996) that were initially proposed to serve as SPS. The product of *sps*2, which is SPS2, is a selenoprotein and can therefore serve as an autoregulator of selenoprotein synthesis (Guimaraes et al. 1996; Kim et al. 1997), as it is indeed the enzyme that synthesizes SeP in mammals (Xu et al. 2006). In studies that further elucidated the roles of SPS1 and SPS2 in mammals, the Sec moiety in SPS2 was mutated to Cys, wherein the mutant was found to have low enzyme activity (Guimaraes et al. 1996; Kim et al. 1997, 1999), but was capable of complementing *selD* minus *E. coli* cells transfected with the mutant mammalian *sps*2 (Kim et al. 1999). Other studies involved complementing *selD* minus *E. coli* cells that had been transfected with either *sps*1 or Sec$^-$ *sps*2, and they suggested that SPS1 has a role in recycling Sec via a selenium salvage pathway, whereas SPS2 was involved in the synthesis of SeP (Tamura et al. 2004). However, these studies did not directly demonstrate the roles of SPS1 and SPS2 in Sec biosynthesis.

To further clarify the roles of SPS1 and SPS2 in Sec biosynthesis, *C. elegans* SPS2, which naturally has Cys instead of Sec at its active site, mouse SPS2 containing a Sec→Cys mutation, *E. coli* SelD and mouse SPS1 were prepared and their abilities to generate SeP from selenide and ATP were determined (Xu et al. 2006). Each SPS synthesized SeP with the exception of mouse SPS1 demonstrating that SPS2, and not SPS1, was SPS in higher animals.

It should first be noted, however, that none of the earlier studies had shown that SeP could serve directly as the selenium donor which would unequivocally demonstrate that SeP was the active selenium donor. SeP was therefore synthesized chemically and added to Sec biosynthesis reactions (Xu et al. 2006). SeP and *O*-phosphoseryl-tRNA$^{[Ser]Sec}$ incubated in the presence of mouse SecS did indeed generate Sec. Reactions containing seryl-tRNA$^{[Ser]Sec}$ in place of *O*-phosphoseryl-tRNA$^{[Ser]Sec}$, with or without SeP, or containing another protein in place of SecS, did not form Sec. These reactions unambiguously proved that SeP is the active selenium donor in Sec biosynthesis (Xu et al. 2006). Reactions containing mouse SecS, *O*-phosphoseryl-tRNA$^{[Ser]Sec}$, mouse mutant Sec→Cys SPS2, selenide and ATP produced selenocysteyl-tRNA$^{\{Ser\}Sec}$, but seryl-tRNA$^{[Ser]Sec}$ in place of *O*-phosphoseryl-tRNA$^{[Ser]Sec}$, or mouse SPS1 in place of SPS2, did not. Thus, SPS2 synthesizes SeP and SPS1 must have another role that may or may not be related to selenoprotein biosynthesis (Lobanov, Hatfield and Gladyshev 2008a). In addition to unequivocally demonstrating that SeP is the active donor and that SPS2, and not SPS1, is the SPS in higher animals, the above in vitro studies showed that SLA is the mammalian SecS and *O*-phosphoseryl-tRNA$^{[Ser]Sec}$ is the correct intermediate in the pathway. At the same time these studies on elucidating the Sec biosynthetic pathway were being carried out, the archaeal SLA gene (*SecS*) was cloned, expressed, and the gene product shown to convert *O*-phosphoseryl-tRNA$^{[Ser]Sec}$ to selenocysteyl-tRNA$^{[Ser]Sec}$ (Yuan et al. 2006).

The roles of SPS1 and SPS2 were further elucidated intracellularly in a complementary study (Xu et al. 2007). SPS1 and SPS2 were knocked down using RNA interference technology in NIH 3T3 cells and the effect of their loss on selenoprotein synthesis examined. Selenoprotein synthesis was abolished completely by the removal of SPS2, but was unaffected by removal of SPS1. The knockdown cells were then used for transfection with SPS2, SelD, or SPS1. Either SPS2 or SelD complemented the loss of SPS2, but SPS1 did not. These "in vivo" studies showed that SPS2, which synthesizes SeP (Xu et al. 2006), is essential for selenoprotein biosynthesis, but SPS1 is not (Xu et al. 2007). Furthermore, SPS1 has been found to occur in animals in which the SPS2 and the other Sec insertion machinery have been lost providing additional evidence that SPS1 has roles other than in Sec biosynthesis and its insertion into protein (Lobanov et al. 2008a).

1.2.6 The Sec biosynthetic pathway

The entire Sec biosynthetic pathway in eukaryotes is shown in Fig. 1.3. The pathway begins with the aminoacylation of tRNA$^{[Ser]Sec}$ with serine by SerS (Sunde and Evenson, 1987; Lee et al. 1989a; Leinfelder et al. 1989). PSTK phosphorylates the serine moiety to form *O*-phosphoseryl-tRNA$^{[Ser]Sec}$ (Carlson et al. 2004a) which

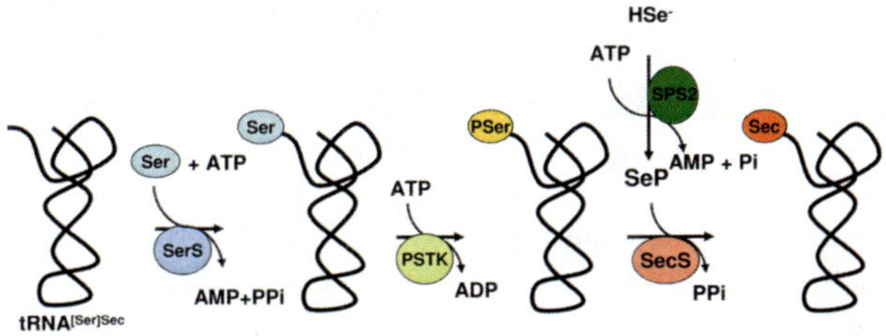

Fig. 1.3 Biosynthesis of Sec in eukaryotes and archaea. Abbreviations of the factors involved in Sec biosynthesis are defined in the text

then serves as a substrate for SecS that hydrolyzes the phosphate group to form the acceptor molecule for SeP, likely dehydroalanyl-tRNA$^{[Ser]Sec}$, that remains bound to SecS (Xu et al. 2006). SPS2 synthesizes SeP, the active selenium donor, using selenide and ATP as substrates and with the addition of SeP to the intermediate attached to SecS, the synthesis of Sec is complete. This pathway established how the 21st and last known eukaryotic amino acid in the genetic code whose biosynthesis had not been established, is synthesized.

Although PSTK is not found in eubacteria, SelA can use *O*-phosphoseryl-tRNA$^{[Ser]Sec}$ as a substrate (Xu et al. 2006). The major difference in the biosynthesis of Sec in eubacteria and in eukaryotes and archaea is the extra step involving *O*-phosphoseryl-tRNA$^{[Ser]Sec}$ which is synthesized using seryl-tRNA$^{[Ser]Sec}$ and ATP by PSTK. *O*-phosphoseryl-tRNA$^{[Ser]Sec}$ then serves as a substrate for SecS in eukaryotes and archaea, whereas seryl-tRNA$^{[Ser]Sec}$ is a substrate for eubacterial SecS. Selenocysteyl-tRNA$^{[Ser]Sec}$ is now poised to be incorporated into selenoproteins (see Chapter 2 by Berry and Howard).

1.3 Identification of Selenoproteins in Sequence Databases

The major form of selenium in cells is Sec residues in proteins. This is illustrated, for example, by the analyses of mice, in which the tRNA$^{[Ser]Sec}$ gene is disrupted in liver (Carlson et al. 2004b). In these animals, liver selenium content is significantly reduced. Similarly, selenoproteins account for most of selenium in body fluids. For example, the main selenoprotein in plasma of mammals is selenoprotein P (SelP), which accounts for more than half of selenium in plasma (Burk and Hill 2005). Selenium may also occur in selenoproteins in the form of a cofactor. For example, in several bacterial selenium-containing molybdoenzymes, such as xanthine dehydrogenase and nicotinic acid hydroxylase, a Se-Mo cofactor is formed in the active site (Gladyshev et al. 1994). This labile cofactor is easily destroyed releasing both elements. The possibility that similar enzymes exist in eukaryotes has not been addressed.

Sec-containing proteins are often misannotated in sequence databases. This is because their TGA codons are interpreted as stop signals by available annotation tools (Gladyshev et al. 2004). It is obviously impossible to identify selenoprotein genes by only searching for TGA codons. However, selenoprotein genes have an RNA structure known as the Sec insertion sequence (SECIS) element (see Chapter 2 for details). SECIS elements are highly specific for selenoprotein genes and possess a sufficiently complex secondary structure (Chapter 2). Initial bioinformatics analyses of selenoprotein genes focused on SECIS elements. In these studies, selenoprotein genes were identified using the following strategy: (1) detection of SECIS elements by searching for conserved stem-loop structures satisfying SECIS consensus sequence and structure; (2) analyzing regions upstream of SECIS elements for coding regions of selenoprotein genes; and (3) computational and experimental analyses of candidate selenoproteins (Kryukov et al. 1999; Lescure et al. 1999; Castellano et al. 2001). This strategy immediately resulted in the identification of several novel selenoproteins. Subsequently, it was applied to entire genomes, identifying full sets of selenoproteins (selenoproteomes) in a variety of organisms. For large and complex genomes, searches were carried with pairs of closely related genomes (e.g., *D. melanogaster* and *D. pseudoobscura*, or human and mouse genomes) by detecting conserved pairs of SECIS elements located upstream of a pair of selenoprotein orthologs (Kryukov et al. 2003). In particular, this strategy was useful in the analysis of the human genome: these analyses were assisted by the availability of mouse and rat sequenced genomes.

A second strategy was also developed wherein selenoproteins can be identified by searching for cysteine (Cys) homologs (Kryukov et al. 2003, 2004; Fomenko et al. 2007). This strategy is based on the observation that most selenoprotein genes have homologs, in which Cys replaces Sec. Thus, protein sequence databases (e.g., NCBI protein database, and ORFs from genome and environmental genome projects) were searched against large nucleotide sequence databases (genomes, ESTs, metagenomics projects, etc.) to identify nucleotide sequences containing an in-frame TGA codon, which, when translated, aligned with Cys-containing protein homologs such that the resulting Sec/Cys pairs were flanked by conserved sequences. It should be noted that such Cys/Sec homology strategy is completely independent of the searches for SECIS elements and thus provided a SECIS-independent tool for selenoprotein detection. In addition, since both strategies (i.e., SECIS based and Sec/Cys pair based) identified identical or nearly identical sets of selenoprotein genes in various genomes, both tools should be viewed as satisfactory and complementary for selenoprotein analyses in sequence databases. Moreover, this observation suggested that the two procedures can identify nearly all or all selenoproteins in sequence databases as well as in completely sequenced genomes.

1.4 Selenoproteins

While the first selenoproteins were discovered in 1973, until recently only a handful of such proteins were known. In fact, the majority of known selenoproteins have been discovered within the last 6 years. Currently, more than 50 selenoprotein families are known (Fig. 1.4). Our laboratories have described many of these proteins and

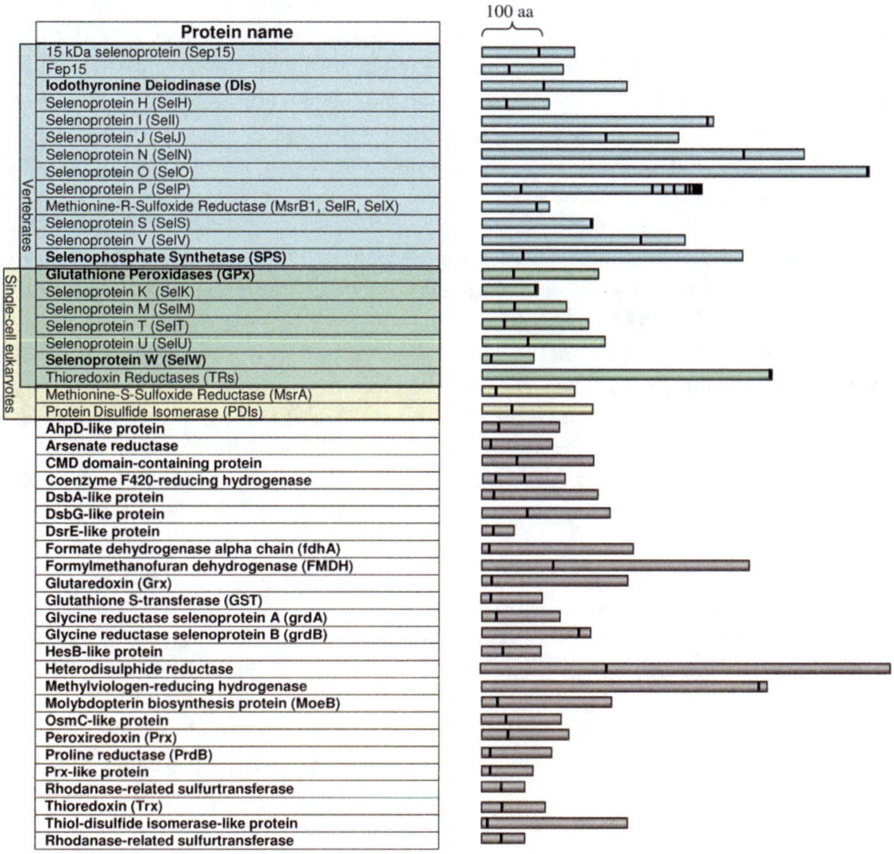

Fig. 1.4 Selenoprotein families. Selenoproteins in vertebrate and single-celled eukaryotes are highlighted, and those selenoproteins shown in bold are also present in bacteria. Other selenoproteins (*lower part of the figure*) are prokaryotic. On the *right side* of the figure, relative lengths of selenoproteins and location of Sec are shown

the reader is referred to the corresponding primary literature (Martin-Romero et al. 2001; Kryukov et al. 2003; Lobanov et al. 2006b, c, 2007; Mix et al. 2007; Lobanov et al. 2008a, b). As several detailed reviews covering selenoproteins and selenoprotein functions have been published recently (Gromer et al. 2005; Schweizer and Schomburg 2005; Hatfield et al. 2006; Holmgren 2006; Moghadaszadeh and Beggs 2006; Papp et al. 2007; Brigelius-Flohe 2008; Gromadzinska et al. 2008; Margis et al. 2008; Schweizer et al. 2008), we do not cover individual selenoproteins here.

1.4.1 Overview of Selenoprotein Functions

Those selenoproteins for which functions have been established are oxidoreductases with Sec located in catalytic sites and serving redox function (Kryukov et al. 2004; Zhang and Gladyshev 2008). By analogy, it may be predicted that many selenoproteins with unknown functions are also oxidoreductases (Fomenko et al. 2007). Sec offers certain catalytic advantages compared with chemically similar Cys residues, including stronger nucleophilicity, lower p*K*a, and lower redox potential. Although Cys homologs are known for the majority of selenoproteins, it appears that the catalytic properties of Cys are insufficient for some proteins and this residue is then replaced with a better catalyst, Sec. On the other hand, Sec may be too reactive and toxic even at very low levels. Thus, there is also a selective pressure for cells to evolve without selenoproteins and Sec. Overall, there appears to be a balance wherein Sec and Cys residues are inter-replaceable depending on protein, environment, substrates, and other factors (Kim and Gladyshev 2005; Lobanov et al. 2007; Zhang and Gladyshev 2008).

Another observation that emerged from the description of individual selenoproteins is that selenoproteins may be loosely classified into three distinct protein groups. In the first group, which is also the most abundant (e.g., 15 out of 25 human selenoproteins belong to this group), selenoproteins have a thioredoxin or thioredoxin-like fold. In addition, nearly all thioredoxin-fold proteins with redox function have selenoprotein homologs. In the second group, Sec is located in the C-terminal sequences, most often in the C-terminal penultimate positions. Such proteins occur in eukaryotes and include selenoproteins K, S, O, I, and TRs. In this group, the role of Sec has been established only for TRs. In the third group, selenoproteins utilize Sec to coordinate redox metals (molybdenum, tungsten, nickel, and perhaps iron) in enzyme active sites. This protein class includes hydrogenase, formate dehydrogenase, formylmethanofuran dehydrogenase, and possibly HesB-like proteins. Finally, several selenoproteins, such as MsrB and GrdB, do not fit the three selenoprotein groups; however, redox catalysis again emerges as the common feature of selenoproteins.

1.5 Selenoproteomes

Efficient tools for detection of selenoproteins in sequence databases, including genomes and metagenomics projects, provided an opportunity for the identification of entire sets of these proteins in organisms (selenoproteomes). In turn, selenoproteomes allowed, for the first time, addressing the roles of selenium in biology at system-wide and organismal levels. These analyses were also used to link individual selenoproteins with dietary effects of selenium and identify new roles of selenium in biology and medicine.

Selenoprotein searches revealed a great variety in size and composition of selenoproteomes. First, there is no single organism that has all known selenoprotein families, even if the three domains of life (eukaryotes, archaea, and bacteria) are separately considered. The largest selenoproteomes have been found in fish, green algae, and some symbiotic bacteria (30–60 selenoproteins) (Lobanov et al. 2007; Lobanov et al. 2008a, b; Zhang and Gladyshev 2007, 2008). However, there are also organisms in each domain of life that lack selenoproteins. These organisms also lost the Sec biosynthesis and insertion system. Once lost, this system is impossible to replace (e.g., via horizontal gene transfer). As a result, three quarters of prokaryotes and about half of eukaryotes, as judged from the genomes which have been sequenced, lack selenoproteins. Second, eukaryotic and prokaryotic selenoproteins are largely different (Fig. 1.4).

Even within animals, selenoproteome varies from zero (e.g., silkworm) to more than 30 (e.g., zebrafish) selenoproteins. It appears that aquatic lifestyle correlates with increased utilization of Sec, whereas terrestrial organisms tend to lose selenoproteins. Environmental factors responsible for these phenotypes are not known. Evolutionary analyses showed that selenoproteins were independently lost in insects (compared to aquatic arthropods), nematodes (compared to other worms), fungi (lost all selenoproteins), higher plants (compared to green algae), and stramenopiles (compared to their distant aquatic relatives). Moreover, even in mammals, there appears to be a trend for reduced utilization of Sec.

1.6 Thioredoxin Reductase and Cancer

When selenium is considered in regard to human health, the best described effect is the role of selenium in cancer prevention. Among selenoproteins implicated in this effect, TR1 plays a particularly enigmatic role as it has been described as both pro- and anti-tumorigenic. Knockdown of TR1 expression in cancer cells using RNAi technology has elucidated its role as a driver of malignancy and further substantiated it as a prime target for cancer therapy (Yoo et al. 2006, 2007). These studies not only demonstrated that many of the cancer-related properties could be reversed in a mouse lung cancer cell line (Yoo et al. 2006), but reduction of TR1 activity in another mouse cancer cell line caused these malignant cells to lose self-sufficiency of growth, manifest a defective progression in their S phase, and manifest a decreased expression of DNA polymerase α (Yoo et al. 2007). Several earlier studies elucidating the function of TR1 in normal and malignant mammalian cells and tissues also set the stage for understanding the role of this protein in cancer. For example, TR1 was known as one of the major antioxidant and redox regulators in mammalian cells (e.g., see Gromer et al. 2005; Lu and Holmgren 2008) and was reported to be an essential selenoprotein (Conrad et al. 2004) that is expressed in all cell types and organs (Behne and Kyriakopoulos 2001; Gromer et al. 2005). In addition, TR1 was observed to be overexpressed in many malignant cells and tissues and its inhibition by various potent cancer drugs was shown to alter the malignant phenotypes of a number of tumor and malignant cell types suggesting that this selenoenzyme was a target for cancer therapy (Rundlöf and Arnér 2006, Biaglow and Miller 2005; Arner and Holmgren 2006; Fujino et al. 2006; Nguyen et al. 2006; Lu et al. 2007). Alternatively, TR1-supported p53 function was known to have other tumor suppressor activities and to be a target for carcinogenic, electrophilic compounds (Moos et al. 2003). Overall, current data argue that TR1 has opposing effects in malignancy development implicating it both in cancer prevention (e.g., see Urig and Becker 2006) and in cancer promotion (Rundlöf and Arnér 2006, Biaglow and Miller 2005; Arner and Holmgren 2006). The data also suggest that TR1 serves as an anti-cancer and antioxidant protein, but once a tumor is formed, it is essential to sustain tumor growth (Hatfield 2007).

1.7 Selenoprotein Knockout Mouse Models

Several selenoproteins have been characterized by targeting their gene removal using genetic engineering techniques primarily adapted for the mouse genome (e.g., see reviews by Schweizer and Schomburg, 2005; Hatfield et al. 2006). Knockout of several selenoproteins has shown that some are essential to development and survival, while others appear to be non-essential. Those selenoproteins that are indispensable for survival have been designated as housekeeping selenoproteins (e.g., TR1) and those that are dispensable have been designated stress-related selenoproteins (e.g., GPx1) (Carlson et al. 2005a). TR1 (Jakupoglu et al. 2005), TR3 (TrxR2) (Conrad et al. 2004), and GPx4 (PHGPX) (Yant et al. 2003) are all essential selenoproteins as their knockout was embryonic lethal. Conversely, knockout of GPx1 (Ho et al. 1997) or GPx2 (Esworthy et al. 2001) had little or no effect on phenotype and these two selenoproteins are therefore non-essential to survival.

1.8 Sec tRNA Knockout and Transgenic Mouse Models

The fact that selenoprotein expression is uniquely dependent on a single tRNA, $tRNA^{[Ser]Sec}$, which is present in single copy in the genomes of mammals, also provides a means of elucidating the function of selenoproteins and their role in health and development (Hatfield et al. 2006). Altering the expression of $tRNA^{[Ser]Sec}$ can in turn alter the expression of selenoproteins. Several mouse models employing $tRNA^{[Ser]Sec}$ for assessing the functions of selenoproteins and their roles in health and development have been generated wherein the mice carry (1) *Trsp* wild-type or mutant transgenes, (2) *Trsp* wild-type or mutant transgenes and a standard or conditional knockout*Trsp*, or (3) *Trsp* conditional knockout. Total knockout of *Trsp* is embryonic lethal (Bosl et al. 1997; Kumaraswamy et al. 2003) and such animals can only survive if they are rescued by introducing a wild-type or mutant transgenes (Carlson et al. 2005a, b). Studies involving *Trsp* transgenic mouse models, *Trsp* transgenic/*Trsp* standard or conditional mouse models, and *Trsp* conditional knockout mouse models using *loxP-Cre* technology are summarized in Tables 1.1, 1.2, and 1.3, respectively.

Mutant *Trsp* transgenes used in mouse models prepared thus far contain a mutation at position 37 (A→G; designated i^6A^-) or at position 34 (T → A; designated A34) (Table 1.1). Mutations at either position yield a *Trsp* product that lacks Um34 (Kim et al. 2000). The initial study involving mutant *Trsp* transgenic mice carried the i^6A^- transgene and resulted in the levels of some selenoproteins being dramatically reduced (e.g., GPx1), while others were unaffected or slightly increased (e.g., TR3), and selenoprotein expression was most and least affected in liver and

Table 1.1 Mutant *Trsp* transgenic mouse models

Transgene[a]	Model description	Major findings	Authors
A37→G37 (i^6A^-)	Mice encode a mutant i6A$^-$ transgene in all tissues and organs.	Levels of stress-related selenoproteins decreased in a protein- and tissue-specific manner in mice expressing the mutant i^6A^- tRNA$^{[Ser]Sec}$ isoform. GPx1 and TR3 were the most and least affected selenoproteins, while selenoprotein expression was most and least affected in the liver and testes, respectively.	Moustafa et al. (2001)
A37→G37 (i^6A^-)	Mice encode a mutant i6A$^-$ transgene in all tissues and organs.	Mice manifest exercise-induced growth following synergist ablation.	Hornberger et al. (2003)
A37→G37 (i^6A^-)	Mice encode a mutant i6A$^-$ transgene in all tissues and organs. Colon is targeted with azoxymethane.	Mice had more azoxymethane-induced aberrant crypt formation (a preneoplastic lesion for colon cancer). First demonstration that low molecular weight selenocompounds and selenoproteins reduce colon cancer incidence.	Irons et al. (2006)
A37→G37 (i^6A^-)	Mice encode a mutant i6A$^-$ transgene in all tissues and organs.	Mutant mice exhibited accelerated development of lesions associated with prostate cancer progression.	Diwadkar-Navsariwala et al. (2006)
A37→G37 (i^6A^-)	Mice encode a mutant i6A$^-$ transgene in all tissues and organs. Lung is targeted by administration of influenza virus.	Although immune system changes were observed following influenza viral infection, lung pathology was similar in i^6A^- and WT mice.	Sheridan et al. (2007)
A37→G37 (i^6A^-)	Mice encode a mutant i6A$^-$ transgene in all tissues and organs.	Selenoproteins have a role in protecting DNA from damage.	Baliga et al. (2008)

[a]Transgene – the numbers in brackets refer to the number of transgene copies inserted in the genome.

testes, respectively (Moustafa et al. 2001). Since an isoform containing Um34 was not synthesized from the mutant i^6A^- transgene and the tRNA$^{[Ser]Sec}$ population was enriched at the expense of the mcm^{5}Um isoform, this study provided the first direct evidence that two tRNA$^{[Ser]Sec}$ isoforms are used in synthesizing different subclasses of selenoproteins that were subsequently called stress-related selenoproteins (e.g., GPx1 and GPx3) and housekeeping selenoproteins (e.g., TR1 and

TR3) (Carlson et al. 2005a). Further studies confirmed the selective use of the two tRNA$^{[Ser]Sec}$ isoforms in selenoprotein synthesis. *Trsp* knockout mice that were rescued with the i^6A^- transgene synthesized stress-related selenoproteins poorly

Table 1.2 Mutant *Trsp* transgenic/conditional or standard knockout mouse models

Transgene[a]	Model description	Major findings	Authors
A37→G37 (i^6A^-)	All tissues lack a wild-type *Trsp* gene and are rescued with a mutant i6A$^-$ transgene.	The absence of Um34 plays a major role in the expression of stress-related selenoproteins, but not housekeeping selenoproteins.	Carlson et al. (2005a,b)
T34→A34 (mcmU$^-$ and Um34$^-$); A37→G37 (i^6A^-)	*Trsp* is knocked out in liver and mouse encodes either mutant T34→A34 or i^6A transgene.	Both mutant tRNAs lacked Um34, and both supported expression of housekeeping selenoproteins (e.g. TR1), but stress-related proteins (e.g. GPx 1) poorly. Um34 is responsible for synthesis of a select group of selenoproteins, the stress-related selenoproteins, in liver rather than the entire selenoprotein population.	Carlson et al. (2007)
T34→A34 (mcmU$^-$ and Um34$^-$); A37→G37 (i^6A^-)	*Trsp* is knocked out in liver and mouse encodes either mutant T34→A34 or i^6A^-transgene.	In *Trsp* mutant mouse lines, the expression of ApoE, as well as genes involved in cholesterol biosynthesis, metabolism, and transport were similar to those observed in wild-type mice indicating for the first time that housekeeping selenoproteins have a role in regulating lipoprotein biosynthesis and metabolism.	Sengupta et al. (2008)
T34→A34 (mcmU$^-$ and Um34$^-$); A37→G37 (i^6A^-)	*Trsp* is knocked out in liver and mouse encodes either mutant T34→A34 or i^6A transgene.	The loss of selenoproteins in liver was compensated for by an enhanced expression of several phase II response genes and their corresponding gene products. The replacement of selenoprotein synthesis in mice carrying mutant *Trsp* transgenes led to normal expression of phase II response genes. **Provides evidence for a functional link between housekeeping selenoproteins and phase II enzymes.**	Sengupta et al. (2008)

[a]Transgene – the numbers in brackets refer to the number of transgene copies inserted in the genome.

Table 1.3 Conditional Trsp knockout mouse models

Cre	Targeted organ	Major findings	Authors
MMTV-Cre	Mammary gland	First description of the *Trsp* conditional knockout mouse.	Kumaraswamy et al. (2003)
Alb-Cre	Liver	Death between 1 and 3 months of age due to severe hepatocellular degeneration and necrosis. **Selenoproteins have a role in proper liver function.**	Carlson et al. (2004b)
TieTek2-Cre	Endothelial cell	14.5 dpc embryos were smaller in size, more fragile, had a poorly developed vascular system, underdeveloped limbs and tails and smaller heads. **Selenoproteins have a role in endothelial cell function.**	Shrimali et al. (2007)
MCK-Cre	Heart and skeletal muscle	Died from acute myocardial failure day 12 after birth. **Selenoproteins have a role in preventing heart disease.**	Shrimali et al. (2007)
LysM-Cre	Macrophage	Elevated oxidative stress and transcriptional induction of cytoprotective antioxidant and detoxification enzyme genes.	Suzuki et al. (2008)
Alb-Cre	Liver	Compensatory induction of cytoprotective antioxidant and detoxification enzyme genes by *Nrf2*.	Suzuki et al. (2008)
NPHS2-Cre	Kidney	Loss of podocyte selenoproteins does not lead to increased oxidative stress nor worsening nephropathy.	Blauwkamp et al. (2008)
LCK-Cre	T cells	Decreased pools of mature T cells and a defect in T cell-dependent antibody responses. Antioxidant hyperproduction in T cells and thereby suppression of T cell proliferation in response to T cell receptor stimulation. **Selenoproteins have a role in immune function**	Shrimali et al. (2008)
Tα1-Cre	Neuron Specific	Enhanced neuronal excitation followed by massive neurodegeneration of the hippocampus. Cerebellar hypoplasia was associated with degeneration of Purkinje and granule cells. **Selenoproteins have a role in neuronal function.**	Wirth et al.
Col2a1-Cre	Osteo-chondroprogenitor	Post-natal growth retardation, chondrodysplasia, chondronecrosis and delayed skeletal ossification characteristic of Kashin-Beck disease. **Model for Kashin-Beck disease.**	Jirik et al.
LysM-Cre	Macrophage	Accumulation of ROS levels and impaired invasiveness. Altered expression of several extracellular matrix and fibrosis-associated genes. **Selenoproteins have a role in immune function.**	Carlson et al.[a]

Table 1.3 (continued)

Cre	Targeted organ	Major findings	Authors
K14-Cre	Skin	Runt phenotype, premature death, alopecia along with a flaky and fragile skin, epidermal hyperplasia along with changes in hair follicle appearance wherein the hair cycle was disturbed with an early regression of hair follicles. **Selenoproteins have a role in skin and hair follicle development.**	Sengupta et al.[a]
MMTV-Cre	Mammary gland	Mice develop tumors more rapidly following DMBA treatment or with a cancer driver gene.	Hudson et al.[a]

[a]In preparation (see text).

(Carlson et al. 2005a, b), and *Trsp* conditional knockout mice, wherein the selenoprotein population was partially replaced with either the i^6A^- or the A34 transgene, also synthesized stress-related selenoproteins poorly (see Carlson et al. 2007 and Table 1.2). The common denominator between the two *Trsp* mutant $tRNA^{[Ser]Sec}$s is that they both lacked Um34. The most obvious parameters that would be expected to influence how the two $tRNA^{[Ser]Sec}$ isoforms are selectively used in synthesizing subclasses of selenoproteins have been considered in detail (Hatfield et al. 2006) and none of these appear to play a role. Our attention more recently has focused on the Um34 methylase as the most likely factor playing a major role in how mcm^5U and mcm^5Um are selectively used (D.L. Hatfield and V.N. Gladyshev, unpublished data).

The i^6A^- transgenic mouse model has also proven to be a useful model in elucidating the role of selenoproteins in health and development. For example, this model has been used to show that (1) both selenoproteins and low molecular weight selenocompounds play a role in reducing colon cancer incidence (Irons et al. 2006), (2) selenoproteins play a role in reducing prostate cancer incidence (Diwadkar-Navsariwala et al. 2006), (3) plantaris muscles from mice in i^6A^- mice exhibited enhanced exercise-induced growth following synergist ablation (Hornberger et al. 2003), (4) influenza virally infected i^6A^- mice manifest an altered immune response that did not affect lung pathology (Sheridan et al. 2007), and (5) selenoproteins play a role in protecting DNA from damage (Baliga et al. 2008).

Conditional *Trsp* knockout mouse models that result in abolishing selenoprotein expression in various targeted tissues and organs have also provided insights into the role of this class of proteins in a variety of health and developmental issues (Table 1.3). The initial study that knocked out *Trsp* in epithelial cells of mammary tissue manifested virtually no phenotypic changes, but a selective reduction of some selenoproteins such as Sep15 and GPx1 was observed (Kumaraswamy et al. 2003). Challenging these mice with the carcinogen, DMBA, or with a cancer driver fusion gene, C3(1)/SV40 Tag, however, showed enhanced tumor formation in breast tissue of the knockout mice (T. Hudson, B.A. Carlson, D.L. Hatfield, J.E. Green, unpublished data). The targeted removal of *Trsp* in endothelial cells demonstrated a role of selenoproteins in their development as the selenoproteinless animals did not develop beyond the embryonic stage and at 14.5 dpc embryos had poorly developed vascular systems and smaller heads and bodies than their normal siblings (Shrimali et al. 2007). Mice in which *Trsp* was deleted in heart and skeletal muscle died abruptly of acute myocardial failure on day 12 after birth demonstrating that selenoproteins have a role in heart disease prevention (Shrimali et al. 2007).

In addition, the role of selenoproteins in immune function has been examined. Selenoprotein expression was abolished in T cells by *Trsp* deletion and the resulting selenoproteinless cells exhibited decreased pools of mature T cells, a defect in T-cell-dependent antibody responses and an oxidant hyperproduction that is likely responsible for suppressing T-cell proliferation in response to T-cell simulation (Shrimali et al. 2008). Knockout of *Trsp* in macrophages (and in liver) resulted in increased oxidative stress and in the induction of cytoprotective antioxidant and detoxification enzymes (Suzuki et al. 2008) and an accumulation of ROS levels and impaired invasiveness (B.A. Carlson, M.-H. Yoo, R. Irons, V.N. Gladyshev, J.M. Park, D.L. Hatfield, unpublished data). The osteo-chondroprogenitor-specific deletion of *Trsp* exhibited decreased growth, chondronecrosis, chondrodysplasia, and reduced skeletal ossification (Downey et al. 2009). Since patients of Kashin–Beck disease manifest similar characteristics of stunted growth, chondronecrosis, chondrodysplasia, and reduced skeletal ossification as the *Trsp* knockout osteo-chondroprogenitor mice, these mice likely provide an excellent model for studying this disease (F.R. Jirik, personal communication). The neuron-specific deletion of *Trsp* generated a mouse line with enhanced neuronal excitation followed by massive neurodegeneration in the cerebral cortex and hippocampus and death in the second week after birth (Wirth et al. 2009). Finally, *Trsp* removal in the epidermis of skin has generated mice-manifesting retarded growth, alopecia with flaky and fragile skin, epidermal hyperplasia that is accompanied by changes in hair follicle appearance, and early death, usually by day 9 (A. Sengupta, U. Lichti, B.A. Carlson, A. Ryscavage, V.N. Gladyshev, S. Yuspa, D.L. Hatfield, unpublished data).

Although the targeted removal of *Trsp* in specific tissues and organs has provided many new insights and detected novel roles of selenoproteins in development and disease, a limitation in this approach is that the effects cannot be attributed to a single selenoprotein or selenoproteins. This can partially be rectified by targeting the removal of specific selenoprotein genes such as TR1 (Conrad et al. 2004) and GPx4 (Jadupoglu et al. 2005) in tissues and organs in which *Trsp* has been removed. Interestingly, deletion of TR1 or GPx4 in neurons has been carried out and TR1 loss had no apparent effect, while GPx4 loss exhibited many of the same deleterious effects as the removal of *Trsp* (Wirth et al. 2009).

Acknowledgments The authors express their sincere appreciation to Bradley A. Carlson for his help with the figures and proofreading of the manuscript, Joyce Ore for proofreading, and Alexey V. Lobanov and Dmitri E. Fomenko for help with the figures. This research was supported by

NIH grants to VNG and by the intramural Research Program of the National Institutes of Health, National Cancer Institute, Center for Cancer Research, to DLH.

References

Arnér ES, Holmgren A (2006) The thioredoxin system in cancer. Semin Cancer Biol 16:420–426

Behne D, Kyriakopoulos A (2001) Mammalian selenium-containing proteins. Annu Rev Nutr 21:453–473

Baliga MS, Diwadkar-Navsariwala V, Koh T, Fayad R, Fantuzzi G, Diamond AM (2008) Selenoprotein deficiency enhances radiation-induced micronuclei formation. Mol Nutr Food Res 52:1300–1304

Bell SD, Jackson SP (1998) Transcription and translation in Archaea: a mosaic of eukaryal and bacterial features. Trends Microbiol 6:222–228

Biaglow, JE Miller, RA (2005) The thioredoxin reductase/thioredoxin system: novel redox targets for cancer therapy. Cancer Biol Ther 4:6–13

Blauwkamp MN, Yu J, Schin MA, Burke KA, Berry MJ, Carlson BA, Brosius FC 3rd, Koenig RJ (2008) Podocyte specific knock out of selenoproteins does not enhance nephropathy in streptozotocin diabetic C57BL/6 mice. BMC Nephrol 9:7

Böck, A Forchhammer, K Heider, J Baron, C (1991) Selenoprotein synthesis: an expansion of the genetic code. Trends Biochem Sci 16:463–467

Bosl MR, Takaku K, Oshima M, Nishimura S, Taketo MM (1997) Early embryonic lethality caused by targeted disruption of the mouse selenocysteine tRNA gene (Trsp). Proc Natl Acad Sci USA 94:5531–5534

Brigelius-Flohe R (2008) Selenium compounds and selenoproteins in cancer. Chem Biodiversity 5:389–395

Burk RF, Hill, KE (2005) Selenoprotein P: an extracellular protein with unique physical characteristics and a role in selenium homeostasis. Annu Rev Nutr 25:215–235

Carlson BA, Xu X.-M, Kryukov GV, Rao M, Berry MJ, Gladyshev VN, Hatfield DL (2004a) Identification and characterization of phosphoseryl-tRNA[Ser]Sec kinase. Proc Natl Acad Sci USA 101:12848–12853

Carlson BA, Novoselov SV, Kumaraswamy E, Lee BJ, Anver MR, Gladyshev VN, Hatfield DL (2004b) Specific excision of the selenocysteine tRNA[Ser]Sec (Trsp) gene in mouse liver demonstrates an essential role of selenoproteins in liver function. J Biol Chem 279:8011–8017

Carlson BA, Xu X-M, Gladyshev VN, Hatfield DL (2005a) Um34 in selenocysteine tRNA is required for the expression of stress-related selenoproteins in mammals. In: Grosjean H (ed) Topics in current genetics. Springer-Verlag, Berlin-Heidelberg, pp 431–438

Carlson BA, Xu X-M, Gladyshev VN, Hatfield DL (2005b) Selective rescue of selenoprotein expression in mice lacking a highly specialized methyl group in Sec $tRNA^{[Ser]Sec}$. J Biol Chem 280:5542–5548

Carlson B, Moustafa M, Sengupta A, Schweizer U, Shrimali R, Rao M, Zhong N, Wang S, Feigenbaum L, Lee B, Gladyshev V, Hatfield D (2007) Selective restoration of the selenoprotein population in a mouse hepatocyte selenoproteinless background with different selenocysteine tRNAs lacking Um34. J Biol Chem 282:32591–32602

Castellano S, Morozova N, Morey M, Berry MJ, Serras F, Corominas M, Guigo R (2001) In silico identification of novel selenoproteins in the Drosophila melanogaster genome. EMBO Rep 2:697–702

Chambers I, Frampton J, Goldfarb P, Affara N, McBain W, Harrison PR (1986) The structure of the mouse glutathione peroxidase gene: The selenocysteine in the active site is encoded by the "termination" codon, TGA. EMBO J 5:1221–1227

Chittum HS, Hill KE, Carlson BA, Lee BJ, Burk RF, Hatfield DL (1997) Replenishment of selenium deficient rats with selenium results in redistribution of the selenocysteine tRNA population in a tissue specific manner. Biochim Biophys Acta 1359:25–34

Conrad M, Jakupoglu C, Moreno SG, Lippl S, Banjac A, Schneider M, Beck H, Hatzopoulos AK, Just U, Sinowatz F, Schmahl W, Chien KR et al (2004) Essential role for mitochondrial thioredoxin reductase in hematopoiesis, heart development, and heart function. Mol Cell Biol 24:9414–9423

Diwadkar-Navsariwala V, Prins GS, Swanson SM, Birch LA, Ray VH, Hedayat S, Lantvit DL, Diamond AM (2006) Selenoprotein deficiency accelerates prostate carcinogenesis in a transgenic model. Proc Natl Acad Sci USA 103:8179–8184

Downey CM, Horton CR, Carlson BA, Parsons TE, Hatfield DL, Hallgrimsson B, Jirik F (2009) Osteo-chondroprogenitor-specific deletion of the selenocysteine tRNA gene, *Trsp*, leads to chondronecrosis and abnormal skeletal development: a putative model for Kashin-Beck disease. PLoS Genetics In Press

Esworthy RS, Aranda R, Martin MG, Doroshow JH, Binder SW, Chu FF (2001) Mice with combined disruption of Gpx1 and Gpx2 genes have colitis. Am J Physiol Gastrointest Liver Physiol 281:G848–G855

Fomenko DE, Xing W, Adair BM, Thomas DJ, Gladyshev VN (2007) High-throughput identification of catalytic redox-active cysteine residues. Science 315:387–389

Fujino G, Noguchi T, Takeda K, Ichijo H (2006) Thioredoxin and protein kinases in redox signaling. Semin Cancer Biol 16:427–435

Gelpi C, Sontheimer EJ, Rodriguez-Sanchez JL (1992) Autoantibodies against a serine tRNA-protein complex implicated in cotranslational selenocysteine insertion. Proc Natl Acad Sci USA 89:9739–9743

Gladyshev VN, Khangulov SV, Stadtman TC (1994) Nicotinic acid hydroxylase from Clostridium barkeri: Electron paramagnetic resonance studies show that selenium is coordinated with molybdenum in the catalytically active selenium-dependent enzyme. Proc Natl Acad Sci 91:232–236

Gladyshev VN, Kryukov GV, Fomenko DE, Hatfield DL (2004) Identification of trace element-containing proteins in genomic databases. Annu Rev Nutr 24:579–596

Glass RS, Singh WP, Jung W, Veres Z, Scholz TD, Stadtman TC (1993) Monoselenophosphate: synthesis, characterization, and identity with the prokaryotic biological selenium donor, compound SePX. Biochemistry 32:12555–12559

Gromadzinska J, Reszka I, Bruzelius K, Wasowicz W, Akesson B (2008) Selenium and cancer: biomarkers of selenium status and molecular action of selenium supplements. Eur J Nutri 47:29–50

Gromer S, Eubel JK, Lee BL, Jacob J (2005) Human selenoproteins at a glance. Cell Mol Life Sci 62:2414–2437

Guimaraes, MJ Peterson, D Vicari, A Cocks, BG Copeland, NG Gilbert, DJ, Jenkins NA, Ferrick DA, Kastelein RA, Bazan JF, Zlotnik A (1996) Identification of a novel selD homolog from eukaryotes, bacteria, and archaea: Is there an autoregulatory mechanism in selenocysteine metabolism? Proc Natl Acad Sci USA 93:15086–15091

Hao B, Gong W, Ferguson TK, James CM, Krzycki JA, Chan MK (2002) A new UAG-encoded residue in the structure of methanogen methyltransferase. Science 296:1462–1466

Hatfield, D Portugal FH (1970) Seryl-tRNA in mammalian tissues: Chromatographic differences in brain and liver and a specific response to the codon, UGA. Proc Natl Acad Sci USA 67: 1200–1206

Hatfield DL, Diamond A, Dudock B (1982) Opal suppressor serine tRNAs from bovine liver form phosphoseryl-tRNA. Proc Natl Acad Sci USA 79:6215–6219

Hatfield DL, Choi IS, Ohama T, Jung J-E, Diamond AM (1994) Selenocysteine tRNA(Ser)sec isoacceptors as central components in selenoprotein biosynthesis in eukaryotes. In: Burk RF (ed), Selenium in biology and human health. Springer-Verlag, New York, pp 25–44

Hatfield DL, Gladyshev VN, Park JM, Park SI, Chittum HS, Huh JH, Carlson BA, Kim M, Moustafa ME, Lee BJ (1999) Biosynthesis of selenocysteine and its incorporation into protein as the 21st amino acid. In: Kelly JW (ed), Comprehensive natural products chemistry, Vol. 4. Elsevier Science, Oxford, England, pp 353–380

Hatfield DL, Gladyshev VN (2002) How selenium has altered our understanding of the genetic code. Mol Cell Biol 22:3565–3576

Hatfield D, Xu X. -M, Carlson BA, Zhong N, Gladyshev, VN (2006) Selenocysteine incorporation machinery and the role of selenoproteins in health. Prog Nucl Acid Res Mol Biol 81:97–142

Hatfield DL (2007) Thioredoxin reductase 1. A double-edged sword in cancer prevention and promotion. CCR Frontiers in Science 6:8–10

Heider J, Bock A (1993) Selenium metabolism in microorganisms. Adv Microb Physiol 35:71–109

Ho YS, Magnenat JL, Bronson RT, Cao J, Gargano M, Sugawara M, Funk CD (1997) Mice deficient in cellular glutathione peroxidase develop normally and show no increased sensitivity to hyperoxia. J Biol Chem 272:16644–16651

Holmgren A (2006) Selenoproteins of the thioredoxin system. In: Hatfield DL, Berry MJ, Gladyshev VN (eds), Selenium: its molecular biology and role in human health, 2nd ed. Springer Science+Business Media, New York, pp 183–194

Hornberger TA, McLoughlin TJ, Leszczynski JK, Armstrong DD, Jameson RR, Bowen PE, Hwang ES, Hou H, Moustafa ME, Carlson BA, Hatfield DL, Diamond AM, Esser KA (2003) Selenoprotein-deficient transgenic mice exhibit enhanced exercise-induced muscle growth. J Nutr 133:3091–3097

Hubert N, Sturchler C, Westhof E, Carbon P, Krol A (1998) The 9/4 secondary structure of eukaryotic selenocysteine tRNA: More pieces of evidence. RNA 4:1029–1033

Irons R, Carlson BA, Hatfield DL, Davis CD (2006) Both selenoproteins and low molecular weight selenocompounds reduce colon cancer risk in mice with genetically impaired selenoprotein expression. J Nutri 136:1311–1317

Jukes TH, Osawa S (1990) The genetic code in mitochondria and chloroplasts. Experientia 46:1117–1126

Jakupoglu C, Przemeck GK, Schneider M, Moreno SG, Mayr N, Hatzopoulos AK, de Angelis MH Wurst W, Bornkamm GW, Brielmeier M, Conrad M (2005) Cytoplasmic thioredoxin reductase is essential for embryogenesis but dispensable for cardiac development. Mol Cell Biol 25: 1980–1988

Kernebeck T, Lohse AW, Grotzinger J (2001) A bioinformatical approach suggests the function of the autoimmunehepatitis target antigen soluble liver antigen/liver pancreas. Hepatology 34:230–233

Kim IY, Stadtman TC (1995) Selenophosphate synthetase: Detection in extracts of rat tissues by immunoblot assay and partial purification of the enzyme from the archaean Methanococcus vannielii. Proc Natl Acad Sci USA 92:7710–7713

Kim IY, Guimaraes MJ, Zlotnik A, Bazan JF, Stadtman TC (1997) Fetal mouse selenophosphate synthetase 2 (SPS2): characterization of the cysteine mutant form overproduced in a baculovirus-insect cell system. Proc Natl Acad Sci USA 94: 418–421

Kim TS, Yu MH, Chung YW, Kim J, Choi EJ, Ahn K, Kim IY (1999) Fetal mouse selenophosphate synthetase 2 (SPS2): biological activities of mutant forms in *Escherichia coli*. Mol Cells 9: 422–428

Kim LK, Matsufuji T, Matsufuji S, Carlson BA, Kim SS, Hatfield DL, Lee BJ (2000) Methylation of the ribosyl moiety at position 34 of selenocysteine tRNA[Ser]Sec is governed by both primary and tertiary structure. RNA 6:1306–1315

Kim HY, Gladyshev VN (2005) Different catalytic mechanisms in mammalian selenocysteine- and cysteine-containing methionine-R-sulfoxide reductases. PLoS Biol 3:e375

Kryukov, GV Kryukov, VM Gladyshev, VN (1999) New mammalian selenocysteine-containing proteins identified with an algorithm that searches for selenocysteine insertion sequence elements. J Biol Chem 274:33888–33897

Kryukov GV, Castellano S, Novoselov SV, Lobanov AV, Zehtab O, Guigo R, Gladyshev VN (2003) Characterization of mammalian selenoproteomes. Science 300:1439–1443

Kryukov, GV Gladyshev, VN (2004) The prokaryotic selenoproteome. EMBO Rep 5:538–543

Kumaraswamy E, Carlson BA, Morgan F, Miyoshi K, Robinson G, Su D, Wang S, Southon E, Tessarollo L, Lee BJ, Gladyshev VN, Hennighausen L, Hatfield, DL (2003) Selective removal of the selenocysteine $tRNA^{[Ser]Sec}$ gene (*Trsp*) in mouse mammary epithelium. Mol Cell Biol 23:1477–1488

Lee BJ, de la Pena P, Tobian JA, Zasloff M, Hatfield D (1987). Unique pathway of expression of an opal suppressor phosphoserine tRNA. Proc Natl Acad Sci USA 84:6384–6388

Lee BJ, Worland PJ, Davis JN, Stadtman TC, Hatfield DL (1989a) Identification of a selenocysteyl-tRNA(Ser) in mammalian cells that recognizes the nonsense codon, UGAJ. Biol Chem 264:9724–9727

Lee BJ, Kang SK, Hatfield D (1989b) Transcription of Xenopus selenocysteine $tRNA^{Ser}$ (formerly designated opal suppressor phosphoserine tRNA) is directed by multiple 5′ extragenic regulatory elements. J Biol Chem 264:9696–9702

Leinfelder W, Stadtman TC, Bock A (1989) Occurrence in vivo of selenocysteyl-tRNA(SERUCA) in *Escherichia coli*. Effect of sel mutations. J Biol Chem 264:9720–9723

Lescure A, Gautheret D, Carbon P, Krol A (1999) Novel selenoproteins identified in silico and in vivo by using a conserved RNA structural motif. J Biol Chem 274:38147–38154

Lobanov AV, Kryukov GV, Hatfield DL, Gladyshev VN (2006a) Is there a 23rd amino acid in the genetic code? Trends Genet 22:357–360

Lobanov AV, Delgado C, Rahlfs S, Novoselov SV, Kryukov GV, Gromer S, Hatfield DL, Becker K, Gladyshev VN (2006b) The plasmodium selenoproteome. Nucl Acids Res 34: 496–505

Lobanov AV, Gromer S, Salinas G, Gladyshev VN (2006c) Selenium metabolism in Trypanosoma: characterization of selenoproteomes and identification of a Kinetoplastida-specific selenoprotein. Nucleic Acids Res 34:4012–4024

Lobanov AV, Fomenko DE, Zhang Y, Sengupta A, Hatfield DL, Gladyshev VN (2007) Evolutionary dynamics of eukaryotic selenoproteomes: large selenoproteomes may associate with aquatic and small with terrestrial life. Genome Biol 8:R198

Lobanov AV, Hatfield DL, Gladyshev VN (2008a) Selenoproteinless animals: selenophophate synthetase SPS1 functions in a pathway unrelated to selenocysteine biosynthesis. Protein Sci 17:176–182

Lobanov AV, Hatfield DL, Gladyshev VN (2008b) Reduced reliance on the trace element selenium during evolution of mammals. Genome Biol 9:R62

Low SC, Harney JW, Berry MJ (1995) Cloning and functional characterization of human selenophosphate synthetase, an essential component of selenoprotein synthesis. J Biol Chem 270:21659–21664

Lu J, Chew EH, Holmgren A (2007) Targeting thioredoxin reductase is a basis for cancer therapy by arsenic trioxide. Proc Natl Acad Sci USA 104:12288–12293

Lu, J, Holmgren A (2008) Selenoproteins. J Biol Chem 284:723–727

Maenpaa PH, Bernfield MR (1970) A specific hepatic transfer RNA for phosphoserine. Proc Natl Acad Sci USA 67:688–695

Margis R, Dunand C, Teixeria FK, Margis-Pinheiro M (2008) Glutathione peroxidase family – an evolutionary overview. FEBS J 275:3959–3870

Martin-Romero FJ, Kryukov GV, Lobanov AV, Carlson BA, Lee BJ, Gladyshev VN, Hatfield DL (2001) Selenium metabolism in Drosophila: selenoproteins, selenoprotein mRNA expression, fertility and mortality. J Biol Chem 276:29798–29804

Meyer F, Schmidt HJ, Plümper E, Hasilik A, Mersmann G, Meyer HE, Engström A, Heckmann K (1991) UGA is translated as cysteine in pheromone 3 of *Euplotes octocarinatus*. Proc Natl Acad Sci USA 88:3758–3762

Mix H, Lobanov AV, Gladyshev VN (2007) SECIS elements in the coding regions of selenoprotein transcripts are functional in higher eukaryotes. Nucleic Acids Res 35:414–423

Moos PJ, Edes K, Cassidy P, Massuda E, Fitzpatrick FA (2003) Electrophilic prostaglandins and lipid aldehydes repress redox-sensitive transcription factors p53 and hypoxia-inducible factor by impairing the selenoprotein thioredoxin reductase. J Biol Chem 278:745–750

Moghadaszadeh B, Beggs AH, (2006) Selenoproteins and their impact on human health through diverse physiological pathways. Physiology 21:307–315

Mourier, T Pain, A Barrell B, Griffiths-Jones S (2005) A selenocysteine tRNA and SECIS element in Plasmodium falciparum. RNA 11:119–122

Moustafa ME, Carlson BA, El-Saadani MA, Kryukov GV, Sun QA, Harney JW, Hill KE, Combs GF, Feigenbaum L, Mansur DB, Burk RF, Berry MJ, Diamond AM, Lee BJ, Gladyshev VN, Hatfield DL (2001) Selective inhibition of selenocysteine tRNA maturation and selenoprotein synthesis in transgenic mice expressing isopentenyladenosine-deficient selenocysteine tRNA. Mol Cell Biol 21:3840–3852

Nguyen P, Awwad RT, Smart DD, Spitz DR, Gius D (2006) Thioredoxin reductase as a novel molecular target for cancer therapy. Cancer Lett 236:164–174

Nirenberg M, Caskey T, Marshall R, Brimacombe R, Kellogg D, Doctor B, Hatfield DL, Levin J, Rottman F, Pestka S, Wilcox M, Anderson F (1966) The RNA code and protein synthesis. Cold Spring Harbor Symposium on Quant Biol 31:11–24

Papp LV, Lu J, Holmgren A, Khanna KK (2007) From selenium to selenoproteins: synthesis, identity, their role in human health. Antioxid Redox Signal 9:775–806

Pesole G, Lotti M, Alberghina L, Saccone C (1995) Evolutionary origin of non-universal CUG(Ser) codon in some *Candida* species as inferred from a molecular phylogeny. Genetics 141:903–907

Rundlöf AK, Arnér ES (2006) Regulation of the mammalian selenoprotein thioredoxin reductase 1 in relation to cellular phenotype, growth, signaling events. Antioxid. Redox Signal 6:41–52

Schweizer U, Schomburg L (2005) New insights into the physiological actions of selenoproteins from genetically modified mice. IUBMB Life 57:737–744

Schweizer U, Chiu J, Köhrle J (2008) Peroxides and peroxide-degrading enzymes in the thyroid. Antioxid Redox Signal 10:1577–1591

Sengupta A, Carlson BA, Weaver JA, Novoselov SV, Fomenko DE, Gladyshev VN, Hatfield DL (2008) A functional link between housekeeping selenoproteins and phase II enzymes. Biochem J 413:151–161

Sheridan PA, Zhong N, Carlson BA, Perella CM, Hatfield DL, Beck MA (2007) Decreased selenoprotein expression alters the immune response during influenza virus infection in mice. J Nutr 137:1466–1471

Shrimali RK, Weaver JA, Miller GF, Carlson BA, Novoselov SN, Kumaraswamy E, Gladyshev VN, Hatfield DL (2007) Selenoprotein expression is essential in endothelial development and cardiac muscle function demonstrating a direct link between loss of selenoprotein expression and cardiovascular disease. Neuromuscular Disorders 17:135–142

Shrimali RK, Irons RD, Carlson BA, Sano Y, Gladyshev VN, Park JM, Hatfield DL (2008) Selenoproteins mediate T cell immunity through an antioxidant mechanism. J Biol Chem 283:20181–20185

Small-Howard A, Morozova N, Stoytcheva Z, Forry EP, Mansell JB, Harney JW, Carlson BA, Xu X-M, Hatfield DL, Berry M.J (2006) A supra-molecular complex mediates selenocysteine incorporation in vivo. Mol Cell Biol 26:2337–2346

Srinivasan G, James CM, Krzycki JA (2002) Pyrrolisine encoded by UAG in archaea: charging of a UAG-decoding specialized tRNA. Science 296:1459–1462

Sunde RA and Evenson JK (1987) Serine incorporation into the selenocysteine moiety of glutathione peroxidase. J Biol Chem 262:933–937

Suzuki T, Kelly VP, Motohashi H, Nakajima O, Takahashi S, Nishimura S, Yamamoto M (2008) Deletion of the selenocysteine tRNA gene in macrophages and liver results in compensatory gene induction of cytoprotective enzymes by Nrf2. JBC 283:2021–2080

Tamura T, Yamamoto S, Takahata M, Sakaguchi H, Tanaka H, Stadtman, TC Inagaki, K (2004) Selenophosphate synthetase genes from lung adenocarcinoma cells: Sps1 for recycling L-selenocysteine and Sps2 for selenite assimilation. Proc Natl Acad Sci USA 101: 16162–16167

Turanov AA, Lobanov AV, Fomenko DE, Morrison HG, Sogin ML, Klobutcher LA, Hatfield DL, Gladyshev VN (2009) Genetic code supports targeted insertion of two amino acids by one codon. 323:259–261

Urig S, Becker K (2006) On the potential of thioredoxin reductase inhibitors for cancer therapy. Semin Cancer Biol 16:452–465

Wirth EK, Conrad M, Winterer J, Wozny C, Carlson BA, Roth S, Schmitz D, Bornkamm GW, Coppola V, Tessarollo L, Schomburg L, Kohrle J, Hatfield DL, Schweizer U (2009) Neuronal selenoprotein expression leads is required for interneuron development and prevents seizures and neurodegeneration. FASEB J (In Press)

Xu X-M, Carlson BA, Grabowski PJ, Gladyshev VN, Berry MJ, Hatfield DL (2005) Evidence for direct roles of two additional factors, SECp43 and soluble liver antigen, in the selenoprotein synthesis machinery. J Biol Chem 280:41568–41575

Xu X-M, Carlson BA, Mix H, Zhang Y, Saira K, Glass RS, Berry MJ, Gladyshev VN, Hatfield, DL (2006) Biosynthesis of selenocysteine on its tRNA in eukaryotes. PLoS Biol 5:e4

Xu X-M, Carlson BA, Irons, RA Mix H, Zhong N, Gladyshev VN, Hatfield DL (2007) Selenophosphate synthetase 2 is essential for selenoprotein biosynthesis. Biochem J 404: 115–120

Yamao F, Muto A, Kawauchi Y, Iwami M, Iwagami S, Azumi Y, Osawa S (1985) UGA is read as tryptophane in *Mycoplasma capricolum*. Proc Natl Acad Sci USA 82:2306–2309

Yant LJ, Ran Q, Rao L, Van Remmen H, Shibatani T, Belter JG, Motta L, Richardson A, Prolla TA (2003) The selenoprotein GPX4 is essential for mouse development and protects from radiation and oxidative damage insults. Free Radic Biol Med 34:496–502

Yokobori S, Suzuki T, Watanabe K (2001) Genetic code variations in mitochondria: tRNA as a major determinant of genetic code plasticity. J Mol Evol 53:314–326

Yoo M-H, Xu X-M, Carlson BA, Gladyshev VN, Hatfield, DL (2006) Selenoprotein thioredoxin reductase 1 deficiency reverses tumor phenotype and tumorigenicity of lung carcinoma cells. J Biol Chem 281:13005–13008

Yoo M-H, Xu X-M, Carlson BA, Patterson AD, Gladyshev VN, Hatfield DL (2007) Targeting thioredoxin reductase 1 reduction in cancer cells inhibits self-sufficient growth and DNA replication. PLoS ONE 2:e1112

Yuan J, Palioura S, Salazar JC, Su D, O' Donoghue P, Hohn MJ, Cardoso AM, Whitman WB, Söll D (2006) RNA-dependent conversion of phosphoserine forms selenocysteine in eukaryotes and archaea. Proc Natl Acad Sci USA 103:18923–18927

Zhang Y, Gladyshev VN (2007) High content of proteins containing 21st and 22nd amino acids, selenocysteine and pyrrolysine, in a symbiotic deltaproteobacterium of gutless worm Olavius algarvensis. Nucleic Acids Res 35:4952–4963

Zhang Y, Gladyshev VN (2008) Trends in selenium utilization in marine microbial world revealed through the analysis of the global ocean sampling (GOS) project. PLoS Genet 4:e1000095

Zinoni F, Birkmann A, Stadtman TC, Bock A (1986) Nucleotide sequence and expression of the selenocysteine-containing polypeptide of formate dehydrogenase (formatehydrogenlyase-linked) from *Escherichia coli*. Proc Natl Acad Sci USA 83:4650–4654

Chapter 2
Reprogramming the Ribosome for Selenoprotein Expression: RNA Elements and Protein Factors

Marla J. Berry and Michael T. Howard

Abstract Many of the benefits of the antioxidant selenium can be attributed to its incorporation into selenoenzymes as the 21st amino acid, selenocysteine. Selenocysteine incorporation occurs cotranslationally at UGA codons in a subset of messages in prokaryotes, eukaryotes, and archaea. UGA codons are recoded to specify selenocysteine, rather than termination, by the presence of specialized *cis*- and *trans*-acting factors. Here we discuss the mechanism of selenocysteine insertion, the factors which affect efficiency of incorporation, and regulation of mRNA levels. Although much remains to be learned about the multiple factors affecting gene and tissue-specific regulation of the selenoenzymes, significant advances in this regard have been made in understanding the role of selenium status, the expression and selective modification of specific *trans*-acting factors, and the *cis*-acting sequences associated with each selenoenzyme message.

Contents

M.J. Berry (✉)
Department of Cell and Molecular Biology, John A. Burns School of Medicine, University of Hawaii at Manoa, Honolulu, HI 96813, USA
e-mail: mberry@hawaii.edu

J.F. Atkins, R.F. Gesteland (eds.), *Recoding: Expansion of Decoding Rules Enriches Gene Expression*, Nucleic Acids and Molecular Biology 24,
DOI 10.1007/978-0-387-89382-2_2,

2.1 Selenium, Selenocysteine, and Selenoproteins

Selenium has long been known for its antioxidant properties, but it has only in recent years come to light that the beneficial effects of this trace element in our diet are attributable to selenoenzymes. In animals approximately 80% of selenium is covalently associated with proteins in the form of the 21st amino acid selenocysteine (Hawkes et al., 1985). This amino acid has a lower p*K*a than cysteine, producing a highly reactive group at physiological pH which is often responsible for catalyzing reduction/oxidation reactions. The known functions of selenoenzymes include protecting cell membranes, proteins, and nucleic acids from cumulative oxidative damage. These functions are carried out by the glutathione peroxidases, enzymes that break down hydroperoxides and lipid peroxides, the thioredoxin reductases, which catalyze regeneration of the essential thiol cofactor, thioredoxin, and other recently identified selenoproteins. Selenoenzymes function in preserving mammalian sperm integrity and in thyroid hormone homeostasis, highlighting essential roles for the trace element in development and metabolism. Selenium deficiency has been linked to cardiovascular disease in deficient regions of rural China, and cumulative oxidative damage has been implicated in the pathogenesis of cancers, diabetes, Alzheimer's and Parkinson's diseases. Further, the oxidative stress caused by selenium deficiency has been shown in experimental animals to increase susceptibility to infection by influenza and other viruses.

2.2 The Mechanism of Selenocysteine Incorporation in Eukaryotes

The mechanism of selenocysteine incorporation in eukaryotes has, for the last ~15 years, been assumed to be inherently different from that in prokaryotes due to differences in the architecture of selenoprotein mRNAs and in the factors catalyzing selenocysteine biosynthesis and incorporation. After extensive efforts spanning the same time frame, many of the essential differences in these mechanisms are being revealed through identification of the *cis*- and *trans*-acting factors catalyzing selenocysteine biosynthesis and its cotranslational insertion in eukaryotes. Additional insights into the efficiency of selenoprotein synthesis are being unveiled through studies of the interactions among these factors.

2.2.1 Identification of Cis-Acting Factors in Eukaryotes

Selenocysteine incorporation occurs cotranslationally at UGA codons in a subset of messages in prokaryotes, eukaryotes, and archaea. UGA codons are recoded to specify selenocysteine, rather than termination, by the presence of specific secondary structures in selenoprotein mRNAs termed selenocysteine insertion sequences, or SECIS, elements. In prokaryotes, SECIS elements are located in the coding region, immediately downstream of the UGA codons they serve (Fig. 2.1A).

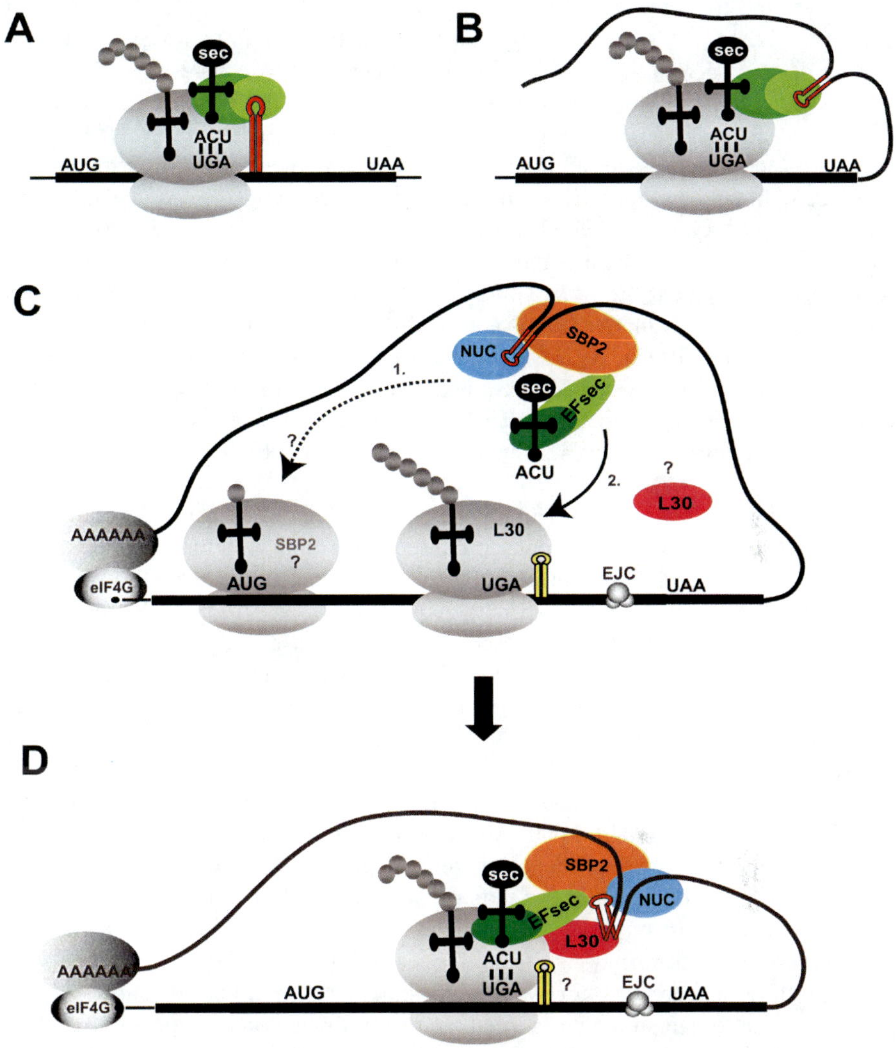

Fig. 2.1 Models for selenocysteine insertion in bacteria, archaea, and eukaryotes. (**A**) The bacterial selenocysteine elongation factor (*green*) binds the Sec-tRNA and also binds directly to the bacterial SECIS element (*red*) located adjacent to and downstream of the UGA codon to deliver the Sec-tRNA to the ribosome. (**B**) Similarly, the archaeal elongation factor binds to the Sec-tRNA and interacts with the 3′ UTR SECIS element analogous to the situation in eukaryotes. (**C**) In eukaryotes the SECIS element binds to SBP2 (*orange*) which binds to Sec-tRNA-bound EFsec. SBP2 also binds to the ribosome. Consequently it is unclear if the ribosome is loaded with SBP2 and possibly other selenocysteine insertion factors prior to decoding the UGA codon (1) or if the factors assemble during decoding of the UGA codon (2). L30 (*magenta*) exists bound to the ribosome and in a free form. A structure downstream of the UGA codon (*yellow*) stimulates selenocysteine insertion by a yet to be determined mechanism. L30 can compete with SBP2 for binding to the SECIS element under conditions which favor the kink-turn conformation at the GA:AG quartet (**D**). It has been suggested that this may trigger conformational changes which allow delivery of the Sec-tRNA to the A-site by EFsec. Decoding of the UGA codon is required to remove the exon junction complex (EJC) downstream to protect selenoprotein messages from nonsense-mediated decay (see Fig. 2.2)

Selenocysteine incorporation occurs via a bifunctional protein, SELB, consisting of a Sec-tRNA$^{[Ser]Sec}$-specific elongation factor (EF) domain and a SECIS RNA-binding domain (Kromayer et al., 1996). In archaea and eukaryotes (Fig. 2.1B and C, respectively), SECIS elements are typically located in the 3′ untranslated region (UTR), but at least one SECIS element has been identified in the 5′ UTR in an archaea selenoprotein gene (Wilting et al., 1997). In eukaryotes, SECIS elements have been shown to recode the entire message, functioning for any upstream in-frame UGA (Berry et al., 1993; Hill et al., 1993; Shen et al., 1993), provided a minimal spacing requirement is met (Martin et al., 1996). In addition, information is encoded locally near the UGA codon which influences the efficiency of selenocysteine insertion (Grundner-Culemann et al., 2001; Gupta and Copeland, 2007; McCaughan et al., 1995). At least a subset of eukaryotic selenoprotein messages contain a highly conserved RNA secondary structure, referred to as the selenocysteine codon redefinition element or SRE, which resides just downstream of the UGA codon and modulates selenocysteine insertion efficiency (Howard et al., 2005, 2007).

Eukaryotic SECIS elements: Eukaryotic SECIS elements consist of a stem-loop structure that contains several conserved sequence and structural features. The sequence features initially identified include AUGA and GA at the 5′ and 3′ bases of the stem, respectively, and a conserved AAR motif in a loop at the top of the stem (Berry et al., 1991, 1993). The sequences at the base of the stem were shown to form a quartet of non-Watson–Crick base pairs, with a central tandem of sheared G.A pairs (Walczak et al., 1996). The stem separating the SECIS core from the conserved adenosines is typically fixed at 9–11 base pairs (Grundner-Culemann et al., 1999). An open loop below the quartet and an additional helix below this were subsequently delineated. As additional selenoprotein sequences were elucidated, compilation revealed variation in the conserved features, including substitution of G for the first A at the 5′ base of the upper stem (Buettner et al., 1999), the presence of the AAR motif in an internal bulge rather than an apical loop (Grundner-Culemann et al., 1999), and substitution of C's for A's in the AAR motif (Kryukov et al., 2003). Nonetheless, with the variations and subsequent refinements, these features allowed the generation of search programs for SECIS elements such that the entire selenoproteome of an organism could be predicted from the genome sequence (Kryukov et al., 2003). Delineation of the conserved or semiconserved features also proved essential in identifying cognate binding proteins, as discussed below.

The SRE and UGA codon context: Although the distal 3′ UTR SECIS element is sufficient for UGA to encode selenocysteine, the efficiency of selenocysteine insertion varies substantially depending upon the codon context. One explanation is that some UGA codons are decoded with greater efficiency than others and that ribosome pausing, competition with termination, and RNA elements near the UGA codon play an active role in determining the efficiency of selenocysteine insertion.

Factors known to influence the efficiency of termination of translation include the sequence context of the stop codon, with the nucleotide following the stop codon having a strong influence. In most cases, C in this position results in more readthrough by near cognate tRNAs than the other three nucleotides (Beier and Grimm, 2001; Howard et al., 2000; Li and Rice, 1993; Manuvakhova et al., 2000; Martin et al., 1993). However, the sequence context effect is complex and can extend to the following six nucleotides and adjacent codons upstream of the stop codon as well (Harrell et al., 2002; Mottagui-Tabar et al., 1994, 1998; Namy et al., 2001) preventing prediction of termination efficiency based simply on examination of the sequence context alone (Bidou et al., 2004). In direct studies of the effect of adjacent sequence context on selenocysteine insertion efficiency, it was shown that the nucleotides immediately upstream and downstream of the UGA codon affect selenocysteine insertion efficiency (Grundner-Culemann et al., 2001b; Gupta and Copeland, 2007; McCaughan et al., 1995). In some but not all cases, contexts favorable for termination result in lowered amounts of selenocysteine insertion. A likely explanation is that the competition between termination and selenocysteine insertion is determined by a larger sequence context which can affect termination and/or the selenocysteine insertion machinery directly to determine the ratio of truncated to full-length protein.

In cases of stop codon redefinition where standard near cognate tRNAs are used to decode stop codons, RNA pseudoknot structures (ten Dam et al., 1990) have been shown to directly readthrough in several mammalian retroviruses (Wills et al., 1991; Feng et al., 1992). A well-studied example is *gag-pol* expression in the murine leukemia virus (MuLV) where the *gag* UAG stop codon is redefined with approximately 5–10% efficiency (Philipson et al., 1978; Yoshinaka et al., 1985). Another example of regulatory stop codon redefinition comes from studies of *kelch* expression during *Drosophila* development (Robinson and Cooley, 1997). In this study, the ratio of the termination to readthrough product was suggested to be regulated in a tissue-specific manner. These findings illustrate that not only can redefinition levels be specified by local sequence context for proper gene expression but also in some cases readthrough efficiency is dynamically regulated to achieve optimal gene expression.

The occurrence of downstream RNA secondary structures associated with other cases of stop codon redefinition, as well as the location of the bacterial SECIS element downstream of the UGA-Sec codon, prompted a re-evaluation of the extended sequence context of selenocysteine UGA codons in eukaryotes for the presence of downstream RNA structures (Howard et al., 2005). Phylogenetic and mutagenic analysis identified one such element downstream from the *SEPN1* selenocysteine UGA codon which was designated the Selenocysteine codon Redefinition Element, or SRE. The functional *SEPN1* SRE consists of upstream sequences and a highly conserved stem-loop structure that starts six nucleotides downstream of the UGA codon. Experimental evidence illustrated that the SRE alone was sufficient to cause high-level UGA readthrough by near cognate tRNAs (Howard et al., 2005, 2007). In the presence of the SECIS element, the SRE was not required for selenocysteine insertion but had a significant stimulatory effect. The upstream sequence, the stem-loop structure, and the length and sequence of the spacer separating it from the

UGA-Sec codon were important for stimulation of selenocysteine incorporation. Interestingly the same RNA secondary structures were independently identified in *SEPN1* and *SelT* in a genome-wide search for deeply conserved functional RNA structures (Pedersen et al., 2006).

Phylogenetic and experimental analysis indicates that in a subset of selenoprotein mRNAs, there is the potential for stable and conserved downstream RNA structures (unpublished data MTH). An intriguing possibility is that the eukaryotic SREs interact directly with components of the selenocysteine insertion machinery to facilitate selenocysteine insertion at the upstream UGA-Sec codon. In addition, the SRE elements may influence selenoprotein message levels by affecting nonsense-mediated decay (NMD) under limiting selenocysteine conditions due to its ability to induce near cognate tRNA decoding. However, definitive answers to the mechanism(s) of SRE action, extent of their occurrence in selenoproteins, and role in the dynamic regulation of selenoprotein expression await further studies.

Mutations in SEPN1 cis-acting elements provide insight into mechanism: Mutations in *SEPN1* result in *SEPN1*-related myopathy consisting of four autosomal recessive disorders originally considered to be separate entities: rigid spine muscular dystrophy (RSMD1) (Flanigan et al., 2000; Moghadaszadeh et al., 2001), the classical form of multiminicore disease (Ferreiro et al., 2002), desmin-related myopathy with Mallory body-like inclusions (Ferreiro et al., 2004), and congenital fiber-type disproportion (Clarke et al., 2006). All are clinically characterized by poor axial muscle strength, scoliosis and neck weakness, and a variable degree of spinal rigidity. Recent studies demonstrate that SelN protein can affect the redox state and is physically associated with the ryanodine receptor intracellular calcium release channel (RyR) (Jurynec et al., 2008). The simplest interpretation is that SelN modifies the regulation of RyR-mediated calcium mobilization required for normal muscle development and that disruption of this process results in the congenital myopathies described above.

Recent studies have identified mutations in the *SEPN1* gene which cause disease by interfering with the selenocysteine insertion mechanism during translation of SelN. A single homozygous disease-causing point mutation was identified in the *SEPN1* 3′ UTR SECIS of a patient with RSMD1 (Allamand et al., 2006). This mutation is sufficient to prevent SBP2 binding and selenocysteine incorporation, and significantly reduces both SelN mRNA and protein levels. A second study analyzed four disease-causing missense mutations identified in the SRE element of *SEPN1* (Maiti et al., 2009). One of these mutations, c.1397G>A, which results in a C:A mismatch near the base of the SRE stem-loop, was shown to significantly reduce selenocysteine insertion efficiency and likewise resulted in negligible levels of *SEPN1* mRNA or protein in the patients muscle. It is notable in both cases that not only was selenocysteine insertion impaired but also messages levels were substantially reduced. These studies highlight the importance of both the SECIS and SRE in maintaining the stability of the message and the selenocysteine insertion pathway in vivo.

2.2.2 Identification of Trans-Acting Factors in Eukaryotes

In 2000, two *trans*-acting factors essential for selenocysteine incorporation were identified in eukaryotes, SBP2 and EFsec (Copeland et al., 2000; Fagegaltier et al., 2000; Tujebajeva et al., 2000). SBP2 was shown to specifically interact with SECIS elements and to be required for selenocysteine incorporation in vitro, but unlike bacterial SELB, SBP2 does not contain elongation factor homology or activity. This activity resides in EFsec, a Sec-tRNA$^{[Ser]Sec}$-specific elongation factor, that lacks SECIS-binding activity but contains a C-terminal extension that interacts with SBP2. The interaction between these two factors was shown to be strongly stimulated by the presence of Sec-tRNA$^{[Ser]Sec}$ (Zavacki et al., 2003). Two additional proteins previously known for other functions were identified as components of the selenocysteine incorporation machinery. The first of these, ribosomal protein L30 (Chavatte et al., 2005), was shown to interact with SECIS elements at a site overlapping the SBP2-binding site, and the second, nucleolin, a major component of the nucleolus has also been identified as a SECIS-binding protein (Wu et al., 2000). These factors and their roles in selenocysteine insertion are discussed in more detail below (see Fig. 2.1C).

Finally, recent studies have shed light on the roles of two additional factors that had been implicated in the selenoprotein biosynthesis pathway. These are SECp43, identified in a degenerate PCR screen for RNA-binding proteins (Ding and Grabowski, 1999), and SLA/LP, identified as an autoantigen in chronic autoimmune hepatitis (Gelpi et al., 1992). Both proteins were shown to bind Sec-tRNA$^{[Ser]Sec}$. Recent studies provide evidence that SECp43 plays a role in Sec-tRNA$^{[Ser]Sec}$ methylation and that SLA/LP is the Sec-tRNA$^{[Ser]Sec}$ synthase (Xu et al., 2005).

With the availability of these factors, some of the crucial questions concerning the mechanism of selenocysteine insertion in eukaryotes could begin to be addressed. The location of most eukaryotic SECIS elements in the 3′ UTR and the assembly of decoding complexes there, resulting in recoding from distances up to ~5 kilobases, might be predicted to decrease incorporation efficiency. In addition, many selenoprotein genes encode one or more introns downstream of the UGA codon(s), marking these codons as premature termination codons if not decoded efficiently. The ability of ribosomes to initiate translation on mRNAs while they are still undergoing export through the nuclear pore (Mehlin et al., 1992) suggests that decoding complexes might need to be assembled on the mRNA prior to export, such that they would be in place before the first ribosome reached the first UGA codon (Fig. 2.2). Otherwise, the UGA codon would be recognized as a premature termination codon and the mRNA degraded.

EFsec: EFsec was identified through homology searches based in part on what was known about SELB and the mechanism of selenocysteine incorporation in prokaryotes. Searches focused on homology to EF1, the canonical eukaryotic elongation factor that delivers most amino acyl-tRNAs to the ribosomal A-site, with the

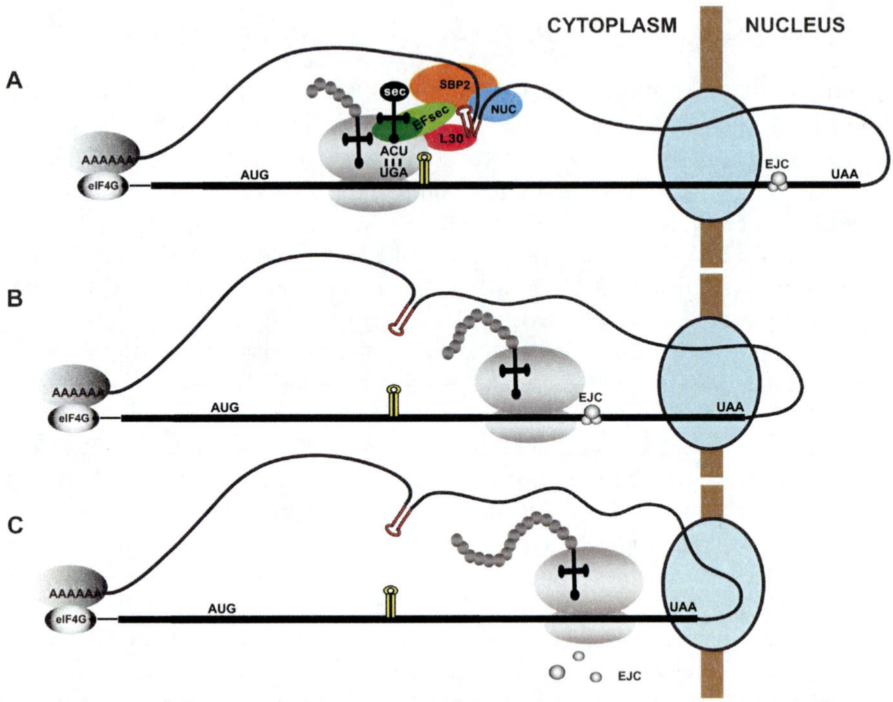

Fig. 2.2 Selenocysteine incorporation complexes may assemble on selenoprotein mRNAs prior to or during nucleocytoplasmic transport. (**A**) The ability of ribosomes to initiate translation on mRNAs while they are undergoing export through the nuclear pore (*light blue*) (Mehlin et al., 1992) suggests that decoding complexes might need to be assembled prior to export, such that they would be in place before the first ribosome reached the first UGA codon. Otherwise, the UGA codon would be recognized as a premature termination codon and the mRNA would be degraded. The decoding complex consists of EFsec (*green*), SBP2 (*orange*), L30 (*magenta*), and the SECIS element. (**B**) Decoding of UGA as selenocysteine allows the ribosome to proceed downstream and (**C**) to remove the EJC, circumventing nonsense-mediated decay

additional condition that a C-terminal extension might be present to interact with SECIS elements. Candidate factors were identified in several genomes, with efforts from two groups focusing on characterization of the murine factor (Fagegaltier et al., 2000 Tujebajeva et al., 2000). The N-terminal elongation factor domain was shown to recognize Sec-tRNA$^{[Ser]Sec}$ but not the Ser-tRNA$^{[Ser]Sec}$ precursor. Two isoforms of Sec-tRNA$^{[Ser]Sec}$, distinguished by the absence or the presence of a wobble base methylation, had previously been characterized. EFsec does not appear to distinguish between the two in binding, but interactions at the ribosome have not been reported.

The selenocysteine elongation factors reveal differences from the standard eukaryotic elongation factor which delivers all other known amino-acylated tRNAs to the ribosome. These differences include its specificity for the selenocysteine-charged tRNA, a C-terminal extension with unknown function, its ability to bind SBP2 as discussed above, and interestingly its higher affinity for GTP than GDP (Fagegaltier et al., 2000; Hilgenfeld et al., 1996). The latter result suggests it may not need a recycling factor to replace GDP with GTP following Sec-tRNA$^{[Ser]Sec}$ delivery to the ribosome. Recently, a GTPase-activating protein GAPSec was identified as a protein which interacts with the *Drosophila* EFSec protein. The protein is conserved in worms, mice, and humans and is highly expressed early in development. Surprisingly, although readthrough of UGA codons in reporter genes is SECIS dependent and GAPsec binds to EFSec, mutants do not appear to effect selenocysteine insertion or the expression of at least some selenoproteins in flies. Although further studies are needed, this protein may be involved in a developmentally regulated SECIS-dependent UGA redefinition pathway through its interactions with EFsec and GTP hydrolysis (Hirosawa-Takamori et al., 2009).

The identification of SBP2 and demonstration of its RNA-binding specificity suggested that SECIS binding by EFsec might not be required for function. Instead, EFsec was shown to be recruited to selenoprotein mRNAs via interaction of its C-terminal domain with SBP2. A crucial mechanistic insight came with the demonstration that the interaction between these two factors is strongly stimulated by the presence of Sec-tRNA$^{[Ser]Sec}$ bound to the N-terminal domain of EFsec (Zavacki et al., 2003). Strikingly, binding of the C-terminal region of EFsec to SBP2 is also increased upon deletion of the N-terminal domain, indicating that an empty elongation factor domain may hinder binding to SBP2. These findings provide a mechanism whereby SBP2 would only recruit EFsec carrying Sec-tRNA$^{[Ser]Sec}$ and would dissociate from the factor upon delivery of the Sec-tRNA$^{[Ser]Sec}$ to the ribosome.

SBP2: SBP2 was identified and purified using SECIS elements as ligand in affinity purification, followed by functional characterization of the recombinant protein in reticulocyte in vitro translation reactions (Copeland et al., 2000). Initial studies showed that the N-terminal half of the protein was dispensable for both SECIS binding and selenoprotein synthesis. Subsequent studies delineated a central domain that is required for selenocysteine incorporation and mapped the SECIS RNA-binding domain to a C-terminal region of the protein (Copeland et al., 2001. Within the SBP2 RNA-binding domain is a canonical L7Ae RNA-binding motif found in several proteins known to interact specifically with kink-turns. Mapping of the SBP2-binding site on several SECIS RNAs showed that binding is limited to the region that includes the conserved G.A/A.G tandem of non-Watson–Crick base pairs (Fletcher et al., 2001). This region was predicted in earlier studies to form a kink-turn (Walczak et al., 1996), an RNA helical structure first identified in ribosomal RNAs. SBP2 has also been shown to stably associate with ribosomes in transfected cells and in vitro possibly through interactions with the L7Ae region and 28S rRNA (Copeland, Stepanik and Driscoll 2001). Mutagenesis of conserved amino acids in the L7Ae region identified a core motif required for SECIS RNA and ribosome binding and for selenocysteine incorporation, whereas additional mutations separated SECIS binding from the other two activities (Caban et al., 2007). The boundaries of the essential RNA-binding domain have been further mapped by deletion analysis to a 235 amino acid region (Bubenik and Driscoll, 2007). Two smaller regions of between 70 and 90 amino acids were found to be highly conserved in vertebrates with the second containing the L7Ae motif discussed above, and both are required for selenocysteine insertion activity. The intervening amino acids were not conserved and found to be dispensable for selenocysteine insertion in vitro. In fact deletions of the intervening sequences increased specific binding affinity for the GPx4 SECIS element. The functional requirement for the two RNA-binding motifs, the role of an apparently inhibitory intervening sequence, as well as the N-terminus of SBP2 remain to be clarified.

L30: Ribosomal protein L30 was identified as a SECIS-binding protein through a similar approach to that used in identifying SBP2 (Chavatte et al., 2005). L30 belongs to the ribosomal protein L7Ae family, of which SBP2 is a member – as discussed above. In vitro binding studies showed that L30 required the same G.A/A.G tandem as SBP2 and further revealed competition between the two proteins for SECIS binding. L30 was further shown to enhance UGA recoding and to bind to SECIS elements in vivo. Magnesium was found to play a crucial role in the competition between SBP2 and L30 binding. Prior studies showed that magnesium and other divalent metal ions induce formation of the kink-turn structure in RNAs that contain two tandem G.A pairs. Chavatte et al. (2005) showed that magnesium addition decreased the SBP2–SECIS interaction in favor of the L30–SECIS interaction. L30 exists in both ribosome-associated and free forms, and the ribosome-associated form was shown to exhibit a higher affinity for SECIS elements than the free recombinant protein, leading to speculation that L30 may adopt a more favorable conformation for SECIS binding when part of the ribosome and/or other ribosomal components may facilitate the L30–SECIS interaction. A possible model proposed by these investigators envisions SBP2 as the initial SECIS selectivity factor, recruiting EFsec and Sec-tRNA$^{[Ser]Sec}$. Once associated with the ribosome, SBP2 would be transiently displaced by L30, which may function in anchoring and/or positioning the complex at the ribosomal A-site. The model includes speculation on simultaneous interactions of L30 with ribosomal RNA and the SECIS element through two RNA-binding interfaces. Finally, they suggest that L30 may induce conformational transitions that function in GTP hydrolysis and Sec-tRNA$^{[Ser]Sec}$ delivery.

Nucleolin: Nucleolin has been identified as an additional SECIS-binding protein (Wu et al., 2000). Mutation of the region where the highly conserved G.A/A.G is conserved eliminated binding. In contrast to the differential affinity of SBP2 to specific SECIS elements (discussed in detail below), nucleolin was found to bind most selenoprotein mRNAs similarly although the role this protein plays in selenocysteine insertion has not been investigated (Squires et al., 2007).

2.3 Efficiency of Selenocysteine Incorporation in Eukaryotes

Two intriguing questions in the field of eukaryotic selenoprotein synthesis are how efficient is selenoprotein synthesis in vivo and to what extent do selenocysteine incorporation and termination compete at any given UGA codon. Selenocysteine incorporation has been reported to be inefficient in all systems studied. Termination occurs in *Escherichia coli* selenoproteins, in rabbit reticulocyte in vitro translation reactions (Berry et al., 1991; Jung et al., 1994), in transiently transfected mammalian cells (Nasim et al., 2000; Tujebajeva et al., 2000), and in baculovirus–insect cell expression systems (Kim et al., 1997). In mammalian cells, overexpression of selenoprotein mRNAs by transfection of increasing amounts of selenoprotein-encoding plasmid increases the ratio of termination product to full-length protein (Berry et al., 1994; Grundner-Culemann et al., 2001). Cotransfection of some components of the selenocysteine incorporation pathway, including $tRNA^{[Ser]Sec}$ (Berry et al., 1994), selenophosphate synthetase (Low et al., 1995), or SBP2 partially reverses this effect, increasing selenocysteine incorporation (de Jesus et al., 2006). Selenium supplementation also increases incorporation (Berry et al., 1994; Brigelius-Flohe et al., 1997). These findings suggest that one or more of these factors may be limiting in some cell types or conditions. Other components of the machinery, such as EFsec, do not appear to be limiting (de Jesus et al., 2006). However, even with overexpression of multiple limiting factors, the levels of full-length selenoprotein do not approach those of the corresponding cysteine-mutant proteins under any of these conditions, implying that selenocysteine incorporation may be inherently inefficient. Attempts at overexpression might exacerbate any inherent inefficiency in this process. Termination at selenocysteine codons has also been observed in intact animals. Purification of selenoprotein P (Sel P) from rat plasma revealed multiple isoforms of the protein. These isoforms were shown by carboxypeptidase sequencing (Himeno et al., 1996) and mass spectrometry (Ma et al., 2002) to comprise full-length and prematurely UGA-terminated species. The amounts of truncated products increased upon dietary selenium limitation, but premature termination was even observed in animals maintained on a selenium-sufficient diet.

Sel P may be a special case as production of full-length protein requires readthrough of multiple UGA codons, and their incorporation is directed by two SECIS elements. The number of selenocysteines predicted by Sel P sequences ranges from 10 in humans and rodents to 28 in sea urchin (Lobanov et al., 2008). This invites the question, with the possibility of ribosomes positioned at multiple UGA codons simultaneously, how do two SECIS elements recode multiple UGAs?

One possibility is that ribosomes may be "reprogrammed" upon encountering the first UGA codon, such that they are now more competent to decode subsequent UGAs. By analogy with translation initiation, where ribosomes remain competent to re-initiate translation for a period of time following initiation due to the continued association of initiation factors, a similar phenomenon may occur with selenocysteine incorporation. For example, with the first selenocysteine incorporation event, the ribosome may undergo a conformational change that favors decoding by EFsec–Sec-tRNA complex over termination by eRF1. The conformational change could be acquisition or loss of L30, SBP2, or nucleolin by the ribosome, or more global ribosomal rearrangements involving the A-site. It is noteworthy that the majority of the UGA codons in selenoprotein P genes are clustered near the 3′ end of the coding sequence. Consequently, the putative rearrangement(s) may be transient lasting only long enough to translate through the closely positioned UGA codons or more permanent such that circularization of the message could allow for reprogrammed ribosomes to be recycled back onto the same message. Even if only a subset of ribosomes is reprogrammed to be processive, this would result in a mixture of full-length and premature termination products as has been reported in rodent studies.

Another possible contributing factor is the concept that the first UGA may serve as a checkpoint for the presence of the factors required for selenocysteine incorporation. If the necessary factors are present, selenocysteine is incorporated, and if they are not, termination ensues. Thus, the rate at which elongating ribosomes progress toward the second UGA would be controlled by inefficient decoding at the first UGA. After the first UGA codon, most of the remaining UGA codons are found close together in a UXU or UXUXU organization, where U is selenocysteine and X is any amino acid. This configuration decreases the number of ribosomes simultaneously decoding UGA codons as several UGAs would be covered by a single ribosome at any given time. A combination of reduced numbers of ribosomes and enrichment for those associated with selenocysteine incorporation factors may favor processive translation to the natural termination codon. The scenarios presented above are speculative and the mechanism by which multiple UGA codons are decoded on a single message is under investigation.

As a first step toward this goal, studies were undertaken to investigate the functions of the two SECIS elements in decoding the UGA codons in Sel P (Stoytcheva et al., 2006). Early studies showed that the first SECIS element exhibited about threefold higher selenocysteine incorporation activity than the second element when linked to the same reporter (Berry et al., 1993). Subsequent sequence alignments reveal the first SECIS to be highly conserved, whereas the second is much less so (unpublished, MTH). Mutation or deletion of the first SECIS element resulted in complete loss of detectable full-length Sel P and a corresponding increase in termination at the first and second UGA codons. This indicates that the first SECIS element is required for production of full-length Sel P, serving the second UGA codon and beyond. In contrast, and quite surprisingly, mutation or deletion of the second SECIS element was found to have minimal effects on selenocysteine incorporation. The effects of swapping the positions of the two elements, duplicating one element and deleting the other, or introducing additional elements were also assessed. These studies show that the first SECIS element is required for efficient incorporation, regardless of its position, whereas the second element, even when duplicated, is unable to confer the ability to produce full-length protein. This result confirms the essential function of the first SECIS, indicating that the two elements are functionally distinct. In further support of this notion, polysome loading on messages containing wild-type or mutant SECIS elements revealed a shift to lighter polysomes only with deletion of the first but not the second SECIS. SBP2 was subsequently shown to preferentially bind to the first versus the second SECIS element in vivo, providing a possible mechanistic basis for the differential functions of the two (Squires et al., 2007).

2.4 Hierarchy of Selenoprotein Synthesis

Selenoproteins exhibit differential priority for available selenium stores, in what has come to be referred to as a hierarchy of selenoprotein synthesis. That is, when selenium is limiting, certain selenoproteins appear to preferentially utilize the selenium that is available at the expense of other selenoproteins. Interestingly, the selenoproteins that appear to have preference coincide with those found through targeted gene disruption studies to be the most essential for viability. The cellular mechanisms contributing to the differential efficiency of selenoprotein synthesis have been under investigation for a number of years, but for the most part have remained elusive.

Several published studies have shown differing selenium retention in different tissues. For example, testes have been shown to retain their selenium stores approximately 20-fold better than liver or heart upon dietary selenium limitation (Behne et al., 1998). Testes also exhibit the highest levels of SBP2 and glutathione peroxidase 4 (GPX4) mRNAs and proteins of any tissue examined (Copeland et al., 2000). The high level of GPX4 in the sperm mitochondrial capsid has been shown to be crucial for sperm integrity and motility and thus to male fertility (Ursini et al., 1999). In addition, a hierarchy for synthesis of different selenoproteins within a single tissue, as well as in different tissues and cell lines, has been observed (Behne et al., 1988; Hill et al., 1992; Lei et al., 1995; Mitchell et al., 1997). As examples of this, glutathione peroxidase 1 (GPX1) activity was reduced to 1% of normal levels in liver and to about 4–9% in kidney, heart, and lung of selenium deficient rats. GPX4 activity was decreased to 25–50% in these same tissues but was unaffected by selenium deprivation in testes. The dramatic decline in GPX1 activity upon selenium deprivation is due in large part to rapid turnover of the mRNA for this protein, most likely via the nonsense-mediated decay (NMD) pathway (Christensen and Burgener, 1992; Lei et al., 1995; Saedi et al., 1988).

Nonsense-Mediated Decay (NMD): In addition to the direct contribution of the ratio of selenocysteine insertion to termination on the expression of selenoproteins, the efficiency of this process may influence message levels by activating or preventing mRNA decay pathways such as nonsense-mediated decay or no-go decay. These pathways are designed to eliminate messenger RNAs with premature stop codons or stalled ribosomes, respectively [Review; Isken and Maquat, 2007]. mRNAs containing premature nonsense codons are eliminated from most cells via the NMD pathway (Hentze and Kulozik, 1999; Nagy and Maquat, 1998). NMD typically occurs during nucleocytoplasmic export of mRNAs, and targeting of mRNAs containing premature termination codons for NMD has been shown to require translation (Thermann et al., 1998), typically via ribosomes initiating on the cytoplasmic side of the nuclear pore complex (Fig. 2.2). A critical feature in discrimination between physiological and premature termination codons in mammalian cells is the position of the last intron in the pre-mRNA relative to the termination codon. According to a recent analysis of the human genome, the termination codon is found in the last exon in ~98.7% of all human genes (Hong et al., 2006). A termination codon upstream of the last exon will typically be recognized as premature, marking the mRNA for NMD (Nagy and Maquat, 1998; Thermann et al., 1998). Thus, selenoprotein mRNAs whose pre-mRNAs contain introns downstream of the selenocysteine codon should be targeted for NMD when selenocysteine incorporation is inefficient. This was shown to be the case for GPX1 mRNA (Moriarty et al., 1998; Weiss and Sunde, 1998). In contrast, GPX4 mRNA is much less sensitive to NMD, despite the presence of appropriately spaced introns in its pre-mRNA (Lei et al., 1995; Weiss and Sunde, 1998).

SBP2 as a limiting determinant for NMD sensitivity? Demonstration that overexpression of SBP2 increases selenocysteine incorporation implies a possible role for this factor in the hierarchy of selenoprotein synthesis and possibly in sensitivity to NMD. To investigate this, the effects of knocking down or overexpressing SBP2 on expression of selenoprotein mRNAs were recently investigated and found to result in hierarchical effects (de Jesus et al., 2006). Transient and stable knockdowns of SBP2 expression decreased SBP2 mRNA levels in each case to ~30% of control levels. In the transient knockdowns, SelH and Gpx1 mRNAs showed the greatest decreases, whereas Gpx4, Trxr2, and Trxr3 mRNAs, among others, were relatively unchanged. In the stable knockdown cell line, Gpx4, Trxr2, and Trxr3 mRNAs exhibited the greatest decreases, while Gpx1 was unchanged. The reasons for these differences are not known, but may be due to changes in transcription, RNA turnover, or both. This may in turn relate to differences in the level of oxidative stress in cells undergoing transient versus stable inhibition of selenoprotein synthesis.

Binding of SBP2 to selenoprotein mRNAs in vivo was examined via immunoprecipitation of the protein and real-time RT-PCR to quantitate bound RNA (Squires et al., 2007). These studies revealed widely differing specificities for different selenoprotein mRNAs. SelW mRNA was precipitated with the highest affinity, followed by Gpx4, Sep15, and SelH, whereas Gpx1 exhibited much lower enrichment in the immunoprecipitates. In vitro binding studies using the SBP2 RNA-binding domain confirmed a significantly higher affinity for the Gpx4 SECIS compared to that of Gpx1 (Bubenik et al., 2007). The resistance of Gpx4 to NMD has been documented in several prior studies of the effects of selenium deficiency (Moriarty et al., 1998; Weiss and Sunde, 1998).

SBP2 mutations provide insights into hierarchy: Intriguing insights into the consequences of impaired SBP2 function were provided with the identification of a homozygous missense mutation in SBP2 in several siblings who presented with abnormal thyroid function tests (Dumitrescu et al., 2005). Investigation of the underlying cause failed to map the defects to members of the iodothyronine deiodinase family of selenoproteins, and components of the selenoprotein synthesis machinery were investigated. The SBP2 mutation was identified in the affected siblings who were subsequently shown to exhibit decreased Gpx activity in serum and fibroblasts and decreased Sel P and total selenium in serum. Quantitation of effects on other selenoproteins was not feasible due to their tissue localization. However, as targeted disruption of some selenoprotein genes, including Gpx4 and Trxr1, has been shown to result in embryonic lethality in rodents, the inference is that the expression of these genes was not significantly impaired. In vivo binding studies showed a reduced affinity for the two Sel P SECIS elements (Squires et al., 2007) which likely explains the reduction in Sel P and defects in selenium transport. In vitro binding studies showed that the mutation alters SBP2 RNA-binding affinity such that interaction with GPx1, Dio1, or Dio2 SECIS elements is not detected in electrophoretic mobility shift assays, whereas binding to Gpx4 and Trxr1 SECIS elements is observed. Further, the mutation reduced the ability of Dio2 SECIS to compete with the GPx4 SECIS in SBP2 binding (Bubenik et al., 2007). Thus, this mutation appears to differentially affect binding to different SECIS elements. These findings suggest a role for SBP2 in conferring resistance or sensitivity to NMD and thus in regulating levels of selenoprotein mRNAs. Understanding the underlying reasons for differences in NMD sensitivity is prerequisite to investigating the consequences for mRNA turnover, selenoprotein expression levels, and the hierarchy of selenoprotein synthesis.

2.5 Other Factors Effecting Differential Selenoprotein Expression

Evidence demonstrates that two isoforms of Sec-tRNA$^{Ser[Sec]}$ exist in higher vertebrates and that the relative abundance of these isoforms plays a role in regulating selenoprotein expression (Chittum et al., 1997; Jameson and Diamond, 2004; Moustafa et al., 2001). The two Sec-tRNA$^{Ser[Sec]}$ isoforms differ by a single methyl group ribosyl moiety of the anticodon wobble base, methylcarboxylmethyluridine (mcm^{5}U), or methylcarboxymethyluridine-2′-*O*-methylribose (mcm^{5}Um). Methylation of the 2′-*O*-hydroxyl is the last step in tRNA maturation and is influenced by selenium status (Hatfield and Gladyshev, 2002). The abundance of the methylated form is reduced under conditions of selenium deficiency and enhanced when selenium levels are sufficient (Hatfield and Gladyshev, 2002). Of relevance to differential selenoprotein expression is the observation that the abundance of a subset of selenoproteins is strongly affected by alterations in the ratio of the Um34-modified isoform to the unmethylated isoform (Carlson et al., 2005, 2007). In these studies, an increase in the unmethylated isoform strongly reduced expression of selenoproteins involved in stress response (e.g., GPx1), whereas other selenoproteins (GPx4, SelT, TR1, TR3) were less affected or even revealed increased expression levels.

In addition, recent evidence indicates that the eukaryotic initiation factor 4a3 (eIF4a3) binds with varying affinity to SECIS elements in competition with SBP2 and can selectively inhibit selenocysteine incorporation (Budiman et al., 2009). Binding affinities were examined for several selenoprotein SECIS elements. Higher binding affinities were found for those known to be affected by selenium status, such as GPx1, consistent with eIF4a3 playing a role in the heirarchy of selenoprotein expression. This information combined with the observation that eIF4a3 levels are increased under conditions of selenium insufficiency strongly suggests that it may be yet another factor influencing differential selenoprotein expression.

It is apparent that multiple mechanisms contribute to the differential efficiency of synthesis of each selenoprotein. As discussed above, these include at least the tissue levels of selenium and factors involved in selenoprotein synthesis, differential interactions with these factors, and differences in sequences and secondary structures in the coding region and 3′ UTR of the selenoprotein mRNAs.

2.6 Where do Selenoprotein mRNA Decoding Complexes Assemble?

Many selenoprotein genes encode one or more introns downstream of the UGA codon(s), marking these codons as premature termination codons if not decoded efficiently. The ability of ribosomes to initiate translation on mRNAs while they are still undergoing export through the nuclear pore (Mehlin et al., 1992) suggests that decoding complexes might need to be assembled early in the life of the mRNA, perhaps even prior to export, such that they would be in place before the first ribosome reached the first UGA codon. Otherwise, the UGA codon would be recognized as a premature termination codon and the mRNA would be degraded. Immunofluorescence and confocal microscopy were used to investigate the levels of SBP2 and the subcellular localization of SBP2 and EFsec (de Jesus et al., 2006). In HEK-293 cells, endogenous SBP2 cannot be detected by immunofluorescence. Following transfection, the protein is easily detected and localizes primarily to the cytoplasm. In three other cell lines, Hep-G2, HT22, and MSTO-211, the endogenous levels of SBP2 are higher and are easily detected by immunofluorescence. These cell lines expressed significant levels of endogenous selenoproteins. Strikingly, much of the SBP2 protein in these cells is found in the nucleus. Nuclear retention of a fraction of SBP2 may be due to recruitment to SECIS elements on newly transcribed mRNAs, and this may function in protecting these mRNAs from NMD. Nuclear localization and nuclear export signals are predicted in the SBP2 protein sequence, and heterokaryon studies showed that the minimal functional domain of the protein shuttles between the nucleus and cytoplasm. Subcellular localization of EFsec has also been examined using epitope-tagged constructs and antibodies. These studies revealed a pattern of predominantly cytoplasmic localization in transfected HEK-293 cells but both nuclear and cytoplasmic localization in HEP-G2, HT22, and MSTO-211 cells. Cotransfection of EFsec and SBP2 revealed the intriguing finding that SBP2 appears to either cotransport EFsec into the nucleus or increase nuclear retention of shuttling EFsec.

Subsequent studies demonstrated the striking finding that nuclear localization of SBP2 is significantly increased in response to cellular stresses, including H_2O_2-induced oxidative stress or UV exposure (Papp et al., 2006). Oxidation of a redox-sensitive cluster of cysteine residues in the C-terminus of SBP2 was implicated in increased nuclear localization, linking cellular redox state to ongoing selenoprotein synthesis. These modifications were efficiently reversed in vitro by human thioredoxin and glutaredoxin, suggesting that these antioxidant systems might regulate redox status of SBP2 in vivo. These results suggest that oxidative stress functions in regulating SBP2 function and thus selenoprotein synthesis.

The subcellular localization and association of factors implicated in generating mature Sec-tRNA$^{[Ser]Sec}$, including selenophosphate synthetase 1, Sec-tRNA$^{[Ser]Sec}$ synthase (SLA), and Sec-tRNA$^{[Ser]Sec}$ methylase (SECp43), were also investigated (Small-Howard et al., 2006). These studies showed that the three enzymes coimmunoprecipitated, and when coexpressed, exhibited nuclear localization. This localization may contribute to ensuring that all the necessary components are present in the nucleus for assembly of decoding complexes concurrent with export. As discussed above, nuclear assembly of decoding complexes may in turn be a key factor in allowing selenoprotein mRNAs to circumvent NMD.

2.7 Elucidating the Functions of Selenoproteins

The selenoproteins whose functions are best understood are, not surprisingly, those whose enzymatic activities were described or characterized independent of their identification as selenoproteins. These include the glutathione peroxidase, iodothyronine deiodinase, and thioredoxin reductase families, selenophosphate synthetase 2, and methionine sulfoxide reductase B. Progress in elucidating the functions of other selenoproteins has relied on traditional biochemical and molecular biological approaches, bioinformatics tools to identify structural motifs, e.g., the identification of Sel I as a diacylglycerol ethanolamine/choline phosphotransferase, and the rare genetic mapping of inherited disorders to selenoprotein genes, such as Sel N. Using a combination of knockdown experiments in zebra fish and biochemical analysis of protein interactions and function in normal muscle and disease tissue, SelN was shown to affect normal muscle development by altering activity of the ryanodine receptor calcium release channel (Jurynec et al., 2008). A combination of experimental and bioinformatics approaches has provided new insights into the functions of Sel H (Novoselov et al., 2007; Panee et al., 2007). A classic nuclear localization signal was identified in the Sel H sequence, followed by experimental confirmation of nuclear/nucleolar location of the protein (Novoselov et al., 2007; Panee et al., 2007). Overexpression and knockdown studies provided support for an antioxidant/redox role of the protein, consistent with identification of a thioredoxin fold (Ben Jilani et al., 2007) (Novoselov et al., 2007; Panee et al., 2007). SelH was found to upregulate expression of the two subunits of gamma glutamyl cysteine synthase, leading to bioinformatics analysis resulting in subsequent identification of the AT-hook DNA-binding motif (Panee, Stoytcheva et al., 2007). Chromatin immunoprecipitation assays confirmed binding of Sel H to stress response and heat shock response elements as are found in the promoters of the two gamma glutamyl cysteine synthase subunits. Using these combinatorial approaches, the functions of other recently identified selenoproteins are currently under investigation in a number of laboratories.

2.8 Summary

The discovery that the *cis*-acting SECIS element resides in the 3′ UTR along with characterization of the important structural and sequence elements required to recruit the selenocysteine insertion machinery has allowed for the identification of most if not all selenoprotein genes in organisms whose genomes have been sequenced. Although the presence of other genes utilizing unique sequence elements or accessory factors to incorporate selenocysteine cannot be ruled out, to date there is no evidence for their existence. Extensive efforts are being made to determine the biological function of this interesting class of selenium-containing proteins.

Many studies now indicate that selenoproteins are expressed in a differential manner depending on selenium status, tissue, and developmental stage. NMD is clearly involved in controlling message levels of some selenoproteins and is one factor in determining the hierarchy of selenoprotein expression. Clarification of how selenocysteine messages escape NMD and the mechanism that determines the degree of sensitivity to NMD is an important line of research in answering this question. However, the full answer to how the expression of each selenoprotein is regulated is certain to be more complicated involving multiple factors including tissue levels of selenium, the expression and selective modification of specific *trans*-acting factors, and the *cis*-acting sequences associated with each selenoprotein message.

While significant advances have been made over the past 15 years in our understanding of the selenocysteine insertion mechanism and its role in regulating selenoprotein expression in eukaryotes, fundamental questions remain to be answered. How is the information contained far downstream of the UGA codon in the 3′ UTR conveyed to reprogram the ribosome during decoding of the UGA codon? Is this via a looping mechanism whereby the SECIS element interacts with the ribosome via L30, SBP2, or yet to be identified factors? Does recruitment of the selenocysteine insertion machinery to the ribosome occur during decoding of the UGA codon perhaps facilitated by ribosome pausing or factor binding to the SRE? Alternatively, functional circularization of mRNAs through interactions with polyA-binding protein and initiation factors suggests that the SECIS elements may be in position to reprogram ribosomes for UGA redefinition prior to decoding of the UGA codon. Thus, suggesting the possibility of a tracking model where the SECIS element and associated factors translocate along the message with the ribosome or that the ribosome is reprogrammed early in translation and maintains an altered state without continued association of the SECIS; now competent for redefinition of the UGA codon. In the absence of the SECIS and associated factors, EFSec is not able to deliver Sec-tRNA$^{[SerSec]}$ to the ribosomal A-site. This implies that conformational changes must occur to either the ribosome or the elongation factor during UGA redefinition to allow access. Structural studies of the ribosome are advancing rapidly and the technology to address this question is now available.

Many questions remain to be answered in our understanding of both the biology and regulated synthesis of selenoproteins and we look forward to many new discoveries in the search to understand the use of selenocysteine, the 21st amino acid.

Acknowledgments This work was supported by grants from the National Institutes of Health to MJB and MTH.

References

Allamand V, Richard P, Lescure A, Ledeuil C, Desjardin D, Petit N, Gartioux C, Ferreiro A, Krol A, Pellegrini N, et al (2006) A single homozygous point mutation in a 3′ untranslated region motif of selenoprotein N mRNA causes SEPN1-related myopathy. EMBO Rep 7:450–454

Behne D, Hammel C, Pfeifer H, Rothlein D, Gessner H, Kyriakopoulos A. (1998) Speciation of selenium in the mammalian organism. Analyst 123:871–873

Behne D, Hilmert H, Scheid S, Gessner H, Elger W (1988) Evidence for specific selenium target tissues and new biologically important selenoproteins. Biochim Biophys Acta 966:12–21

Beier H, Grimm M. (2001) Misreading of termination codons in eukaryotes by natural nonsense suppressor tRNAs. Nucleic Acids Res 29:4767–4782

Ben Jilani KE, Panee J, He Q, Berry MJ, Li PA (2007) Overexpression of selenoprotein H reduces Ht22 neuronal cell death after UVB irradiation by preventing superoxide formation. Int J Biol Sci 3:198–204

Berry MJ, Banu L, Chen YY, Mandel SJ, Kieffer JD, Harney JW, Larsen PR (1991) Recognition of UGA as a selenocysteine codon in type I deiodinase requires sequences in the 3′ untranslated region. Nature 353:273–276

Berry M.J, Banu L, Harney JW, Larsen PR (1993) Functional characterization of the eukaryotic SECIS elements which direct selenocysteine insertion at UGA codons. EMBO J 12: 3315–3322

Berry MJ, Harney JW, Ohama T, Hatfield DL (1994) Selenocysteine insertion or termination: factors affecting UGA codon fate and complementary anticodon: codon mutations. Nucleic Acids Res 22:3753–3759

Bidou L, Hatin I, Perez N, Allamand V, Panthier JJ, Rousset JP (2004) Premature stop codons involved in muscular dystrophies show a broad spectrum of readthrough efficiencies in response to gentamicin treatment. Gene Ther 11:619–627

Brigelius-Flohe R, Friedrichs B, Maurer S, Streicher R. (1997) Determinants of PHGPx expression in a cultured endothelial cell line. Biomed Environ Sci 10:163–176

Bubenik JL, Driscoll DM (2007) Altered RNA binding activity underlies abnormal thyroid hormone metabolism linked to a mutation in selenocysteine insertion sequence-binding protein 2. J Biol Chem 282:34653–34662

Budiman ME, Bubenik JL, Miniard AC, Middleton LM, Gerber CA, Cash A, Driscoll DM (2009) Eukaryotic initiation factor 4a3 is a selenium-regulated RNA-binding protein that selectively inhibits selenocysteine incorporation. Mol Cell 35:479–489

Buettner C, Harney JW, Berry MJ (1999) The *Caenorhabditis elegans* homologue of thioredoxin reductase contains a selenocysteine insertion sequence (SECIS) element that differs from mammalian SECIS elements but directs selenocysteine incorporation. J Biol Chem 274:21598–21602

Caban K, Kinzy SA, Copeland PR (2007) The L7Ae RNA binding motif is a multifunctional domain required for the ribosome-dependent Sec incorporation activity of Sec insertion sequence binding protein 2. Mol Cell Biol 27:6350–6360

Carlson BA, Moustafa ME, Sengupta A, Schweizer U, Shrimali R, Rao M, Zhong N, Wang S, Feigenbaum L, Lee BJ et al (2007) Selective restoration of the selenoprotein population in a mouse hepatocyte selenoproteinless background with different mutant selenocysteine tRNAs lacking Um34. J Biol Chem 282:32591–32602

Carlson BA, Xu XM, Gladyshev VN, Hatfield DL (2005) Selective rescue of selenoprotein expression in mice lacking a highly specialized methyl group in selenocysteine tRNA. J Biol Chem 280:5542–5548

Chavatte L, Brown BA, Driscoll DM (2005) Ribosomal protein L30 is a component of the UGA selenocysteine recoding machinery in eukaryotes. Nat Struct Mol Biol 12:408–416

Chittum HS, Hill KE, Carlson BA, Lee BJ, Burk RF, Hatfield DL (1997) Replenishment of selenium deficient rats with selenium results in redistribution of the selenocysteine tRNA population in a tissue specific manner. Biochim Biophys Acta 1359:25–34

Christensen MJ, Burgener KW (1992) Dietary selenium stabilizes glutathione peroxidase mRNA in rat liver. J Nutr 122:1620–1626

Clarke NF, Kidson W, Quijano-Roy S, Estournet B, Ferreiro A, Guicheney P, Manson JI, Kornberg AJ, Shield LK, North KN (2006) SEPN1: associated with congenital fiber-type disproportion and insulin resistance. Ann Neurol 59:546–552

Copeland PR, Fletcher JE, Carlson BA, Hatfield DL, Driscoll DM (2000) A novel RNA binding protein, SBP2, is required for the translation of mammalian selenoprotein mRNAs. EMBO J 19:306–314

Copeland PR, Stepanik VA, Driscoll DM (2001) Insight into mammalian selenocysteine insertion: domain structure and ribosome binding properties of Sec insertion sequence binding protein 2. Mol Cell Biol 21:1491–1498

de Jesus LA, Hoffmann PR, Michaud T, Forry EP, Small-Howard A, Stillwell RJ, Morozova N, Harney JW, Berry MJ (2006) Nuclear assembly of UGA decoding complexes on selenoprotein mRNAs: a mechanism for eluding nonsense mediated decay? Mol Cell Biol 26:1795–1805

Ding F, Grabowski PJ (1999) Identification of a protein component of a mammalian tRNA(Sec) complex implicated in the decoding of UGA as selenocysteine. RNA 5:1561–1569

Dumitrescu AM, Liao XH, Abdullah MS, Lado-Abeal J, Majed FA, Moeller LC, Boran G, Schomburg L, Weiss RE, Refetoff S (2005) Mutations in SECISSBP2 result in abnormal thyroid hormone metabolism. Nat Genet 37:1247–1252

Fagegaltier D, Hubert N, Yamada K, Mizutani T, Carbon P, Krol A (2000) Characterization of mSelB, a novel mammalian elongation factor for selenoprotein translation. EMBO J 19:4796– 4805

Feng YX, Yuan H, Rein A, Levin JG (1992) Bipartite signal for read-through suppression in murine leukemia virus mRNA: an eight-nucleotide purine-rich sequence immediately downstream of the *gag* termination codon followed by an RNA pseudoknot. J Virol 66:5127–5132

Ferreiro A, Ceuterick-de Groote C, Marks JJ, Goemans N, Schreiber G, Hanefeld F, Fardeau M, Martin JJ, Goebel HH, Richard P and others (2004) Desmin-related myopathy with Mallory body-like inclusions is caused by mutations of the selenoprotein N gene. Ann Neurol 55: 676–686

Ferreiro A, Quijano-Roy S, Pichereau C, Moghadaszadeh B, Goemans N, Bonnemann C, Jungbluth H, Straub V, Villanova M, Leroy JP et al. (2002) Mutations of the selenoprotein N gene, which is implicated in rigid spine muscular dystrophy, cause the classical phenotype of multiminicore disease: reassessing the nosology of early-onset myopathies. Am J Hum Genet 71:739–749

Flanigan KM, Kerr L, Bromberg MB, Leonard C, Tsuruda J, Zhang P, Gonzalez-Gomez I, Cohn R, Campbell KP, Leppert M (2000) Congenital muscular dystrophy with rigid spine syndrome: a clinical, pathological, radiological, and genetic study. Ann Neurol 47:152–161

Fletcher JE, Copeland PR, Driscoll DM, Krol A (2001) The selenocysteine incorporation machinery: interactions between the SECIS RNA and the SECIS-binding protein SBP2. RNA 7:1442–1453

Gelpi C, Sontheimer EJ, Rodriguez-Sanchez JL (1992) Autoantibodies against a serine tRNA-protein complex implicated in cotranslational selenocysteine insertion. Proc Natl Acad Sci USA 89:9739–9743

Grundner-Culemann E, Martin GW 3rd, Harney JW, Berry MJ (1999) Two distinct SECIS structures capable of directing selenocysteine incorporation in eukaryotes. RNA 5:625–635

Grundner-Culemann E, Martin GW 3rd, Tujebajeva R, Harney JW, Berry MJ (2001) Interplay between termination and translation machinery in eukaryotic selenoprotein synthesis. J Mol Biol 310:699–707

Gupta M, Copeland PR (2007) Functional analysis of the interplay between translation termination, selenocysteine codon context, and selenocysteine insertion sequence-binding protein 2. J Biol Chem 282:36797–36807

Harrell L, Melcher U, Atkins JF (2002) Predominance of six different hexanucleotide recoding signals 3′ of read-through stop codons. Nucleic Acids Res 30:2011–2017

Hatfield DL, Gladyshev VN (2002) How selenium has altered our understanding of the genetic code. Mol Cell Biol 22:3565–3576

Hawkes WC, Wilhelmsen EC, Tappel AL (1985) Abundance and tissue distribution of selenocysteine-containing proteins in the rat. J Inorg Biochem 23:77–92

Hentze MW, Kulozik AE (1999) A perfect message: RNA surveillance and nonsense-mediated decay. Cell 96:307–310

Hilgenfeld R, Böck A, Wilting R (1996) Structural model for the selenocysteine-specific elongation factor SelB. Biochimie 78:971–978

Hill KE, Lloyd RS, Burk RF (1993) Conserved nucleotide sequences in the open reading frame and 3′ untranslated region of selenoprotein P mRNA. Proc Natl Acad Sci USA 90:537–541

Hill KE, Lyons PR, Burk RF (1992) Differential regulation of rat liver selenoprotein mRNAs in selenium deficiency. Biochem Biophys Res Commun 185:260–263

Himeno S, Chittum HS, Burk RF (1996) Isoforms of selenoprotein P in rat plasma. Evidence for a full-length form and another form that terminates at the second UGA in the open reading frame. J Biol Chem 271:15769–15775

Hirosawa-Takamori M, Ossipov D, Novoselov SV, Turanov AA, Zhang Y, Gladyshev VN, Krol A, Vorbruggen G, Jackle H. (2009) A novel stem loop control element-dependent UGA read-through system without translational selenocysteine incorporation in *Drosophila*. FASEB J 23:107–113

Hong X, Scofield DG, Lynch M. (2006) Intron size, abundance and distribution within untranslated regions of genes. Mol Biol Evol 23:2392–2404

Howard MT, Aggarwal G, Anderson CB, Khatri S, Flanigan KM, Atkins JF (2005) Recoding elements located adjacent to a subset of eukaryal selenocysteine-specifying UGA codons. EMBO J 24:1596–1607

Howard MT, Moyle MW, Aggarwal G, Carlson BA, Anderson CB (2007) A recoding element that stimulates decoding of UGA codons by Sec tRNA[Ser]Sec. RNA 13:912–920

Howard MT, Shirts BH, Petros LM, Flanigan KM, Gesteland RF, Atkins JF (2000) Sequence specificity of aminoglycoside-induced stop codon readthrough: potential implications for treatment of Duchenne muscular dystrophy. Ann Neurol 48:164–169

Isken O, Maquat LE (2007) Quality control of eukaryotic mRNA: safeguarding cells from abnormal mRNA function. Genes Dev 21:1833–1856

Jameson RR, Diamond AM (2004) A regulatory role for Sec tRNA[Ser]Sec in selenoprotein synthesis. RNA 10:1142–1152

Jung JE, Karoor V, Sandbaken MG, Lee BJ, Ohama T, Gesteland RF, Atkins JF, Mullenbach GT, Hill KE, Wahba AJ and others (1994) Utilization of selenocysteyl-tRNA[Ser]Sec and seryl-tRNA[Ser]Sec in protein synthesis. J Biol Chem 269:29739–29745

Jurynec MJ, Xia R, Mackrill JJ, Gunther D, Crawford T, Flanigan KM, Abramson JJ, Howard MT, Grunwald DJ (2008) Selenoprotein N is required for ryanodine receptor calcium release channel activity in human and zebrafish muscle. Proc Natl Acad Sci USA 105: 12485–12490

Kim IY, Guimaraes MJ, Zlotnik A, Bazan JF, Stadtman TC (1997) Fetal mouse selenophosphate synthetase 2 (SPS2): characterization of the cysteine mutant form overproduced in a baculovirus-insect cell system. Proc Natl Acad Sci USA 94:418–421

Kromayer M, Wilting R, Tormay P, Böck A (1996) Domain structure of the prokaryotic selenocysteine-specific elongation factor SelB. J Mol Biol 262:413–420

Kryukov GV, Castellano S, Novoselov SV, Lobanov AV, Zehtab O, Guigo R, Gladyshev VN (2003) Characterization of mammalian selenoproteomes. Science 300:1439–1443

Lei XG, Evenson JK, Thompson KM, Sunde RA (1995) Glutathione peroxidase and phospholipid hydroperoxide glutathione peroxidase are differentially regulated in rats by dietary selenium. J. Nutr. 125:1438–1446

Li G, Rice CM (1993) The signal for translational readthrough of a UGA codon in Sindbis virus RNA involves a single cytidine residue immediately downstream of the termination codon. J Virol 67:5062–5067

Lobanov AV, Hatfield DL, Gladyshev VN (2008) Reduced reliance on the trace element selenium during evolution of mammals. Genome Biol 9:R62

Low SC, Harney JW, Berry MJ (1995) Cloning and functional characterization of human selenophosphate synthetase, an essential component of selenoprotein synthesis. J Biol Chem 270:21659–21664

Ma S, Hill KE, Caprioli RM, Burk RF (2002) Mass spectrometric characterization of full-length rat selenoprotein P and three isoforms shortened at the C terminus. Evidence that three UGA codons in the mRNA open reading frame have alternative functions of specifying selenocysteine insertion or translation termination. J Biol Chem 277:12749–12754

Maiti B, Arbogast S, Allamand V, Moyle MW, Anderson CB, Richard P, Guicheney P, Ferreiro A, Flanigan KM, Howard MT (2009) A mutation in the SEPN1 selenocysteine redefinition element (SRE) reduces selenocysteine incorporation and leads to SEPN1-related myopathy. Hum Mutat 30:411–416.

Manuvakhova M, Keeling K, Bedwell DM (2000) Aminoglycoside antibiotics mediate context-dependent suppression of termination codons in a mammalian translation system. RNA 6: 1044–1055

Martin GW 3rd, Harney JW, Berry MJ (1996) Selenocysteine incorporation in eukaryotes: insights into mechanism and efficiency from sequence, structure, and spacing proximity studies of the type 1 deiodinase SECIS element. RNA 2:171–182

Martin R, Phillips-Jones MK, Watson FJ, Hill LS (1993) Codon context effects on nonsense suppression in human cells. Biochem Soc Trans 21:846–851

McCaughan KK, Brown CM, Dalphin ME, Berry MJ, Tate WP (1995) Translational termination efficiency in mammals is influenced by the base following the stop codon. Proc Natl Acad Sci USA 92:5431–5435

Mehlin H, Daneholt B, Skoglund U. (1992) Translocation of a specific premessenger ribonucleoprotein particle through the nuclear pore studied with electron microscope tomography. Cell 69:605–613

Mitchell JH, Nicol, F, Beckett GJ, Arthur JR (1997) Selenium and iodine deficiencies: effects on brain and brown adipose tissue selenoenzyme activity and expression. J Endocrinol 155: 255–263

Moghadaszadeh B, Petit N, Jaillard C, Brockington M, Roy SQ, Merlini L, Romero N, Estournet B, Desguerre I, Chaigne D, and others (2001) Mutations in SEPN1 cause congenital muscular dystrophy with spinal rigidity and restrictive respiratory syndrome. Nat Genet 29:17–18

Moriarty PM, Reddy CC, Maquat LE (1998) Selenium deficiency reduces the abundance of mRNA for Se-dependent glutathione peroxidase 1 by a UGA-dependent mechanism likely to be nonsense codon-mediated decay of cytoplasmic mRNA. Mol Cell Biol 18: 2932–2939

Mottagui-Tabar S, Björnsson A, Isaksson LA (1994) The second to last amino acid in the nascent peptide as a codon context determinant. EMBO J 13:249–257

Mottagui-Tabar S, Tuite MF, Isaksson LA. (1998) The influence of 5′ codon context on translation termination in *Saccharomyces cerevisiae*. Eur J Biochem 257:249–254

Moustafa ME, Carlson BA, El-Saadani MA, Kryukov GV, Sun QA, Harney JW, Hill KE, Combs GF, Feigenbaum L, Mansur DB and others (2001) Selective inhibition of selenocysteine tRNA maturation and selenoprotein synthesis in transgenic mice expressing isopentenyladenosine-deficient selenocysteine tRNA. Mol Cell Biol 21:3840–3852

Nagy E, Maquat LE (1998) A rule for termination-codon position within intron-containing genes: when nonsense affects RNA abundance. Trends Biochem Sci 23:198–199

Namy O, Hatin I, Rousset JP (2001) Impact of the six nucleotides downstream of the stop codon on translation termination. EMBO Rep 2:787–793

Nasim MT, Jaenecke S, Belduz A, Kollmus H, Flohe L, McCarthy JE (2000) Eukaryotic selenocysteine incorporation follows a nonprocessive mechanism that competes with translational termination. J Biol Chem 275:14846–14852

Novoselov SV, Kryukov GV, Xu XM, Carlson BA, Hatfield DL, Gladyshev VN (2007) Selenoprotein H is a nucleolar thioredoxin-like protein with a unique expression pattern. J Biol Chem 282:11960–11968

Panee J, Stoytcheva Z, Liu W, Berry M (2007) Selenoprotein H is a redox-sensing HMG family DNA-binding protein that upregulates genes involved in glutathione synthesis and phase II detoxification. J Biol Chem 282:23759–23765

Papp LV, Lu J, Striebel F, Kennedy D, Holmgren A, Khanna KK (2006) The redox state of SECIS binding protein 2 controls its localization and selenocysteine incorporation function. Mol Cell Biol 26:4895–4910

Pedersen JS, Bejerano G, Siepel A, Rosenbloom K, Lindblad-Toh K, Lander ES, Kent J, Miller W, Haussler D (2006) Identification and classification of conserved RNA secondary structures in the human genome. PLoS Comput Biol 2:e33

Philipson L, Andersson P, Olshevsky U, Weinberg R, Baltimore D, Gesteland R (1978) Translation of MuLV and MSV RNAs in nuclease-treated reticulocyte extracts: enhancement of the Gag-Pol polypeptide with yeast suppressor tRNA. Cell 13:189–199

Robinson DN, Cooley L. (1997) Examination of the function of two kelch proteins generated by stop codon suppression. Development 124:1405–1417

Saedi MS, Smith CG, Frampton J, Chambers I, Harrison PR, Sunde RA (1988) Effect of selenium status on mRNA levels for glutathione peroxidase in rat liver. Biochem Biophys Res Commun 153:855–861

Shen Q, Chu FF, Newburger PE (1993) Sequences in the 3′-untranslated region of the human cellular glutathione peroxidase gene are necessary and sufficient for selenocysteine incorporation at the UGA codon. J Biol Chem 268:11463–11469

Small-Howard A, Morozova N, Stoytcheva Z, Forry EP, Mansell JB, Harney JW, Carlson BA, Xu SM, Hatfield DL, Berry MJ (2006) Supramolecular complexes mediate selenocysteine incorporation in vivo. Mol Cell Biol 26:2337–2346

Squires JE, Stoytchev I, Forry EP, Berry MJ (2007) SBP2 binding affinity is a major determinant in differential selenoprotein mRNA translation and sensitivity to nonsense-mediated decay. Mol Cell Biol 27:7848–7855

Stoytcheva Z, Tujebajeva RM, Harney JW, Berry MJ (2006) Efficient incorporation of multiple selenocysteines involves an inefficient decoding step serving as a potential translational checkpoint and ribosome bottleneck. Mol Cell Biol 26:9177–9184

ten Dam EB, Pleij CW, Bosch L (1990) RNA pseudoknots: translational frameshifting and readthrough on viral RNAs. Virus Genes 4:121–136

Thermann R, Neu-Yilik G, Deters A, Frede U, Wehr K, Hagemeier C, Hentze MW, Kulozik AE. (1998) Binary specification of nonsense codons by splicing and cytoplasmic translation. EMBO J 17:3484–3494

Tujebajeva RM, Copeland PR, Xu XM, Carlson BA, Harney JW, Driscoll DM, Hatfield DL, Berry MJ (2000) Decoding apparatus for eukaryotic selenocysteine incorporation. EMBO Rep 2: 158–163

Ursini F, Heim S, Kiess M, Maiorino M, Roveri A, Wissing J, Flohe L (1999) Dual function of the selenoprotein PHGPx during sperm maturation. Science 285:1393–1396

Walczak R, Westhof E, Carbon P, Krol A (1996) A novel RNA structural motif in the selenocysteine insertion element of eukaryotic selenoprotein mRNAs. RNA 2:367–379

Weiss SL, Sunde RA (1998) *Cis*-acting elements are required for selenium regulation of glutathione peroxidase-1 mRNA levels. RNA 4:816–827

Wills NM, Gesteland RF, Atkins JF (1991) Evidence that a downstream pseudoknot is required for translational read-through of the Moloney murine leukemia virus *gag* stop codon. Proc Natl Acad Sci USA 88:6991–6995

Wilting R, Schorling S, Persson BC, Böck A (1997) Selenoprotein synthesis in archaea: identification of an mRNA element of *Methanococcus jannaschii* probably directing selenocysteine insertion. J Mol Biol 266:637–641

Wu R, Shen Q, Newburger PE (2000) Recognition and binding of the human selenocysteine insertion sequence by nucleolin. J Cell Biochem 77:507–516

Xu XM, Mix H, Carlson BA, Grabowski PJ, Gladyshev VN, Berry MJ, Hatfield DL (2005) Evidence for direct roles of two additional factors, SECp43 and SLA, in the selenoprotein synthesis machinery. J Biol Chem 280:41568–41575

Yoshinaka Y, Katoh I, Copeland TD, Oroszlan S (1985) Murine leukemia virus protease is encoded by the gag-pol gene and is synthesized through suppression of an amber termination codon. Proc Natl Acad Sci USA 82:1618–1622

Zavacki AM, Mansell JB, Chung M, Klimovitsky B, Harney JW, Berry MJ (2003) Coupled $tRNA^{Sec}$ dependent assembly of the selenocysteine decoding apparatus. Mol Cell 11:773–781

Chapter 3
Translation of UAG as Pyrrolysine

Joseph A. Krzycki

Abstract Pyrrolysine followed selenocysteine in order of discovery. While both atypical amino acids are encoded by canonical stop codons, the mechanisms by which they are inserted into protein are very different. Pyrrolysine is carried to the ribosome by $tRNA^{Pyl}$ (encoded by *pylT*) whose unusual structure possesses the CUA anticodon needed to decode UAG. A pyrrolysyl-tRNA synthetase (product of *pylS*) ligates pyrrolysine to $tRNA^{Pyl}$. Pyrrolysine is made by the products of the *pylBCD* genes without the need for $tRNA^{Pyl}$, contrasting with selenocysteine synthesis on $tRNA^{Sec}$. Isolated examples of the *pylTSBCD* genes, often in a single cluster, have been found in genomes of methanogenic Archaea, G+ Bacteria, and δ-proteobacteria. *Escherichia coli* transformed with *pyl* genes translates UAG as endogenously synthesized pyrrolysine. The ease of the lateral transfer of the genetic encoding of pyrrolysine is now being exploited for tailoring recombinant proteins. Pyrrolysine incorporation appears to occur to some extent by amber suppression on a genome-wide basis in methanogenic Archaea. With some methylamine methyltransferase transcripts, a putative pyrrolysine insertion sequence (PYLIS) forms an in-frame stem-loop 3′ to the translated UAG, analogous to such loops required in Bacteria for translation of UGA as selenocysteine. PYLIS sequences are not found in all types of methylamine methyltransferases. Unlike the precedent of selenocysteine, after deletion of PYLIS, significant UAG translation remains with a marked increase in UAG-directed termination, suggesting some part of the PYLIS sequence functions in enhancing amber suppression. Some methanogen genomes encode additional homologs of elongation and release factors, however, their limited distribution suggests at best a nonessential role in enhancing UAG translation as pyrrolysine.

J.A. Krzycki (✉)
Department of Microbiology, The Ohio State University, Columbus, OH, USA
e-mail: krzycki.1@osu.edu

J.F. Atkins, R.F. Gesteland (eds.), *Recoding: Expansion of Decoding Rules Enriches Gene Expression*, Nucleic Acids and Molecular Biology 24,
DOI 10.1007/978-0-387-89382-2_3,

Contents

3.1 Introduction

Twenty protein residues are found in common among all organisms on earth. Each of these familiar and abundant amino acids has been represented in the genetic code with one to six of the 61 sense codons, leaving three nonsense codons to signal the end of translation. However, in some organisms, nonsense codons have come to also signal the co-translational incorporation of the two atypical residues selenocysteine and pyrrolysine (Böck et al. 2004; Krzycki 2005). Selenocysteine was discovered in 1976 as a residue of clostridial glycine reductase (Cone et al. 1976). Ten years elapsed until the realization that selenocysteine was inserted under the direction of an opal (TGA=UGA) codon during the synthesis of *E. coli* formate dehydrogenase (Zinoni et al. 1986) and mammalian glutathione reductase (Chambers et al. 1986). Pyrrolysine was found some 16 years later (Hao et al. 2002; Srinivasan et al. 2002). In contrast to the discovery of selenocysteine, pyrrolysine was presaged by discovery of an in-frame amber (TAG=UAG) codon within the genes encoding the different methylamine methyltransferases of select methanogenic Archaea (Burke and Krzycki 1998; Paul et al. 2000).

In deference to their entrance into the genetic code, selenocysteine and pyrrolysine have been called the 21st and 22nd amino acids (Bock et al. 1991b; Atkins and Gesteland 2002). The two residues have continued to provide a study in contrasts and similarities as to how amino acids can enter the genetic code. For example, UGA is recoded as selenocysteine on a gene-by-gene basis, while pyrrolysine appears to be inserted by a combination of genome-wide amber suppression, combined with local context effects that amplify the efficiency of pyrrolysine incorporation.

3.2 The Discovery and Biological Context of Pyrrolysine

The methanogens are the charter members of the Archaea, being the first group identified by Woese whose 16S ribosomal RNA sequences supported a phylogeny distinct from both Bacteria and Eucarya (Woese 1977, 1990). Most methanogens are capable of growth by reduction of CO_2 to methane, with little other recourse for cellular energy or carbon. However, some families, such as many within the order Methanosarcinales, are also capable of utilizing compounds such as acetate, methanol, as well as a few types of methylthiols or methylamines (Thauer 1998; Ferry 1999; Krzycki 2004). The latter category includes trimethylamine (TMA), dimethylamine (DMA), or monomethylamine (MMA).

Characterization of the enzymes initiating methylamine metabolism in *Methanosarcina barkeri* MS revealed distinct methyltransferases specific for TMA, DMA, or MMA (Burke and Krzycki 1997; Ferguson and Krzycki 1997; Ferguson et al. 2000). These proteins methylate a small cognate corrinoid-binding protein that is then used to generate major precursors of methane and carbon assimilation. As each methyltransferase is highly abundant, it was of considerable surprise when *mtmB1*, the gene encoding the predominant MMA methyltransferase (MtmB), was found to contain an in-frame amber codon (Burke et al. 1998). An ORF encoding the N-terminus of MtmB and ending with a TAA codon was identified with a single mid-frame amber codon in two *M. barkeri* strains. The UAG codon was present in the *mtmB1* transcript. Numerous stops were detected in both other reading frames, indicating successive frameshifts of opposite polarity, or in-frame ribosome hopping, would be necessary to bypass the UAG codon. Both scenarios were unprecedented, leaving a reasonable probability of direct translation, possibly as a specialized catalytic residue (Burke et al. 1998).

The sequence of genes encoding the TMA methyltransferase (*mttB*) and DMA methyltransferase (*mtbB*) from *M. barkeri* demonstrated that all three nonhomologous methylamine methyltransferase genes possess an in-frame amber codon (Paul et al. 2000). Two additional *mtbB* genes in the same genome average 95% identity, each with an in-frame amber codon (Paul et al. 2000). The C-terminal sequence of the isolated DMA methyltransferase confirmed that the UAA codon, and not the in-frame amber codon, signaled the termination of translation of the mature protein (Paul et al. 2000). The TMA methyltransferase from *Methanosarcina thermophila* also possesses a conserved in-frame amber codon (Paul et al. 2000). Analysis of the transcript from the TMA methyltransferase gene *mttB* indicated that the UAG codon is represented in the nonedited transcript. Numerous stops in both other reading frames again left the possibilities of a ribosome-hopping event or successive compensating frameshifts or UAG translation.

The sequencing of the genomes of *M. barkeri* Fusaro, *Methanosarcina acetivorans*, *Methanosarcina mazei*, and *Methanococcoides burtonii* (also a member of the family Methanosarcinaceae) has now shown that all instances of the *mttB*, *mtbB*, and *mtmB* genes contained in-frame amber codons, the position being completely conserved in each type of methyltransferase (Deppenmeier et al. 2002; Galagan et al. 2002; Maeder et al. 2006; Goodchild et al. 2004; Zhang et al. 2005). The phenomenon of multiple, nearly identical, copies of each methyltransferase gene is common among the Methanosarcinaceae. *M. acetivorans*, for example, has two to three copies of each type of methylamine methyltransferase gene, each one possessing a conserved in-frame amber codon (Galagan et al. 2002). These copies are named numerically, e.g., *mtmB1* and *mtmB2*. The isolated MtmB protein is produced from *mtmB1* in *M. barkeri* (Hao et al. 2002; Soares et al. 2005).

Both Edman degradation and mass spectrometry of a tryptic peptide of MtmB confirmed that translation continued through the amber codon to the terminal UAA codon (James et al. 2001). The efficiency of UAG translation appeared very high, in that little of a possible amber-termination product from the *mtmB* genes could be identified by immunoblotting *M. barkeri* extracts. Both mass spectrometry and Edman degradation indicated lysine was at the amber codon position. However, given the tryptic fragment had been isolated with extended exposure to acidic conditions, the stated possibility remained that the UAG codon could encode a labile lysine derivative (James et al. 2001).

Crystallography of MtmB by Bing Hao and Michael Chan revealed a lysine derivative was indeed present at the UAG-encoded position (Hao et al. 2002). The 1.55 Å structure showed the N^{ε} of the UAG-encoded lysine was in amide linkage with 4-substituted pyrroline-5-carboxylate. Initially, the identity of the 4-substiutent could not be deduced, however, subsequent crystallography of the protein in which the residue was derivatized with hydroxylamine or sulfite allowed assignment as a methyl group (Hao et al. 2004), providing a structure for pyrrolysine (Fig. 3.1). Accurate mass determination of the UAG-encoded residue in the MMA methyltransferase confirmed the proposed empirical formula and that pyrrolysine was present in the DMA and TMA methyltransferases as well (Soares et al. 2005). Thus, UAG stands for pyrrolysine in the products from all three of the methyltransferase genes, in spite of the fact that the TMA, DMA, or MMA methyltransferases have no identifiable primary sequence similarity.

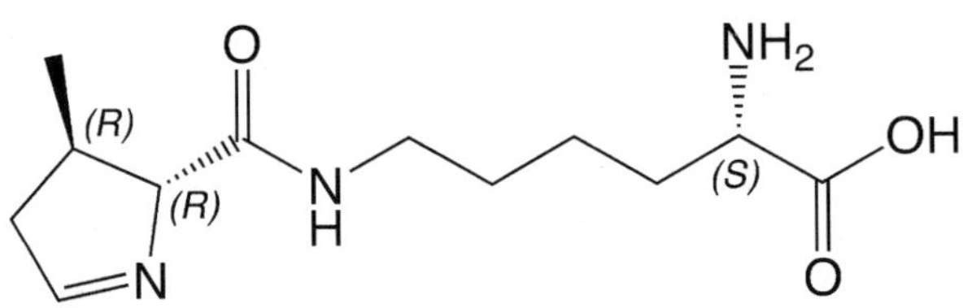

Fig. 3.1 The structure of pyrrolysine

Coincident with the discovery of pyrrolysine, the *pylT* gene was found near a cluster of methylamine methyltransferase genes in *M. barkeri* (Srinivasan et al. 2002). The *pylT* gene encodes an amber-decoding tRNA that participates in UAG translation. Adjacent to *pylT* is *pylS*, whose product is homologous to class II aminoacyl-tRNA synthetases. It was proposed that PylS is instrumental in charging $tRNA_{CUA}$ for decoding UAG (Srinivasan et al. 2002). It is now known that PylS is a pyrrolysyl-tRNA synthetase that can charge $tRNA_{CUA}$ with pyrrolysine, and thus, $tRNA_{CUA}$ is $tRNA^{Pyl}$ (Blight et al. 2004; Polycarpo et al. 2004; Schimmel and Beebe 2004). Close homologs of *pylS* and *pylT* genes have also been found in *Desulfitobacterium hafniense* (Srinivasan et al. 2002) and a symbiotic δ-proteobacterium (Woyke et al. 2006; Zhang and Gladyshev 2007), as well as *M. mazei* (Deppenmeier et al. 2002), *M. acetivorans* (Galagan et al. 2002), and *M. burtonii* (Goodchild et al. 2004).

3.3 Novel Functionality Underlies Pyrrolysine Addition to the Genetic Code

Every sequenced genome having *pylT* and *pylS* also has examples of *mttB*, *mtbB*, or *mtmB* homologs with conserved amber codon (Deppenmeier et al. 2002; Galagan et al. 2002; Srinivasan et al. 2002; Goodchild et al. 2004; Zhang et al. 2005; Woyke et al. 2006; Zhang and Gladyshev 2007). This continues to suggest that UAG translation as pyrrolysine is closely associated with methylamine-dependent methyltransferase function. Indeed, an *M. acetivorans* Δ*ppylT* mutant containing a deletion of *pylT* and the *pyl* promoter cannot metabolize MMA, DMA, or TMA for either methane production or nitrogen assimilation. The Δ*ppylT* strain has normal growth rates on acetate or methanol (Mahapatra et al. 2006).

The imine bond of pyrrolysine brings to the genetic code an electrophilic functionality (Hao et al. 2002) not observed without modification of the canonical 20 residues (Retey 2003). Several crystal structures of MtmB with sulfite, ammonia, or hydroxylamine substitutions at the C-2 position of the pyrrolysine ring illustrate the reactivity of the imine bond of pyrrolysine (Hao et al. 2002, 2004) and fueled the proposal that pyrrolysine plays a unique catalytic role in MtmB, in which methylamine substitutes at the C-2 position prior to methyl group transfer to the corrinoid protein (Hao et al. 2002), see also Krzycki (2004).

Homologs of the methanogen methylamine methyltransferases without amber codons are found in the genomes of many Bacteria and a few non-methanogenic Archaea (Srinivasan et al. 2002; Zhang et al. 2005; Nonaka et al. 2006; Woyke et al. 2006; Atkins and Baranov 2007; Zhang and Gladyshev 2007). Genes encoding such presumably pyrrolysine-free homologs of the TMA methyltransferase are most prevalent, and BLAST searches will readily retrieve such homologs predominantly from various α-proteobacteria, as well as in the crenarchaeaote *Thermofilum pendens*, and *Bacteroides* spp. The functions of these genes are unknown, but significantly *pylS* and *pylT* are not found in genomes unless the in-frame amber codon is conserved in at least one of the methyltransferase gene homologs.

3.4 The *pyl* Gene Cluster

Selenocysteinyl-$tRNA^{Sec}$ is made from seryl-$tRNA^{Sec}$ (Bock et al. 1991a). Prior to the demonstration that PylS is a pyrrolysyl-tRNA synthetase (Blight et al. 2004; Polycarpo et al. 2004), lysyl-$tRNA^{Pyl}$ was analogously considered a likely intermediate toward the formation of pyrrolysyl-tRNA^{Pyl}. However, the lysyl-tRNA^{Pyl} synthetase activity attributed to PylS could not be observed with a more precise assay that directly monitors charging of tRNA^{Pyl} (Srinivasan et al. 2002; Polycarpo et al. 2003; Blight et al. 2004). An alternative means of charging tRNA^{Pyl} with lysine by concerted action of the *M. barkeri* class I and class II lysyl-tRNA synthetases (LysK and LysS) was proposed (Polycarpo et al. 2003). Deletion of either *lysS* or *lysK* from *M. acetivorans* does not effect the formation of pyrrolysyl proteins such as MtmB nor diminish levels of charged tRNA^{Pyl} in the cell (Mahapatra et al. 2007). Therefore, this path to lysyl-tRNA^{Pyl} is unlikely to be a major route to pyrrolysyl-tRNA^{Pyl}.

With the first description of *pylT*, the possibility that tRNA^{Pyl} could be charged directly with pyrrolysine was proposed, but the lack of the amino acid as a chemically synthesized compound left this an untested hypothesis (Srinivasan et al. 2002). The difficult synthesis of pyrrolysine by the Chan group surmounted the problem (Hao et al. 2004), and with this substrate it was shown that PylS could activate pyrrolysine to the pyrrolysyl-adenylate, as well as ligate-charged tRNA^{Pyl} with pyrrolysine (Blight et al. 2004). Similar activities were observed with a pyrrolysine analog (Polycarpo et al. 2004) in vitro.

In an important test of the specificity of PylS for pyrrolysine and tRNA^{Pyl}, the *pylT* and *pylS* genes were transformed into *E. coli* (Blight et al. 2004). The resultant strain could incorporate exogenous chemically synthesized pyrrolysine into MtmB into the UAG-encoded position. The in vivo and in vitro evidence support the entrance of pyrrolysine into the genetic code via the first aminoacyl-tRNA synthetase known to have specificity for an amino acid beyond than the original 20 (Krzycki 2005).

In *M. acetivorans*, *M. mazei*, two *M. barkeri* strains, and *M. burtonii*, three additional genes, *pylB*, *pylC*, and *pylD*, form an apparent transcriptional unit with *pylT* and *pylS* (Fig. 3.2). This was directly demonstrated for *pylC*, *pylB*, and *pylS* in *M. barkeri* MS (Srinivasan et al. 2002). The *pylT* gene can be detected on this same transcript, as well as the mature 72 bp species in the extracted tRNA pool. The association of *pylT* and *pylS* with *pylBCD* was dramatically illustrated by the genome of the gram-positive anaerobic Bacteria, *D. hafniense* (Srinivasan et al. 2002). *D. hafniense* possesses the capacity to produce tRNA^{Pyl}, but in this organism, *pylS* has been split into two genes (Fig. 3.2). A homolog of the catalytic domain of archaeal PylS is encoded by *pylSc* and is found adjacent to *pylT*, but a homolog of the more degenerate N-terminal domain of archaeal PylS is encoded by *pylSn*, which is found downstream of *pylSc*. Between *pylSc* and *pylSn*, homologs of *pylB*, *pylC*, and *pylD* are present. Thus, in this gram-positive organism, the *pylTScBCDSn* genes are found in an apparent transcriptional unit whose arrangement is nearly identical to that found in unrelated methanogenic Archaea but for the presence of a split *pylS* gene (Srinivasan et al. 2002; Zhang et al. 2005).

Aside from Methanosarcinaceae and *D. hafniense*, only one other complete set of *pyl* genes has been found; this is the uncultivated methyltransferase-replete δ-proteobacterium described in a metagenomic study of *Olavius* symbionts (Woyke

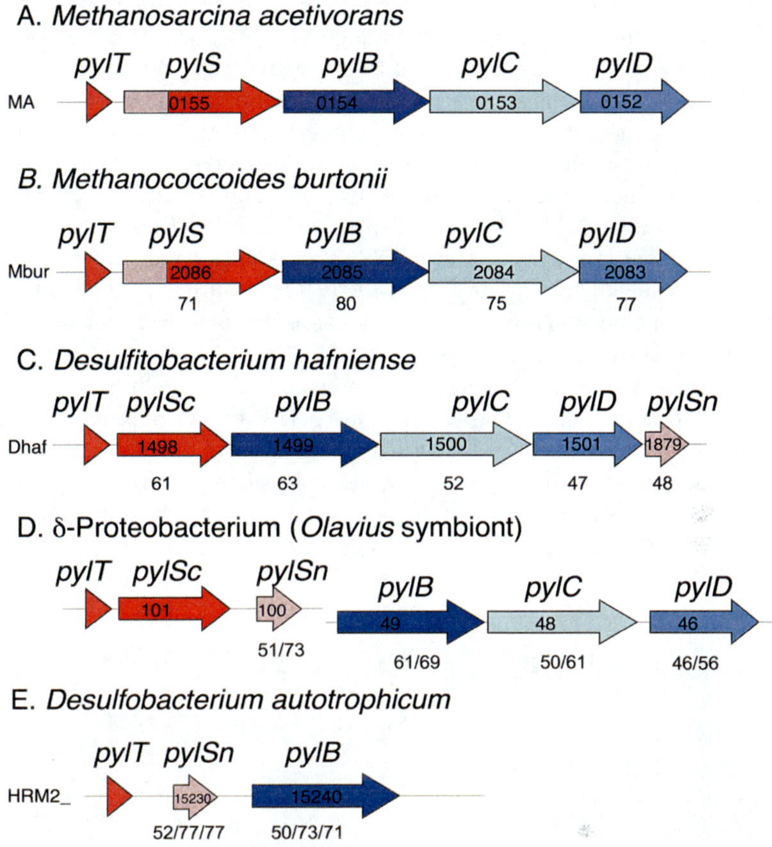

Fig. 3.2 Known examples of *pyl* gene clusters. Colors indicate homologous genes or gene regions. The *pyl* gene designations are found above each ORF, within each protein encoding gene is a number corresponding to the ORF designation from the genomic sequencing project, for brevity the letter designations preceding each gene cluster are shown left of the gene cluster, e.g., MA0155 corresponds to the *pylS* from *M. acetivorans*. In the case of the *Olavius* symbiont the designations are the ORF numbers found on the relevant contigs. (**A**) The *pyl* gene cluster in *M. acetivorans* and related *Methanosarcina* spp. was the first shown to encode proteins required to make and genetically encode pyrrolysine in vivo. The Pyl proteins from *M. barkeri* or *M. mazei* range from 86 to 95% similar to those of *M. acetivorans*. All Methanosarcinacea including *Mc. burtonii* (**B**) share a similar *pyl* gene order, which is dramatically altered in gram-positive Bacteria (**C**). In (**B**) and (**C**) the numbers below each gene refer to the percent similarity to the homologous gene or gene region in *M. acetivorans*. In the gutless worm δ-proteobacterial symbiont (**D**), the split of *pylS* into *pylSn* and *pylSc* is also observed, but the pyrrolysine biosynthetic genes are found on a separate portion of the genome. The numbers below each symbiont gene refer, respectively, to the percent similarity of each gene to those in *M. acetivorans* or *D. hafniense* DCB-1. Finally, a disrupted *pyl* gene cluster is present in the recently sequenced genome of a free-living δ-proteobacteria, *D. autotrophicum*. The percent similarities shown beneath each *D. autotrophicum* gene correspond to the homologous genes in *M. acetivorans*, *D. hafniense*, or the *Olavius* symbiont, respectively. References are found in the text

et al. 2006; Atkins and Baranov 2007; Zhang and Gladyshev 2007). The bacterial strategy for encoding PylS has been maintained in this gram-negative bacterium with the pyrrolysyl-tRNA synthetase encoded in two separate genes associated with *pylT* (Fig. 3.2). In contrast to *D. hafniense*, the proteobacterial *pylBCD* genes are found in a separate gene cluster (and contig) from the *pylTScSn* genes, but nonetheless binning with the same symbiont in the metagenomic analysis. This arrangement indicates richer diversity of *pyl* gene clusters in nature. Indeed, a recent metagenomic study of genes associated with aerobic methane oxidation in freshwater sediment (Kalyuzhnaya et al. 2008) sequenced an environmental DNA fragment upon which homologs of *pylSc* directly adjacent to *pylC* can be identified. Recently, the genome of *Desulfobacterium autotrophicum*, a δ-proteobacterium, was sequenced (Strittmatter, et al. 2009). BLAST searches of this genome revealed a cluster of genes that are not annotated as *pyl* genes, but are highly similar to *pylT*, *pylSn*, and *pylB* (Figs. 3.2 and 3.3). Other *pyl* genes are not present. The nearby transposase gene fragments suggest the *pyl* gene cluster in this organism has been disrupted, leading to loss of the ability to genetically encode pyrrolysine. Although *D. autotrophicum* has a number of methylamine methyltransferase gene homologs, none of these have an in-frame amber codon.

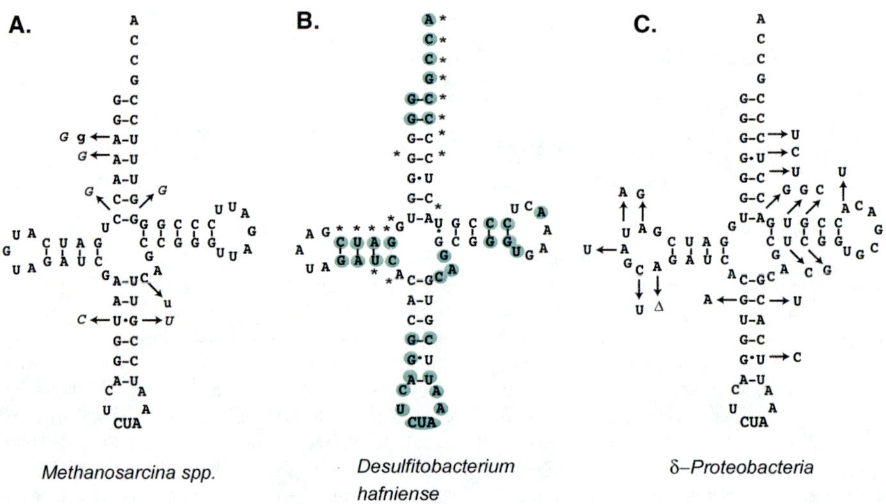

Fig. 3.3 Known examples of tRNA^{Pyl} in the Archaea and Bacteria reveal conservation of residues and unique secondary structure features. The tRNA^{Pyl} from *M. barkeri* Fusaro (**A**) is identical to that from *M. acetivorans* and *M. mazei*. Base changes in *M. barkeri* MS (*small letters*) or in *Mc. burtonii* (*capital italics*) are indicated by *arrows*. The example from *D. hafniense* (**B**) is the only one sequenced from G+ Bacteria, but maintains a number of residues (*shaded in turquoise*) conserved in tRNA^{Pyl} from Archaea and δ-proteobacteria. Residues that contact PylSc are indicated (*asterisks*). The tRNA^{Pyl} from a symbiotic δ-proteobacteria (**C**) has an expanded D-loop but maintains the 6 bp acceptor stem and small variable loop. The base variations relative to the symbiont tRNA^{Pyl} found in the putative tRNA^{Pyl} from *D. autotrophicum* are indicated by *arrows*

3.5 Structure and Binding of tRNA^{Pyl} by PylS

The addition of pyrrolysine to the genetic code of *E. coli* by transformation with *pylS* and *pylT* illustrates the centrality of their gene products in programming UAG codons to also act as pyrrolysine codons. This requires the precise recognition of tRNA^{Pyl} by PylS.

While tRNA^{Pyl} from methanogenic Archaea are nearly identical, *D. hafniense* and δ-proteobacterial tRNA^{Pyl} are, respectively, 68 or 62% identical to *M. barkeri* tRNA^{Pyl} (Fig. 3.3). Nonetheless, the bacterial examples retain the unusual properties first described with *M. barkeri* tRNA^{Pyl} (Srinivasan et al. 2002). The nearly universal GG sequence in the D-loop and the TψC loop in the T-sequence are missing, suggesting atypical loop interaction in the final tertiary structure. Generally, the D-loop is small. The anticodon stem forms with six, rather than five, base pairs constraining the variable loop to only three nucleotides, not the typical four. One, not two, base lies between the D-stem and the acceptor stem. An alternative folding with a shorter anticodon stem was proposed (Polycarpo et al. 2003). However, a subsequent structure probing study supported the extended anticodon stem, small variable loop, and overall resemblance of secondary structure to that of bovine mitochrondrial tRNA^{Ser} (Théobald-Dietrich et al. 2004) that has the typical L-shaped tertiary structure (Hayashi et al. 1998).

Most recently, the structure of the *D. hafniense* tRNA^{Pyl} complexed with *D. hafniense* PylSc (Nozawa et al. 2009) confirmed the atypical secondary structure of tRNA^{Pyl} and revealed that the tertiary structure was similar to canonical tRNA, but with variations. For example, only one nucleotide between the anticodon- and D-stems, and small D- and variable loops, lead to a compact core for tRNA^{Pyl}. Alternative base pairings compensate for lack of the conserved sequences in the D- and T-loops noted above.

The C-terminal portion of the *M. mazei* PylS protein has been crystallized (Yanagisawa et al. 2006) and the structure solved (Kavran et al. 2007; Yanagisawa et al. 2008b). The first 184 residues were deleted to increase protein stability; the remainder represents the catalytic domain of PylS. The structure of the *D. hafniense* PylSc alone (Lee et al. 2008) and in complex with tRNA^{Pyl} (Nozawa et al. 2009) have also been obtained. PylSc is active as a pyrrolysyl-tRNA synthetase (Herring et al. 2007b), in spite of the lack of the N-terminal domain found in the *M. mazei* PylS (encoded separately by *pylSn*), and is essentially equivalent to the crystallized *M. mazei* PylS catalytic domain (Krzycki 2005; Lee et al. 2008; Nozawa et al. 2009).

Each monomer of the homodimeric *D. hafniense* PylSc binds one tRNA^{Pyl} by interacting with the major groove of *D. hafniense* tRNA^{Pyl} (Nozawa et al. 2009). Thirty-one PylSc residues were shown to directly contact either the acceptor stem or the compact tRNA^{Pyl} core, with little direct contact of the T-loop or anticodon stem and loop. Most PylSc contacts to tRNA are located in the N-terminus of PylSc and a C-terminal tail that form a complementary surface for the compact core of tRNA^{Pyl}. Neither sequence is conserved in other aminoacyl-tRNA synthetases leading to the cognate relationship of PylS and tRNA^{Pyl} (Nozawa et al. 2009). Previous modeling of the interaction of *M. mazei* PylS catalytic domain and tRNA^{Pyl} also predicted significant interaction between the N- and the C-termini of the PylS fragment with the D-loop and acceptor stem, but not with the T-arm or anticodon stem (Yanagisawa et al. 2008b).

Unlike archaeal PylS, PylSc has limited capacity to support in vivo UAG translation in *E. coli* monitored with reporter genes such as *lacZ* (Herring et al. 2007a) or *mtmB1* (Jiang and Krzycki, unpublished). However, *pylSc* does support low levels of UAG translation in extremely sensitive tests of amber suppression (Nozawa et al. 2009). The inefficiency of PylSc in vivo may result from the poorer affinity of PylSc for $tRNA^{Pyl}$ relative to archaeal PylS (Herring et al. 2007a) that may be due to the lack in PylSc of a region homologous to the N-terminal domain of archaeal PylS. In *D. hafniense* the N-terminal PylS domain is encoded by the separate *pylSn* gene product. PylSn binds $tRNA^{Pyl}$ with high affinity (Jiang and Krzycki, manuscript in preparation) and may bind $tRNA^{Pyl}$ on the T-stem or anticodon loop thereby enhancing PylSc binding. These were identity elements for the archaeal PylS (Ambrogelly et al. 2007), but PylSc did not contact these regions of the tRNA in the crystal structure.

3.6 Pyrrolysine Recognition by PylS and PylSc

Pyrrolysine is the only known physiological substrate of PylS, and methylamine methyltransferases have no other detectable residue at the UAG-encoded position (Soares et al. 2005). However, a number of unnatural amino acids are substrates of PylS, such as N^{ε}-cyclopentyloxycarbonyl-L-lysine (Cyc) (Polycarpo et al. 2006), N^{ε}-(tert-butoxycarbonyl)-l-lysine (Yanagisawa et al. 2008a), 2-amino-6-(cyclopentanecarboxamino)hexanoic acid, and 2-amino-6-((*R*)-tetrahydrofuran-2-carboxamido)hexanoic acid (2Thf-lys) (Li et al. 2009). The basis for recognition of pyrrolysine and non-natural substrates is evident in the structure of the catalytic domains of the *M. mazei* and *D. hafniense* enzymes (Kavran et al. 2007; Lee et al. 2008; Nozawa et al. 2009; Yanagisawa et al. 2008a). Pyrrolysyl-adenylate binds within a deep groove upon the surface of PylS with the aminoacyl moiety contacted by surprisingly few H-bonding residues. In the *M. mazei* PylS fragment, Asn346 forms an indirect H-bond with the pyrrolysine α-amino group via a bound water, and directly to the primary carbonyl oxygen. Arg330 also H-bonds to the carboxylate moiety (Kavran et al. 2007). The pyrroline ring is surrounded by a hydrophobic pocket formed by tyrosine, tryptophan, cysteine, and valine residues. A mobile loop between strands β7 and β8 bears Y384 that closes the hydrophobic cavity and introduces an additional H-bonding contact with the pyrrolysine imine nitrogen (Kavran et al. 2007; Yanagisawa et al. 2008b). The β7/β8 hairpin assumes a number of conformations, and in the apo-enzyme, or in complexes with pyrrolysine or the Cyc analog, the hairpin is an open conformation that does not close off the hydrophobic cavity (Kavran et al. 2007; Yanagisawa et al. 2008b). Mutation of Y384 yields an active enzyme (Yanagisawa et al. 2008b). Nonetheless, the influence of H-bonding with the imine nitrogen of pyrrolysine is illustrated by the favorable kinetics of amino acid activation by PylS with pyrrolysine analogs that have an electronegative group at the imine nitrogen position versus those that do not (Li et al. 2009). Such substrates might reveal distinctions in how the mobile loop bearing the tyrosine influences substrate binding and subsequent amino acid activation and $tRNA^{Pyl}$ charging.

The lack of a strict requirement for PylS residue interaction with the pyrrolysine imine allows recognition of a number of other lysine amides in which the pyrroline ring of pyrrolysine is replaced by a moiety that fits the hydrophobic pocket of PylS. The structures of PylS with various lysine N^{ε}-amides suggest that a hydrophilic group adjacent to a hydrophobic regions can H-bond Asn346 and position the hydrophobic group into the pocket that normally accommodates the pyrroline ring of pyrrolysine (Yanagisawa et al. 2008a). The need for a bulky hydrophobic group for wild-type PylS is further illustrated by very poor reactivity with acetyl-lysine (Polycarpo et al. 2006; Li et al. 2009).

The *D. hafniense* PylSc averages 60% similarity to methanogen PylS proteins (Krzycki 2005), and the catalytic site of the G+ protein is somewhat different. Many H-bonding and hydrophobic residues contacting pyrrolysine are conserved, but other changes lead to a smaller hydrophobic pocket. PylSc may prove more highly selective for its substrates than the archaeal enzymes, especially with reference to the bulkiness of the hydrophobic group of the pyrrolysine analog (Lee et al. 2008).

3.7 PylS and $tRNA^{Pyl}$-Based Amber Suppression in *E. coli* as a Tool for Biotechnology

The ability of PylS and $tRNA^{Pyl}$ to partially reprogram UAG as pyrrolysine in *E. coli* (Blight et al. 2004) is highlighted by recent efforts to employ the two molecules as an orthogonal pair. Orthogonal pairs of aminoacyl-tRNA synthetase and cognate tRNA have been exploited as amber suppressors in *E. coli* and other systems for site-specific incorporation of unnatural amino acids into recombinant proteins (Wang et al. 2006). At significant levels, other aminoacyl-tRNA synthetases do not appear to recognize $tRNA^{Pyl}$, $tRNA^{Pyl}$ does not recognize codons other than UAG, and other tRNA species are not significantly recognized by PylS in vivo (Blight et al. 2004; Neumann et al. 2008). PylS and $tRNA^{Pyl}$ thus function as an orthogonal pair in recombinant organisms. Importantly, this activity is independent of whether the recombinant gene with a translated amber codon naturally encodes a pyrrolysyl protein.

One of the first efforts to exploit the potential of PylS and $tRNA^{Pyl}$ achieved significant reactivity of PylS with acetyl-N^{ε}-lysine by mutation of the pyrroline-binding pocket with resultant acetyl-lysine incorporation into superoxide dismutase (Neumann et al. 2008). More recently, the utility of PylS and $tRNA^{Pyl}$ for incorporation of chemically modifiable residues into recombinant proteins has been demonstrated. Mutagenesis of *M. mazei* PylS resulted in increased activity for

N^{ε}-(*o*-azidobenzyloxycarbonyl)-L-lysine, allowing specific incorporation into protein and tagging with a fluorescein derivative (Yanagisawa et al. 2008a). Wild-type PylS recognized a derivative of the 2THF-lys pyrrolysine analog bearing an alkyne group at the 4-position, allowing the UAG-encoded residue of a protein to be modified with azidocoumarin for FRET analysis (Fekner et al. 2009). PylS can also recognize high concentrations of pyrrolysine analogs that lack an α-amine group, such as an α-hydroxy acid derivative. Recombinant $tRNA^{Pyl}$ can be charged with such a derivative and participate in ribosomal protein synthesis (Kobayashi et al. 2009). This provides a means by which an ester bond can be directly incorporated into the backbone of a recombinant protein for backbone mutagenesis and chemical cleavage.

3.8 Transmissible Biosynthesis and Genetic Encoding of Pyrrolysine

While the function of *pylS* and *pylT* has begun to be understood, the precise reactions catalyzed by the products of the other *pyl* genes remain elusive. The association of *pylBCD* with *pylT* and *pylS* homologs in Bacteria and Archaea (Srinivasan et al. 2002) indicated an important role in pyrrolysine metabolism. A general role has now been demonstrated, *pylBCD* are essential for pyrrolysine biosynthesis (Longstaff et al. 2007b).

In the absence of exogenous pyrrolysine, *E. coli*-bearing *pylTSBCD* can translate amber codons in *uidA*, encoding β-glucuronidase, (GUS), or *mtmB1* encoding the MMA methyltransferase (Longstaff et al. 2007b). The UAG-encoded residue has the mass of natively produced pyrrolysine. Transformation of *E. coli* with the *pylBCD* genes leads to intracellular pyrrolysine production, as the amino acid pool now includes a substrate for in vitro PylS-mediated reactions (Longstaff et al. 2007b). The amino acid produced in *E. coli*-bearing *pylBCD* co-migrates with chemically synthesized pyrrolysine in thin layer chromatography (M. Thalhoffer and J. Krzycki, unpublished data). Production of either pyrrolysine as a free amino acid or as a protein residue required the presence of *pylB*, *pylC*, and *pylD*.

Therefore, PylB, PylC, and PylD comprise unique gene products branching toward pyrrolysine (or a PylS substrate) from common metabolites found in Archaea or Bacteria. A biosynthetic role was first suggested for *pylBCD* based on their similarities to gene families whose products function in the biosynthesis of amino acids and vitamins (Srinivasan et al. 2002; Krzycki 2004, 2005; Longstaff et al. 2007a). These similarities inform possible biosynthetic routes (Krzycki 2004, 2005; Longstaff et al. 2007a). For example, the *pylB* gene product displays signatures of the Radical SAM family whose members catalyze various intramolecular rearrangements, reductions, and methylation reactions (Frey et al. 2008). PylB may catalyze methylation of the pyrroline ring precursor, or an intramolecular rearrangement that leads to pyrroline ring formation. The source of the pyrroline ring precursor is made more problematic as two chiral centers in the ring have the *R* configurations (Fig. 3.1), suggesting the need for a racemase yielding an *R*-carboxylate as a ring precursor. The *pylC* gene product is related to carbamoyl-phosphate synthetase, and the D-alanine–D-alanine ligase superfamily, suggesting a role in formation of the amide bond of pyrrolysine from lysine and a carboxylated ring precursor. PylD has an NAD-binding signature, and in PSI-BLAST searches is most often aligned with proteins involved in various dehydrogenations associated with amino acid metabolism. This protein might be involved in steps that lead to ring formation and/or imine bond formation.

The transformation of *E. coli* with the five *pyl* genes from *M. acetivorans* illustrates that addition of pyrrolysine to the genetic code of a naïve organism can happen with some ease and with little obvious detriment to the recipient (Longstaff et al. 2007b). UAG introduced into transcripts of an otherwise non-pyrrolysyl protein can encode pyrrolysine. It is notable that such similar arrangements of the *pyl* gene cluster are found in both Archaea and G+ Bacteria. These properties suggest that natural lateral transfer of the *pyl* genes would result in the ability to biosynthesize and decode UAG as pyrrolysine on a genome-wide basis. The mechanism underlying this would essentially be amber suppression, and amber-directed termination would be expected to continue at levels from 50 to 90% of amber translation.

3.9 Predictions of UAG as Sense and Stop Codon in *pyl*-Containing Organisms

As it became clear that pyrrolysine is encoded by UAG in methanogenic Archaea, it was a key question if UAG translation was a genome-wide phenomenon in organisms naturally containing the *pyl* operon or if UAG meaning as a stop codon was recoded only at the level of individual genes encoding pyrrolysyl proteins. This question entails to what extent UAG is a stop codon in methanogenic archaea, and if a sense codon, how widespread its use might be. As methanogens are notoriously difficult to culture and have limited genetics, most of the initial efforts to answer these questions were bioinformatics approaches.

As the first methylamine methyltransferase gene was sequenced, the limited database available indicated an unusually small percentage of sequenced genes from *Methanosarcina* spp. annotated to end with UAG (Burke et al. 1998). The first genomic sequencing efforts confirmed this trend, and *M. mazei* and *M. acetivorans* were found to, respectively, have only 3 and 5% of total ORFs ending with UAG, which was attributed to adaptation to UAG as a sense codon (Deppenmeier et al. 2002; Galagan et al. 2002; Zhang et al. 2005). *M. burtonii* and *M. barkeri* Fusaro also have depressed numbers of UAG as terminators, the former has only 2% of total ORFs apparently ending with UAG (Goodchild et al. 2004; Maeder et al. 2006). In comparison, closely related Archaea such as *Archaeoglobus fulgidis* ended 19% of their ORFs with amber codons, while *M. jannaschii* approaches 10%. However, this picture was clouded by comparison to *D. hafniense*; this G+ bacterium terminates over 22% of ORFs with UAG (Zhang et al. 2005); indicating that adaptation to life with pyrrolysine in the genetic code does not require a decrease in UAG utilization as a stop codon. Overall, *D. hafniense* seems to bucking a trend. Discrimination against UAG or UGA as a stop codon is, respectively, exaggerated in several organisms known to encode UAG as pyrrolysine or UGA as selenocysteine, with *D. hafniense* a notable exception (Fujita et al. 2007). Rationales for this discrepancy with reference to pyrrolysine are discussed further below.

Although a number of instances of UAG serving as a conserved substitution for UAA or UGA are observed when homologs from other organisms are compared to genes from *D. hafniense*, a study of *Methanosarcina* genomes found no clear instances of UAG serving as a stop codon (Zhang et al. 2005). Our own examination of the *M. acetivorans* genome indeed revealed very few instances where UAG did appear to function as a stop codon. One example is particularly interesting, as it is involved in pyrrolysine biosynthesis. The *pylB* gene ends with a TAG in *M. barkeri* MS, *M. acetivorans*, and *M. mazei* but in *M. barkeri* Fusaro, TAG is replaced with TAA (Longstaff et al. 2007b). The cloned *M. acetivorans pylB* gene is functional in *E. coli* for pyrrolysine biosynthesis when the TAG is mutated to TAA (Longstaff et al. 2007b).

Those *pylB* genes that end with TAG are followed within a short stretch of codons by TAA or TGA. This is a relatively common condition for amber-ending ORFs in methanogens (Zhang et al. 2005). In an unpublished analysis by our laboratory, 230 ORFs in the genome of *M. acetivorans* that were annotated as ending in an amber codon were treated as though UAG was a sense codon, leaving TAA or TGA as the only stop codons. The median relative placement of the presumed in-frame UAG was 94% of the putative ORF, suggesting that placement of such "in-frame" amber codons is always near the end of the ORF. Of the 112 ORFs above the median, approximately 25 had a UAA or UGA within three codons of the putative in-frame UAG codon. This data would indicate that pyrrolysine is preferentially incorporated into the extreme C-termini of putative pyrrolysyl proteins, or more reasonably, that UAG can serve as one of a tandem pair of stop codons.

The percentage of genes that would overlap if any one of the three canonical stop codons were converted completely to sense codon was predicted in *M. mazei* and *M. acetivorans* (Zhang et al. 2005). The absolute percent increase in overlaps was similar for the three stop codons, but these overlaps tended to be shorter when UAG was treated as the sense codon, rather than UGA or UAA.

A reasonable number of ORFs in methanogenic Archaea have in-frame amber codons at positions removed from the C-terminus. These ORFs provide some candidates for proteins produced via UAG translation as pyrrolysine. The original genome descriptions of *M. mazei* and *M. acetivorans*, as well as several directed searches, identified a few genes that might contain translated amber codons (Deppenmeier et al. 2002; Galagan et al. 2002; Chaudhuri and Yeates 2005; Zhang et al. 2005). Most of these genes were identified by their homology to known genes before and after the suspected in-frame amber codon. The majority of candidates are only found in a single species of *Methanosarcina*, suggesting they acquired the in-frame amber codon by mutation, rather than evolution of a class of proteins toward a functional adaptive role for the incorporation of pyrrolysine. Examples include genes encoding methylcobamide:CoM methyltransferase, several SAM-dependent methyltransferases, and CobN; these are found only in *M. acetivorans*.

However, two *tetR* homologs with conserved in-frame amber codon were identified in both *M. acetivorans* (MA0354) and *M. barkeri* (Mbar3297/6) (Galagan et al. 2002; Fujita et al. 2007). MA0354 is one of 15 *tetR* family members in the genome, yet is most closely related to Mbar3297/3296. The *tetR(amber)* genes are approximately 90% similar, but are not adjacent to homologous genes in the two genomes. The two highly similar *tetR* genes suggest translation and functionality of the gene products in spite of the amber codon.

Multiple copies of transposases whose genes contain single in-frame amber codons were identified during the sequencing of the *M. mazei* and *M. acetivorans* genomes (Deppenmeier et al. 2002; Galagan et al. 2002) that are highly similar to those associated with the ISBst12 family of insertion sequences (Filee et al. 2007). In *M. mazei* a total of 18 copies were originally identified, while in *M. acetivorans* only four copies were found. All can be readily aligned and shown to maintain the amber codon at a conserved position. In *M. acetivorans*, three of the transposases are nearly identical, while the fourth (MA1425) is dissimilar and clusters much more closely with a group of the "amber transposases" from *M. mazei* (Fig. 3.4). MA1425 may have been involved in a gene transfer event between *M. mazei* and *M. acetivorans*, and as the transposase itself is likely to have a role in such a transfer, this is strong evidence that amber translation renders these transposases functional. Multiple transposase genes having conserved amber codons also indicate active transposase products. Interestingly, in *M. burtonii*, which also possesses the *pyl* genes, multiple ISBst12-associated transposase genes are also present, but a glutamate codon substitutes for the in-frame UAG. The amber codon may have been acquired during introduction into the *M. mazei* and *M. acetivorans* lineages.

3.10 UAG *Is* Both Stop and Sense in *M. acetivorans*

Following the discovery of pyrrolysine, one of the more complete bioinformatics analysis of *pyl*-containing genomes concluded that UAG conversion to a sense codon had occurred in *Methanosarcina* spp. but the extent to which it remained a stop codon could not be reliably estimated (Zhang et al. 2005). Several scenarios were seen as possible. The UAG might be completely converted to sense, which would explain the low usage of UAG as a stop codon in *Methanosarcina* spp. Another was that UAG was partially converted to global sense codon, and that under conditions where pyrrolysine insertion was paramount the level of incorporation was modified by either individual gene context or environmental factors favoring translation as pyrrolysine. Fortunately, recent advances in methanogen genetics had made possible empirical investigation of these alternatives.

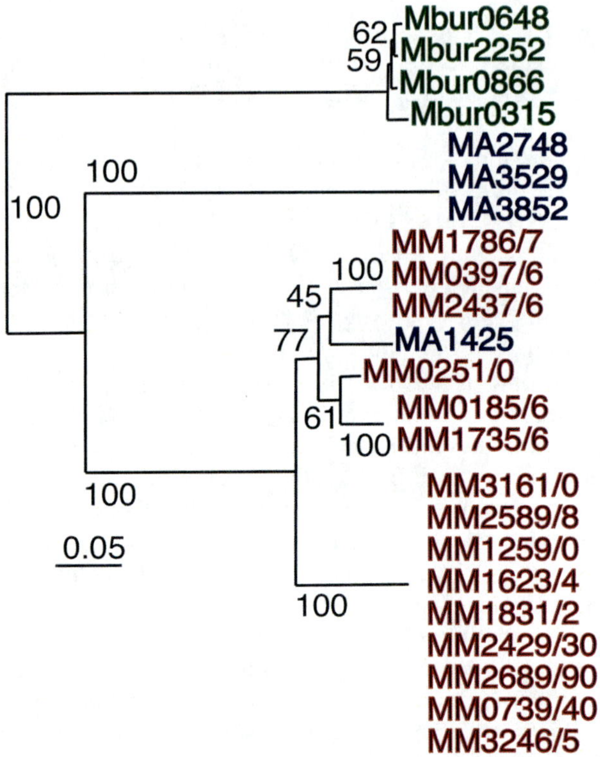

Fig. 3.4 Phylogenetic analysis of transposases from *M. mazei* (*orange*), *M. acetivorans* (*blue*), and *Mc. burtonii* (*green*). Transposases were aligned with Clustal W2 (Larkins et al., 2007) and a phylogenetic tree generated with BioNJ (Gascuel, 1997). The percentage values supporting each node for 500 boostrap replicates are indicated. In the original annotation of *M. acetivorans* the transposase genes were annotated as single ORFs containing translated amber codons. In *M. mazei* the transposases were annotated as two separate genes due to the in-frame amber codon, although the annotation recognized these two ORFs likely formed one protein. The homologous transposases encoded in the *Mc. burtonii* genome lack the conserved amber codon found in the *Methanosarcina* spp. The close relationship of MA1425 to several *M. mazei* clades is evident, suggesting the transposase gene products are functional due to UAG translation as pyrrolysine

The *E. coli uidA* gene encoding β-glucuronidase (GUS) was inserted into chromosome at the *hpt* locus in *M. acetivorans* to act as a reporter of translation or termination directed by an in-frame TAG or a TAA introduced at codon 286 (replacing a lysine codon). Results were compared to expression of the wild-type *uidA* gene at the same chromosomal location (Longstaff et al. 2007a). Activity assays and immunoblots indicated that introduction of a TAA codon into *uidA* led to accumulation of inactive TAA-terminated truncated GUS, consistent with TAA predominance as a stop codon in *M. acetivorans*. However, the archaeal strain bearing *uidA* with an introduced TAG codon in the same position displayed 30% of the GUS activity of the strain bearing wild-type *uidA*. Immunoblots revealed approximately 20% UAG readthrough to produce the full-length GUS protein. Mass spectrometry confirmed that the UAG in *uidA* transcript was translated as pyrrolysine. Translation of UAG continued to occur, albeit at various levels, when the amber codon was moved to various locations in *uidA* (Longstaff et al. 2007a).

This simple yet laborious experiment demonstrated that UAG translation could occur in *M. acetivorans* in the absence of any evolved *cis*-acting signals within a particular transcript. However, termination still appears to occur with a frequency exceeding that of translation by as much as four to five times in *uidA* containing an in-frame amber codon. These results indicate that many of the genes predicted to be pyrrolysyl proteins in methanogens are probably made as such, though undoubtedly with variable levels of efficiency. It further suggests that UAG can function as a stop codon, though an extraordinarily leaky one, explaining the high instance of UAG codons in close proximity to a TAA or TGA in *Methanosarcina* genomes.

The ability of *M. acetivorans* to support UAG translation in *uidA* is similar to the well-known phenomenon of amber suppression, which underlies the translation of UAG in *E. coli* transformed with the *pyl* genes. As expected for amber suppression in that system, dependence on $tRNA^{Pyl}$ is demonstrable (Blight et al. 2004). UAG translation versus termination increases as a function of the pyrrolysine present in the medium (Li et al. 2009). Mutation of the anticodon of $tRNA^{Pyl}$ to UCA obliterates suppression in a reporter gene with an in-frame amber codon, but can be recovered if the reporter gene amber codon is converted to UGA (Ambrogelly et al. 2007). This, coupled with the demonstration of that *E. coli* EF-TU will bind charged $tRNA^{Pyl}$ is further evidence that in *E. coli* $tRNA^{Pyl}$ acts as an amber-suppressor tRNA (Théobald-Dietrich et al. 2004). Amber suppression also appears to operate with *pylT* and *pylS* genes in mammalian cell lines (Mukai et al. 2008).

We suggest that in *M. barkeri*, as well as in other organisms that naturally contain the *pyl* gene cluster(s), amber suppression underlies global UAG translation as pyrrolysine in genes lacking any evolved *cis*-acting sequences, but importantly, at a fraction of UAG function as a terminator. Several important tests of this concept remain. For example, the amount of suppression at in-frame amber codons in different genes in *Methanosarcina* spp. should be measured in the presence and in the absence of clean deletions of the *pylT* gene, as well as in different reporter genes.

If the presence of *pyl* genes leads to a generalized level of amber suppression in an organism, adaptation toward lowered use of UAG as a terminator may not be a prerequisite for use of pyrrolysine as a genetically encoded amino acid. *E. coli* has approximately 10% of ORFs ending with UAG (Blattner et al. 1997), yet amber suppressor strains have been identified at relatively high frequency in natural population of *E. coli* (Robeson et al. 1980). The ability of these strains to not only tolerate amber suppression, but successfully compete in natural environments suggests little detriment is imparted by amber suppression. Indeed, *E. coli*-bearing *pylTSBCD* does not display notable growth defects (Longstaff et al. 2007b). *Bacillus subtilus* can tolerate induction of an amber suppressor tRNA with 10% apparent readthrough of UAG in a reporter gene (Grundy and Henkin 1994). This may explain the large discrepancy of the apparent usage of UAG as a terminator between *D. hafniense* and the methanogenic Archaea that possess *pyl* genes. Amber suppression resulting from the *pyl* genes might not dictate a decrease in UAG usage as a terminator, such as seen in *D. hafniense*. Why then have the pyl-containing methanogens appeared to decrease their UAG usage as a stop codon? The energetics of methanogenesis are very poor (Deppenmeier and Muller 2008), and the resultant growth rates slow. This would provide a strong driving force for eliminating ambiguous interpretation of the lengths of reading frames by decreasing UAG codons. Members of *Desulfitobacterium* genus, on the other hand, carry out a diverse number of anaerobic respirations with relatively favorable energetics (Villemur et al. 2006). The *pylT* gene seems to only be obligately required for growth on methylamine (Mahapatra et al. 2006), and such metabolism is a specialty of *Methanosarcina* spp. and relatives, as evidenced by multiple copies of *mttB*, *mtbB*, and *mtmB*. Few other classes of substrates are known for these methanogenic Archaea. In contrast, *D. hafniense* has a single pyrrolysine-dependent methylamine methyltransferase gene. The ability of *Desulfitobacterium* spp. to utilize a number of alternative non-methylamine substrates would weaken the driving force to limit UAG usage as a terminator, yet such choices are few for the methanogen. Further, the *pyl* operon is not apparently regulated during growth on non-methylamine substrate in methanogens (Veit et al. 2006), suggesting a constant pressure to minimize damage from ambiguous interpretation of UAG as a sense codon. It is unknown how pyrrolysine metabolism is regulated in *Desulfitobacterium* spp. The extent of adaptation to UAG as a sense codon in *Methanosarcina* spp. may also reflect a longer period in which this lineage has maintained pyrrolysine in the genetic code. It may be a recent adaptation in *D. hafniense* acquired by lateral transfer.

If UAG translation as pyrrolysine is a global trait (albeit a relatively inefficient one) in an organism with the *pyl* genes, why then is UAG not accumulated in a number of genes as a tolerated sense codon in *Methanosarcina* spp.? Translation of UAG in *uidA* in *M. acetivorans* indicates inefficient translation as pyrrolysine, leading to probable drop in expression levels of most genes acquiring amber mutations. Further, pyrrolysine is large and bulky, and the ring nitrogen would likely introduce a positive charge into proteins at physiological pH, making it a difficult substitution for many amino acids. Beyond this, pyrrolysine itself is chemically reactive. An imine bond, such as found in the pyrrolysine ring, is reversibly hydrolyzed in aqueous solution to the amine and aldehyde form. For example, the proline precursor pyrroline-5-carboxylate in water is in equilibrium with 0.05% glutamate semialdehyde (Bearne and Wolfenden 1995). Pyrrolysine opening to the aldehyde/amine form would expose a protein to a number of unfavorable side reactions, an additional selection against accumulation of UAG in genes encoding typical proteins.

3.11 Amber Suppression May Not Be Enough for Methanogenic Archaea

Methanogens produce high amounts of their catabolic proteins. The methylamine methyltransferases, which directly provide the substrate to the methyl-CoM reductase from TMA, DMA, or MMA, were individually estimated to comprise 2–3%

of total soluble protein (Burke and Krzycki 1997; Ferguson and Krzycki 1997; Ferguson et al. 2000). Each methyltransferase must be synthesized during growth on TMA, and up to 10% of the cellular protein would involve UAG translation as pyrrolysine. If the efficiency of translation averaged 20%, as observed by immunoblotting strains expressing *uidA* with an in-frame amber codon (Longstaff et al. 2007a), this quantity of full-length methyltransferases would require a third of the soluble protein be made as amber-termination products. Again, given the meager energetics of methanogenesis, selective pressure would be present in methanogens to correct the problem, either at the level of individual gene or genome.

Direct examination of the cells for the amber-termination product of the *mtmB* genes revealed only trace amounts of UAG-terminated MtmB in either stationary or log phase cells during growth on MMA (James et al. 2001). This indicates that in *mtmB* amber translation is very efficient or that any UAG-termination product is rapidly degraded. In order to address this question, the *M. barkeri mtmB1* gene was introduced on a plasmid into *M. acetivorans* under control of *Pmcr*, a strong constitutive promoter (Longstaff et al. 2007a). Little or no UAG-termination product of the introduced *mtmB1* gene could be detected, although over 1% of total protein was produced as the His-tagged *M. barkeri* MtmB. Substitution of the TAG codon with TAA led to a loss of the C-terminally His-tagged MtmB protein and to accumulation of the amber-termination product.

This result suggested that UAG translation must be more efficient during *mtmB1* expression than during expression of *uidA* genes containing amber codons in *M. acetivorans*. Such an effect could be due to the context surrounding the UAG codon in the *mtmB1* gene or to the environmental triggers causing increased efficiency of UAG translation. However, one possible environmental trigger, the presence of methylamines, does not affect the efficiency of UAG translation. The translation of UAG introduced into *uidA* does not change substantially regardless if *M. acetivorans* is grown on trimethylamine, methanol, or acetate (Longstaff et al. 2007a). This leaves the possibility that transcript context near the UAG codon influences the efficiency of translation. However, such context is unlikely to lie in untranslated regions of the *M. barkeri mtmB1* transcript, as these are not necessary for expression of high amounts of *mtmB1* with minimal detectable amber-termination product (Longstaff et al. 2007a).

3.12 A Putative Pyrrolysine Insertion Sequence

As the *mttB*, *mtbB*, and *mtmB* genes were first sequenced, a putative structure having conserved sequence elements was apparent that might form following the in-frame amber codon of *mtmB* and *mttB* transcripts. This was initially considered a potential player in UAG readthrough during translation, but the apparent significance was diminished with the acquisition of the first *mtbB* sequences. A stem-loop following the UAG codon in that strain lacked such features. Furthermore, the sequence of a *M. thermophila mttB* gene did not provide covariance support for the proposed structures (Paul 1999).

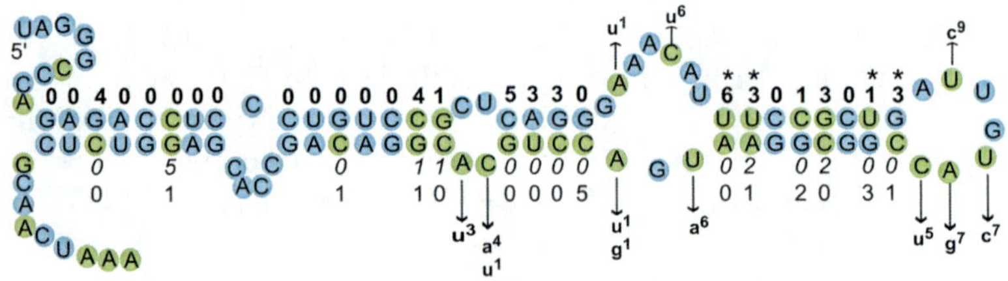

Fig. 3.5 The proposed PYLIS structure as seen in the *mtmB1* gene from M. *barkeri* MS. *Blue-shaded bases* are those that are conserved in the 10 sequences of *mtmB* genes from *M. barkeri* strains MS or Fusaro, *M. acetivorans*, *M. mazei*, and *Mc. burtonii*. *Green-shaded bases* are not conserved. Above the stem-loop is tallied in *bold* the number of times a variation is observed in the 10 *mtmB* genes that disrupts the base pair below. The *italicized number* directly below each base pair is the number of times covariant changes were observed in the 10 *mtmB* genes for that base pair. The number in plain text below each base pair indicates the number of times a single base variation is observed in the 10 sequences that still maintains the base pair above. Base variations in loops within PYLIS seen in the different sequences are indicated by *arrows*; the *superscript* indicates the number of times a particular base was seen. Finally, the structure as shown represents that as originally proposed in Namy et al. (2004), the asterisks mark bases that were shown to be melted in in vitro structure probing (Theobald-Dietrich et al. 2005)

Following the discovery of pyrrolysine, portions of this structure were independently found by Namy et al. in the *mtmB1* and *mtmB2* genes from *M. mazei* and *M. barkeri* (Namy et al. 2004). This was hypothesized to be a "pyrrolysine insertion sequence" (PYLIS) that might be important for UAG translation as pyrrolysine (Fig. 3.5). Alignment and covariance analysis of PYLIS in these different organisms generally supported the PYLIS structure, although a number of nucleotide substitutions not supporting covariance could also be found (Theobald-Dietrich et al. 2005). Structure probing of the in vitro transcribed 86 bases following the UAG codon in *M. barkeri mtmB1* demonstrated the PYLIS did form the proposed structure, with the variation that several base pairs at the bottom of the apical stem were melted (Theobald-Dietrich et al. 2005). An element suggested to have PYLIS-like features was also proposed downstream of the UAG codon in the *mttB* transcript from *D. hafniense* (Ibba and Söll 2004).

However, the similarity of the PYLIS element in *mtmB* to the *mttB* or *mtbB* genes was disputed in a later analysis of existing methylamine methyltransferase genes (Zhang et al. 2005). These workers could find no clear similarities in downstream elements of different methyltransferase genes and emphasized the lack of an apparent PYLIS-like element in the *mtbB* methylamine methyltransferase genes. The dissimilarity between stem-loops downstream of the UAG codon, coupled with the translation of UAG introduced into *uidA* genes in *M. acetivorans*, led us to empirically test the function of the PYLIS within the *M. barkeri mtmB1* transcript when introduced into *M. acetivorans*.

Replacement of the PYLIS with dissimilar sequence from a gene encoding a structural homolog of MtmB decreased the total expression of the modified *mtmB1* gene by fivefold relative to wild-type (Longstaff et al. 2007a). Nonetheless, in keeping with the translation of UAG introduced into *uidA*, mass spectrometric sequencing of the PYLIS-less *mtmB1* UAG-translation product revealed pyrrolysine at the UAG-encoded position, followed by the sequence with which pyrrolysine had been replaced. Thus, translation of UAG occurs in the absence of the PYLIS, but significantly, an increased amount of UAG-termination product was also observed. The sequence, or portions of the sequence, termed "PYLIS" in some fashion acts to enhance the efficiency of UAG translation. Several mutations introduced to disrupt different stems in the PYLIS structure modestly increased the amount of UAG-terminated *mtmB1* product and led to a 35% decrease on average in total abundance of *mtmB1* (Longstaff et al. 2007a). It must be emphasized, however, that the need for the PYLIS structure, or the complete length of the PYLIS, has not yet been demonstrated. Indeed, only limited covariance support for the structure can be seen in the10 known *mtmB* gene sequences from Methanosarcinaceae (Fig. 3.5).

Little if any UAG-termination products are detectable from either the DMA or the TMA methyltransferase gene in *M. acetivorans* (Lee and Krzycki, unpublished data). Future experiments will test the effects of the downstream stem-loops on translation of UAG in the various methylamine methyltransferase transcripts, as well as the surrounding context of each in-frame amber codon. It is well known that the surrounding bases can influence the efficiency of stop codons (Beier and Grimm 2001); for example, in yeast sequences upstream and downstream of a stop codon can influence termination over 100-fold (von der Haar and Tuite 2007). Therefore, mutational analysis of nearby bases must also be undertaken. A key development is required, however, and that is the development of a tractable UAG-translation reporter gene system that can be used in the methanogenic Archaea. Although experiment with the intact methyltransferase genes are important starting points, understanding how the efficiency of UAG translation is modulated in individual genes will require an accurate method to quantitate translation and termination in these Archaea.

3.13 Multiple Termination and Elongation Factors in *Methanosarcina* spp

In Bacteria, the specialized elongation factor SelB specifically recognizes selenocysteinyl-tRNA^{Sec}, which is not bound by EF-TU, the standard bacterial elongation factor. *Methanosarcina* spp. and *M. burtonii* possess a gene encoding a truncated SelB-like molecule that has been proposed as a possible participant in translating UAG as pyrrolysine (Ibba and Söll 2004). Each of these species also possesses another gene encoding an EF-1α homolog, more typical of Archaea. The presence of the SelB-like protein in these methanogens is unusual, given that selenocysteine does not appear to present in *Methanosarcina* spp. The methanogen SelB-like protein lacks the RNA-binding domain of bacterial SelB that would typically bind the SECIS element following a UGA codon translatable as selenocysteine (Ibba and Söll 2004). The SelB-like protein is thus unlikely to directly interact with any sequence near the UAG codon, such as the PYLIS, though it may interact via a second protein. BLAST searches reveal that *Methanosarcina* spp. and *M. burtonii* genes encode SelB-like proteins that are very similar to one another, but are also closely related to a series of putative elongation factors found in other methanogenic Archaea that lack pyrrolysine. A SelB homolog is found in *D. hafniense*, which without doubt participates in the selenocysteine metabolism this organism is known to possess (Zhang et al. 2005), but this factor is more distantly related to the *Methanosarcina* spp. SelB-like protein.

The ability of pyl-tRNA^{Pyl} to bind bacterial EF-TU (Blight et al. 2004; Théobald-Dietrich et al. 2004) would seem to obviate the need for a specialized elongation factor for binding pyl-tRNA^{Pyl} in Bacteria. Since tRNA^{Pyl} functions as an amber suppressor in mammalian cells (Mukai et al. 2008) it also seems likely that tRNA^{Pyl} will be recognized by the archaeal EF-1α. Still, an attractive possibility is that the methanogen SelB-like protein plays a role in more efficient translation of UAG as pyrrolysine; as there may be differences in affinity for tRNA^{Pyl}-binding by the SelB-like and the archaeal EF1-α elongation factor.

Both *M. barkeri* and *M. acetivorans* encode two release factors in their genomes, and it has been postulated these might have different affinities for recognition of UAG (Zhang et al. 2005). It would be a worthwhile line of investigation to test the efficiency with which these two factors recognize the three stop codons within *M. acetivorans*, especially in the presence and in the absence of mutations in the *pyl* operon. However, if one of these proteins does a play a role in limiting recognition of UAG as a stop codon, this is not a likely widespread solution to increasing UAG translation as pyrrolysine, as *M. mazei* and *M. burtonii* have only a single release factor encoded in their genomes.

3.14 Beyond Pyrrolysine

Following the discovery of pyrrolysine, a natural question arose: Are there more genetically encoded amino acids than the current known set? Two approaches have been taken to examine existing sequenced genomes for the potential to encode novel amino acids. The first surveyed genomes using several tRNA search algorithms to find tRNA species whose anticodons would predict decoding of stop codons, but which were not similar to known suppressor tRNAs (Lobanov et al. 2006). Another study surveyed 191 prokaryotic genomes and identified ORFs based solely on their conservation between organisms and then examined such ORFs for in-frame stop codon that could signify a new amino acid (Fujita et al. 2007).

Neither approach yielded promising candidates for new amino acids. Both groups of authors admitted the limitations of their studies. Novel tRNA species could be discarded because of extremity of structure. Novel tRNAs or ORFs with in-frame stop codons having narrow phylogenetic distributions would be missed. Both searches also relied on the decoding of a stop codon as a novel amino acid, however, since some degenerate codons for certain amino acids are used to a very small extent in some genomes, even a rare sense codon might be used for an usual amino acid in a small group of organisms.

Such searches should certainly continue. Given the diversity of microbial metabolism and species, it is the author's belief that an amino acid beyond the 22nd will be found. Pyrrolysine brings a precedent that genetically encoded amino acids of limited distribution and specialized function exist. Pyrrolysine was discovered in proteins for which no close homologs were immediately apparent when first sequenced; and even now, homologs with amber codons are of limited distribution in nature. The recent discovery through metagenomic sequencing of *pyl* genes in a new kingdom of Bacteria suggests that we do not have a full idea of pyrrolysine's distribution as yet. It is a simple matter to imagine that as we uncover one more instance of a now known, but recently discovered, amino acid in the Proteobacteria, that we will eventually discover the first instance of a novel amino acid elsewhere. However, to find it will no doubt require continued attention to the nuances of metabolism, and the conviction that we yet to see all the variations inherent in the diversity of life.

Acknowledgments Research in the author's laboratory is supported by the Department of Energy (DEFG020291ER200042) and the National Institute of Health (GM070663).

References

Ambrogelly A, Gundllapalli S, Herring S, Polycarpo C, Frauer C, Söll, D (2007) Proc Natl Acad Sci USA 104:3141–3146
Atkins JF, Baranov PV (2007) Nature 448:1004–1005
Atkins JF, Gesteland, R (2002) Science 296:1409–1411
Bearne SL, Wolfenden, R (1995) Biochemistry 34:11515–11520
Beier H, Grimm, M (2001) Nucleic Acids Res 29:4767–4782
Blattner FR, Plunkett G 3rd, Bloch CA, Perna NT, Burland V, Riley M, et al (1997) Science 277:1453–1474
Blight SK, Larue RC, Mahapatra A, Longstaff DG, Chang E, Zhao G, et al (2004) Nature 431: 333–335
Böck A, Forchhammer K, Heider J, Baron C (1991a) Trends Biochem. Sci. 16:463–467
Böck A, Forchhammer K, Hëider J, Leinfelder W, Sawers G, Veprek B, et al (1991b) Mol Microbiol 5:515–520
Böck A, Thanbichler M, Rother M, Resch, A (2004) In: Ibba M, Francklyn C, Cusack S (eds), *The Aminoacyl-tRNA synthetases*. Landes Bioscience
Burke SA, Krzycki JA (1997) J Biol Chem 272:16570–16577
Burke SA, Lo SL, Krzycki JA (1998) J Bacteriol 180:3432–3440
Chambers I, Frampton J, Goldfarb P, Affara N, McBain W, Harrison PR (1986) EMBO J 5: 1221–1227
Chaudhuri BN, Yeates TO (2005) Genome Biol 6:R79
Cone JE, Del Rio RM, Davis JN, Stadtman TC (1976) Proc Natl Acad Sci USA 73:2659–2663
Deppenmeier U, Johann A, Hartsch T, Merkl R, Schmitz RA, Martinez-Arias R, et al (2002) J Mol Microbiol Biotechnol 4:453–461
Deppenmeier U, Müller, V (2008) Results Probl Cell Differ 45:123–152
Fekner T, Li X, Lee MM, Chan MK (2009) Angew Chem Int Ed Engl 48:1633–1635
Ferguson DJ Jr, Gorlatova N, Grahame DA, Krzycki JA (2000) J Biol Chem 275:29053–29060
Ferguson DJ Jr, Krzycki JA (1997) J Bacteriol 179:846–852
Ferry JG (1999) FEMS Microbiol Rev 23:13–38

Filee J, Siguier P, Chandler, M (2007) Microbiol Mol Biol Rev 71:121–157
Frey PA, Hegeman AD, Ruzicka FJ (2008) Crit Rev Biochem Mol Biol 43:63–88
Fujita M, Mihara H, Goto S, Esaki N, Kanehisa, M (2007) BMC Bioinformatics 8:225
Galagan JE, Nusbaum C, Roy A, Endrizzi MG, Macdonald P, FitzHugh W, et al (2002) Genome Res 12:532–542
Gascuel, O (1997) Mol Biol Evol 14:685–695
Goodchild A, Saunders NF, Ertan H, Raftery M, Guilhaus M, Curmi PMG, et al (2004) Mol Microbiol 53:309–321
Grundy FJ, Henkin TM (1994) J Bacteriol 176:2108–2110
Hao B, Gong W, Ferguson TK, James CM, Krzycki JA, Chan MK (2002) Science 296:1462–1466
Hao B, Zhao G, Kang P, Soares J, Ferguson T, Gallucci J, et al (2004) Chem. Biol. 11: 1317–1324
Hayashi I, Kawai G, Watanabe, K (1998) J Mol Biol 284:57–69
Herring S, Ambrogelly A, Gundllapalli S, O'Donoghue P, Polycarpo CR, Soll D (2007a) FEBS Lett 581:3197–3203
Herring S, Ambrogelly A, Polycarpo CR, Soll D (2007b) Nucleic Acids Res 35:1270–1278
Ibba M, Söll, D (2004) Gen Dev 18:731–738
James CM, Ferguson TK, Leykam JF, Krzycki JA (2001) J Biol Chem 276:34252–34258
Kalyuzhnaya MG, Lapidus A, Ivanova N, Copeland AC, McHardy AC, Szeto E, et al (2008) Nat Biotechnol 26:1029–1034
Kavran JM, Gundllapalli S, O'Donoghue P, Englert M, Soll D, Steitz TA (2007) Proc Natl Acad Sci USA 104:11268–11273
Kobayashi T, Yanagisawa T, Sakamoto K, Yokoyama, S (2009) J Mol Biol 385:1352–1360
Krzycki JA (2004) Curr Opin Chem Biol 8:484–491
Krzycki JA (2005) Curr Opin Microbiol 8:706–712
Larkins MA, Blackshields G, Brown NP, Chenna R, McGettigan PA, McWilliam H, et al (2007) Bioinformatics 23:2947–2948
Lee MM, Jiang R, Jain R, Larue RC, Krzycki J, Chan MK (2008) Biochem Biophys Res Commun 374:470–474
Li WT, Mahapatra A, Longstaff DG, Bechtel J, Zhao G, Kang PT, et al (2009) J Mol Biol 385:1156–1164
Lobanov AV, Kryukov GV, Hatfield DL, Gladyshev VN (2006) Trends Genet 22:357–360
Longstaff DG, Blight SK, Zhang L, Green-Church KB, Krzycki JA (2007a) Mol Microbiol 63: 229–241
Longstaff DG, Larue RC, Faust JE, Mahapatra A, Zhang L, Green-Church KB, et al (2007b) Proc Natl Acad Sci USA 104:1021–1026
Maeder DL, Anderson I, Brettin TS, Bruce DC, Gilna P, Han CS, et al (2006) J Bacteriol 188: 7922–7931
Mahapatra A, Patel A, Soares JA, Larue RC, Zhang JK, Metcalf WW, et al (2006) Mol Microbiol 59:56–66
Mahapatra A, Srinivasan G, Richter KB, Meyer A, Lienard T, Zhang JK, et al (2007) Mol Microbiol 64:1306–1318
Mukai T, Kobayashi T, Hino N, Yanagisawa T, Sakamoto K, Yokoyama, S (2008) Biochem Biophys Res Commun 371:818–822
Namy O, Rousset JP, Napthine S, Brierley, I (2004) Mol Cell 13:157–168
Neumann H, Peak-Chew SY, Chin JW (2008) Nat Chem Biol doi:10.1038
Nonaka H, Keresztes G, Shinoda Y, Ikenaga Y, Abe M, Naito K, et al (2006) J Bacteriol 188: 2262–2274
Nozawa K, O'Donoghue P, Gundllapalli S, Araiso Y, Ishitani R, Umehara T, et al (2009) Nature 457:1163–1167
Paul, L (1999) Ph.D. Dissertation. Ohio State University, Columbus, Ohio
Paul L, Ferguson DJ, Krzycki JA (2000) J. Bacteriol. 182:2520–2529

Polycarpo C, Ambrogelly A, Berube A, Winbush SM, McCloskey JA, Crain PF, et al (2004) Proc Natl Acad Sci USA 101:12450–12454
Polycarpo C, Ambrogelly A, Ruan B, Tumbula-Hansen D, Ataide SF, Ishitani R, et al (2003) Mol Cell 12:287–294
Polycarpo CR, Herring S, Berube A, Wood JL, Soll D, Ambrogelly, A (2006) FEBS Letts 580:6695–6700
Retey, J (2003) Biochim Biophys Acta 1647:179–184
Robeson JP, Goldschmidt RM, Curtiss R 3rd (1980) Nature 283:104–106
Schimmel P, Beebe, K (2004) Nature 431:257–258
Soares JA, Zhang L, Pitsch RL, Kleinholz NM, Jones RB, Wolff JJ, et al (2005) J Biol Chem 280:36962–36969
Srinivasan G, James CM, Krzycki JA (2002) Science 296:1459–1462
Strittmatter AW, Liesegang H, Rabus R, Decker I, Amann J, Andres S, et al. (2009) Environ Microbiol doi:10.1111/j.1462–2920.2008.01825.x
Thauer RK (1998) Microbiology 144:2377–2406
Théobald-Dietrich A, Frugier M, Giegé R, Rudinger-Thirion, J (2004) Nucl Acid Res 32: 1091–1096
Théobald-Dietrich A, Giegé R, Rudinger-Thirion, J (2005) Biochimie 87:813–817
Veit K, Ehlers C, Ehrenreich A, Salmon K, Hovey R, Gunsalus RP, et al (2006) Mol. Genet. Genomics 276:41–55
Villemur R, Lanthier M, Beaudet R, Lepine, F (2006) FEMS Microbiol Rev 30:706–733
von der Haar T, Tuite MF (2007) Trends Microbiol 15:78–86
Wang L, Xie J, Schultz PG (2006) Annu Rev Biophys Biomol Struct 35:225–249
Woese CR, Fox GE (1977) Proc Natl Acad Sci USA 74:5088–5090
Woese CR, Kandler O, Wheelis ML (1990) Proc Natl Acad Sci USA 87:4576–4579
Woyke T, Teeling H, Ivanova NN, Huntemann M, Richter M, Gloeckner FO, et al (2006) Nature 443:950–955
Yanagisawa T, Ishii R, Fukunaga R, Kobayashi T, Sakamoto K, Yokoyama S (2008a) Chem Biol 15:187–1197
Yanagisawa T, Ishii R, Fukunaga R, Kobayashi T, Sakamoto K, Yokoyama S (2008b) J Mol Biol 378:634–652
Yanagisawa T, Ishii R, Fukunaga R, Nureki O, Yokoyama, S (2006) Acta Crystallogr Sect F Struct Biol Cryst Commun 62:1031–1033
Zhang Y, Baranov PV, Atkins JF, Gladyshev VN (2005) J Biol Chem 280:20740–20751
Zhang Y, Gladyshev VN (2007) Nucleic Acids Res 35:4952–4963
Zinoni F, Birkmann A, Stadtman TC, Böck, A (1986) Proc Natl Acad USA 83:4650–4654

Chapter 4
Specification of Standard Amino Acids by Stop Codons

Olivier Namy and Jean-Pierre Rousset

Abstract Translation termination is usually a very efficient process. When a stop codon enters the ribosomal A-site it is recognized by the termination complex which promotes release of the polypeptide and dissociation of the ribosome. However, the efficiency of termination depends of the local context of the stop codon. In a number of cases, programmed stop codon readthrough occurs allowing the synthesis of two polypeptides from the same mRNA. These events have been identified both in viral and in cellular genes. In cells, either standard or specialized amino acids (selenocystein, pyrrolysine) can be incorporated at the stop codon by near cognate or cognate tRNAs, respectively. In this chapter, we focus on readthrough events involving incorporation of standard amino acids. In addition to their biological relevance, stop codon readthrough sites are useful tools to study translation termination mechanisms, especially in eukaryotes where they are less understood. We present an overview of this field discussing the mechanisms involved and how new readthrough sites can be identified in databases. Finally we propose further directions to better understand termination and readthrough mechanisms.

Contents

J.-P. Rousset (✉)
IGM, CNRS, UMR 8621, Orsay, F 91405 France, Univ Paris-Sud, Orsay, F 91405, France
e-mail: jean-pierre.rousset@igmors.u-psud.fr

J.F. Atkins, R.F. Gesteland (eds.), *Recoding: Expansion of Decoding Rules Enriches Gene Expression*, Nucleic Acids and Molecular Biology 24,
DOI 10.1007/978-0-387-89382-2_4,

4.1 Introduction

For organisms using a standard genetic code, translation termination occurs when one of the three stop codons, UAA, UGA, and UAG, enters the ribosomal A-site. In contrast to the recognition of sense codons by tRNA, stop codons are recognized by extra-ribosomal proteins called class I release factors. The efficiency of translation termination depends on competition between the recognition of the stop codon by a class I release factors and the decoding of the stop codon by a near-cognate tRNA. In wild-type situations there is no cognate tRNA for decoding the stop codon; near-cognate tRNAs recognize stop codons with low efficiency. This leads to a very low background of stop codon suppression. However, a number of viruses (mostly plant RNA viruses) and several cellular genes display a high level of natural suppression at particular sequences, allowing two related polypeptides to be produced from a single mRNA. This phenomenon is called "programmed stop codon readthrough" and will be hereafter referred to as "readthrough." Readthrough depends on particular mRNA sequences and structures. In most cases, the elements involved in readthrough efficiency are located close to the stop, although at least one element lying hundreds of nucleotides downstream from the stop has been described (see below). Elucidating the precise mechanisms by which the sequence context surrounding the stop codon can influence termination is a major challenge in understanding readthrough. There are two main ways that readthrough can be increased: either by stimulating stop codon decoding by a near-cognate tRNA or by limiting the access of release factors to the stop codon. Finally, the difference between "standard stop codon readthrough" and incorporation of selenocysteine and pyrrolysine should be noted (see Chapters 1, 2, and 3). Incorporation of selenocysteine and pyrrolysine requires a specific cognate tRNA to read the stop codon, helped by a secondary structure downstream from the stop codon and several dedicated factors.

4.2 Translation Termination and Programmed Stop Codon Readthrough

In eukaryotes, two release factors eRF1 and eRF3 mediate translation termination. Either full or partial X-ray structures are available for both proteins and provide interesting insight into their function (Fig. 4.1) (Kong et al., 2004; Song et al., 2000). The overall shape of human eRF1 is similar to a tRNA, with functional motifs in both the peptidyl transferase center and the decoding site. eRF1 recognizes all three stop codons through the NIKS motif in its N-terminal domain. It triggers peptidyl–tRNA hydrolysis by activating the peptidyl transferase center of the ribosome through the highly conserved GGQ motif in the eRF1 central domain. The C-terminal domain of eRF1 is involved in binding eRF3, which is essential for in vivo termination (Eurwilaichitr et al., 1999). The molecular mechanisms underlying this process remain unclear in eukaryotes. It is possible that the binding of eRF1 to eRF3 induces a conformational change in eRF3 to stabilize

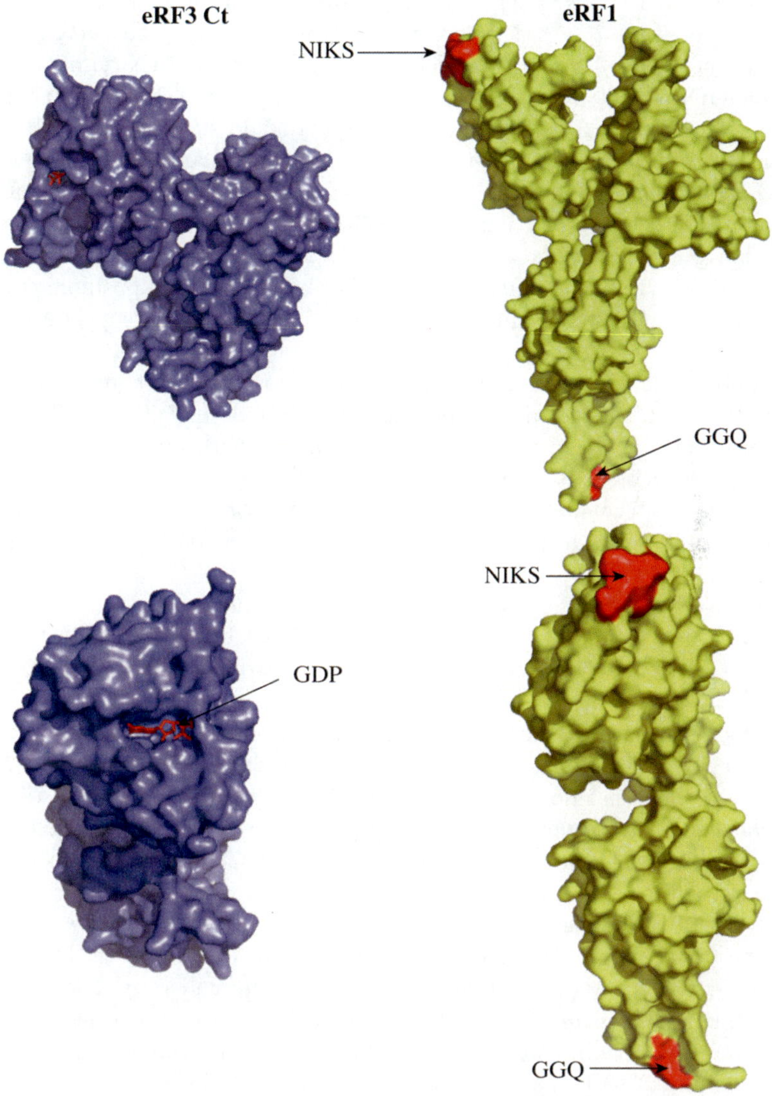

Fig. 4.1 Structures of both eukaryotic release factors. *Left panel* shows the structure of the carboxy-terminal domain of eRF3 (PDB 1R5N) bound to GDP (in *red*). *Right panel* corresponds to the structure of the human eRF1 (PDB: 1DT9). Two important motifs highly conserved between species are shown in *red*. The NIKS motif can be seen at the top of domain 1, which is thought to directly interact with the stop codon, and the GGQ motif is at the end of domain 2, involved in cleavage of the last peptidyl–tRNA. The lower part of the figure represents a rotation of 90° of the upper region. The pocket where GDP binds eRF3 is clearly visible

the binding of GTP to eRF3 (Pisareva et al., 2006). In *Saccharomyces cerevisiae*, eRF1 and eRF3 are encoded by the genes *SUP45* and *SUP35*, respectively. The eRF3 protein is composed of two different domains. Its N-terminal domain (called NM), specific to *S. cerevisiae*, is not necessary for termination activity but is a major determinant of eRF3 aggregation, involved in the formation of prion-like polymers known as [*PSI*$^+$]. The C-terminal domain is highly conserved through evolution and is involved in both GTP and eRF1 binding. The affinity of free eRF3 for GDP is 100 times greater than its affinity for GTP; eRF1 binding to eRF3 is thus required for GTP binding. A recent study suggested that eRF3 acts as a proofreading factor for the termination process (Salas-Marco and Bedwell, 2004). The ternary complex eRF1:eRF3:GTP may bind the ribosomal A-site, but the binding of GTP to eRF3 prevents eRF1 from catalyzing termination. If a stop codon is located in the A-site, a conformational rearrangement occurs and activates eRF3 GTPase activity. GTP hydrolysis allows proper positioning of the eRF1 GGQ motif in the peptidyl transferase center to catalyze peptidyl–tRNA cleavage.

The only available structure of eRF1 is unlikely to represent its functional form. Indeed, the distance between the NIKS and the GGQ motifs is 97.5 Å, 22 Å larger than the distance between the anticodon and the CCA acceptor stem in tRNAs. Thus, in this form eRF1 is too large to enter the ribosomal A-site. A bioinformatics model of the conformational changes induced upon ribosomal binding has been proposed, fitting more closely to the structural constraints of the ribosome (Trobro and Aqvist, 2007).

In prokaryotes, three factors are involved in the termination process: two class I factors, RF1 and RF2, which recognize UAG and UAA or UGA and UAA, respectively, and one class II factor called RF3. Many structural (Petry et al., 2005; Rawat et al., 2003) and mutational studies (Ito et al., 2000; Mora et al., 2003; Trobro and Aqvist, 2007; Zavialov et al., 2002) have provided essential information about the translation termination process. Recently, high resolution crystal structures of the translation termination complex bound to the bacterial ribosome was published (Laurberg et al., 2008; Weixlbaumer et al., 2008; Korostelev et al., 2008). These allows detailed observation of release factors frozen in the act of termination (Fig. 4.2a). In a structure illustrated, the stop codon is bound in a pocket formed by interactions between the conserved PxT motif (the equivalent of eukaryotic NIKS motif) and conserved nucleotides of the 16S rRNA decoding site (Fig. 4.2b). The codon and the 30S subunit A-site undergo induced-fit binding resulting in stabilization of an RF1 conformation that promotes the positioning of Gln 230 (of the GGQ motif) in the peptidyl transferase centre of the ribosome to allow its direct participation in peptidyl–tRNA hydrolysis (Fig. 4.2c). Before this recent publication, the only high-resolution structure of a prokaryotic release factor (RF2) described did not fit the cryoEM density observed with RF2 bound to the ribosome (Klaholz et al., 2003; Vestergaard et al., 2001), making it difficult to compare the structures of prokaryotic and eukaryotic class I release factors. The high resolution structure of the terminating ribosome allows this comparison to be made. Figure. 4.3 shows both structures aligned by their conserved GGQ motifs. The overall shape of the two factors is clearly very different; however, both functional motifs (NIKS (green) and GGQ (blue)) of the closed conformation

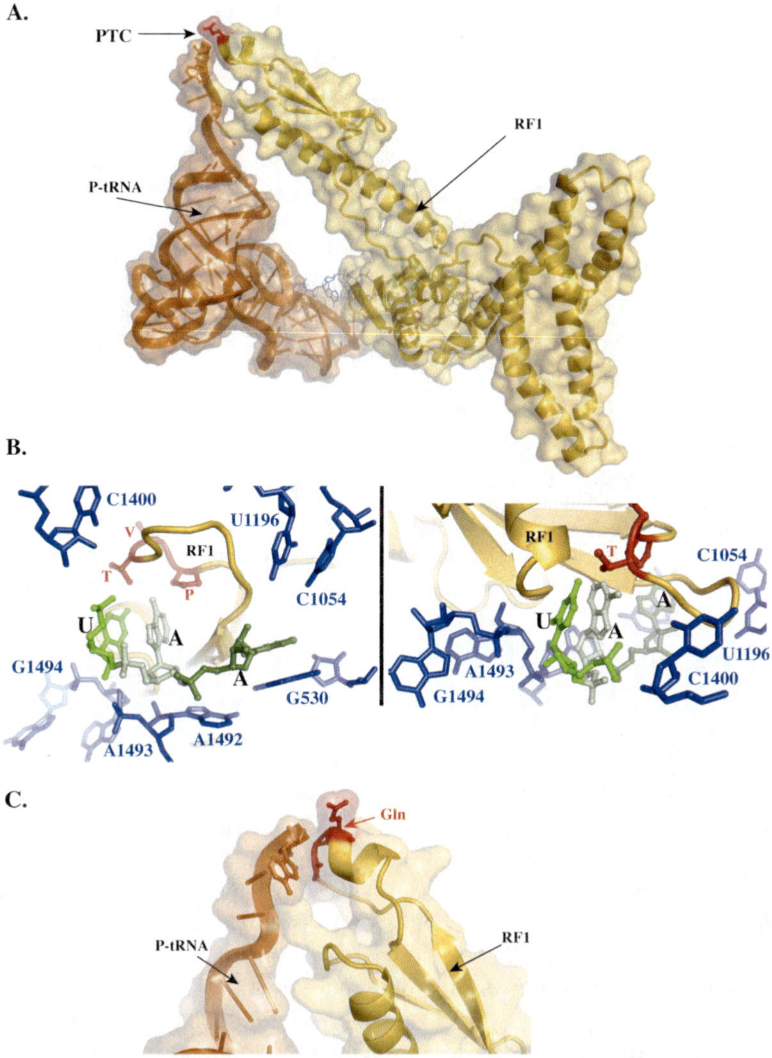

Fig. 4.2 Structure of prokaryotic RF1 bound to the ribosome. (**a**) the P-site tRNA is indicated in orange and RF1 in *yellow*. As in the previous figure, the two highly conserved motifs are in red. The close proximity of the ends of the tRNA and RF1 in the peptidyl transferase center is clearly seen. (**b**) two close views of the interactions between RF1, the stop codon (*green*) and the rRNA (*dark blue*) The PVT motif of RF1 is indicated *red*, as well as the identity of the important nucleotides of the rRNA. (**c**) the glutamine of the GGQ motif is shown in red; the last adenosine of the CCA t-RNA domain is also shown. This close view of this region also highlights the various interactions between RF1 and this region of the tRNA

proposed by Tobro and colleagues display a similar orientation of PXT/GGQ to RF1, whereas these motifs have a very different orientation in the x-ray structure (Fig. 4.3).

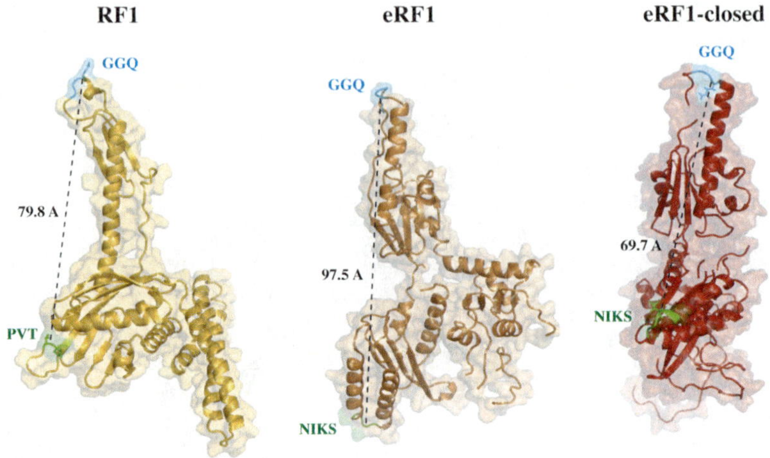

Fig. 4.3 Comparison between prokaryotic and eukaryotic class I release factors. *From left to right*: prokaryotic RF1 (*yellow*), human eRF1 (*orange*), and a bioinformatics-derived model of a closed conformation of eRF1 (*red*). The distances between the GGQ (*cyan*) and NIKS/PVT (*green*) motifs are indicated for each structure. By comparison, the distance between the anticodon and the tRNA CCA motif is 75 Å. All the structures are aligned to allow direct comparison of GGQ domains

4.3 Readthrough in Viruses and Phages

The majority of recoding events identified so far (including frameshifting) has been found in viruses or phages. The aim of this review is not to provide an exhaustive view of all the known viral readthrough sites, but to provide insights into stop codon readthrough mechanisms and into the biological relevance of readthrough products. Many readthrough sites have been identified simply by sequence analysis (presence of an in-frame stop codon); we will therefore only focus on well-characterized readthrough sites, discussing their biological function where known.

One of the earliest examples of programmed readthrough comes from bacteriophage Qβ. This single-stranded positive-strand RNA phage encodes four viral proteins from its own genome. These proteins (coat protein, replicase, and maturation/lysis) are all involved in viral replication. However, only the maturase and the coat protein are found in the mature particle, with just one protein – rather than two separate proteins – exhibiting maturation and lysis activity. Qβ also encodes a fourth protein obtained by readthrough of the UGA stop codon of the coat protein. The molecular weight of this recoded protein, called II*b*, is 22 kDa greater than the coat protein and accounts for about 5% of the amount of coat protein produced (Weiner and Weber, 1971). Trp tRNA probably reads this UGA. Readthrough results in a considerably elongated product that is incorporated into the virion and is essential for infectivity (Hofstetter et al., 1974). This minor form of the coat protein could thus play a role in the assembly of the mature particle, like the frameshifted products involved in tail assembly in phage lambda (Levin et al., 1993) (see Chapter 11).

The simplest readthrough motif is found in the Sindbis virus RNA. Indeed, a single cytidine residue immediately downstream from the termination codon is needed to give a readthrough efficiency of 10% (Li and Rice, 1993). However, the next residue (+2) also plays a role in readthrough efficiency, in both mammalian and insect cells (JPR and John F. Atkins, unpublished results).

Around one hundred plant RNA viruses use readthrough to express the replicase domain of their genome. The archetypal representative of this class is the tobacco mosaic virus (TMV), an RNA virus that infects more than 150 types of herbaceous, dicotyledonous plants including many vegetables, flowers, and weeds. The virus damages the leaves, flowers, and fruits leading to stunted plant growth. This virus uses UAG readthrough to produce its RNA-dependent RNA polymerase (Pelham, 1978) essential for its replication. A region of six nucleotides downstream from the stop codon promotes a high level of stop codon readthrough. Mutational analysis in plant cells has shown that this small unstructured region has the consensus motif CAR-YYA (Skuzeski et al., 1991). This motif can induce readthrough at all three stop codons. The near-cognate tRNA likely to mediate UAG readthrough in plants is Tyr tRNA. Tyrosine, lysine, and tryptophan (at a ratio of 4:2:1) have been found as products of a closely related readthrough site created by a nonsense mutation in the *S. cerevisiae* gene *STE6* (Fearon et al., 1994). Several natural tRNAs are thus able to read the UAG stop codon in this sequence context. Other plant viruses use slightly different readthrough signals, most of which are not fully defined (Dreher and Miller, 2006). A second motif found in most plant RNA viruses displaying a readthrough event consists of two adenines located just 5′ of the stop. This motif is also associated with low termination efficiency in yeast (Tork et al., 2004). In the Luteoviridae, the *cis*-acting signals involved in readthrough are composed of two elements: a cytidine-rich repeat (CCNNNN)8–16 beginning about 20 nt downstream from the stop codon and a distal sequence found 700 nt downstream (Brown et al., 1996), which does not seem to exhibit significant secondary structure. This is a unique example of a stimulatory element located at such a distance from the site at which readthrough takes place. Deciphering the precise mechanisms involved would reveal the ways in which the ribosome can be forced to perform unconventional decoding.

More than a dozen animal viruses also use readthrough for replication (recode db). The Moloney murine leukemia virus (MoMulV), a retrovirus, produces the Gag-Pol polyprotein by readthrough of a UAG (Philipson et al., 1978). A frameshifting site is usually found at this position in most retroviruses (see Chapters 7 and 8). A glutamine tRNA reads the UAG codon with an efficiency of 5% (Yoshinaka et al., 1985). This maintains a precise ratio of Gag to Gag-pol protein for viral replication and in particular for nucleocapsid formation. Replacing the UAG codon by a sense codon leads to a defective viral cycle, whereas replacing UAG by another stop codon does not affect viral propagation (Jones et al., 1989).

The signal stimulating UAG readthrough in the MoMulV is far more complex than those found in either Qβ or TMV. Indeed, the sequence surrounding the UAG stop codon displays two alternative secondary structures: a pseudoknot and a stem-loop (Alam et al., 1999; Wills et al., 1991; Feng et al., 1992; Wills et al., 1994).

The stop codon is located at the top of the stem-loop structure and comprises a 9 nt loop and a 14 nt stem. The stem overlaps the potential pseudoknot by 10 nt. The pseudoknot, but not the stem, has a stimulatory effect on readthrough, so when the stem-loop and the pseudoknot are present the level of readthrough is reduced by increasing the proportion of the stem-loop relative to that of the pseudoknot. This suggests that these two structures are in competition, but it is unknown if the proportion of time one structure forms at the expense of the other varies during the infective cycle with consequent modulation of readthrough efficiency and that a swap mechanism may regulate stop codon readthrough efficiency (Fig. 4.4a). The importance of this pseudoknot for UAG recoding is clearly established; in particular, nucleotides from loop 2 and from the spacer are important for readthrough efficiency (Wills et al., 1994). However, how the pseudoknot stimulates readthrough remains largely unknown. The pseudoknot probably induces a pause and/or a conformational change in the ribosome, but it is unclear how this is related to stop codon readthrough. A recent cryoEM analysis of a ribosome pausing at a −1 frameshifter pseudoknot (from IBV) provided structural information on the mechanisms of action of the pseudoknot. The pseudoknot blocks the ribosome during translocation, preventing a complete step to occur (Namy et al., 2006). The mRNA and tRNA are thus placed under tension, leading to a displacement and distortion of the translocating tRNA. These observations can be used to provide a model for the role of the MoMulV pseudoknot in stop codon readthrough. If this pseudoknot induces a pause in translocation at the same step as the IBV pseudoknot, the alteration of the ribosome structure would prevent eRF1 from entering the A-site, allowing more time for a near-cognate tRNA to read the stop codon. Obviously this model needs to be tested, but it could explain the importance of the pseudoknot in MoMulV readthrough. eRF1 may interact with the viral reverse transcriptase (RT) (product of the pol gene) (Orlova et al., 2003). This sequestering of eRF1 by the RT would stimulate stop codon readthrough, creating a positive feedback loop and increasing RT production (Fig. 4.4b). However, given that stop codon readthrough level is not modified during infection (i.e., whether in the presence or absence of RT) (Berteaux et al., 1991), the biological relevance of this observation remains unclear.

As discussed above, viruses often use new strategies to regulate the level of their own proteins. Initially found frequently in viral decoding, recoding events are now known to occur in all the kingdoms of life. The biological function of readthrough has been investigated in a few cases, but unfortunately remains a mystery in many cases.

Fig. 4.4 (continued) This could be part of a swap mechanism controlling stop codon readthrough efficiency. (**b**) simplistic representation of the MoMulV replication cycle involving GAG and POL protein production. When ribosomes terminate at the stop codon, only the GAG protein is synthesized, whereas stop codon readthrough allows synthesis of the GAG-POL fusion protein. In turn, the POL protein can interact with eRF1, depleting release factors and thus stimulating readthrough and increasing GAG-POL protein synthesis

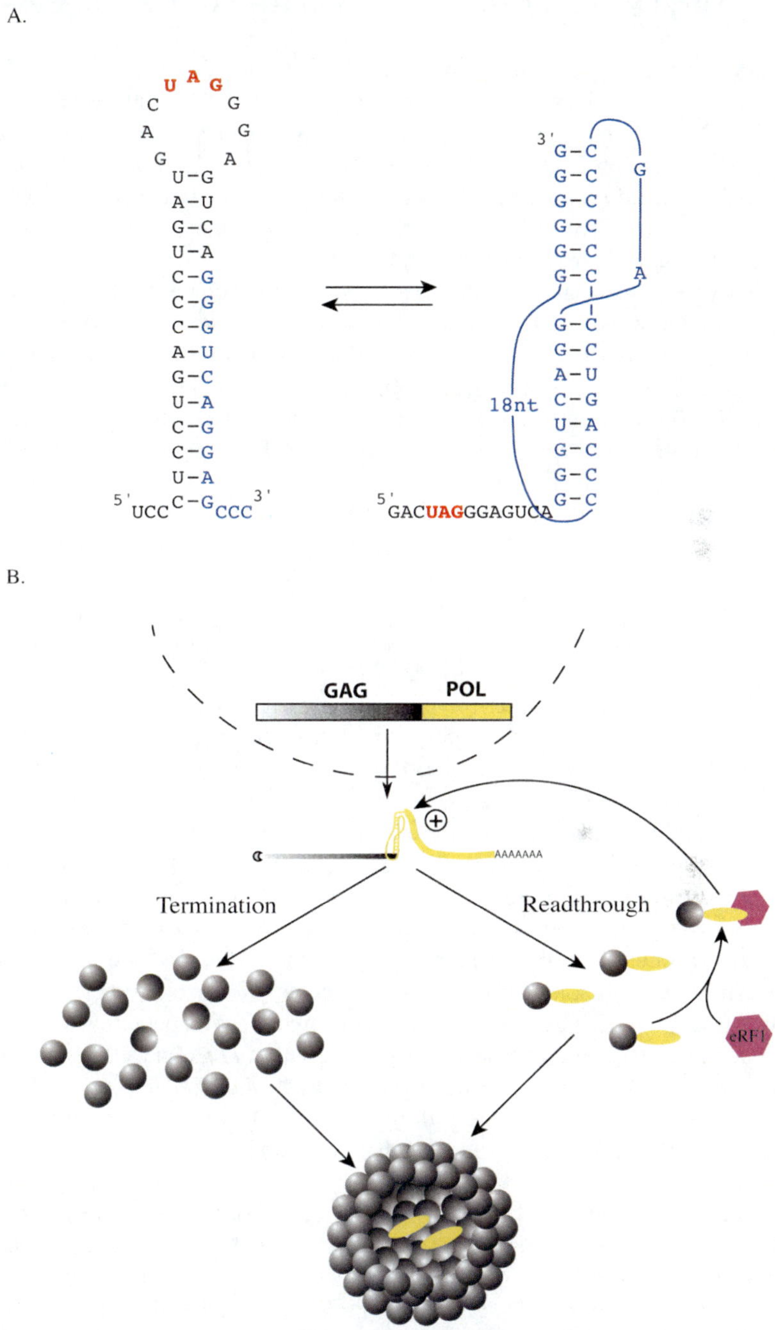

Fig. 4.4 Stop codon readthrough of MoMulV. (**a**) representation of the two potential structures; the stem-loop encompassing the stop codon, and the downstream pseudoknot. Sequences involved in the pseudoknot formation are shown in *blue*. The 3′ part of the stem can also form the pseudoknot.

4.4 Biological Relevance of Stop Codon Readthrough in Cells

In prokaryotes, the most common programmed readthrough event involves the insertion of the non-standard amino acid selenocysteine at the UGA codon (see Chapters 1 and 2). The only reported case of incorporation of a standard amino acid in place of a stop codon is the synthesis of the adherent CS3 pilus of the enterotoxigenic *Escherichia. coli* CFA/II strain, requiring the production of a 104 kDa protein by stop codon readthrough. A suppressor glutamine tRNA is necessary for full expression at this site (Jalajakumari et al., 1989).

In eukaryotes, stop codon readthrough, other than selenocysteine insertion, has been described in yeast and Drosophila. A few years ago, we reported that the yeast gene *PDE2*, encoding the high affinity cAMP-phosphodiesterase, is subjected to stop codon readthrough. The readthrough product is targeted to the 26S proteasome (Namy et al., 2002). The stop codon readthrough is influenced both by the nucleotide context of the stop codon and by the environment. Readthrough levels are increased either in the absence of glucose or in the presence of the prion [PSI^+]. One phenotype associated with this increased level of stop codon readthrough is increased thermosensitivity of the cells. This suggests that stop codon readthrough of *PDE2* can modify yeast fitness under stress conditions. Given that cAMP is a major secondary messenger in the cell, readthrough of the *PDE2* stop codon may also affect other biological functions. Notably, the presence of the prion [PSI^+] affects the regulation of gene expression in yeast (see below).

In *Drosophila melanogaster*, the expression of at least three genes is subject to readthrough. All three corresponding proteins are involved in developmental processes. This raises interesting questions about the origin of these genes. The *oaf* gene (out at first) contains a first open reading frame, which encodes a protein of 322 amino acids, separated from a second ORF by a UGA stop codon. Readthrough of the UGA codon will add 155 amino acids to the product of the first ORF, resulting in a protein of 477 amino acids. OAF is produced in nurse cells during oogenesis and is widely distributed throughout embryonic development (Bergstrom et al., 1995). This protein is found in the CNS but mutant larvae do not exhibit any obvious nervous system defects. However, some homozygous *oaf* mutants display peripheral nervous system defects with a penetrance depending on the mutant tested. These observations suggest that *oaf* is necessary for proper neuronal development and hatching. Homozygous mutants show a lethal phenotype late in embryogenesis or early during the first larval instar, including those showing no CNS defects. This suggests that additional roles of OAF remain to be identified.

The *hdc* gene (headcase) has a 3241 nt ORF interrupted by an in-frame UAA codon at position 2981 (Steneberg et al., 1998). The short and long polypeptides are both efficient in inhibiting terminal branching in the trachea. However, the longer product is more efficient than the peptide from ORF1 alone (Steneberg and Samakovlis, 2001). A secondary structure is predicted downstream from the UAA codon. This structure is sufficient to induce readthrough whatever the identity of the stop codon (Steneberg and Samakovlis, 2001). This, together with MoMulV, is the only known example with secondary structure involved in stop codon readthrough. This stem-loop structure has a similar shape to the PYLIS structure involved in the incorporation of pyrrolysine in archaea (Namy et al., 2004; Namy et al., 2007) (and see Chapter 3). It could be speculated that a rare amino acid would be inserted at this position in the hdc protein.

Kelch is an essential protein for the production of viable eggs in *Drosophila* ovaries. It is a structural component of the ring canals that act as intercellular conduits through which cytoplasm is transported from nurse cells to the oocyte in an egg chamber. The gene encodes a 76 kDa protein from one open reading frame (ORF1; 689 aa) and a 160 kDa product (ORF1 + ORF2; 782 aa) from the same mRNA. This stop codon readthrough is conserved among *Drosophila* species. The ratio of long to short product is regulated during embryonic development to give a maximum ratio during metamorphosis (Robinson and Cooley, 1997). The amino acid incorporated at the UGA codon is not known, but the reduced activity of a mutant generated by a deletion of the in-frame UGA indicates that this amino acid is important for the proper function of the protein. However, expression of the ORF1 is sufficient for *Drosophila* development. By contrast to *hdc*, the UGA stop codon is essential and cannot be replaced by another stop codon. This strongly suggests that both genes use different readthrough mechanisms to suppress the stop codon. It has been proposed that a selenocysteine is incorporated at the site of the Kelch readthrough event. However, this is unlikely given that no selenocysteine insertion sequence (SECIS) is found in the 3′ UTR and no ^{75}Se incorporation has been demonstrated (Robinson and Cooley, 1997). Alternatively, mRNA editing could take place at this stop codon. RNA editing involves the adenosine deaminase (ADAR) enzyme, which catalyzes the deamination of adenosine to inosine (Bass, 2002). Due to its base-pairing properties, inosine is recognized as a guanosine. This modifies the genetic information carried by the mRNA, changing a stop codon to a sense codon.

As we can see, the biological relevance of stop codon readthrough in *Drosophila* remains to be clearly determined.

4.5 Identification of Readthrough Sites in Genomes

Several genomes have been screened extensively for recoding events, namely frameshifting and selenocysteine incorporation (Castellano et al., 2008; Kryukov et al., 1999; Lescure et al., 1999; Mix et al., 2007). There is no doubt that incorporation of a standard amino acid during stop codon readthrough can be used as a regulatory mechanism (Fujita et al., 2007). However, such events have received much less attention, probably because they are far more difficult to identify due to the absence of a consensus motif.

Two approaches are commonly used to identify readthrough sites:

(i) searching for readthrough motifs within a particular genome. In *S. cerevisiae*, this approach led to the identification of genes with inefficient stop codons (Namy et al., 2002; Williams et al., 2004). A given 3′ readthrough motif is probably specific to a subset of organisms, so the nature of this motif is likely to change from an organism to another. This method of identifying new programmed readthrough genes is powerful but needs prior systematic analysis to identify inefficient termination sequence context.

(ii) searching for genes with expression controlled by stop codon suppression, without a priori knowledge of the sequences involved. This approach has been developed for the *S. cerevisiae* genome (Namy et al., 2003) and inefficient termination signals have been searched for in the *Oryza sativa* genome (Liu and Xue, 2004). Other methods developed to identify rare amino acids incorporated at a stop codon could also be used to identify programmed readthrough sites (Fujita et al., 2007). These methods do not require prior knowledge about the motifs or the mechanisms involved. It is therefore an efficient method of identifying new recoding events involving a stop codon. However, the main challenge remains the biological validation of the candidates identified by bioinformatics methods. Indeed, many factors can influence the quality of the results. In prokaryotes, the presence of operons with many ORFs separated by a single stop codon leads to a high number of false-positive candidates. In eukaryotes, the complexity of genome readout is a major challenge, with alternative splicing making it difficult to search for recoding sites. A clear example is provided by the large-scale analysis of 12 *Drosophila* genomes (Lin et al., 2007). This study identified 149 new candidates for stop codon readthrough. The analysis took into account conservation of amino acids downstream from the stop. Indeed, sequences that do not encode a protein tend to be less conserved. Similarly, the ENCODE project identified a number of potential candidate readthrough sites (Birney et al., 2007). However, it is difficult to determine the biological significance of these observations without functional validation. A large proportion of these genes may use alternative splicing, mRNA editing or a different recoding event (i.e. hopping?). For all these approaches, the results obtained must be considered with caution until the genes identified are characterized further.

Despite these considerations, the main problem, common to all genomes, in identifying readthrough sites seems to be the quality of the DNA sequence. Indeed, a search for readthrough sites cannot rely only on the presence of an in-frame stop codon because of the high frequency of sequencing errors. These sequencing errors are almost indistinguishable from a bona-fide programmed readthrough site in a coding sequence. Even incorporating other criteria in the analysis like protein motifs or coding sequence probability (HMM) – does not solve the problem.

As mentioned above, the identification and characterization of programmed readthrough sites give significant insight into cell physiology and could lead to the discovery of new pathways regulating translation. However, readthrough sites and recoding events in general are also interesting, providing powerful tools for the study of ribosome function.

4.6 Programmed Readthrough as a Tool to Study Translation Termination

A powerful approach for deciphering the molecular mechanisms underlying a phenomenon is to identify "exceptions to the rule" and look for defects in the regulation and function of regular components behaving in a non conventional way. Programmed stop codon readthrough is very useful in studying translation termination because it decreases the natural high termination efficiency. As mentioned above, the analysis of *cis*-acting sequences involved in readthrough has allowed further characterization of the role of surrounding nucleotide sequence in controlling termination efficiency of the stop codon.

The analysis of stop codon context in a large number of organisms clearly shows a bias. In the sequence 5′ of the stop codon, this effect may depend on the nature of the P-site tRNA pairing with the adjacent codon, on the amino acid present on this tRNA, or on the nucleotides present in the mRNA. It is difficult to distinguish between these possibilities due to the intricacies of these signals. One study has shown P-tRNA structure to influence stop codon readthrough efficiency (Smith and Yarus, 1989). The authors suggest that interactions take place with the release factor through the anticodon loop of the P-tRNA. This is consistent with recent structural data showing direct contact between RF1 and the P-tRNA (Figs. 4.2a and b). However, the effect exerted by the 5′ nucleotide sequence is not limited to interactions between the P-tRNA and RF1. In *E. coli*, the last amino acid on the P-tRNA can interfere with the termination process (Björnsson et al., 1996; Mottagui-Tabar and Isaksson, 1997; Mottagui-Tabar and Isaksson, 1998). These initial observations in *E. coli* have been extended to other organisms (Arkov et al., 1995; Bonetti et al., 1995; McCaughan et al., 1995).

The role of 3′ nucleotides remains unclear despite many attempts to elucidate their potential function. A strong bias has been found at position +1 following the stop codon in both prokaryotes and eukaryotes (Brown et al., 1990; Cridge et al., 2006; Poole et al., 1995). The importance of the 3′ nucleotide sequence is not limited to the nucleotide immediately following the stop codon; the signal can be extended up to six nucleotides downstream from the stop. To identify efficient readthrough sequences, we set up a SELEX-like screen in yeast. This allowed us identify the motif CAR-NBA as the readthrough consensus 3′ motif. This general motif (including the plant CAR-YYA motif), first identified in the TMV readthrough site, is functional in *S. cerevisiae*. It is also sufficient to drive a high level of stop codon readthrough in mammalian cells (Cassan and Rousset, 2001). Many hypotheses have been proposed to explain the role of the 3′ sequence (Bonetti et al., 1995; Cassan and Rousset, 2001; Namy et al., 2001; Skuzeski et al., 1991; Tate and Mannering, 1996) but none of them have been confirmed.

Since readthrough sites are generally located in the middle of the mRNA, the stop codon is equivalent to a non-sense mutation appearing in a normal gene; the stop can thus be viewed as a premature termination codon. Premature termination codons have recently received considerable interest due to their being potential therapeutic targets (see Chapter 6). Underlying this interest in therapeutic benefit, a more fundamental issue needs to be addressed, to determine whether a premature termination codon behaves like a standard stop codon. In eukaryotes, when a ribosome encounters a premature termination codon, the "Nonsense mRNA decay" (NMD) pathway is activated, leading to the decapping and rapid degradation of the mRNA, whereas a stop codon in its normal position does not induce mRNA decapping (Frischmeyer et al., 2002; Mitchell and Tollervey, 2003). Recent studies have shown that eRF3 can bind either UPF proteins or PABP bound to the polyA tail (Ivanov et al., 2008; Singh et al., 2008). Thus, two termination complexes should exist, one of them able to activate the NMD pathway. In this case, we would expect signals to be present on the mRNA, indicating whether the stop codon is a premature termination codon or not. Moreover, if the nucleotide context of stop codons is under selective pressure to confer a high termination efficiency, the premature termination codon is unlikely to be in an appropriate nucleotide context for termination. One major factor determining NMD activation is the distance between the premature termination codon and the 3′UTR/ polyA tail (Amrani et al., 2004; Silva et al., 2008). However, as mRNAs are always highly folded, it is unclear how the distance between the premature termination codon and the 3′UTR/polyA tail is measured by the cell. Stop codon readthrough directly affects NMD efficiency; indeed, a threshold level of stop codon readthrough antagonizes NMD in yeast (Keeling et al., 2004) and in human cells (Allamand et al., 2008).

Identifying the precise mechanisms by which the local nucleotide context can influence the balance between release factors and suppressor tRNA remains a major challenge. The analysis of readthrough signals will be very helpful in elucidating the function of these nucleotides.

Programmed readthrough is also useful in deciphering the role of trans-acting factors in termination. One major question rarely addressed is the identity of the tRNA reading the stop codon. In eukaryotes, several naturally occurring cytoplasmic tRNAs have been shown to recognize stop codons involved in programmed translational readthrough events (Beier and Grimm, 2001; Lecointe et al., 2002). In all cases, stop codon recognition implies non-orthodox base pairing between the second or the third base of the anticodon and the first or second base of the codon. The ability of these tRNAs to compete with release factors by reading a stop codon depends largely on their modified nucleotide content, particularly in their anticodon branch (Beier and Grimm, 2001). The study of viruses has allowed the identification of several new tRNAs that use different modified nucleotides for decoding. For example, the plant $tRNA^{Arg}$ is a natural suppressor of UGA in the PEMV (pea enation mosaic virus) (Baum and Beier, 1998). As discussed above, the $tRNA^{Tyr}$ decodes the UAG codon in the TMV. Two modified nucleotides have been shown to play an important role for UAG readthrough: (i) Y35 in the middle of the anticodon is a major determinant for UAG readthrough by $tRNA^{Tyr}$; (ii) The queosine modification at position 34 of the same tRNA counteracts the effect of Y35 (Zerfass and Beier, 1992). Surprisingly, the absence of pseudouridinylation at positions 38 or 39 of the anticodon branch decreases readthrough efficiency of a programmed stop codon in *S. cerevisiae* (Lecointe et al., 2002). Indeed, deletion of the *PUS3* gene responsible for these modifications affects readthrough of all three stop codons in the TMV. Interestingly, all three known near-cognate tRNAs able to decode the stop codon in *S. cerevisiae* ($tRNA^{Trp}$, $tRNA^{Tyr}$and $tRNA^{Lys}$) harbor a pseudouridine at position 39. It is thus possible that this modification allows a stronger interaction between the codon and the anticodon, which is particularly important for decoding a stop codon. This modification also increases +1 translational frameshifting efficiency (Lecointe et al., 2002).

As described above, stop codon readthrough can be stimulated by modifying the decoding efficiency of the suppressor tRNAs. Alternatively, readthrough can be increased by decreasing the efficiency of release factors. One way to achieve this is to modify the concentration of either eRF1 or eRF3. In *S. cerevisiae*, [*PSI*$^+$] is the prion form of eRF3. The conformational change impairs eRF3′s termination activity through aggregation of free active molecules to form inactive polymers. This consequently increases stop codon readthrough, resulting in the production of proteins with carboxy-terminal extensions (Serio and Lindquist, 1999). Although termination is affected at all the stop codons, stop codons in a weak termination context are more sensitive to the presence of [*PSI*$^+$]. These conditions thus provide a weak termination background in which to study the incorporation of near-cognate tRNAs. Moreover, [*PSI*$^+$] has interesting physiological functions. Indeed, many phenotypes, dependent on the genetic background of the host, are associated with the [*PSI*$^+$] status of the cell (Namy et al., 2008; True and Lindquist, 2000). These phenotypes reflect the broad effects of modifying termination efficiency on yeast physiology. Indeed, these phenotypes may be connected to the general disruption of the yeast proteome. However, we have recently shown that [*PSI*$^+$] increases frameshifting at the programmed shifty stop present in the ornithine decarboxylase antizyme-encoding gene (Namy et al., 2008). Frameshifting in turn stimulates the expression of the antizyme, which negatively regulates the ornithine decarboxylase, leading to a general decrease of polyamines in cells. This reduction of polyamine concentration is responsible for about half of the [*PSI*$^+$]-related phenotypes.

4.7 What's Next? Remaining Questions and Objectives

As mentioned above, the identification of programmed readthrough sites using bioinformatics is very challenging, because it is very difficult to distinguish bona-fide readthrough sites from sequencing errors. Although the number of sequenced genomes in the database is rapidly increasing, there are unfortunately no efficient approaches in place to identify these sites. Moreover, in the most favorable cases, automatic annotations systematically indicate such recoding events as pseudogenes and in some cases the sequence is corrected to get rid of the stop codon. Correction of the sequence in such cases leads to a loss of the primary information, making it impossible to study the potential existence of a translational recoding event. A better understanding of the mechanisms of stop codon readthrough would help to identify these sites. Clearly, in addition to searching for in-frame stop codons, other factors such as the presence of protein motifs or conservation among different species also

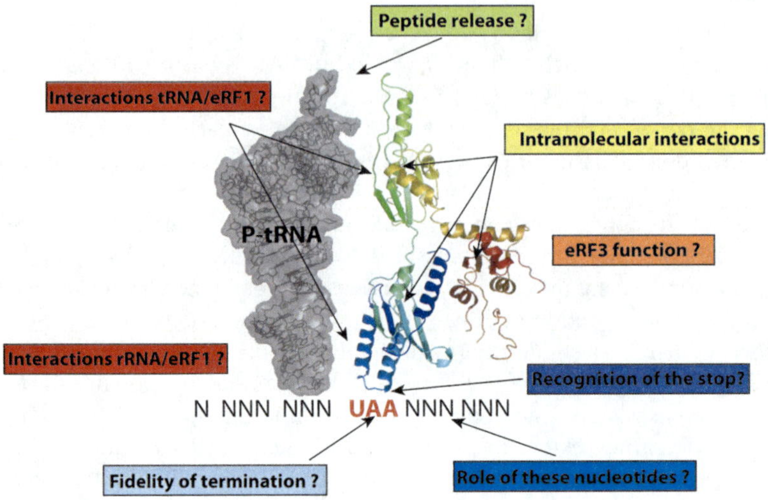

Fig. 4.5 The remaining questions on translation termination. Schematic representation of a P-site tRNA and an A-site stop codon being decoded by eRF1. The main questions are summarized in the boxes

need to be considered; however, this does not allow the specific identification of translational readthrough events.

The study of readthrough will improve our understanding of translation termination mechanisms, which remain largely unknown (summarized in Fig. 4.5). The role of the GGQ motif is still a matter of debate. Indeed, it has been suggested that the glutamine coordinates a water molecule in the peptidyl transferase center (Heurgue-Hamard et al., 2005; Song et al., 2000). However, glutamine may play a more direct role in the hydrolysis of the last peptidyl–tRNA (Seit-Nebi et al., 2001). This glutamine residue is methylated both in prokaryotes and in eukaryotes (Figaro et al., 2008). Although this modification is necessary for regulating termination efficiency in prokaryotes, its role in eukaryotes is unclear. Very little is known about the intramolecular interactions between the different domains and their role in the activity of class I release factors. As discussed above, it is highly likely that eRF1 undergoes a large conformational change upon binding to the ribosome. Recent high-resolution NMR structure shows the overall folding of eRF1 to be similar in solution and in the crystal. However, there are noticeable differences in the GGQ loop between the two structures (Ivanova et al., 2007). It is therefore possible that the crystal conformation of eRF1 does not represent a biologically relevant conformation.

In prokaryotes, RF3 binding induces conformational changes in the ribosome, breaking the interactions between RF1/2 and the decoding and peptidyl transferase centers and thus leading to the release of the class I release factor (Gao et al., 2007). The function of eRF3 is unclear. This factor seems to play a very different role from its prokaryotic counterpart. As recently suggested, the binding of eRF1 to eRF3 may induce a conformational change in eRF3 to stabilize the binding of GTP to eRF3 (Pisareva et al., 2006). eRF3 might also trigger eRF1 conformational changes to couple stop codon recognition and peptide release (Fan-Minogue et al., 2008).

Recognition of the stop codon by release factors has been the subject of many studies. A major challenge was to understand how these factors distinguish between a stop codon like UGA and a sense codon (UGG) with such high efficiency (Chavatte et al., 2003). Organisms that use an alternative genetic code have been very useful in studies of eRF1 regions involved in distinguishing between the different stop codons (Alkalaeva et al., 2006; Kervestin et al., 2001; Lekomtsev et al., 2007; Salas-Marco et al., 2006). Several models have been proposed to explain how the release factor binds the stop codon. The first model proposed was the tripeptide anticodon, based on the similarity between class I release factors and tRNA (Nakamura and Ito, 2002). A cavity model was later proposed by Stansfield's group (Bertram et al., 2000) and has gained support more recently (Fan-Minogue et al., 2008). Moreover, this proposal is consistent with the recently published structure of RF2 bound to the ribosome (Laurberg et al., 2008) (see Fig. 4.2b).

Termination efficiency is modulated by the local nucleotide context of the stop codon. However, the molecular mechanisms underlying these observations have not been determined. This is currently a major challenge facing researchers in the field. These nucleotides also modulate the action of aminoglycoside antibiotics, strongly limiting their use in therapeutics protocols for "stop codon diseases" (see Chapter 6). We believe that the identification of new natural readthrough sites, together with further analysis of the role of the nucleotides surrounding the stop codon, will help to understand these striking observations. Last but not least, interactions between class I release factors with rRNA or with the P-tRNA probably play an important role in stop codon readthrough (Poole et al., 2007). Currently, this translational step requires eukaryotic structural data to complement the elegant and numerous biochemical and genetic analyses performed so far. The analysis of stop codon readthrough mechanisms should be very useful in addressing all these issues.

References

Alam SL, Wills NM, Ingram JA, Atkins JF, Gesteland RF (1999) Structural studies of the RNA pseudoknot required for readthrough of the gag-termination codon of murine leukemia virus. J Mol Biol 288:837–852

Alkalaeva EZ, Pisarev AV, Frolova LY, Kisselev LL, Pestova TV (2006) In vitro reconstitution of eukaryotic translation reveals cooperativity between release factors eRF1 and eRF3. Cell 125:1125–1136

Allamand V, Bidou L, Arakawa M, Floquet C, Shiozuka M, Paturneau-Jouas M, Gartioux C, Bulter-Browne GS, Mouly V, Rousset JP, Matsuda R, Ikeda D, Guicheney p (2008) Drug-induced readthrough of premature stop codons leads to the stabilization of laminin alpha2 chain mRNA in CMD myotubes, J Gene Med 10:217–224

Amrani N, Ganesan R, Kervestin S, Mangus D.A, Ghosh S, Jacobson A (2004) A faux 3′-UTR promotes aberrant termination and triggers nonsense-mediated mRNA decay. Nature 432: 112–118

Arkov AL, Korolev SV, Kisselev LL (1995) 5′ contexts of *Escherichia coli* and human termination codons are similar. Nucleic Acids Res 23:4712–4716

Bass B.L, (2002) RNA editing by adenosine deaminases that act on RNA. Annu Rev Biochem 71:817–846

Baum M, Beier H (1998) Wheat cytoplasmic arginine tRNA isoacceptor with a U∗CG anticodon is an efficient UGA suppressor in vitro. Nucleic Acids Res 26:1390–1395

Beier H, Grimm M (2001) Misreading of termination codons in eukaryotes by natural nonsense suppressor tRNAs. Nucleic Acids Res 29:4767–4782

Bergstrom DE, Merli CA, Cygan JA, Shelby R, Blackman RK (1995) Regulatory autonomy and molecular characterization of the *Drosophila* out at first gene. Genetics 139: 1331–1346

Berteaux V, Rousset JP, Cassan M (1991) UAG readthrough is not increased in vivo by Moloney murine leukemia virus infection. Biochimie 73:1291–1293

Bertram G, Bell H.A, Ritchie D.W, Fullerton G, Stansfield I (2000) Terminating eukaryote translation: domain 1 of release factor eRF1 functions in stop codon recognition. RNA 6:1236–1247

Birney E, Stamatoyannopoulos JA, Dutta A, Guigo R, Gingeras TR, Margulies EH, Weng Z, Snyder M, Dermitzakis ET, Thurman RE, Kuehn MS, Taylor CM, Neph S, Koch CM, Asthana S, Malhotra A, Adzhubei I, Greenbaum JA, Andrews RM, Flicek P, Boyle PJ, Cao H, Carter NP, Clelland GK, Davis S, Day N, Dhami P, Dillon SC, et al. (2007) Identification and analysis of functional elements in 1% of the human genome by the ENCODE pilot project. Nature 447:799–816

Björnsson A, Mottagui-Tabar S, Isaksson LA (1996) Structure of the C-terminal end of the nascent peptide influences translation termination. EMBO J 15:1696–1704

Bonetti B, Fu L, Moon J, Bedwell DM (1995) The efficiency of translation termination is determined by a synergistic interplay between upstream and downstream sequences in *Saccharomyces cerevisiae*. J Mol Biol 251:334–345

Brown CM, Dinesh-Kumar SP, Miller WA (1996) Local and distant sequences are required for efficient readthrough of the barley yellow dwarf virus PAV coat protein gene stop codon. J Virol 70:5884–5892

Brown CM, Stockwell PA, Trotman CN, Tate WP (1990) Sequence analysis suggests that tetranucleotides signal the termination of protein synthesis in eukaryotes. Nucleic Acids Res 18:6339–6345

Cassan M, Rousset JP (2001) UAG readthrough in mammalian cells: effect of upstream and downstream stop codon contexts reveal different signals. BMC Mol Biol 2:3

Castellano S, Gladyshev VN, Guigo R, Berry MJ (2008) SelenoDB 1.0: a database of selenoprotein genes, proteins and SECIS elements. Nucleic Acids Res 36:D332–D338

Chavatte L, Kervestin S, Favre A, Jean-Jean O (2003) Stop codon selection in eukaryotic translation termination: comparison of the discriminating potential between human and ciliate eRF1 s. EMBO J 22:1644–1653

Cridge AG, Major LL, Mahagaonkar AA, Poole ES, Isaksson LA, Tate WP (2006) Comparison of characteristics and function of translation termination signals between and within prokaryotic and eukaryotic organisms. Nucleic Acids Res 34:1959–1973

Dreher TW, Miller WA (2006) Translational control in positive strand RNA plant viruses. Virology 344:185–197

Eurwilaichitr L, Graves FM, Stansfield I, Tuite MF (1999) The C-terminus of eRF1 defines a functionally important domain for translation termination in *Saccharomyces cerevisiae*. Mol Microbiol 32:485–496

Fan-Minogue H, Du M, Pisarev AV, Kallmeyer AK, Salas-Marco J, Keeling KM, Thompson SR, Pestova TV, Bedwell DM (2008) Distinct eRF3 requirements suggest alternate eRF1 conformations mediate peptide release during eukaryotic translation termination. Mol Cell 30:599–609

Fearon K, McClendon V, Bonetti B, Bedwell DM (1994) Premature translation termination mutations are efficiently suppressed in a highly conserved region of yeast Ste6p, a member of the ATP-binding cassette (ABC) transporter family. J Biol Chem 269:17802–17808

Feng YX, Yuan H, Rein A, Levin JG (1992) Bipartite signal for read-through suppression in murine leukemia virus mRNA: an eight-nucleotide purine-rich sequence immediately downstream of the gag termination codon followed by an RNA pseudoknot. J Virol 66: 5127–5132

Figaro S, Scrima N, Buckingham RH, Heurgue-Hamard V (2008) HemK2 protein, encoded on human chromosome 21, methylates translation termination factor eRF1. FEBS Lett 582: 2352–2356

Frischmeyer PA, van Hoof A, O'Donnell K, Guerrerio AL, Parker R, Dietz HC (2002) An mRNA surveillance mechanism that eliminates transcripts lacking termination codons. Science 295:2258–2261

Fujita M, Mihara H, Goto S, Esaki N, Kanehisa M (2007) Mining prokaryotic genomes for unknown amino acids: a stop-codon-based approach. BMC Bioinformatics 8:225

Gao H, Zhou Z, Rawat U, Huang C, Bouakaz L, Wang C, Cheng Z, Liu Y, Zavialov A, Gursky R, Sanyal S, Ehrenberg M, Frank J, Song H (2007) RF3 induces ribosomal conformational changes responsible for dissociation of class I release factors. Cell 129:929–941

Heurgue-Hamard V, Champ S, Mora L, Merkulova-Rainon T, Kisselev LL, Buckingham RH (2005) The glutamine residue of the conserved GGQ motif in *Saccharomyces cerevisiae* release factor eRF1 is methylated by the product of the YDR140w gene. J Biol Chem 280: 2439–2445

Hofstetter H, Monstein HJ, Weissmann C (1974) The readthrough protein A1 is essential for the formation of viable Q beta particles. Biochim Biophys Acta 374:238–251

Ito K, Uno M, Nakamura Y (2000) A tripeptide 'anticodon' deciphers stop codons in messenger RNA. Nature 403:680–684

Ivanov PV, Gehring NH, Kunz JB, Hentze MW, Kulozik AE (2008) Interactions between UPF1, eRFs, PABP and the exon junction complex suggest an integrated model for mammalian NMD pathways. EMBO J 27:736–747

Ivanova EV, Kolosov PM, Birdsall B, Kelly G, Pastore A, Kisselev LL, Polshakov VI (2007) Eukaryotic class 1 translation termination factor eRF1–the NMR structure and dynamics of the middle domain involved in triggering ribosome-dependent peptidyl-tRNA hydrolysis. FEBS J 274:4223–4237

Jalajakumari MB, Thomas CJ, Halter R, Manning PA (1989) Genes for biosynthesis and assembly of CS3 pili of CFA/II enterotoxigenic *Escherichia coli*: novel regulation of pilus production by bypassing an amber codon. Mol Microbiol 3:1685–1695

Jones DS, Nemoto F, Kuchino Y, Masuda M, Yoshikura H, Nishimura S (1989) The effect of specific mutations at and around the gag-pol gene junction of Moloney murine leukaemia virus. Nucleic Acids Res 17:5933–5945

Keeling KM, Lanier J, Du M, Salas-Marco J, Gao L, Kaenjak-Angeletti A, Bedwell DM (2004) Leaky termination at premature stop codons antagonizes nonsense-mediated mRNA decay in *S. cerevisiae*. RNA 10:691–703

Kervestin S, Frolova L, Kisselev L, Jean-Jean O (2001) Stop codon recognition in ciliates: *Euplotes* release factor does not respond to reassigned UGA codon. EMBO Rep 2:680–684.

Klaholz BP, Pape T, Zavialov AV, Myasnikov AG, Orlova EV, Vestergaard B, Ehrenberg M, van Heel M (2003) Structure of the *Escherichia coli* ribosomal termination complex with release factor 2. Nature 421:90–94

Kong C, Ito K, Walsh MA, Wada M, Liu Y, Kumar S, Barford D, Nakamura Y, Song H (2004) Crystal structure and functional analysis of the eukaryotic class II release factor eRF3 from *S. pombe*. Mol Cell 14:233–245

Korostelev A, Asahara H, Lancaster L, Laurberg M, Hirschi A, Zhu J, Trakhanov S, Scott WG, Noller HF (2008). Crystal structure of a translation termination complex formed with release factor RF2. Proc Natl Acad Sci USA 105:19684–19689.

Kryukov GV, Kryukov VM, Gladyshev VN (1999) New mammalian selenocysteine-containing proteins identified with an algorithm that searches for selenocysteine insertion sequence elements. J Biol Chem 274:33888–33897

Laurberg M, Asahara H, Korostelev A, Zhu J, Trakhanov S, Noller HF (2008) Structural basis for translation termination on the 70S ribosome. Nature 454:852–857

Lecointe F, Namy O, Hatin I, Simos G, Rousset JP, Grosjean H (2002) Lack of pseudouridine 38/39 in the anticodon arm of yeast cytoplasmic tRNA decreases in vivo recoding efficiency. J Biol Chem 277:30445–30453

Lekomtsev S, Kolosov P, Bidou L, Frolova L, Rousset JP, Kisselev L (2007) Different modes of stop codon restriction by the *Stylonychia* and *Paramecium* eRF1 translation termination factors. Proc Natl Acad Sci USA 104:10824–10829

Lescure A, Gautheret D, Carbon P, Krol A (1999) Novel selenoproteins identified in silico and in vivo by using a conserved RNA structural motif. J Biol Chem 274:38147–38154

Levin ME, Hendrix RW, Casjens SR (1993) A programmed translational frameshift is required for the synthesis of a bacteriophage lambda tail assembly protein. J Mol Biol 234:124–139

Li G, Rice CM (1993) The signal for translational readthrough of a UGA codon in Sindbis virus RNA involves a single cytidine residue immediately downstream of the termination codon. J Virol 67:5062–5067

Lin MF, Carlson JW, Crosby MA, Matthews BB, Yu C, Park S, Wan KH, Schroeder AJ, Gramates LS, St Pierre SE, Roark M, Wiley KL Jr, Kulathinal RJ, Zhang P, Myrick KV, Antone JV, Celniker SE, Gelbart WM, Kellis M (2007) Revisiting the protein-coding gene catalog of *Drosophila melanogaster* using 12 fly genomes. Genome Res 17:1823–1836

Liu Q, Xue Q (2004) Computational identification and sequence analysis of stop codon readthrough genes in *Oryza sativa*. Biosystems 77:33–39

McCaughan KK, Brown CM, Dalphin ME, Berry MJ, Tate WP (1995) Translational termination efficiency in mammals is influenced by the base following the stop codon. Proc Natl Acad Sci USA 92:5431–5435

Mitchell P, Tollervey D (2003) An NMD pathway in yeast involving accelerated deadenylation and exosome-mediated $3' \rightarrow 5'$ degradation. Mol Cell 11:1405–1413

Mix H, Lobanov AV, Gladyshev VN (2007) SECIS elements in the coding regions of selenoprotein transcripts are functional in higher eukaryotes. Nucleic Acids Res 35:414–423

Mora L, Heurgue-Hamard V, Champ S, Ehrenberg M, Kisselev LL, Buckingham RH (2003) The essential role of the invariant GGQ motif in the function and stability in vivo of bacterial release factors RF1 and RF2. Mol. Microbiol 47:267–275

Mottagui-Tabar S, Isaksson LA (1997) Only the last amino acids in the nascent peptide influence translation termination in *Escherichia coli* genes. FEBS Lett 414:165–170

Mottagui-Tabar S, Isaksson LA (1998) The influence of the 5′ codon context on translation termination in *Bacillus subtilis* and *Escherichia coli* is similar but different from *Salmonella typhimurium*. Gene 212:189–196

Nakamura Y, Ito K (2002) A tripeptide discriminator for stop codon recognition. FEBS Lett 514:30–33

Namy O, Duchateau-Nguyen G, Hatin I, Hermann-Le Denmat S, Termier M, Rousset JP (2003) Identification of stop codon readthrough genes in *Saccharomyces cerevisiae*. Nucleic Acids Res 31:2289–2296

Namy O, Duchateau-Nguyen G, Rousset JP (2002) Translational readthrough of the PDE2 stop codon modulates cAMP levels in *Saccharomyces cerevisiae*. Mol Microbiol 43:641–652

Namy O, Galopier A, Martini C, Matsufuji S, Fabret C, Rousset JP (2008) Epigenetic control of polyamines by the prion [PSI(+)]. Nat Cell Biol 10:1069–1075

Namy O, Hatin I, Rousset JP 2001. Impact of the six nucleotides downstream of the stop codon on translation termination. EMBO Rep 2:787–793

Namy O, Moran SJ, Stuart DI, Gilbert RJ, Brierley I (2006) A mechanical explanation of RNA pseudoknot function in programmed ribosomal frameshifting. Nature 441:244–247

Namy O, Rousset JP, Napthine S, Brierley I (2004) Reprogrammed genetic decoding in cellular gene expression. Mol Cell 13:157–168

Namy O, Zhou Y, Gundllapalli S, Polycarpo CR, Denise A, Rousset JP, Soll D, Ambrogelly A (2007) Adding pyrrolysine to the *Escherichia coli* genetic code. FEBS Lett 581:5282–5288

Orlova M, Yueh A, Leung J, Goff SP (2003) Reverse transcriptase of Moloney murine leukemia virus binds to eukaryotic release factor 1 to modulate suppression of translational termination. Cell 115:319–331

Pelham HR (1978) Leaky UAG termination codon in tobacco mosaic virus RNA. Nature 272: 469–471

Petry S, Brodersen DE, Murphy FV, Dunham CM, Selmer M, Tarry MJ, Kelley AC, Ramakrishnan V (2005) Crystal structures of the ribosome in complex with release factors RF1 and RF2 bound to a cognate stop codon. Cell 123:1255–1266

Philipson L, Andersson P, Olshevsky U, Weinberg R, Baltimore D, Gesteland R (1978) Translation of MuLV and MSV RNAs in nuclease-treated reticulocyte extracts: enhancement of the gag-pol polypeptide with yeast suppressor tRNA. Cell 13:189–199

Pisareva VP, Pisarev AV, Hellen CU, Rodnina MV, Pestova TV (2006) Kinetic analysis of interaction of eukaryotic release factor 3 with guanine nucleotides. J Biol Chem 281: 40224–40235

Poole ES, Brown CM, Tate WP (1995) The identity of the base following the stop codon determines the efficiency of in vivo translational termination in *Escherichia coli*. EMBO J 14: 151–158

Poole ES, Young DJ, Askarian-Amiri ME, Scarlett DJ, Tate WP (2007) Accommodating the bacterial decoding release factor as an alien protein among the RNAs at the active site of the ribosome. Cell Res 17:591–607

Rawat UB, Zavialov AV, Sengupta J, Valle M, Grassucci RA, Linde J, Vestergaard B, Ehrenberg M, Frank J (2003) A cryo-electron microscopic study of ribosome-bound termination factor RF2. Nature 421:87–90

Robinson DN, Cooley L (1997) Examination of the function of two kelch proteins generated by stop codon suppression. Development 124:1405–1417

Salas-Marco J, Bedwell DM (2004) GTP hydrolysis by eRF3 facilitates stop codon decoding during eukaryotic translation termination. Mol Cell Biol 24:7769–7778

Salas-Marco J, Fan-Minogue H, Kallmeyer AK, Klobutcher LA, Farabaugh PJ, Bedwell DM (2006) Distinct paths to stop codon reassignment by the variant-code organisms *Tetrahymena* and *Euplotes*. Mol Cell Biol 26:438–447

Seit-Nebi A, Frolova L,, Justesen J, Kisselev L (2001) Class-1 translation termination factors: invariant GGQ minidomain is essential for release activity and ribosome binding but not for stop codon recognition. Nucleic Acids Res 29:3982–3987

Serio TR, Lindquist SL (1999) [PSI+]: an epigenetic modulator of translation termination efficiency. Annu Rev Cell Dev Biol 15:661–703

Silva AL, Ribeiro P, Inacio A, Liebhaber SA, Romao L (2008) Proximity of the poly(A)-binding protein to a premature termination codon inhibits mammalian nonsense-mediated mRNA decay. RNA 14:563–576

Singh G, Rebbapragada I, Lykke-Andersen J (2008) A competition between stimulators and antagonists of Upf complex recruitment governs human nonsense-mediated mRNA decay. PLoS Biol 6:e111

Skuzeski JM, Nichols LM, Gesteland RF, Atkins JF (1991) The signal for a leaky UAG stop codon in several plant viruses includes the two downstream codons. J Mol Biol 218:365–373

Smith D, Yarus M (1989) tRNA-tRNA interactions within cellular ribosomes. Proc Natl Acad Sci USA 86:4397–4401

Song H, Mugnier P, Das AK, Webb HM, Evans DR, Tuite MF, Hemmings BA, Barford D (2000) The crystal structure of human eukaryotic release factor eRF1–mechanism of stop codon recognition and peptidyl-tRNA hydrolysis. Cell 100:311–321

Steneberg P, Englund C, Kronhamn J, Weaver TA, Samakovlis C (1998) Translational readthrough in the hdc mRNA generates a novel branching inhibitor in the *Drosophila* trachea. Genes Dev 12:956–967

Steneberg P, Samakovlis C (2001) A novel stop codon readthrough mechanism produces functional Headcase protein in Drosophila trachea. EMBO Rep 2:593–597

Tate WP, Mannering SA (1996) Three, four or more: the translational stop signal at length. Mol Microbiol 21:213–219

Tork S, Hatin I, Rousset JP, Fabret C (2004) The major 5′ determinant in stop codon read-through involves two adjacent adenines. Nucleic Acids Res 32:415–421

Trobro S, Aqvist J (2007) A model for how ribosomal release factors induce peptidyl-tRNA cleavage in termination of protein synthesis. Mol Cell 27:758–766

True HL, Lindquist SL (2000) A yeast prion provides a mechanism for genetic variation and phenotypic diversity. Nature 407:477–483

Vestergaard B, Van LB, Andersen GR, Nyborg J, Buckingham RH, Kjeldgaard M (2001) Bacterial polypeptide release factor RF2 is structurally distinct from eukaryotic eRF1. Mol Cell 8: 1375–1382

Weiner AM, Weber K (1971) Natural read-through at the UGA termination signal of Q-beta coat protein cistron. Nat New Biol 234:206–209

Weixlbaumer A, Jin H, Neubauer C, Voorhees RM, Petry S, Kelley AC, Ramakrishnan V (2008). Insights into translational termination from the structure of RF2 bound to the ribosome. Science 322:953–956.

Williams I, Richardson J, Starkey A, Stansfield I (2004) Genome-wide prediction of stop codon readthrough during translation in the yeast *Saccharomyces cerevisiae*. Nucleic Acids Res 32:6605–6616

Wills NM, Gesteland RF, Atkins JF (1991) Evidence that a downstream pseudoknot is required for translational read-through of the Moloney murine leukemia virus *gag* stop codon. Proc Natl Acad Sci USA 88:6991–6995

Wills NM, Gesteland RF, Atkins JF (1994) Pseudoknot-dependent read-through of retroviral gag termination codons: importance of sequences in the spacer and loop 2. EMBO J 13:4137–4144

Yoshinaka Y, Katoh I, Copeland TD, Oroszlan S (1985) Murine leukemia virus protease is encoded by the *gag-pol* gene and is synthesized through suppression of an amber termination codon. Proc Natl Acad Sci USA 82:1618–1622

Zavialov AV, Mora L, Buckingham RH, Ehrenberg M (2002) Release of peptide promoted by the GGQ motif of class 1 release factors regulates the GTPase activity of RF3. Mol Cell 10: 789–798

Zerfass K, Beier H (1992) Pseudouridine in the anticodon G psi A of plant cytoplasmic tRNA(Tyr) is required for UAG and UAA suppression in the TMV-specific context. Nucleic Acids Res 20:5911–5918

Chapter 5
Ribosome "Skipping": "Stop-Carry On" or "StopGo" Translation

Jeremy D. Brown and Martin D. Ryan

Abstract 2A and 2A-like "CHYSEL" sequences ("2As") are oligopeptides that specify ribosome "skipping" (also referred to as "Stop-Carry On" or "StopGo" translation). In this form of recoding event, the synthesis of a specific peptide bond (between the C-terminal glycine of 2A and the N-terminal proline of the downstream peptide) is skipped. We have proposed a model in which the nascent 2A oligopeptide interacts with the exit tunnel of the ribosome to stall, or pause, processivity. We propose this interaction also inhibits the mechanism of peptide bond formation and that the nascent protein is released (in a stop codon-independent manner) by release factors 1 and 3. Translation may then "pseudo-reinitiate" at the proline codon such that two discrete translation products are formed. Although first identified within positive-stranded mammalian RNA viruses (picornaviruses), 2As are also found in a wide range of insect positive-stranded RNA viruses plus mammalian, insect and crustacean double-stranded RNA viruses. Cellular protein biogenesis may also be controlled by 2As: such sequences are also found within non-LTR retrotransposons in the genomes of trypanosomes and the purple sea urchin. 2A appears to form the N-terminal region of open reading frames of sea urchin genes encoding CATERPILLER proteins (involved in the innate immune response). Indeed, ~50% of such genes commence with 2A. It appears, therefore, that this form of recoding plays a role in controlling protein biogenesis both in pathogens and in the innate immune system.

Contents

M.D. Ryan (✉)
Centre for Biomolecular Sciences, Biomolecular Sciences Building, North Haugh University of St. Andrews, St. Andrews KY16 9ST, UK
e-mail: martin.ryan@st-and.ac.uk

J.F. Atkins, R.F. Gesteland (eds.), *Recoding: Expansion of Decoding Rules Enriches Gene Expression*, Nucleic Acids and Molecular Biology 24,
DOI 10.1007/978-0-387-89382-2_5,

5.1 Picornavirus 2A Sequences

The family *Picornaviridae* comprises a number of genera including enterovirus (e.g. poliovirus), rhinovirus (common cold viruses), cardiovirus (e.g. encephalomyocarditis virus, EMCV; Theiler's murine encephalitis virus, TMEV) and aphthovirus (e.g. foot-and-mouth disease virus, FMDV). They all possess single-stranded, positive sense (+ve) RNA genomes of some 7,500–8,500 nts. Early studies on poliovirus infected cells produced a major conundrum: the sum of the molecular masses of the many different poliovirus proteins greatly exceeded that which could be encoded by the genome. The solution lay in the discovery of proteolytic processing of virus proteins. Large, precursor, forms were processed via intermediates into "mature" proteins. Later work in the early 1980s resulted in determination of the complete genome sequence of poliovirus. Along with biochemical analyses, this produced a fairly clear picture: the virus possessed a single, long, open reading frame (ORF) encoding a polyprotein. The polyprotein included two proteinase domains ($2A^{pro}$ and $3C^{pro}$) which cleaved the polyprotein in two co-translational, intramolecular, cleavages in *cis*, generating three products (Fig. 5.1, Panel A; regions P1, P2 and P3). These "primary" processing products then acted as precursors for a series of post-translational cleavages both in *cis* and in *trans* (intermolecular). This same strategy of protein biogenesis was observed for human rhinoviruses.

In the case of two other genera, the cardio- and aphthoviruses, inspection of polyprotein sequences showed that while the 3C proteinase sequences were similar to those of polio- and rhinoviruses, the sequence of the 2A region of their polyproteins bore no similarity. The 2A proteinase of entero- and rhinoviruses cleaved at their own N-termini, while the analogous primary polyprotein cleavage of cardio- and aphthovirus polyproteins occurred at the C-terminus of their 2A proteins. Furthermore, while the 2A protein of cardioviruses was ~150aa, the 2A region of FMDV was only 18aa (Fig. 5.1, Panel B). No sequence similarity

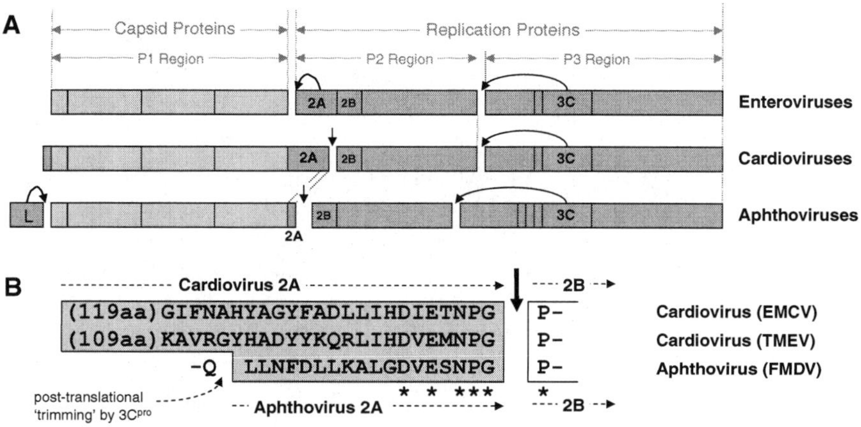

Fig. 5.1 Picornavirus polyproteins. The polyproteins of picornaviruses comprise an N-terminal capsid proteins domain (P1 region) plus two further domains comprising the replication proteins (P2, P3 regions). In the case of the enteroviruses, the "primary" cleavage (curved arrow) between P1 and P2 is mediated by a virus-encoded proteinase ($2A^{pro}$) cleaving at it's own N-terminus. Although also encoding all of their proteins in a single ORF, in many other genera "cleavage" occurs at the C-terminus of 2A – in reality the P1 region is translated as a product separate from P2 and P3 (**Panel A**). Some viruses have short 2As (~20aa; e.g. aphtho-, erbo-, teschoviruses), while others have larger, multifunctional, 2As (e.g. cardioviruses) with the "cleavage" activity mapping to the C-terminal ~20aa (**Panel B**)

was observed between entero-, cardio- and aphthovirus 2A proteins other than a motif (-D(V/I)ExNPG-) conserved at the C-terminus of cardio- and aphthovirus 2A proteins. In addition, the N-terminal proline of the downstream protein, 2B, was conserved completely (Fig. 5.1, Panel B). A major question was, therefore, *what activity was responsible for the co-translational 2A/2B cleavage of the cardio- and aphthovirus polyproteins*? Within infected cells, or within translation systems in vitro programmed with virus RNA (or RNA transcripts derived from clones encoding polyprotein), no "precursor" forms spanning the 2A/2B site could be detected. To map this cleavage activity onto the FMDV polyprotein, a series of recombinant FMDV polyproteins were constructed such that sequences either upstream or downstream of 2A were deleted (maintaining the single ORF). Analyses of the processing properties of these recombinant polyproteins showed that the 2A/2B cleavage was mediated by sequences solely within the 2A oligopeptide region (Ryan et al., 1991).

5.2 Analyses Using Artificial Polyprotein Systems

To demonstrate that 2A was autonomous from other FMDV sequences, artificial polyprotein systems were created. Reporter proteins were used to flank FMDV 2A and, crucially, the N-terminal proline of 2B (for convenience collectively referred to below as "2A"). Initially analyses were performed using a chloramphenicol acetyl-transferase (CAT)-2A-β-glucuronidase (GUS) construct (forming a single ORF) which was used to program translation systems in vitro (Ryan et al., 1994). Here, the only component of the artificial polyprotein from FMDV was 2A (-LLNFDLLKLAGDVESNPGP-; 19aa). High-level cleavage activity was observed, confirming that 2A was not merely the substrate for either the FMDV L or the 3C proteinases, but was an autonomous element. In these systems, translation was monitored by the incorporation of [^{35}S]-methionine and the sequence (number of methionines) was known for all the reporter proteins used. We showed that (i) cleavage occurred at the C-terminus of 2A (as in FMDV polyprotein processing) and (ii) while ~90% was detected in the [CAT-2A] plus GUS cleavage products, ~10% of the radioactivity was incorporated into a [CAT-2A-GUS] uncleaved form, a translation product spanning 2A: a phenomenon not observed in FMDV polyprotein processing. Kinetic analyses showed, however, that there was not a precursor:product relationship between the uncleaved [CAT-2A-GUS] and the cleaved forms ([CAT-2A] and GUS). All translation products were stable upon prolonged incubation: cleavage occurred co-, but not post-translationally. Inserting the C-terminal region of the cardiovirus EMCV and TMEV 2A proteins (corresponding to FMDV 2A; Fig. 5.1, Panel B) between CAT and GUS also gave high-level cleavage – directly comparable to FMDV 2A (Ryan and Drew, 1994).

Although the involvement of virus-encoded proteinase in 2A-mediated cleavage had been eliminated, the possibility remained that 2A was merely the substrate for a hypothetical cellular proteinase tightly coupled to translation. The stability of the [CAT-2A-GUS] uncleaved form in the translation systems in vitro was an argument – a weak argument – against this possibility. Expression of this, and other constructs encoding 2A peptides, showed highly efficient 2A-mediated cleavage within a wide range of mammalian cells, insect cells and plant cell extracts – in fact in all eukaryotic systems tested. Another aspect of the translation profiles we obtained from in vitro translation systems provided a much stronger argument against the involvement of a proteinase. Using phosphorimaging to quantify the distribution of the [^{35}S]-methionine radiolabel between the [CAT-2A] and GUS translation products revealed a substantial molar excess of [CAT-2A] above GUS. We showed this was due neither to different rates of protein degradation for CAT or GUS, nor degradation/cleavage of the transcript RNA used to programme the translation systems. The only explanation left was that this imbalance in accumulation arose from differential rates of *synthesis* of the two products. Specifically, the [CAT-2A] portion of the mRNA was translated more frequently than that encoding GUS, the translation product downstream of 2A – even though they were encoded in a single ORF.

To provide independent confirmation of the findings with the [CAT-2A-GUS] reporter a second reporter system, green fluorescent protein (GFP)-2A-GUS was generated (Donnelly et al., 2001a; Fig. 5.2). In this reporter (unlike in [CAT-2A-GUS]), the (formerly) initiator methionine of GUS was removed. Since the first internal methionine residue in the GUS sequence is over 100 amino acids downstream of 2A, any translation product of the same size as GUS must have arisen from 2A-mediated "cleavage", rather than internal initiation (a formal possibility with

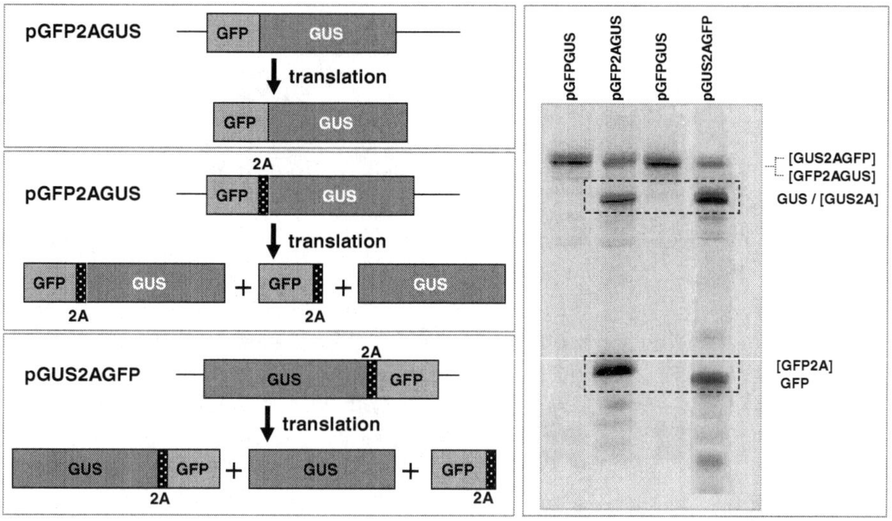

Fig. 5.2 Analyses of 2A-mediated cleavage using artificial polyproteins. Control constructs encoding green fluorescent protein (GFP) and β-glucuronidase (GUS) were assembled into a single ORF (pGFPGUS and pGUSGFP). The FMDV 2A sequence was inserted between the reporter proteins such that the single ORF was maintained. The constructs were then used to program cell-free coupled transcription-translation systems. The de novo protein synthesis was monitored by the incorporation of ^{35}S-methionine and the distribution and quantification of incorporated radiolabel determined by SDS-PAGE and phosphorimaging. The "cleavage" products (GUS + [GFP2A], or, [GUS2A] + GFP) are highlighted within the *dashed boxed areas*

[CAT-2A-GUS]). We also reversed the order of the "genes" flanking 2A, generating [GUS-2A-GFP]. Analyses using [GFP-2A-GUS] again showed the imbalance of translation products with the product upstream of 2A accumulating between 15 and 2.5 fold over that downstream, depending on the translation system used (rabbit reticulocyte lysates/wheatgerm extracts) and also between different batches of lysate/extract. This is clearly demonstrated in Fig. 5.2: although GFP contains only 6 methionines, the band is much more intense than that of GUS (12 methionines). The same effect was observed with [GUS-2A-GFP], and these results clearly showed that the molar excess of the upstream product was not due to some peculiarity in the sequences used, but a feature of the 2A reaction.

Since analysis based upon the 19aa 2A sequence showed the generation of uncleaved polyproteins ([CAT-2A-GUS] or [GFP-2A-GUS]), yet no precursors spanning the 2A/2B boundary were observed in FMDV polyprotein processing, we suspected that viral sequences around 2A may influence the reaction. Confirming this suspicion, in vitro translation of clones encoding the [P1-2ABC] portion of the FMDV polyprotein yielded separated cleavage products in equimolar quantities, with no detectable uncleaved form (Donnelly et al., 2001a). The key part of the viral genome for efficient 2A function was narrowed down by inserting FMDV 1D sequences immediately upstream of 2A into the [GFP-2A-GUS] reporter polyprotein (Donnelly et al., 2001b). As the length of the FMDV sequence present in this construct was increased, cleavage efficiency improved and the imbalance of "cleavage" products disappeared, with 100% efficiency and a 1:1 ratio of products occurring with ~30 amino acids of viral coding sequence. Thus while the length of 2A was defined by self-processing at its own C-terminus and $3C^{pro}$ processing at its N-terminus, the functional length of 2A was somewhat longer. Sequences immediately upstream of the 18aa influenced, but were not critical, for activity. We were fortunate, therefore, in our initial analyses by using a sub-optimal length of sequence which gave vital clues as to the mechanism of "cleavage".

5.3 The Co-translational Model of 2A-Mediated "Cleavage"

The results described above are inconsistent with models invoking proteolysis – either by a virus-encoded or by a host-cell proteinase – of an extant peptide bond, which would predict the generation of equimolar amounts of the upstream and downstream products. However, no described forms of translational recoding accounted for our data. A new, translational, model of 2A-mediated "cleavage" needed to be developed, based solely upon the properties of these oligopeptide sequences, to enable us to predict experimental outcomes that could be tested. Our translational model of "cleavage" was formulated thus;

(i) all ribosomes initiated translation at the start codon of the ORF (N-terminal, upstream of 2A);
(ii) elongation of all ribosomes continued to the C-terminus of FMDV 2A;
(iii) a proportion of ribosomes would release the nascent chain at this site – in the midst of the ORF – and dissociate from the mRNA; and
(iv) the remaining ribosomes would also release the nascent chain at this same site, but could subsequently "re-initiate" translation of downstream sequences, terminating normally at the 3′ end of the entire ORF.

5.3.1 Roles for Conserved and Non-Conserved Portions of 2A

Although the above model provided an explanation of how more of the reporter protein upstream of 2A accumulated over that downstream, it posed many questions as to *how* the oligopeptide sequence could bring about this translational effect. Specifically, how could 2A induce a stop codon-independent termination of translation and mediate a start codon-independent (re)initiation of translation? Comparison of the available 2A sequences from aphtho- and cardioviruses showed that only the C-terminal motif -D(V/I)ExNPGP- was conserved among them. Furthermore, inspection of the nucleotides encoding these peptides showed substantial synonymous variation suggesting that while the amino acid sequence was important for activity, the RNA sequence was not. In this regard, comprehensive analyses using algorithms designed to predict RNA secondary structures failed to show structures with any conservation amongst the 2A coding sequences. In contrast, dynamic molecular modelling of the 2A oligopeptide sequences revealed a common theme amongst the different 2As: a helical structure with a tight turn at the C-terminus (Ryan et al., 1999).

Within the conserved motif at the C-terminus of 2A, the final glycyl-prolyl amino acid pair is absolutely conserved. Mutagenesis results confirmed the critical nature of these amino acids, with changes to them leading to generation of solely the uncleaved translation product (Hahn and Palmenberg, 1996; Donnelly et al., 2001b). In a number of cases, the first amino acid of the downstream proteins produced by the 2A reaction had been shown to correspond to the final proline of the conserved motif (Robertson et al., 1985; Palmenberg et al., 1992; Pringle et al., 1999; Wu et al., 2002; Dodding et al., 2005). More recent data have confirmed that the upstream reaction product indeed ends in glycine, confirming that the reaction is essentially a skip in peptide bond formation (Atkins et al., 2007; Doronina et al., 2008).

Early work showed that amongst aminoacyl-tRNAs, glycyl-tRNA (closely followed by prolyl-tRNA) was the poorest electrophilic centre in the P-site (Rychlík et al., 1970). Proline is unique amongst amino acids in that it has great conformational rigidity since the cyclic structure of the side chain fixes its ϕ backbone dihedral angle at approximately $-75°$. Proline is also a poor nucleophile in the peptidyl transferase reaction compared to most other amino acids. Analogues of the aminonucleoside antibiotic puromycin were synthesised in which the methoxy-phenylalanine moiety was replaced with other amino acids. Analogues were then tested for their ability to terminate protein elongation or to act as acceptors in the "fragment reaction". Amongst all forms tested, the prolyl- and gycyl-substituted analogues were essentially inactive, indicating these were very poor nucleophiles (Nathans and Neidle, 1963; Nathans, 1964). Indeed, recent studies have shown that proline is incorporated significantly more slowly (between 3 and 6 fold) than phenylalanine or alanine (Pavlov et al., 2009). Proline also promotes turns within protein structures and is present in the sequence immediately N-terminal of the cleavage site (-D(V/I)ExNP̲G-) – also predicted to form a tight turn.

As described in other chapters of this book, certain nascent peptides can interact with the exit tunnel of ribosomes to bring about a pause in translation, and, in some cases, to direct subsequent recoding events (Baranov et al., 2003; Tenson and Ehrenberg, 2002). A potential function of the putative helical region of 2A could be, therefore, to interact with the exit tunnel. Atomic structures of ribosomes revealed that exit tunnels can accommodate ~35 amino acids of nascent peptide in a helical structure (Nissen et al., 2000). Indeed, earlier theoretical studies had indicated that a helix was the most probable conformation for nascent peptides within the exit tunnel (Lim and Spirin, 1986). That certain peptides form compact conformations, rather than remaining extended within the exit tunnel, has been shown by the efficient transfer of energy between fluorophores incorporated into nascent chains in FRET experiments (Woolhead et al., 2004, 2006). We had mapped FMDV 2A-mediated "cleavage" activity to ~30 amino acids and predicted a helical conformation for the majority of the peptide. These ribosome structural data were, therefore, consistent with the notion that 2A functioned within the ribosome. That the non-conserved

N-terminal portion of 2A is important in placing the conserved C-terminal motif in the correct context for the reaction was shown by analyses of N-terminally truncated forms of 2A: the conserved -D(V/I)ExNPGP- motif was inactive on its own (Donnelly et al., 2001b). Furthermore, insertion of helix-breaking residues into the non-conserved portion of 2A reduces its activity (Donnelly et al., 2001b; P. Sharma and JDB unpublished data).

5.3.2 Why Is the Glycine-Proline Peptide Bond Not Formed?

We suggested that interactions of 2A with the ribosomal exit tunnel constrained the conformational space of the tight turn at the C-terminus of 2A, ending with the peptidyl(2A)-tRNA^{gly} ester linkage in the P-site of the ribosome (Ryan et al., 1999; Donnelly et al., 2001a). This would then preclude the ester bond from nucleophilic attack specifically by the imide nitrogen of prolyl-tRNA^{pro} in the A-site. Elongation would be paused, possibly "jammed". Since the translation products were formed as discrete entities and we had discounted proteolysis of a peptide bond, we suggested that the nascent protein was liberated by cleavage of the ester linkage between the 2A peptide and the tRNA^{gly}. The $\sim$10% of full-length reporter fusion proteins we observed with sub-optimal lengths of 2A (19 amino acids) would be the result of the same proportion of 2As failing to interact "correctly" with the exit tunnel, conferring a greater degree of conformational space to the peptidyl(2A)-tRNA^{gly} ester linkage, and consequently formation of the peptide bond to prolyl-tRNA^{pro}.

Recent structural and biochemical data have provided a detailed picture of events within the PTC during translation elongation (reviewed in Korostelev and Noller, 2007; Korostelev et al., 2008; Steitz, 2008; Beringer and Rodnina, 2008). The ester group of the peptidyl-tRNA in the P-site is protected from hydrolysis by water by rRNA while the A-site is unoccupied. Entry of aminoacyl-tRNA into the A-site induces conformational changes in the rRNA that expose the carbonyl-carbon of the peptidyl-tRNA for nucleophilic attack by the α-amino group of the aminoacyl-tRNA – an "induced fit" mechanism (Schmeing et al., 2005). Peptidyl(2A)-tRNA^{gly} in the P-site can form peptide bonds with aminoacyl-tRNAs in the P-site – other than prolyl-tRNA. Similarly, prolyl-tRNA in the A-site – in the normal course of events – is able to form peptide bonds with peptidyl-tRNA in the P-site. It seems improbable then, that 2A acts to impair the gating mechanism that normally leads to peptide bond formation. In the context of the 2A reaction, we suggest that the combination of a "poor" electrophile and "poor" nucleophile leads to inhibition of the peptidyl-transferase reaction. "Poor" meaning, perhaps, an aspect in common between two dynamic situations in which both (i) the carbonyl-carbon of 2A peptidyl-tRNA and (ii) the imide nitrogen of prolyl-tRNA exhibit low occupancy of the correct orientation, effectively halting nascent chain elongation. In our original model we proposed that a water molecule present in the PTC could be activated, possibly by the carboxyl groups at the base of the 2A helix, to hydrolyse the ester linkage releasing the nascent peptide. Later experiments suggest a different mechanism (see below).

To continue translation of the downstream sequences, we proposed that the prolyl-tRNA in the A-site was translocated to the P-site, allowing binding of the next aminoacyl-tRNA and thence elongation. An alternative outcome following hydrolysis of the ester bond is that the complex dissociated at this point, terminating translation. This would explain the molar excess of translation product upstream over that downstream of 2A we observed in translation systems in vitro (Fig. 5.2).

5.4 Testing the Co-translational Model

5.4.1 Ribosomal Pausing at 2A

If 2A interacts with the ribosome to generate a pause in elongation at its own C-terminus, we argued that (as long as the pause were at the correct point in the elongation cycle) translation in the presence of low levels of puromycin would produce a "signal" from the incorporation of puromycin at this site above the "noise" from incorporation at random throughout elongation. Translation of pGFP2AGUS in the presence of puromycin and immunoprecipitation of the translation products with anti-puromycin antibodies yielded a species that co-migrated with [GFP-2A] (Donnelly et al., 2001a). The exact position of this pause on the mRNA was examined using the "toe-printing" approach. Here, reverse transcriptase is used to map the position of ribosomes or other complexes on RNA. A translation-specific signal was obtained from mRNAs that included sequences encoding a functional 2A (but not an inactive mutant) (Doronina et al., 2008). The signal indicated that ribosomes pause at the C-terminus of 2A, with the codons encoding the critical glycine 18 and proline 19 (-NP<u>G</u> ↓ <u>P</u>-), in the P- and A-sites, respectively. These data provided significant support for the co-translational model.

With regard to the site of translational arrest, an interesting parallel to our observations is the translational regulation of expression of the *Sec*M gene in *Escherichia coli*. Here, the nascent peptide –FxxxxWIxxxxGIRAGP- has been shown to interact with the ribosome exit tunnel to produce a pause such that peptidyl-tRNA^{Gly165} (underlined; -GIRA<u>G</u>P-) occupies the P-site. Puromycin does not effectively release this arrested peptide since the A-site is occupied by tRNA^{Pro166} (Muto et al., 2006). In both cases the ribosome is paused (by the nascent polypeptide) within an ORF and at a site on the mRNA encoding a glycyl-prolyl pair. In both cases the P-site is occupied by peptidyl-tRNA^{Gly}; however, a significant difference is that in the case of SecM, the A-site is occupied (by tRNA^{pro}) blockading polypeptide release from the complex by puromycin, while in the case of 2A the stalled complex *could* be released effectively by puromycin, indicating the A-site is not similarly blocked (Donnelly et al., 2001a; see below).

5.4.2 The 2A Reaction Takes Place at the Ribosomal Decoding Centre

A key piece of evidence that directly linked the 2A reaction to translation came through analysis of products generated from truncated RNA templates ending in positions across the C-terminus of the 2A peptide (Doronina et al., 2008). RNAs ending at, or before, the glycine 18 codon yielded protein products that remained associated with tRNA. This was expected as ribosomes stall at the 3′ end of RNA templates such as these, with the nascent chain covalently attached to ribosome-associated tRNA. However, an RNA that included the proline 19 codon yielded largely free peptide, indicating that the first part of the 2A reaction – release of the upstream product – had occurred on this template. Incorporation of mutations that inactivated the 2A peptide into this template led to the peptide being retained as a

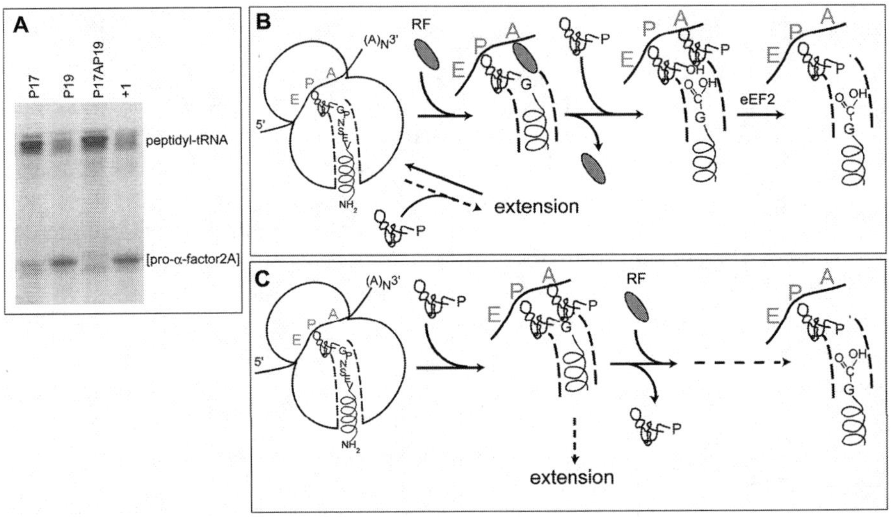

Fig. 5.3 Model of 2A activity. The 2A reaction proceeds with the final glycine and proline codons encoding the 2A peptide in the peptidyl transferase centre. Translation reactions were assembled with wheatgerm lysate and programmed with RNA encoding abbreviated *S. cerevisiae* pro-a-factor followed by sequences encoding the FMDV 2A peptide. DNA templates from which the RNAs were generated were truncated at the positions indicated (including a template bearing the proline 17 to alanine mutation; P17A) using PCR and specific oligonucleotides. Protein synthesis was monitored by incorporation of [^{35}S]-methionine included in the reactions and products examined by SDS-PAGE and phosphorimaging. Peptidyl-tRNA adducts were assigned by the fact that treatment with RNAse A led to their increasing in mobility such that they then migrated at the size expected for the translated polypeptides (not shown) (**Panel A**). Alternative models for the stop-carry on recoding event dictated by CHYSELs (**Panels B and C**). In the first model the conformation of the peptidyl(2A)-ribosome complex would be such that the entry of prolyl-tRNA^{pro} into the A-site is discriminated against, and that the entry of RFs - without the need for stop codon recognition – is favoured. Thus, the majority of ribosomes terminate translation at this point. Subsequent to this the continued interaction of the peptide with the exit tunnel (presumably in a slightly different conformation) promotes entry of prolyl-tRNA^{pro}, which is then moved to the P-site by the action of eEF2, leading to further translation of sequences downstream (**Panel B**). In the alternative model entry of cognate prolyl-tRNA^{pro} into the ribosome, paused at the CHYSEL peptide, occurs as normal. The failure to generate a new peptide bond and subsequent dissociation of prolyl-tRNA^{pro} from the A-site of the ribosome is a pre-requisite, due to the conformation that it leaves behind, for entry or RF into the A-site and hydrolysis of the ester linkage (**Panel C**). Subsequent events are similar to those after RF entry in the pathway shown in **Panel B**. See text for further details

tRNA-adduct, confirming the necessity for an active 2A peptide for nascent chain release (Fig. 5.3A).

5.4.3 Translation Terminating Release Factors and the 2A Reaction

A key feature of our co-translational model of the 2A reaction (above) is hydrolytic release of the nascent chain from tRNA. During the normal course of translation this reaction is catalysed by release factors (RFs) when a termination codon is reached. While non-conventional, we considered the possibility that release factors could catalyse the separation of the nascent chain from the ribosome at 2A, despite the presence of a proline codon in the A-site (Doronina et al., 2008). To probe this hypothesis we set out to examine the products of translation of an ORF including an internal 2A coding sequence (ssDaF-2A-GFP; de Felipe et al., 2003) in a variety of situations in which RF activity was compromised. Eukaryotic RF comprises 2 subunits, eRF1, which decodes all three stop codons and catalyses hydrolysis of the peptidyl-tRNA ester linkage joining the nascent chain to the final tRNA, and eRF3, a GTP binding protein. The conversion of GTP to GDP on eRF3 is intimately linked to the termination reaction, recent models suggesting that it facilitates stop codon decoding by eRF1 (Salas-Marco and Bedwell, 2004; Alkalaeva et al., 2006; Fan-Minogue et al., 2008).

Yeast provided an excellent system for these experiments, as a number of characterised mutations in either eRF1 or eRF3 are available. Further, the epigenetic [PSI+] trait provides a naturally occurring situation in which available eRF3 levels are low due to aggregation of much of the protein into prion-like particle. The strategy that we chose for these studies was to induce high-level expression of the 2A-containing protein from the yeast *GAL1* promoter, pulse-label the cells with [^{35}S]-methionine, and immunoprecipitate the translation products from cell lysates with specific antibodies. Depletion of eRF1 activity was achieved by the use of a thermosensitive allele, *sup45-2* (Stansfield et al., 1997). Consistent with RF directing the 2A reaction, we found that following incubation of cells carrying this mutation at the restrictive temperature, the amount of "unprocessed", full-length ORF translation product of a protein containing an internal 2A peptide was increased relative to the separated upstream and downstream reaction products. A similar effect was also seen in vitro using translation-competent extracts made from yeast cells genetically depleted of eRF1. However, although we consistently noted a decrease in 2A "activity" under these conditions, it was by no means blocked.

A somewhat different situation emerged when we examined the 2A reaction in cells and extracts containing mutant forms of eRF3 with reduced rates of GTP hydrolysis. Here, we found two effects. First, there was a substantial reduction in the proportion of the downstream (discrete) translation product. Second, there was a reduction in the proportion of the full-length translation product. Thus there is a significant *increase* in the proportion of ribosomes that underwent the first part of the 2A reaction, accompanied by a dramatic *decrease* in the proportion that go on to synthesise sequences downstream of 2A. These observations are hard to rationalise in terms of the accepted functions of release factors – in the context of termination at a stop codon, a reduced rate of GTP hydrolysis on eRF3 has been shown to lead to increased readthrough, i.e. a concomitant reduction in the efficiency of stop codon recognition or hydrolysis of the tRNA-nascent chain ester linkage. However, given that the decoding function of RF must be bypassed for the 2A reaction (with a proline codon in the A-site), hydrolysis of the $tRNA^{gly}$-nascent chain ester linkage may be uncoupled from GTP hydrolysis on eRF3. In this context, the delay to GTP hydrolysis in the eRF3 mutant may extend the occupancy of RF on the ribosome, thereby increasing the time window for both ester bond hydrolysis and release of the 2A peptide from the ribosome. This would lead to the observed changes to the outcome of the 2A reaction in the mutant cells and extracts. Importantly, the dramatic changes in products seen in the presence of eRF3 mutants provide strong support for RF being associated with ribosomes at 2A. Cumulatively the data obtained from these studies are consistent with RF catalysing, or at least contributing to the hydrolytic separation of the nascent chain from the ribosome – i.e. that the first part of the reaction is indeed a non-conventional termination reaction.

Surprisingly, while carrying out in vivo experiments to determine the effect of reducing RF levels on the 2A reaction we found that growth of cultures of all the RF mutants tested slowed, or stopped, when transcription of the 2A reporter was induced. This effect was not seen with wild-type strains, nor a number of strains carrying mutations in other translation factors, but was seen when the 2A sequence of FMDV used in the initial construct was replaced by active 2A sequences from either *Thosea asigna* virus (TaV) or Theiler's murine encephalitis virus (TMEV). These results suggested a rather remarkable effect of 2A expression – that somehow it titrated the available RF, already limited in these strains, such that they were no longer able to grow. Supporting this possibility, we found that a "strong" [PSI+] strain, in which a high proportion of eRF3 was in aggregates, was more affected by 2A expression than an isogenic "weak" [PSI+] strain with more available eRF3.

Decreased availability of RF results in decreased efficiency of translation termination at stop codons. We therefore tested whether 2A expression resulted in increased readthrough at stop codons *in trans* using reporters comprising sequences encoding b-galactosidase and luciferase separated by a stop codon. Luciferase activity expressed relative to β-galactosidase indicates the efficiency of termination at the internal stop codon in such reporters, and, consistent with titration of RF, we found that luciferase activity was greatly increased (up to 30-fold in some cases) by 2A expression in cells already limited for RF activity. This effect was also seen in wild-type cells, indicating that, even though it did not affect cell growth, high-level expression of 2A depleted RF activity sufficiently to cause elevated readthrough in these sensitive reporters of stop codon recognition.

Direct inhibition of RF by cytosolic 2A peptide binding RF subunits was excluded as the mechanism for RF titration, since constructs containing co-translationally recognised ER-targeting signal sequences (that direct efficient localisation of the 2A peptide into the lumen of the organelle where it can not bind cytosolic RF) inhibited growth of the RF-limited strains as effectively as proteins that remained in the cytosol. A number of further mechanisms can be suggested to explain how 2A could reduce the effective concentration of RF: (i) RF leaves the ribosome "disabled" in some way, unable to recycle as rapidly as normal for another round of termination, (ii) termination on a sense codon might generate a signal within the cell leading to down-regulation of RF activity or (iii) RF may dwell on the ribosome for an extended period during the translational pause at 2A.

5.5 Refining the Co-translational Model

The identification of eRFs as players in the 2A reaction has prompted refinement of the translational model, and we now consider it a "stop-carry on" recoding event.

5.5.1 Binding/Dissociation of Prolyl-tRNA

We still propose that cognate prolyl-tRNA^{pro} (-NPGP-; underlined) is able to bind at the A-site. For the reasons discussed above, it is reasonable to assume that binding of the prolyl-tRNA^{pro} ternary complex to the cognate A-site codon leads to hydrolysis of GTP by eEF1 and movement of the aminoacylated 3′ end of the tRNA into the peptidyl-transferase centre – "accommodation". The key difference that the involvement of RFs in the 2A reaction makes to the model is that, instead of hydrolysis of the peptidyl(2A)-tRNA^{gly} ester linkage with prolyl-tRNA^{pro} in the A-site, the tRNA cannot be in the ribosome for the release factors to bind. Situations in which a cognate tRNA in the A-site is unable to form a peptide bond are extremely rare and there are no data for K_{off}. Using two methods (puromycin incorporation, toe-printing), we have observed a pause in elongation at this site so there must be at least one "slow" step in the 2A reaction. Dissociation of cognate prolyl-tRNA must be a good candidate for a slow step, though it would not be detected by puromycin.

In bacteria, ribosomal stalling can lead to endonucleolytic cleavage of the mRNA and elimination of the truncated protein through the tmRNA system (Hayes and Sauer, 2003; Sunohara et al., 2004a, b). Endonucleolytic cleavage is also a key feature of mechanisms that remove ribosomes stalled during elongation in eukaryotes and, in at least some instances, is a primary event in nonsense-mediated decay, the pathway through which mRNAs containing premature stop codons are eliminated (Gatfield and Izaurralde, 2004; Doma and Parker, 2007; Huntzinger et al., 2008). Analysis of mRNAs encoding a 2A peptide has not provided any evidence for these transcripts being unstable, and thus ribosomes paused at 2A must avoid degradative pathways.

5.5.2 eRF Activity

Dissociation of prolyl-tRNA^{pro} from the A-site might lead to cycling of aminoacyl-tRNAs and RF into the A-site. However, a model in which RFs have a low probability of terminating translation at any codon, but are more likely to terminate translation at the C-terminus of 2A due to re-iterative entry into the A-site is unattractive. The consequence of such a model for normal translational elongation is that the longer an ORF is the less likely it is that the entire protein will be synthesised. Previously we suggested that the conformation imposed on the PTC and the ribosome as a whole, mimicked that taken when RF is bound, effectively pre-organising it for release (Fig. 5.3B; Doronina et al., 2008). As discussed above, since codons specifying amino acids other than proline at position 19 lead to extension of the nascent chain, it seems unlikely that tRNA binding is disfavoured by the interactions of 2A – which might argue against this model. A further possibility is that, following entry of cognate prolyl-tRNA^{pro} into the A-site and failure to generate a peptide bond, dissociation of the tRNA might leave structural rearrangements that accompanied its accommodation in place (Steitz, 2008). Such an unusual ribosomal conformation, with, e.g. the P-site "open", could be a substrate for productive RF binding and be an intrinsic part of the 2A reaction mechanism (Fig. 5.3C). Specifically, the ribosome's conformation could be sufficiently similar to that which it normally takes on RF/stop codon binding to allow productive association of the catalytic domain of eRF1 with the PTC without a requirement for stop codon recognition.

5.5.3 "Regulation" of the 2A Reaction?

Following the first step of the 2A reaction, i.e. release of the nascent peptide, a number of steps are required for the ribosome to then go on to synthesise the sequences downstream of 2A: (i) exit of RFs from the A-site, (ii) re-entry of (cognate) prolyl-tRNA^{pro} into the A-site, (iii) translocation of prolyl-tRNA^{pro} from the A- to P-site and (iv) ingress of the next aminoacyl-tRNA into the vacant A-site.

As discussed above, decreased eRF3 GTPase activity both increases termination at the C-terminus of 2A and reduces synthesis of sequences downstream of 2A. Regulation of eRF3 activity may then be a means by which the qualitative outcome of the 2A reaction could be regulated.

The RNAse L pathway forms part of the cellular antiviral response mounted when the replication of viruses is detected due to the formation of double-stranded RNA (dsRNA). Accumulation of dsRNA leads to the activation of a family of 2′–5′ oligoadenylate synthetase (OAS) and OAS-like (OASL) proteins. 2′–5′ oligoadenylates synthesized by these proteins bind and activate RNase L, which then degrades single-stranded RNA within infected cells (reviewed in Silverman, 2007). Recently it has been shown that, in addition to 2′–5′ oligoadenylates, RNase L binds eRF3 (Le Roy et al., 2005). Further, this leads to increased read through at stop codons – i.e. RNAse L binding reduces eRF3 activity. Interestingly, the replication of picornaviruses such as FMDV and EMCV is very sensitive to 2′–5′ oligoadenylates or over-expression of OAS or RNase L (Li et al., 1998; Zhou et al., 1998; Marié et al., 1999). As RNase L becomes activated and binds eRF3, this may lead to a similar effect to reduced GTPase activity of eRF3 – decreased synthesis of sequences downstream of 2A in comparison with those upstream of 2A.

Another possible "control" step is suggested by the early observation that translation of EMCV virus RNA (vRNA) using Krebs cell-free extracts showed a *Translational Barrier in Central Region of Encephalomyocarditis Virus Genome* which could be overcome by the addition of purified eEF2 (Svitkin and Agol, 1983). Examination of the translation profiles in these experiments strongly suggests that this "barrier" occurs at the C-terminus of 2A and that supplementation of this cell-free extract with eEF2 promoted the synthesis of sequences downstream of 2A. eEF2 activity is dependent upon its phosphorylation status, regulated by eEF2 kinase (eEF2K) and protein phosphatase 2A (PP2A). The activities of both eEF2K and PP2A are regulated by cellular stress signalling pathways. During 2A-mediated "cleavage", once the nascent peptide is released by eRF1, RFs have left the A-site and prolyl-tRNA^{pro} has re-entered the A-site, the ribosome would contain deacylated tRNAs in the P- and E-sites and prolyl-tRNA^{pro} in the A-site. A necessary step for further protein synthesis is then translocation of the prolyl-tRNA^{pro} into the P- site without the formation of a peptide bond, and this unusual translocation reaction may be particularly sensitive to eEF2 activity over "normal" translocation events. It is conceivable, therefore, that eEF2 activity may also "regulate" the outcome of the 2A reaction – but only if a reduction in the rate of translocation of prolyl-tRNA^{pro} from the A- to P-site leads to increased dissociation of the ribosome from the mRNA.

In viruses which encode 2A/2A-like sequences, 2A is found either (i) forming the boundary between upstream polyprotein domains comprising capsid proteins and downstream domains comprising RNA replication proteins (e.g. picornavirus) or (ii) located towards the N-terminus of separate ORFs encoding replication proteins (e.g. insect discistroviruses, see below). Given the possibilities outlined above for regulation of synthesis of protein encoded downstream of 2A, the intriguing possibility arises that these viruses may have evolved an accommodation with, or even the ability to harness, cellular responses to infection such that they can utilise the diminishing translational resources of the infected cell for the preferential synthesis of capsid proteins over replicative functions. When the virus genome is initially delivered into the cytoplasm the cell is not "stressed" and there is no dsRNA generated by the replication of vRNA. As an infection progresses the cellular translational apparatus is placed under increasing stress and virus-specific dsRNA accumulates. Changes to eRF3 or eEF2 activity may then lead to reduced translation of sequences downstream of 2A. In the case of picornaviruses, since 240 capsid proteins are required to encapsidate a single RNA genome, increased synthesis of capsid proteins would lead to a higher yield of infectious particles.

5.6 "2A-Like" Sequences

As more virus genome sequences became available, it rapidly became apparent that 2A proteins of viruses from picornavirus genera other the aphtho- and cardioviruses (e.g. tescho-, erbo- and certain parechoviruses) directed stop-carry on recoding, rather than being 2A proteinases. Indeed, as more sequences became available, peptides predicted to have such activity were found in many other types of mammalian and insect virus genomes (both +ve ssRNA and dsRNA; Luke et al., 2008). These were tested for activity by insertion into our artificial polyprotein assay system and all were active. This method of controlling virus protein biogenesis was much more wide-spread than we anticipated. We coined the term "CHYSEL" as an acronym (*cis*-acting *hy*drolyase *el*ement) to distinguish aphtho- and cardioviral and other similar peptides that promote stop-carry on recoding from the proteinase-type 2A of entero- and rhinoviruses (de Felipe et al., 2006).

Sheer probability suggested that the short -D(V/I)xNPGP- motif would be present in some cellular proteins and, indeed, this is the case. However, our analyses of N-terminally truncated forms of FMDV 2A showed this motif *alone* did not mediate "cleavage": the motif needed an appropriate upstream context to be active (Ryan and Drew, 1994). As with the viral sequences, we analysed these putative cellular CHYSELs in the artificial polyprotein system and, excepting those we discuss below, found none were active.

Probing databases also revealed the presence of putative CHYSELs in the non-LTR retrotransposons of *Trypanosoma brucei*, *T. cruzi*, *T. vivax* and *T. congolense* (Fig. 5.4, Panel A; Donnelly et al., 2001b, Heras et al., 2006; unpublished observations) – specifically in L1Tc and ingi elements clustering into the ingi clade of non-LTR retrotransposons. The determination of the genome sequence of the purple sea urchin *Strongylocentrotus purpuratus* (The Sea Urchin Sequencing Consortium, 2006) revealed the presence of very many more putative CHYSELs. Representative CHYSELs from both organisms were tested and found to be active (Donnelly et al., 2001b, Heras et al., 2006; unpublished observations).

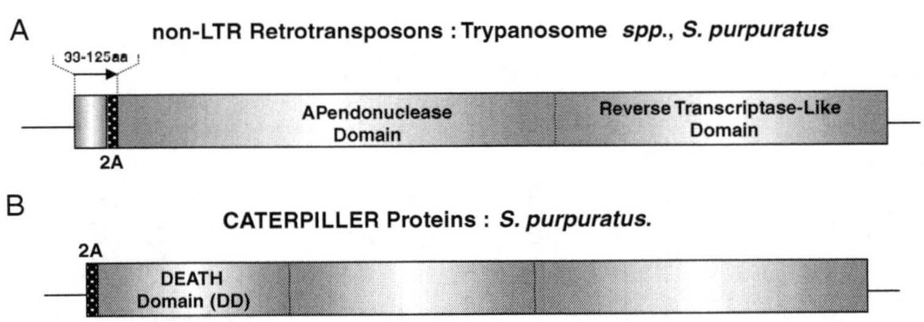

Fig. 5.4 Cellular 2A-like sequences. The 2As present within the non-LTR retrotransposons of trypanosome *spp.* and *S. purpuratus* are located downstream of a short (<125aa) N-terminal domain of unknown function. Downstream of 2A is the typical domain arrangement of CR1-like retrotransposons: AP endonuclease (APE) and reverse transcriptase (RT) domains (**Panel A**). 2A comprises the N-terminus of ~50% of *S. purpuratus* CATERPILER proteins. Downstream of 2A is (**i**) a death domain (DD): related to death effector domains (DED) and caspase-recruitment domains (CARD), (**ii**) a "NACHT" domain: a 300–400aa nucleoside triphosphatase (NTPase) domain and, (**iii**) a leucine–rich repeat (LRR) domain: comprising repeats of 22–28 amino acid motifs involved in binding pathogen-associated molecular patterns, or PAMPs (**Panel B**)

Bioinformatic analyses indicate that *S. purpuratus* cellular sequences harbouring CHYSELs fall into two classes. First, they are found within three different clades of non-LTR retrotransposon, CR1, L2 and Rex1. Like the trypanosome elements, CHYSELs occur towards the N-terminus of the ORF (Fig. 5.4, Panel A; unpublished observations). Their presence within such elements of both trypanosomes and *S. purpuratus* suggests that they play a role – as yet entirely unknown – in the transposition of these elements. Second, CHYSELs are found at the N-terminus of nucleotide binding oligomerization domain (NOD)-like, or CATERPILLER (CARD – caspase-recruitment domain transcription enhancer, R (purine)-binding, pyrin, lots of leucine repeats) proteins (Fig. 5.4, Panel B; unpublished observations). CATERPILLER proteins are cytosolic proteins involved in innate immunity (reviewed in Lich & Ting, 2007; Dostert et al., 2008; Rietdijk et al., 2008). Compared to mammals, the genes encoding the CATERPILLER proteins of *S. purpuratus* comprise a massively expanded family (Rast et al., 2006). Of over 200 such genes, our bioinformatic analyses suggest that ~50% of these ORFs actually begin with a CHYSEL. Preliminary data indicates that some of these peptides may (also) function as signal sequences: "uncleaved" forms enter the exocytic pathway while those translation products in which 2A has mediated "cleavage" become localised to the cytoplasm (unpublished data).

5.7 Concluding Remarks

Despite the progress made towards dissecting the reaction dictated by CHYSELs, significant questions remain. To direct a pause in translation, the helical portion of the CHYSEL presumably interacts with the ribosomal exit tunnel. Many such interactions have been mapped between nascent peptides and the bacterial ribosome exit tunnel: however, CHYSELs do not function in prokaryotic systems. Further experiments to examine the CHYSEL:ribosome interaction must, therefore, be performed using eukaryotic ribosomes – particularly those for which the most detailed atomic structures are available. Yeast provides an attractive system since it is possible to experimentally manipulate ribosomal RNA. Further insight into the "termination" reaction at CHYSELs may be gained through testing further variants of RF to determine which features of eRF1 and 3 are important, or can be dispensed with, for the CHYSEL reaction. One might expect, for example, that residues implicated in the catalytic activity of eRF1 would be necessary, but those important for stop codon "decoding" might not. The second part of the CHYSEL reaction, the "restart" or "pseudo-initiation" of translation on proline 19, remains uncharted territory. Basic questions such as whether initiation factors play a role in the (re-) entry of prolyl-tRNA^{pro} into the A-site, and what permits the translocation of prolyl-tRNA^{pro} from the A- to P-site (while a deacylated tRNA is present in the A-site – a situation not encountered in normal translation) remain to be answered.

Equally intriguing questions pertain as to the biological role(s) of CHYSELs. First, do these sequences represent a novel method of controlling protein biogenesis? Our data on the effect of eRF3 on the outcome of the CHYSEL reaction suggest this may well be the case. The effects of stress or infection may influence the translational outcome of the activity viral CHYSELs, cellular CHYSELs – or quite possibly both. Second, since CHYSELs are found in a range of non-LTR retrotransposons in the genomes of quite different organisms one assumes this sequence is involved in the regulation, or mechanism, of retrotransposition and that CHYSELs may well be found in similar elements within the genomes of other species. Third, can N-terminal CHYSELs be bi-functional, acting essentially as (possibly regulatable) "self-cleaving" signal sequences to produce dual protein targeting? Lastly, the biological "distribution" of CHYSELs raises interesting questions as to the relationships between them. Are CHYSELs "passed" between different types of viruses, passed between virus and cellular sequences, or, since they are so short, do they predominantly arise independently?

What *is* clear, however, is that 2A/CHYSEL sequences are yet another instance of the expansion of the genetic code: a fascinating new addition to the field of "recoding".

References

Alkalaeva EZ, Pisarev AV, Frolova LY, Kisselev LL, Pestova TV (2006) In vitro reconstitution of eukaryotic translation reveals cooperativity between release factors eRF1 and eRF3. Cell 125:1125–1136

Atkins JF, Wills NM, Loughran G, Wu C-Y, Parsawar K, Ryan MD, Wang CH, Nelson CC (2007) A case for "StopGo": Reprogramming translation to augment codon meaning of GGN by promoting unconventional termination (Stop) after addition of glycine and then allowing continued translation (Go). RNA 13:1–8

Baranov PV, Gurvich OL, Hammer AW, Gesteland RF, Atkins JF (2003) RECODE 2003. Nucleic Acids Res 31:87–89

Beringer M, Rodnina MV (2008) The ribosomal peptidyl transferase. Mol Cell 26:311–321

de Felipe P, Hughes LE, Ryan MD, Brown JD (2003). Co-translational, intraribosomal cleavage of polypeptides by the foot-and-mouth disease virus 2A peptide. J Biol Chem 278: 11441–11448

de Felipe P, Luke GA, Hughes LE, Gani D, Halpin C, Ryan MD (2006) *E unum pluribus:* multiple proteins from a self-processing polyprotein. Trends Biotechnol 24:68–75

Dodding MP, Bock M, Yap MW, Stoye JP (2005) Capsid processing requirements for abrogation of Fv1 and Ref1 restriction. J Virol 79:10571–10577

Doma MK, Parker R (2007) RNA quality control in eukaryotes. Cell 131:660–668

Donnelly M, Gani D, Flint M, Monoghan S, Ryan MD (1997) The cleavage activity of aphtho- and cardiovirus 2A proteins. J Gen Virol 78:13–21

Donnelly MLL, Luke G, Mehrotra A, Li X, Hughes LE, Gani D, Ryan MD 2001a. Analysis of the aphthovirus 2A/2B polyprotein 'cleavage' mechanism indicates not a proteolytic reaction, but a novel translational effect: a putative ribosomal 'skip'. J Gen Virol 82:1013–1025

Donnelly MLL, Hughes LE, Luke G, Li X, Mendoza H, ten Dam E, Gani D, Ryan MD (2001b) The 'cleavage' activities of FMDV 2A site-directed mutants and naturally-occurring '2A-Like' sequences. J Gen Virol 82:1027–1041

Doronina VA, Wu C, de Felipe P, Sachs MS, Ryan MD, Brown JD (2008) Site-specific release of nascent chains from ribosomes at a sense codon. Mol Cell Biol 28:4227–4239

Dostert C, Meylan E, Tschopp J (2008) Intracellular pattern-recognition receptors. Advanced Drug Delivery Rev 60:830–840

Fan-Minogue H, Du M, Pisarev AV, Kallmeyer AK, Salas-Marco J, Keeling KM, Thompson SR, Pestova TV, Bedwell DM (2008) Distinct eRF3 requirements suggest alternate eRF1 conformations mediate peptide release during eukaryotic translation termination. Mol. Cell 30:599–609

Gatfield D, Izaurralde E (2004) Nonsense-mediated messenger RNA decay is initiated by endonucleolytic cleavage in *Drosophila*. Nature 429:575–578

Hahn H, Palmenberg AC (1996) Mutational analysis of the encephalomyocarditis virus primary cleavage. J Virol 70:6870–6875

Hayes CS, Sauer RT (2003) Cleavage of the A site mRNA codon during ribosome pausing provides a mechanism for translational quality control. Mol Cell 12:903–911

Heras SR, Thomas MC, García M., de Felipe P. García-Pérez JL, Ryan MD, López MC (2006) L1Tc non-LTR retrotransposons from *Trypanosoma cruzi* contain a functional viral-like self-cleaving 2A sequence in frame with the active proteins they encode. Cell Mol Life Sci 63: 1449–1460

Huntzinger E., Kashima I., Fauser M, Saulière J, Izaurralde E (2008) SMG6 is the catalytic endonuclease that cleaves mRNAs containing nonsense codons in metazoan. RNA 14: 2609–2617

Korostelev A, Noller HF (2007) The ribosome in focus: new structures bring new insights. Trends Biochem Sci 32:434–441

Korostelev A, Ermolenko DN, Noller HF (2008) Structural dynamics of the ribosome. Curr Opin Chem Biol 12: 674–683

Laurberg Mm, Asahara H, Korostelev A, Zhu J, Trakhanov S, Noller HF (2008) Structural basis for translation termination on the 70S ribosome. Nature 454:852–857

Le Roy F, Salehzada T, Bisbal C, Dougherty JP, Peltz SW (2005) A newly discovered function for RNase L in regulating translation termination. Nat Struct Mol Biol 12:505–512

Ledoux S, Uhlenbeck OC (2008) Different aa-tRNAs are selected uniformly on the ribosome. Mol Cell 31:114–123

Li X-L, Blackford JA, Hassel BA (1998) RNase L mediates the antiviral effect of interferon through a selective reduction in viral RNA during encephalomyocarditis virus infection. J Virol 72:2752–2759

Lich JD, Ting JP-Y (2007) CATERPILLER (NLR) family members as positive and negative regulators of inflammatory responses. Proc American Thoracic Soc 4:263–266

Lim VI, Spirin AS (1986) Stereochemical analysis of ribosomal transpeptidation conformation of nascent peptide. J Mol Biol 188:565–574

Luke GA, de Felipe P, Lukashev A, Kallioinen SE, Bruno EA, Ryan MD (2008) The occurrence, function, evolutionary origins of "2A-like" sequences in virus genomes. J Gen Virol 89: 1036–1042

Marié I, Rebouillat D, Hovanessian AG (1999) The expression of both domains of the 69/71 kDa 2′,5′ oligoadenylate synthetase generates a catalytically active enzyme and mediates an antiviral response. Eur J Biochem 262:155–165

Meenakshi KD, Parker R (2006) Endonucleolytic cleavage of eukaryotic mRNAs with stalls in translation elongation. Nature 440:561–564

Muto H, Nakatogawa H, Ito K (2006) Genetically encoded but nonpolypeptide prolyl-tRNA functions in the A site for SecM-mediated ribosomal stall. Mol Cell 22: 545–552

Nathans D, Neidle A (1963) Structural requirements for puromycin inhibition of protein synthesis. Nature 197:1076–1077

Nathans D (1964) Puromycin inhibition of protein synthesis: incorporation of puromycin into peptide chains. Proc Natl Acad Sci USA 51:585–592

Nissen P, Hansen J, Ban N, Moore PE, Steitz TA (2000) The structural basis of ribosome activity in peptide bond synthesis. Science 289:920–930

Palmenberg AC, Parks GD, Hall DJ, Ingraham RH, Seng TW, Pallai PV (1992) Proteolytic processing of the cardioviral P2 region: primary 2A/2B cleavage in clone-derived precursors. Virology 190:754–762

Pavlov MY, Watts RE, Tan X, Cornish VW, Ehrenberg M, Forster AC (2009) Slow peptide bond formation by proline and other N-alkylamino acids in translation. Proc Natl Acad Sci USA 106:50–54

Pringle FM, Gordon KH, Hanzlik TN, Kalmakoff J, Scotti PD, Ward VK (1999) A novel capsid expression strategy for Thosea asigna virus (Tetraviridae). J Gen Virol 80:1855–63

Rast JP, Smith LC, Loza-Coll M, Hibino T, Litman GW (2006) Genomic insights into the immune system of the sea urchin. Science 314:952–956

Rietdijk ST, Burwell T, Bertin J, Coyle AJ (2008) Sensing intracellular pathogens—NOD-like receptors. Curr Op Pharmacol 8:261–266

Robertson BH, Grubman MJ, Weddell GN, Moore DM, Welsh JD, Fischer T, Dowbenko TJ, Yansura DG, Small B, Kleid DG (1985) Nucleotide and amino acid sequence coding for polypeptides of foot-and-mouth disease virus type A12. J Virol 54:651–660

Ryan MD, King AMQ, Thomas GP (1991) Cleavage of foot-and-mouth disease virus polyprotein is mediated by residues located within a 19 amino acid sequence. J Gen Virol 72: 2727–2732

Ryan MD, Drew J (1994) Foot-and-mouth disease virus 2A oligopeptide mediated cleavage of an artificial polyprotein. EMBO J 134:928–933

Ryan MD, Flint M (1997) Virus-encoded proteinases of the picornavirus super-group. J Gen Virol 78:699–723

Ryan MD, Donnelly ML, Lewis A, Mehotra AP, Wilke J. and Gani D (1999) A model for non-stoichiometric, co-translational protein scission in eukaryotic ribosomes. Bioorganic Chem 27:55–79

Rychlík I, Černá J, Chládek S, Pulkrábek P, Žemlička J (1970) Substrate specificity of ribosomal peptidyl transferase. Eur J Biochem 16:136–142

Salas-Marco J, Bedwell DM (2004) GTP hydrolysis by eRF3 facilitates stop codon decoding during eukaryotic translation termination. Mol Cell Biol 24:7769–7778

Schmeing TM, Huang KS, Srobel SA, Steitz TA (2005) An induced-fit mechanism to promote peptide bond formation and excluded hydrolysis of peptidyl-tRNA. Nature 428:520–524

Silverman RH (2007) Viral encounters with 2′,5′-oligoadenylate synthetase and RNase L during the interferon antiviral response. J Virol 81:12720–12729

Simonović M, Steitz TA (2008) Peptidyl-CCA deacylation on the ribosome promoted by induced fit and the O3′-hydroxyl group of A76 of the unacylated A-site tRNA. RNA 14:1–7

Song H, Mugnier P, Das AK, Webb HM, Evans DR, Tuite MF, Hemmings BA, Barford D (2000) The crystal structure of human eukaryotic release factor eRF1 – mechanism of stop codon recognition and peptidyl-tRNA hydrolysis. Cell 100:311–321

Stansfield I, Kushnirov VV, Jones KM, Tuite MF (1997) A conditional-lethal translation termination defect in a *sup45* mutant of the yeast *Saccharomyces cerevisiae*. Eur J Biochem 245:557–563

Steitz TA (2008) A structural understanding of the dynamic ribosome machine. Mol Cell Biol 9:242–253

Sunohara T, Jojima K, Tagami H, Inada T, Aiba H (2004a) Ribosome stalling during translation elongation induces cleavage of mRNA being translated in *Escherichia coli*. J Biol Chem 279:15368–15375

Sunohara T, Jojima K, Yamamoto Y, Inada T, Aiba H (2004b) Nascent-peptide-mediated ribosome stalling at a stop codon induces mRNA cleavage resulting in nonstop mRNA that is recognized by tmRNA. RNA 10:378–386

Svitkin YV, Agol VI (1983) Translational barrier in central region of encephalomyocarditis virus genome. Modulation by elongation factor 2 (eEF2). Eur J Biochem 133:145–154

Tenson T, Ehrenberg M (2002) Regulatory nascent peptides in the ribosomal tunnel. Cell 108: 591–594

The Sea Urchin Sequencing Consortium (2006) The purple sea urchin genome. Science 314: 941–952

Weixlbaumer A, Jin H, Neubauer C, Voorhees RM, Petry S, Kelley AC, Ramakrishnan V (2008) Insights into translational termination from the structure of RF2 bound to the ribosome. Science 322:953–956

Woolhead CA, McCormick PJ, Johnson AE (2004) Nascent membrane and secretroy proteins differ in FRET-detected folding far inside the ribosome and in their exposure to ribosomal proteins. Cell 116:725–736

Woolhead CA, Johnson AE, Bernstein HD (2006) Translation arrest requires two-way communication between a nascent polypeptide and the ribosome. Mol Cell 22:587–598

Wu CY, Lo CF, Huang CJ, Yu HT, Wang, CH (2002) The complete genome sequence of *Perina nuda* picorna-like virus, an insect-infecting RNA virus with a genome organization similar to that of the mammalian picornaviruses. Virology 294:312–323

Zhou A, Paranjape JM, Hassel BA, Nie H, Shah S, Galinski B, Silverman RH (1998) Impact of RNase L overexpression on viral and cellular growth and death. J Interferon Cytokine Res 18:953–961

Chapter 6
Recoding Therapies for Genetic Diseases

Kim M. Keeling and David M. Bedwell

Abstract Strategies aimed at recoding premature stop mutations or frameshift mutations have the potential to treat a genotypic subset of patients afflicted with many different genetic diseases. In this chapter we provide an overview of approaches to promote readthrough of premature stop mutations, including pharmacological agents and suppressor tRNAs. We also describe the use of oligonucleotides to induce differential splicing to exclude disease-causing mutations or to induce site-specific frameshifting as a method of recoding mutations that alter the ribosomal reading frame. Finally, we discuss issues that could complicate the success of these approaches, such as toxicity or nonsense-mediated mRNA decay. Ultimately, these therapies have the potential to be uniquely tailored to a particular patient to optimize the therapeutic effect. Based upon recent progress in this field, one or more of these therapeutic recoding strategies could be used to treat individuals with one or more genetic diseases within the next few years.

Contents

D.M. Bedwell (✉)
Department of Microbiology, BBRB 432/Box 8, 1530 Third Avenue, South, The University of Alabama at Birmingham, Birmingham, AL 35294-2170, USA
e-mail: dbedwell@uab.edu

J.F. Atkins, R.F. Gesteland (eds.), *Recoding: Expansion of Decoding Rules Enriches Gene Expression*, Nucleic Acids and Molecular Biology 24,
DOI 10.1007/978-0-387-89382-2_6,

6.1 Recoding Premature Stop Codons Using Small Molecules

Premature stop mutations are found in a significant fraction of patients with a wide range of genetic diseases. Over the last decade, a novel therapeutic approach has been investigated that uses small molecules to induce recoding of a premature termination codon to a sense codon. The goal of this treatment is to inhibit translation termination at a premature stop mutation, so full-length, functional protein is made. This treatment strategy is often termed "readthrough" or "suppression."

Two proteins, eukaryotic release factor 1 (eRF1) and eukaryotic release factor 3 (eRF3), form a complex to mediate translation termination (Zhouravleva et al., 1995). eRF1 recognizes and binds to any of the three stop codons (UAA, UAG, and UGA) when they are located in the ribosomal A site (Bertram et al., 2001; Kisselev et al., 2003). After stop codon recognition, eRF1 stimulates the ribosomal peptidyl transferase center to trigger release of the nascent polypeptide (Frolova et al., 1999; Seit-Nebi et al., 2001; Song et al., 2000). eRF3 is a GTPase that, upon binding to both eRF1 and ribosome, hydrolyzes GTP and assists eRF1 in both stop codon recognition and polypeptide release (Alkalaeva et al., 2006; Frolova et al., 1996; Mitkevich et al., 2006; Salas-Marco and Bedwell, 2004).

Near-cognate aminoacyl-tRNAs carry an anticodon that can base pair with two of the three nucleotides of a codon. If a near-cognate aminoacyl-tRNA base pairs with a stop codon in the A site, its bound amino acid may become incorporated into the nascent polypeptide. When this occurs, termination is suppressed and translation elongation continues in the correct ribosomal reading frame (Fig. 6.1). Normally, stop codon recognition by eRF1 is very efficient, where incorporation of an amino acid at a stop codon occurs at a rate of only 0.1–0.5% (Bonetti et al., 1995; Loftfield and Vanderjagt, 1972; Mori et al., 1985; Stansfield et al., 1998). However, small

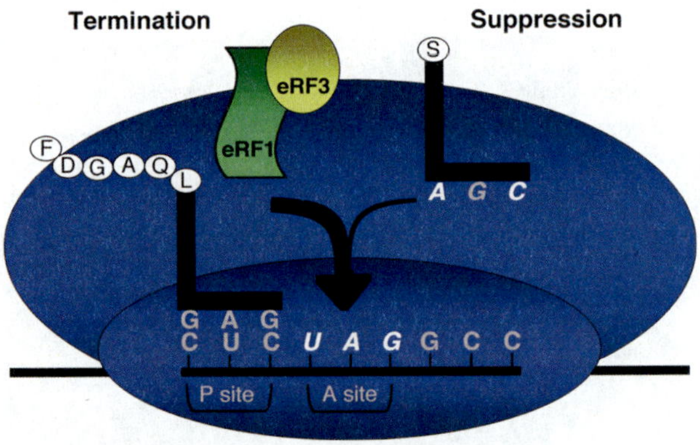

Fig. 6.1 Suppression of premature stop codons. A premature stop codon can be suppressed by a near-cognate aminoacyl-tRNA when the tRNA base pairs with the stop codon in the ribosomal A site and the amino acid carried by the tRNA becomes incorporated into the nascent polypeptide chain. This allows translation elongation to continue in the correct ribosomal reading frame, resulting in the production of a full-length protein

molecules have been discovered that bind to the ribosome and significantly increase the frequency at which translation termination at premature stop codons is suppressed by the incorporation of near-cognate aminoacyl-tRNAs (for a review, see (Keeling and Bedwell, 2005)). Compounds that promote suppression of premature stop codons in mammalian cells include aminoglycosides and aminoglycoside derivatives, negamycin, and PTC124.

6.1.1 Aminoglycosides and Their Derivatives

Aminoglycosides are a group of small molecules that consist of various amino sugars joined to a 2-deoxystreptamine ring by glycosidic linkages (Fig. 6.2). These compounds were originally identified as antibiotics due to their ability to interfere with bacterial protein synthesis. At low concentrations, they induce ribosomal misreading at both sense codons and stop codons by binding to a region of the bacterial 16S ribosomal RNA known as the decoding site (Moazed and Noller, 1986; Purohit and Stern, 1994). This region of the ribosome specifically functions to discriminate cognate from near-cognate aminoacyl-tRNAs within the ribosomal A site (Carter et al., 2000). When aminoglycosides bind to the decoding site, a conformational change occurs in the rRNA that reduces the ability of the ribosome to distinguish cognate from near-cognate aminoacyl-tRNAs (Fourmy et al., 1996, 1998a, b; Recht et al., 1996; Vicens and Westhof, 2001; Yoshizawa et al., 1998). This impairment in ribosomal discrimination can result in the misincorporation of near-cognate amino acids at both sense codons and stop codons.

Aminoglycosides have also been shown to bind the decoding site in the 18S rRNA of the eukaryotic ribosome (Lynch and Puglisi, 2001b). However, minor differences in the nucleotide sequence between the prokaryotic and the eukaryotic decoding site RNAs result in structural variations that substantially reduce the binding affinity of aminoglycosides to the eukaryotic decoding site (Lynch and Puglisi, 2001a, b; Recht et al., 1999). Due to the lower affinity of aminoglycosides for the eukaryotic ribosome, aminoglycoside concentrations that normally inhibit bacterial protein synthesis do not impede eukaryotic protein synthesis. However, these concentrations of aminoglycosides can induce low levels of misreading preferentially at stop codons in eukaryotic cells (Burke and Mogg, 1985; Palmer et al., 1979; Salas-Marco and Bedwell, 2005; Singh et al., 1979), possibly due to differences in the mechanism or kinetics of stop codon vs. sense codon recognition.

Numerous studies have used cultured cells to examine the ability of aminoglycosides to suppress premature stop mutations associated with a wide range of genetic diseases. In these studies, a key objective was to determine whether enough functional protein could be restored by readthrough to alleviate disease phenotypes. A summary of the numerous in vitro studies is shown in Table 6.1. The great majority of these studies revealed that aminoglycosides (and other small molecules) efficiently suppress disease-causing premature stop mutations. The level of suppression achieved among these different disease-causing premature stop mutations presumably varies due to several factors, including the identity of the stop codon, the

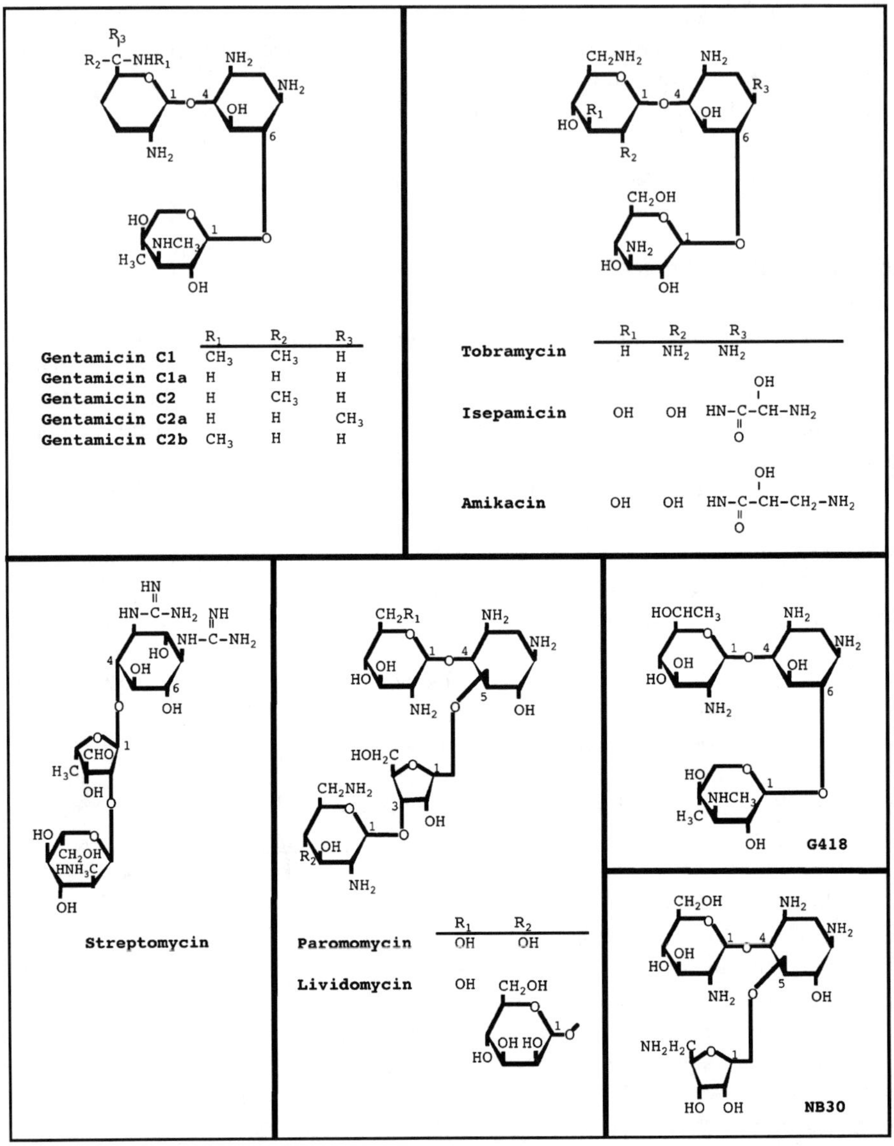

Fig. 6.2 Chemical structures of aminoglycoside compounds that suppress premature termination codons

sequence surrounding the stop codon, the chemical structure of the aminoglycoside examined, the permeability of the cell type for aminoglycosides, the abundance of the mRNA containing the stop mutation, and the amount of functional protein required to normalize a phenotype.

Table 6.1 In vitro suppression of disease-causing premature stop mutations

Gene that carries a premature stop codon	Disease	Drugs analyzed (effectiveness)	Protein function restored	References
ATM	Ataxia telangiectasia; cancer	G418 (+) Gentamicin (+) Paromomycin (+) Tobramycin (–)	Yes	Lai et al. (2004)
ATP2C1	Familial benign pemphigus/Hailey–Hailey disease	Gentamicin (+)	Yes	Kellermayer et al. (2006)
CFTR	Cystic fibrosis	G418 (+) Gentamicin (+) Tobramycin (–) Amikacin (+)	Up to 20% of WT	Bedwell et al. (1997); Howard et al. (1996); Zsembery et al. (2002)
CLNS	Infantile neuronal ceroid lipofuscinosis	Gentamicin (+)	Yes, up to 7% of WT	Sleat et al. (2001)
CTNS	Cystinosis	Gentamicin (+)	Yes	Helip-Wooley et al. (2002)
DMD	Duchenne muscular dystrophy	Gentamicin (+) Amikacin (+) Tobramycin (–) Paromomycin (+) Negamycin (+) PTC124 (+)	Yes, up to 20% of WT	Arakawa et al. (2003); Barton-Davis et al. (1999); Bidou et al. (2004); Howard et al. (2004); Howard et al. (2000); Welch et al. (2007)
F7	Coagulation VII deficiency	G418 (+) Gentamicin (+)	Yes	Pinotti et al. (2006)
IDUA	MPS I-H/Hurler syndrome	Gentamicin (+) Amikacin (+) Tobramycin (+)	Yes, up to 5% of WT	Hein et al. (2004); Keeling and Bedwell (2002); Keeling et al. (2001)
KCNA5	Atrial fibrillation	Gentamicin (+)	Yes	Olson et al. (2006)

Table 6.1 (continued)

Gene that carries a premature stop codon	Disease	Drugs analyzed (effectiveness)	Protein function restored	References
LAMA2	Congenital muscular dystrophy	Gentamicin (+) Amikacin (+) Negamycin (+)	Yes, up to 3% of WT, but not enough to correct phenotype	Allamand et al. (2008)
P53	Li-Fraumeni syndrome; cancer	Gentamicin (+) Amikacin (+) Tobramycin (+)	Yes	Keeling and Bedwell (2002)
PCDH15	Usher syndrome	G418 (+) Gentamicin (+) Paromomycin (+) New compound NB30 (+)	Yes	Rebibo-Sabbah et al. (2007)
PKD2	Autosomal dominant polycystic kidney disease	Gentamicin (+) Isepamicin (+)	Yes	Aguiari et al. (2004)
RP2	X-linked retinitis pigmentosa	Gentamicin (–)	No	Grayson et al. (2002)
SMN1	Recessive spinal muscular atropy	G418 (+) Gentamicin (–) Amikacin (+) Lividomycin (+) Streptomycin (+) Tobramycin (+) New compounds (+)	Yes	Mattis et al. (2006); Sossi et al. (2001); Woolstencroft et al. (2006)

(+) indicates effective suppression; (–) indicates ineffective suppression.

Several studies using mouse models of human diseases have also been carried out to determine the ability of aminoglycosides to suppress premature stop mutations in vivo. A summary of these studies is shown in Table 6.2. In many of these studies, evidence of readthrough of the disease-causing mutation was found and partial correction of the disease phenotype was observed. These findings led to a number of small, preliminary clinical trials using patients that carried premature stop mutations. A summary of these investigations is listed in Table 6.3. The results of these studies are considerably more varied than the animal studies. For example, three of the four clinical trials with cystic fibrosis (CF) patients that carried stop mutations responded to aminoglycoside therapy as shown by increased cAMP-stimulated chloride conductance, indicative of a partial restoration of CFTR protein function (Clancy et al., 2001; Wilschanski et al., 2000, 2003). However, no evidence of restored CFTR function was observed in response to aminoglycoside therapy in one study (Clancy et al., 2007).

The variability observed among these various clinical investigations could be caused by a number of factors. As observed in in vitro experiments, the ability of aminoglycosides to suppress premature stop mutations depends on many variables, including the identity of the stop codon and the sequence context surrounding the stop codon. Due to the low frequency of stop mutations in the CF patient population, each of these studies used patients with different stop mutations in the *CFTR* gene. In addition, other factors such as differential uptake of the compounds or subtle variations in the assays used by different research groups could also potentially complicate the interpretation of in vivo experiments.

Differences in administration protocols could also explain why results differed among the various clinical studies. Some protocols used in clinical trials administer aminoglycosides based on previous strategies that used aminoglycosides to treat bacterial infections via IV administration, while others use more conservative approaches, such as topical drops or nebulization in the nasal cavity. Whatever the reason, the results of various suppression therapy studies showed surprising variability (Table 6.3). A careful survey has not yet been conducted to determine the optimal aminoglycoside administration protocols for suppression therapy.

Toxicity of aminoglycosides and their derivatives – A key disadvantage of aminoglycosides is their relatively high potential for toxic side effects. If extreme care is not shown in their use, aminoglycosides can cause kidney damage and hearing loss. Importantly, many of these side effects appear to be independent of their ability to interact with the translational machinery. Rather, the charged nature of aminoglycosides appears to modulate their association with other cellular constituents that result in toxicity (Mingeot-Leclercq et al., 1999; Nagai and Takano, 2004). Aminoglycosides are taken into cells via megalin, a multi-ligand, endocytic receptor that is particularly abundant in the proximal tubules of the kidney and the hair cells of the inner ear (Nagai and Takano, 2004). Upon entering kidney cells, the positively charged nature of aminoglycosides promotes their binding to acidic phospholipids in the lysosomal membrane (Mingeot-Leclercq et al., 1999; Nagai and Takano, 2004), which alters the activity of a number of enzymes and may lead to the formation of free radical species.

Table 6.2 Suppression of premature stop codons in mouse models of human disease

Disease/mouse model	Mutation†	Treatment	Administration route (optimal)	Dosage (optimal)	Treatment duration	Phenotypic result	References
Cystic fibrosis	*CFTR-G542X* (UGAG) human transgene expressed in *Cftr*^{tm1cam} (knockout) mouse	Gentamicin	Once daily subcutaneous injections	5–34 mg/kg	2 weeks	14–28% of WT protein function restored	Du et al. (2002, 2006)
		Tobramycin	Once daily subcutaneous injections	34 mg/kg	2 weeks	Much lower response than with gentamicin	Du et al. (2002)
		Amikacin	Once daily subcutaneous injections	15–170 mg/kg	2 weeks	22–30% of WT protein function restored	Du et al. (2006)
		PTC124	Once daily subcutaneous injections	15–60 mg/kg	2 weeks	2.1–29% of WT protein function restored	Du et al. (2008)
		PTC124	Oral	0.9 mg/ml	2 weeks	24% of WT protein function restored	Du et al. (2008)
Duchenne muscular dystrophy	*mdx-Q995X* (UAAA) naturally occurring mutation	Gentamicin	Once daily subcutaneous injections	34 mg/kg	2 weeks	10–20% of WT protein restored	Barton-Davis et al. (1999); Loufrani et al. (2004)

Table 6.2 (continued)

Disease/mouse model	Mutation†	Treatment	Administration route (optimal)	Dosage (optimal)	Treatment duration	Phenotypic result	References
		Negamycin	Once daily subcutaneous injections	12–24 μmol/kg	2–4 weeks	≈ 10% of WT protein restored	Arakawa et al. (2001, 2003)
		PTC124	Oral combined with three daily intraperitoneal injections	1.8 mg/ml oral + 33 mg/kg IP	2–8 weeks	20–25% of WT protein restored	Welch et al. (2007)
Hemophilia B	*f9-R338X* (UGAU)	G418	Once daily subcutaneous injections	28 mg/kg	2 days	3.4% of WT activity in R338X mouse	Yang et al. (2007)
	f9-R29X (UGAG) Human transgenes expressed in an *f9* knockout mouse	G418	Once daily subcutaneous injections	28 mg/kg	2 days	2.7% of WT activity in R29X mouse	Yang et al. (2007)
X-linked nephrogenic diabetes insipidus	*avpr2-E242X* (UAGC) Targeted mutation	G418	Once daily intraperitoneal injections	14 mg/kg	1 week	Partial correction of disease phenotype	Sangkuhl et al. (2004)

† The four nucleotide sequence indicates the stop codon and the first nucleotide that follows.

Table 6.3 Suppression of premature stop mutations in human patients

Disease	Gene/mutations†	Treatment	Route	Dosage	Duration	Results	References
Cystic fibrosis	*CFTR* W1282X (UGAA) G542X (UGAG)	Gentamicin	Nasal drops	3 mg/ml; two drops to both nostrils three times daily	14 days	Increased chloride transport in 7/9 patients	Wilschanski et al. (2000)
Cystic fibrosis	*CFTR* G542X (UGAG) W1282X (UGAA) R553X (UGAG)	Gentamicin	Intravenously	2.5 mg/kg every 8 h Peak: 8–10 μg/ml Trough: <2 μg/ml	7 days	Increased chloride transport in 4/5 patients	Clancy et al. (2001)
Cystic fibrosis	*CFTR* W1282X (UGAA) G542X (UGAG)	Gentamicin	Nasal drops	3 mg/ml; two drops to both nostrils three times daily	14 days	Increased chloride transport in 17/19 patients; CFTR protein detected in nasal epithelial cells after treatment	Wilschanski et al. (2003)
Cystic fibrosis	*CFTR* G542X (UGAG) R553X (UGAG) E60X (UAGC) Y1092X (UAGC) R1162X (UGAG) W1282X (UGAA)	Gentamicin Tobramycin	Nebulization	0.3% solution administered in 100 μl sprays three times daily	14 days	No increase in chloride transport was detected; no evidence of CFTR detected in any of the 11 patients	Clancy et al. (2007)
Duchenne muscular dystrophy	*DMD* S757X (UGAG) W3294X (UGAC) Q625X (UAAG) Q2198X (UAGC)	Gentamicin	Intravenously	7.5 mg/kg once daily Peak: 34 μg/ml Trough: <0.4 μg/ml	14 days	No dystrophin protein detected in 4/4 patients; no change in muscle strength; decreased serum creatine kinase levels	Wagner et al. (2001)

Table 6.3 (continued)

Disease	Gene/mutations†	Treatment	Route	Dosage	Duration	Results	References
Duchenne muscular dystrophy	*DMD* *R1314X (UGAC) *R195X (UGAC) E137X (UAAA)	Gentamicin	Intravenously	6 mg/kg once daily 7.5 mg/kg once daily	Two rounds of 6-day treatment with 7-week breaks between treatments	Restoration of dystrophin protein detected in 3/4 patients (no protein found in patient with UAAA mutation); no change in functional tests; decreased serum creatine kinase levels	Politano et al. (2003)
Factor VII deficiency	*F7* K316X (UAGG) W363X (UGAG)	Gentamicin	Injections	3 mg/kg once daily	2 days	Very slight increases in F7 function	Pinotti et al. (2006)
Hailey–Hailey	*ATP2C1* R468X (UGAA)	Gentamicin	Topical	1 mg/ml twice daily	18 days	Early remission of skin eruptions compared to conventional treatments	Kellermayer et al. (2006)
Hemophilia A	*F8* *S1395X (UGAU) R2116X (UGAG) R427X (UGAU)	Gentamicin	Intravenously	7 mg/kg once daily	3 days	1/3 patients showed an increase in F8, and an increase in thrombin	James et al. (2005)
Hemophilia B	*F9* *R252X (UGAG) *R333X (UGAG)	Gentamicin	Intravenously	7 mg/kg once daily	3 days	2/2 patients showed an increase in F9 and increase in thrombin	James et al. (2005)
McArdle disease	*PYGM* R49X (UGAG) R269X (UGAG)	Gentamicin	Intravenously	8 mg/kg once daily peak: 43 μg/ml trough: <0.8 μg/ml	10 days	No significant increase in myophosphorylase activity	Schroers et al. (2006)

* Indicates a positive response for that patient.

† The four nucleotide sequence indicates the stop codon and the first nucleotide that follows.

Several approaches are being undertaken to overcome the potential toxicity of aminoglycosides. First, altering the route and duration of their administration has been shown to reduce aminoglycoside toxicity (Bartal et al., 2003; Beauchamp and Labrecque, 2001). The most common protocols used for the treatment of bacterial infections include either a once daily or three times daily administration of aminoglycosides (Bartal et al., 2003; Beauchamp and Labrecque, 2001). During either treatment regimen, peak and trough serum concentrations are strictly monitored to prevent toxicity. However, these protocols may need to be optimized for suppression therapies to maximize suppression while minimizing toxicity. It is also possible that the coadministration of other compounds with the aminoglycosides may further reduce toxicity. For example, coadministration of antioxidants has been reported to reduce the free radical damage associated with aminoglycosides (Kawamoto et al., 2004; Mazzon et al., 2001; Nakashima et al., 2000; Sener et al., 2002). Other studies have shown that the coadministration of polyanions such as poly-L-aspartate (Beauchamp et al., 1990; Gilbert et al., 1989) or daptomycin (Thibault et al., 1994, 1995) can sequester aminoglycosides from the lysosomal membrane while increasing their level in the cytoplasm, which could theoretically reduce toxicity while improving the level of suppression obtained. Finally, the encapsulation of aminoglycosides in liposomes has been shown to circumvent much of their toxicity by altering their route of clearance from the kidneys to the liver (Schiffelers et al., 2001; Xiong et al., 1999).

Another approach would be to physically separate toxic contaminants from the readthrough agent. Interestingly, many commercial aminoglycosides consist of a mixture of isoforms generated during the manufacturing process. For example, five isoforms of gentamicin, C1, C1a, C2, C2a, and C2b, are commonly found in commercial gentamicin preparations (Deubner et al., 2003) (Fig. 6.2). The relative proportions of these isoforms can vary significantly among different gentamicin lots. Recently it was shown that the C2 form of gentamicin retains significant bactericidal activity compared to other isoforms but exhibits significantly reduced toxicity (Sandoval et al., 2006). This isoform also has the ability to suppress premature stop mutations (our unpublished data). This finding suggests that it may be possible to purify specific aminoglycoside isoforms that will allow efficient suppression of premature stop mutations with reduced toxicity.

Minor alterations in the chemical structure of aminoglycosides also have the potential to significantly alter both their toxicity and their ability to suppress premature stop codons. Preliminary attempts have been made to synthesize new aminoglycoside derivatives that carry the moieties needed for efficient readthrough while omitting those constituent groups that cause toxicity (Nudelman et al., 2006). For example, a synthetic derivative of paromomycin called NB30 has been shown to be 6- to 15-fold less toxic than gentamicin or paromomycin (Rebibo-Sabbah et al., 2007) (Fig. 6.2). In addition, NB30 has been shown to be as effective as gentamicin and more effective than paromomycin in suppressing a premature stop mutation in the *PCDH15* gene that causes type 1 Usher syndrome. The design of new derivatives of aminoglycosides may provide an important method to overcome the toxic side effects that prohibit long-term aminoglycoside use, while increasing their efficacy in suppressing premature stop mutations.

6.1.2 Non-aminoglycoside Compounds

Other compounds that are structurally unrelated to aminoglycosides have also been shown to induce readthrough of premature stop mutations. For example, negamycin, a dipeptide antibiotic (Fig. 6.3), has been found to cause readthrough in mammalian cells. In prokaryotes, negamycin has been shown to inhibit the release of nascent polypeptides and induce misreading (Uehara et al., 1972, 1976).

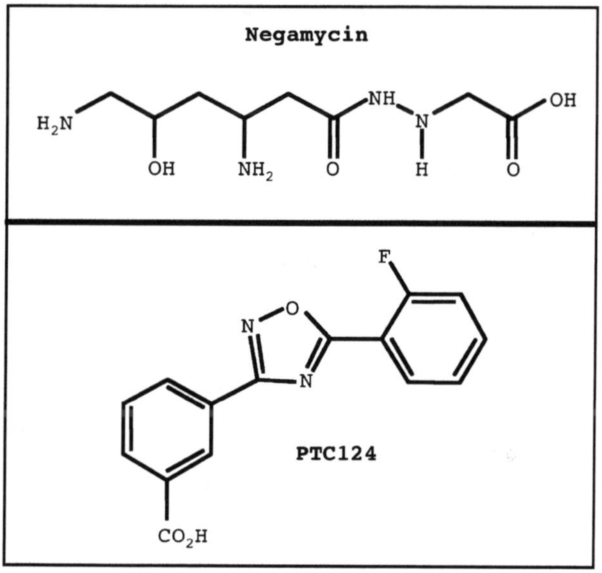

Fig. 6.3 Chemical structure of non-aminoglycoside compounds that suppress premature termination codons

Negamycin has been shown to suppress a UAA premature stop mutation in the gene that codes for the dystrophin protein in the *mdx* mouse, a model of Duchenne muscular dystrophy (Arakawa et al., 2001, 2003). When negamycin was administered to *mdx* mice using the same administration protocol that was previously used to treat *mdx* mice with gentamicin (12–24 μmoles/kg daily via subcutaneous injection for 2 weeks), negamycin restored dystrophin levels to a similar extent as gentamicin (~10% of wild type) (Arakawa et al., 2001, 2003). However, negamycin was reported to be much less toxic than gentamicin in mice.

The ability of negamycin to induce suppression of a C1546X (UGA) premature stop mutation in the *LAMA2* gene (which encodes the α2 chain of laminin-211) has also been investigated. Mutations in the *LAMA2* gene lead to the onset of congenital muscular dystrophy (Allamand et al., 2008). A plasmid expressing a readthrough reporter containing a portion of the *LAMA2* mRNA that contains the C1546X premature stop mutation was injected into skeletal muscle with subsequent electrotransfer. Following daily intramuscular injections of 34 mg/kg negamycin, a 38-fold increase in readthrough of the reporter could be detected. However, when primary human myotubes that carried the *LAMA2-C1546X* mutation were cultured in the presence of negamycin, no restoration of laminin α2-chain expression could be detected by either immunofluorescence or immunoblotting. Interestingly, the same myotubes grown in the presence of negamycin did show an increase in the *LAMA2* mRNA containing the C1546X mutation, suggesting that negamycin might also somehow inhibit degradation of the mRNA by nonsense-mediated mRNA decay.

The mechanism of negamycin-induced readthrough has not yet been fully delineated. Negamycin was shown to bind an RNA oligomer containing a portion of the rRNA from the eukaryotic decoding site, suggesting that negamycin may induce misreading by interacting with the decoding center (Arakawa et al., 2003). However, a crystal structure of negamycin bound to the prokaryotic 50S ribosomal subunit revealed that this compound also binds to the nascent chain exit tunnel, a location consistent with its role in inhibiting polypeptide release (Schroeder et al., 2007). Thus, negamycin may interfere with ribosome function in two distinct ways. Additional studies will be required to better understand how negamycin induces readthrough in eukaryotes.

PTC124 is another non-aminoglycoside compound that can suppress premature stop mutations. This compound does not share any structural similarity to aminoglycosides (Fig. 6.3) and does not possess antibacterial properties. While the mechanism by which PTC124 suppresses termination is currently unknown, PTC124 has been shown to suppress premature stop mutations in animal models of cystic fibrosis and Duchenne muscular dystrophy (Table 6.2). In both studies, PTC124 treatment restored enough functional protein to partially correct the disease phenotype (Du et al., 2008; Welch et al., 2007). A recent study suggested that the initial identification of PTC124 using firefly luciferase-based readthrough reporters might have been based on the inadvertent stabilization of the reporter protein rather than increased readthrough of the stop codon (Auld et al., 2009). While this claim raises questions about its initial discovery and characterization, PTC124 has subsequently been shown to induce levels of readthrough comparable to that of aminoglycosides in animal models of both cystic fibrosis and Duchenne muscular dystrophy (Du et al., 2008; Welch et al., 2007). Those findings provide strong evidence that this compound possesses significant in vivo readthrough activity. Furthermore, PTC124 also appears to have a number of advantages for suppression therapy compared to aminoglycosides. Much lower concentrations of PTC124 are needed to induce the same level of readthrough as aminoglycosides, thereby reducing the potential for nonspecific effects. Consistent with this premise, PTC124 is not associated with any toxic side effects (unlike the aminoglycosides) (Hirawat et al., 2007). Importantly, PTC124 has been shown to preferentially target suppression at premature stop codons, rather than native stop codons at the end of mRNAs (Welch et al., 2007). Finally, PTC124 is readily absorbed through the intestinal lining and it is therefore orally bioavailable. Aminoglycosides are not readily absorbed through the intestines and must be administered via injections. Accordingly, PTC124 has progressed rapidly to phase II clinical trials for cystic fibrosis (Kerem et al., 2008) and Duchenne muscular dystrophy patients who carry premature stop mutations.

6.1.3 Outstanding Questions Regarding Recoding Premature Termination Codons

Each of the compounds described above has shown promise as agents that can efficiently suppress termination at many premature stop mutations. However, none can suppress all premature stop mutations to the same extent. The stop codon and its surrounding sequence context can greatly influence the level of suppression mediated by various pharmacological agents (Cassan and Rousset, 2001; Howard et al., 2000; Keeling and Bedwell, 2002; Manuvakhova et al., 2000; Namy et al., 2001). To address these problems, it may be necessary to routinely screen a battery of compounds for their ability to suppress an individual patient's stop mutation within its unique sequence context. Other issues that could limit the effectiveness of this therapeutic approach are discussed below.

Nonsense-mediated mRNA decay – Differences in the efficiency of suppression observed in patients with the same premature stop mutation in clinical trials could be explained by the presence of different genetic modifier alleles. One likely modifier is the inherent efficiency of nonsense-mediated mRNA decay (NMD), a mechanism that eliminates mRNAs that contain a premature termination codon. NMD is widely conserved among eukaryotes and may have evolved as a quality control mechanism to prevent the synthesis of truncated proteins that may have dominant-negative effects on one or more cellular functions. The process of NMD in mammalian cells has been the subject of intense investigation over the last decade or more, and many of the basic features are now understood (for a recent review, see (Isken and Maquat, 2008)). NMD is a surveillance mechanism that monitors where the termination complex forms relative to exon junction complexes (EJCs) located in the mRNA during the first (or "pioneer") round of translation. If the termination complex forms at a distance ≥50–55 nucleotides upstream of an EJC, the NMD factor Upf1p is recruited and the mRNA is targeted for rapid turnover. As a result of NMD, the abundance of many mutant mRNAs containing a premature termination codon is only 5–10% of normal.

Obviously, this reduction in the level of steady-state mRNA could have a significant effect on the amount of full-length protein that could be restored by suppression of the premature termination codon, and ultimately, the therapeutic effect obtained. A recent study found that the residual steady-state level of *CFTR* mRNA containing a premature stop codon showed a strong correlation with the success of gentamicin-mediated suppression in primary cells derived from CF patients (Linde et al., 2007). This study demonstrated that the abundance of mRNA available for readthrough has a direct effect on the efficacy of suppression therapy. A higher abundance of an mRNA containing a premature stop mutation should increase the effectiveness of suppression therapy, since more mRNA would be available for translation, and consequently, readthrough. For some diseases, combination therapies that target both the NMD process and suppression of the premature stop codon ultimately may be necessary in order to produce a sufficient level of protein for a therapeutic effect.

Targeting premature vs. normal stop codons – A growing body of evidence suggests that premature stop codons may be more susceptible to pharmacological suppression than native stop codons located at the end of open reading frames. The identity of the stop codon as well as several nucleotides surrounding it can influence the efficiency of translation termination, as well as its susceptibility to suppression (Bonetti et al., 1995; Brown et al., 1990; Cassan and Rousset, 2001; Howard et al., 2000; Keeling and Bedwell, 2002; Manuvakhova et al., 2000; McCaughan et al., 1995; Namy et al., 2001). The eRF1 release factor binds to the stop codon and may also bind to a broader termination signal that includes one or more nucleotides downstream of the stop codon. In particular, the nucleotide that directly follows the stop codon plays an important role in determining termination efficiency; it has been proposed that eRF1 may actually recognize a tetranucleotide termination signal (Brown et al., 1990). Evolutionary selection for the most efficient termination signals at the end of frequently expressed mRNAs has been observed in both prokaryotic and eukaryotic systems (Tate et al., 1995). This type of selection for the most efficient translation termination signals would not occur at premature stop codons, possibly making them more susceptible to suppression.

In vitro studies comparing the efficiency of termination at premature vs. normal stop codons have also provided evidence that termination occurs much more efficiently at normal stop codons. For example, ribosomal toeprint assays that monitor how long a termination complex resides at a premature stop codon suggest that termination occurs much more efficiently at native stop codons, suggesting that some feature of the termination complex differs at premature stop codons that may make them more susceptible to suppression (Amrani et al., 2004). Recent studies have suggested that this difference is due to the ability of the termination factor eRF3 to interact with poly(A) binding protein (PABP) bound to the poly(A) tail of mRNAs (Kobayashi et al., 2004). At normal stop codons, translation termination occurs in close proximity to the poly(A) tail (and PABP), thus promoting the interaction between eRF3 and PABP and allowing efficient termination to proceed. In contrast, premature stop codons may be spatially located too far from the poly(A) tail for this interaction to occur, thus reducing the efficiency of the termination process. More recently, it was suggested that the association between eRF3 and PABP is required to prevent the recruitment of Upf1p and the initiation of the NMD process (Ivanov et al., 2008; Singh et al., 2008). Thus, the location of the stop codon in the mRNA may play a key role in determining whether a stop codon is recognized as a premature stop codon or a normal stop codon.

Finally, multiple in-frame stop codons are often observed at the end of mRNAs (Jacobs et al., 2006; Major et al., 2002). These tandem termination signals should ensure efficient termination and prevent the formation of C-terminal protein extensions if high levels of readthrough of the native stop should occur. An investigation into the ability of aminoglycosides to suppress native stop codons of the global transcriptome found evidence that only limited readthrough of normal stop codons occurred under conditions that promoted efficient gentamicin-induced suppression (Keeling et al., 2001).

6.2 Recoding Premature Stop Codons Using Suppressor tRNAs

A second approach to recode premature stop codons as sense codons is to express a suppressor tRNA that carries an anticodon that is complementary to the premature stop codon. A cognate suppressor aminoacyl-tRNA can directly compete with the termination factor eRF1 for stop codon binding in the A site much more effectively than a near-cognate aminoacyl-tRNA, resulting in a significant increase in stop codon suppression.

In mammalian cells, this approach has been shown to suppress premature stop mutations that cause beta-thalassemia (Temple et al., 1982), Duchenne muscular dystrophy (Kiselev et al., 2002), and a collagen VI $\alpha 2$ deficiency known as Ullrich disease (Sako et al., 2006). It has also been shown that the injection of DNA encoding a suppressor tRNA into the skeletal and heart muscles of a transgenic mouse expressing a reporter gene with a premature stop codon resulted in the suppression of the stop codon in vivo (Buvoli et al., 2000).

While this approach can result in much higher levels of suppression than can be obtained using drugs that induce readthrough by near-cognate tRNAs, there are significant drawbacks to this approach. The ability to stably introduce suppressor tRNAs and control their expression in various tissues will undoubtedly prove to be a difficult task. The use of suppressor tRNAs also has some of the same potential disadvantages that occur during the use of small molecules to suppress premature stop codons. These include the context effects of the stop codon and its surrounding sequence and the possible suppression of native stop codons (Bossi and Roth, 1980).

6.3 Recoding Mutations using Antisense Oligonucleotides

In addition to point mutations that lead to premature stop codons, frameshift mutations caused by small nucleotide deletions or insertions also frequently result in the production of truncated proteins due to the occurrence of premature stop codons in the altered reading frame. While small molecules that suppress premature stop mutations created by single nucleotide point mutations have been shown to be effective at partially restoring protein function by recoding the premature stop codon into a sense codon; this approach would not be useful with frameshift mutations since the resulting protein produced by suppression would be translated in the wrong reading frame. Currently, no drugs that are capable of recoding mutations that alter the ribosomal reading frame in eukaryotic cells are available.

However, an approach that can recode frameshift mutations utilizes antisense oligonucleotides to induce specific exon skipping during pre-mRNA splicing.

Antisense oligonucleotides can be designed to hybridize to a specific set of splicing signals in a pre-mRNA to alter the normal splicing pattern. In this way, a specific exon that carries a mutation can be excluded from the pre-mRNA so that the original reading frame (and production of the rest of the protein distal to the removed exon) is restored. While this produces a full-length protein, a region of the protein that was encoded by the skipped exon will be absent from the final polypeptide.

This approach may not be feasible for many diseases, since loss of a portion of the protein may abrogate its function. However, in some diseases, the loss of a portion of the protein may still allow partial (or complete) protein function. One disease model that has been used to extensively investigate the use of antisense oligonucleotides to suppress frameshifting mutations is Duchenne muscular dystrophy (DMD). DMD is caused by loss of a structural protein called dystrophin, which acts to link the cytoskeleton to the extracellular matrix. This connection maintains the stability of muscle fibers during contraction (Matsumura and Campbell, 1994). Without dystrophin, bridging of the cytoskeleton to the extracellular matrix cannot occur, which leads to cumulative damage to muscle fibers. The *DMD* gene is the largest known human gene, spanning 2.4 Mb and containing 79 exons (Roberts et al., 1993). Approximately 70% of mutations that cause DMD are due to deletions of one of more exons that result in a shift in the reading frame and ultimately, protein truncation (Aartsma-Rus et al., 2006). A milder allelic form of this disease known as Becker muscular dystrophy (BMD) is usually caused by in-frame deletions, where the correct ribosomal reading frame is retained. While patients with BMD develop some muscle weakness upon onset of the disease, most BMD patients develop symptoms much later in life and have a significantly longer life expectancy than patients with DMD. This suggests that dystrophin is partially functional even when certain domains are deleted. The unique structure of the dystrophin protein can explain this observation. The N-terminal and C-terminal domains of the dystrophin protein are responsible for linking the cytoskeleton and the extracellular matrix and are thus essential for its function (England et al., 1990; Mirabella et al., 1998). In contrast, portions of the central rod domain of dystrophin are not essential for protein function. Therefore, dystrophin molecules that contain small internal deletions retain partial dystrophin function. These features make it possible to use antisense oligonucleotides to modulate pre-mRNA splicing, so exons that carry frameshift mutations can be excluded from the mature *DMD* mRNA. As a result, production of dystrophin with at least partial function can be restored. A recent review that discusses this approach in the treatment of muscular dystrophy has been published (Muntoni and Wells, 2007).

This approach has been investigated in the *mdx* mouse model of Duchenne muscular dystrophy, where antisense oligonucleotides designed to promote excision of exon 23 that carries an UAA premature stop mutation were injected into muscle tissue (Lu et al., 2003; Mann et al., 2001). A localized restoration of dystrophin protein could be detected at the injection site as demonstrated by immunofluorescence. Functional assays also indicated a partial restoration of dystrophin function. More recently, the *mdx* mouse was treated systemically by tail vein injection of an adeno-associated viral vector that expressed antisense sequences to excise exon 23 (Denti et al., 2006). The antisense oligonucleotide and dystrophin protein were both detected in multiple muscle tissues throughout the body of the mouse. Again, a significant improvement in muscle function was found in these mice.

Preliminary clinical data has also been obtained in a study designed to determine whether the intramuscular injection of an antisense oligonucleotide could remove the dystrophin exon 51 in four patients with Duchenne muscular dystrophy (van Deutekom et al., 2007). Each patient showed specific skipping of exon 51. In addition, dystrophin protein was detected in 64–97% of muscle fibers at the site of injection. Significantly, the amount of dystrophin protein detected in muscle biopsies ranged from 17 to 35% of wild-type levels.

A number of other studies have also tested the ability of antisense oligonucleotides to promote skipping of exons that contain mutations in the mRNAs of various other genes. These include models of dystrophic epidermolysis bullosa caused by mutations in collagen type 7 (*COL7A1*) (Goto et al., 2006); Menkes disease caused by mutations in *ATP7A* (Madsen et al., 2008); a number of cancer models (Mercatante and Kole, 2002; Mercatante et al., 2001; Renshaw et al., 2004; Williams and Kole, 2006); a number of inflammatory disorders (Karras et al., 2001, 2007; Vickers et al., 2006); and atherosclerosis (Khoo et al., 2007). A detailed review of this approach to treat genetic diseases was recently published (Aartsma-Rus and van Ommen, 2007).

In addition to exon skipping, antisense oligonucleotides have also been used to normalize abnormal splicing due to mutations that create cryptic splice sites associated with beta-thalassemia (Dominski and Kole, 1993; Sierakowska et al., 1996), cystic fibrosis (Friedman et al., 1999), and Hutchinson–Gilford progeria syndrome (Scaffidi and Misteli, 2005).

Recent reports have also shown that site-specific ribosomal frameshifting can be stimulated by antisense oligonucleotides. Ribosomal frameshifting requires the presence of a shift-prone sequence that shifts the base paring of the ribosomal P site and A site tRNAs with the mRNA by one nucleotide in either the 3′ direction (a +1 frameshift) or the 5′ direction (−1 frameshift). In addition, the presence of a structural element such as a pseudoknot or stem-loop structure must reside downstream of the shift-prone sequence in order to pause the ribosomal so that tRNA slippage can occur. Antisense oligonucleotides have been designed to bind to a complementary sequence downstream of shift-prone sites to form a simple secondary structure to induce either +1 or −1 ribosomal frameshifting (Howard et al., 2004; Olsthoorn et al., 2004; Henderson et al., 2006). This approach could potentially be developed as a treatment for diseases caused by frameshift mutations. Administration of specifically designed antisense oligonucleotides could be used to revert ribosomes to their original reading frame and restore production of functional protein.

6.4 Concluding Remarks

Recoding premature stop mutations or frameshift mutations has the potential to treat many genetic diseases for which no other treatment is currently available. The general nature of these approaches allows them to be targeted to a genotypic subset of patients with a specific class of mutations found in a broad range of diseases. Due to the targeted nature of this therapeutic approach to specific molecular features of translation or mRNA splicing, it is important that the particular disease-causing mutation carried by each patient is known. This makes the broad utilization of this approach dependent upon the implementation of the "personal genomics" revolution. Already, many patients with genetic diseases such as cystic fibrosis are routinely screened for common mutations in the *CFTR* gene using commercially available kits. Similar efforts by patient groups for other genetic diseases will need to reach this level of sophistication before they can take advantage of these advances. It will also be necessary to determine the consequences of each mutation on mRNA stability, protein expression, and protein function. Ultimately, these therapies have the potential to be uniquely tailored to a particular patient or family. Notably, the compound PTC124 is already in phase II clinical trials for the treatment of cystic fibrosis and Duchenne muscular dystrophy patients with premature termination codons. If this compound is found to be safe and effective in these clinical trials, therapeutic recoding strategies could become commonly used to treat individuals with these genetic diseases within a few years.

References

Aartsma-Rus A, Van Deutekom JC, Fokkema IF, Van Ommen GJ, Den Dunnen JT (2006) Muscle Nerve 34:135–144

Aartsma-Rus A, van Ommen GJ (2007) RNA 13:1609–1624

Aguiari G, Banzi M, Gessi S, Cai Y, Zeggio E, Manzati E, Piva R, Lambertini E, Ferrari L, Peters DJ, et al (2004) FASEB J 18:884–886

Alkalaeva EZ, Pisarev AV, Frolova LY, Kisselev LL, Pestova TV (2006) Cell 125:1125–1136

Allamand V, Bidou L, Arakawa M, Floquet C, Shiozuka M, Paturneau-Jouas M, Gartioux C, Butler-Browne GS, Mouly V, Rousset JP, et al (2008) J Gene Med 10:217–224

Amrani N, Ganesan R, Kervestin S, Mangus DA, Ghosh S, Jacobson A (2004) Nature 432:112–118

Arakawa M, Nakayama Y, Hara T, Shiozuka M, Takeda S, Kaga K, Kondo S, Morita S, Kitamura T, Matsuda R (2001) Acta Myologica 20:154–158

Arakawa M, Shiozuka M, Nakayama Y, Hara T, Hamada M., Kondo S., Ikeda D., Takahashi Y., Sawa R., Nonomura Y., et al (2003) J Biochem Tokyo 134:751–758

Auld DS, Thorne N, Maguire WF, Inglese J (2009) Proc Natl Acad Sci USA 106:3585–3590

Bartal C, Danon A, Schlaeffer F., Reisenberg K, Alkan M, Smoliakov R, Sidi A, Almog Y (2003) Am J Med 114:194–198

Barton-Davis ER, Cordier L., Shoturma DI, Leland SE, Sweeney HL (1999) J Clin Invest 104: 375–381

Beauchamp D, Labrecque G (2001) Curr Opin Crit Care 7:401–408

Beauchamp D, Laurent G, Maldague P, Abid S, Kishore BK, Tulkens PM (1990) J Pharmacol Exp Ther 255:858–866

Bedwell DM, Kaenjak A, Benos DJ, Bebok Z., Bubien JK, Hong J., Tousson A., Clancy JP, Sorscher EJ (1997). Nat Med 3:1280–1284

Bertram G, Innes S, Minella O, Richardson J, Stansfield I (2001) Microbiology 147:255–269

Bidou L, Hatin I, Perez N, Allamand V, Panthier JJ, Rousset JP (2004) Gene Ther 11:619–627

Bonetti B., Fu L., Moon J., Bedwell DM (1995) J Mol Biol 251:334–345

Bossi L, Roth JR (1980) Nature 286:123–127

Brown CM, Stockwell PA, Trotman CN, Tate WP (1990) Nucleic Acids Res 18:6339–6345

Burke JF, Mogg AE (1985) Nucleic Acids Res 13:6265–6272

Buvoli M, Buvoli A, Leinwand LA (2000) Mol Cell Biol 20:3116–3124

Carter AP, Clemons WM, Brodersen DE, Morgan-Warren RJ, Wimberly BT, Ramakrishnan V (2000) Nature 407:340–348

Cassan M, Rousset JP (2001) BMC Mol. Biol. 2:3

Clancy JP, Bebok Z, Ruiz F, King C, Jones J, Walker L, Greer H, Hong J, Wing L, Macaluso M, et al (2001) Am J Respir Crit Care Med 163:1683–1692

Clancy JP, Rowe SM, Bebok Z, Aitken ML, Gibson R, Zeitlin P, Berclaz P, Moss R, Knowles MR, Oster RA, et al (2007) Am J Respir Cell Mol Biol 37:57–66

Denti MA, Rosa A, D′Antona G, Sthandier O, De Angelis FG, Nicoletti C, Allocca M, Pansarasa O, Parente V, Musaro A, et al (2006) Proc Natl Acad Sci USA 103: 3758–3763

Deubner R, Schollmayer C, Wienen F, Holzgrabe U (2003) Magn Reson Chem 41:589–598

Dominski Z, Kole R (1993) Proc Natl Acad Sci USA 90:8673–8677

Du M, Jones JR, Lanier J, Keeling KM, Lindsey JR, Tousson A, Bebok Z, Whitsett JA, Dey CR, Colledge WH, et al (2002) J Mol Med 80:595–604

Du M, Keeling KM, Fan L, Liu X, Kovacs T, Sorscher E, Bedwell DM (2006) J Mol Med 84: 573–582

Du M, Liu X, Welch EM, Hirawat S, Peltz SW, Bedwell DM (2008) Proc Natl Acad Sci USA 105:2064–2069

England SB, Nicholson LV, Johnson MA, Forrest SM, Love DR, Zubrzycka-Gaarn EE, Bulman DE, Harris JB, Davies KE (1990) Nature 343:180–182

Fourmy D, Recht MI, Blanchard SC, Puglisi JD (1996) Science 274:1367–1371

Fourmy D, Recht MI, Puglisi JD (1998a) J Mol Biol 277:347–362

Fourmy D, Yoshizawa S, Puglisi JD (1998b) J Mol Biol 277:333–345

Friedman KJ, Kole J, Cohn JA, Knowles MR, Silverman LM, Kole R (1999) J Biol Chem 274:36193–36199

Frolova L, Le Goff X, Zhouravleva G, Davydova E, Philippe M, Kisselev L (1996) RNA 2:334–341

Frolova LY, Tsivkovskii RY, Sivolobova GF, Oparina NY, Serpinsky OI, Blinov VM, Tatkov SI, Kisselev LL (1999) RNA 5:1014–1020

Gilbert DN, Wood CA, Kohlhepp SJ, KohnenPW, Houghton DC, Finkbeiner HC, Lindsley J, Bennett WM (1989) J Infect Dis 159:945–953

Goto M, Sawamura D, Ito K, Abe M, Nishie W, Sakai K, Shibaki A, Akiyama M, Shimizu H (2006) J Invest Dermatol 126:766–772

Grayson C, Chapple JP, Willison KR, Webster AR, Hardcastle AJ, Cheetham ME (2002) J Med Genet 39:62–67

Hein LK, Bawden M, Muller VJ, Sillence D, Hopwood JJ, Brooks DA (2004) J Mol Biol 338: 453–462

Helip-Wooley A, Park MA, Lemons RM, Thoene JG (2002) Mol Genet Metab 75:128–133

Henderson CM, Anderson CB, Howard MT (2006) Nucleic Acids Res 34: 4302–4310

Hirawat S, Welch EM, Elfring GL, Northcutt VJ, Paushkin S, Hwang S, Leonard EM, Almstead NG, Ju W, Peltz SW, Miller LL (2007) J Clin Pharmacol 47:430–444

Howard M, Frizzell RA, Bedwell DM (1996) Nat Med 2:467–469

Howard MT, Anderson CB, Fass U, Khatri S, Gesteland RF, Atkins JF, Flanigan KM (2004) Ann Neurol 55:422–426

Howard MT, Shirts BH, Petros LM, Flanigan KM, Gesteland RF, Atkins JF (2000) Ann Neurol 48:164–169

Isken O, Maquat LE (2008) Nat Rev Genet 9:699–712

Ivanov IP, Loughran G, Atkins JF (2008) Proc Natl Acad Sci USA 105:10079–10084

Jacobs GH, Stockwell PA, Tate WP, Brown CM (2006) Nucleic Acids Res 34:D37–40

James PD, Raut S, Rivard GE, Poon MC, Warner M, McKenna S, Leggo J, Lillicrap D (2005) Blood 106:3043–3048

Karras JG, Crosby JR, Guha M, Tung D, Miller DA, Gaarde WA, Geary RS, Monia BP, Gregory SA (2007) Am J Respir Cell Mol Biol 36:276–285

Karras JG, Maier MA, Lu T, Watt A, Manoharan M (2001) Biochemistry 40:7853–7859

Kawamoto K, Sha SH, Minoda R, Izumikawa M, Kuriyama H., Schacht J., Raphael Y (2004) Mol Ther 9:173–181
Keeling KM, Bedwell DM (2002) J Mol Med 80:367–376
Keeling KM, Bedwell DM (2005) Current Pharmacogenomics 3:259–269
Keeling KM, Brooks DA, Hopwood JJ, Li P, Thompson JN, Bedwell DM (2001) Hum Mol Genet 10:291–299
Kellermayer R, Szigeti R, Keeling KM, Bedekovics T, Bedwell DM (2006) J Invest Dermatol 126:229–231
Kerem E, Hirawat S, Armoni S, Yaakov Y, Shoseyov D, Cohen M, Nissim-Rafinia M, Blau H, Rivlin J, Aviram M, et al (2008) Lancet 372:719–727
Khoo B, Roca X, Chew SL, Krainer AR (2007) BMC Mol Biol 8:3
Kiselev AV, Ostapenko OV, Rogozhkina EV, Kholod NS, Seit Nebi AS, Baranov AN, Lesina EA, Ivashchenko TE, Sabetskii VA, Shavlovskii MM, et al (2002) Mol Biol Mosk 36:43–47
Kisselev L, Ehrenberg M, Frolova L (2003) EMBO J 22:175–182
Kobayashi T, Funakoshi Y, Hoshino S, Katada T (2004) J Biol Chem 279:45693–45700
Lai CH, Chun HH, Nahas SA, Mitui M, Gamo KM, Du L, Gatti RA (2004) Proc Natl Acad Sci USA 101:15676–15681
Linde L, Boelz S, Nissim-Rafinia M, Oren YS, Wilschanski M, Yaacov Y, Virgilis D, Neu-Yilik G, Kulozik AE, Kerem E, Kerem B (2007) J Clin Invest 117:683–692
Loftfield RB, Vanderjagt D (1972) Biochem J 128:1353–1356
Loufrani L, Dubroca C, You D, Li Z, Levy B, Paulin D, Henrion D (2004) Arterioscler. Thromb Vasc Biol 24:671–676
Lu QL, Mann CJ, Lou F, Bou-Gharios G, Morris GE, Xue SA, Fletcher S, Partridge TA, Wilton SD (2003) Nat Med 9:1009–1014
Lynch SR, Puglisi JD (2001a) J Mol Biol 306:1037–1058
Lynch SR, Puglisi JD (2001b) J Mol Biol 306:1023–1035
Madsen EC, Morcos PA, Mendelsohn BA, Gitlin JD (2008) Proc Natl Acad Sci USA 105: 3909–3914
Major LL, Edgar TD, Yee Yip P, Isaksson LA, Tate WP (2002) FEBS Lett 514:84–89
Mann CJ, Honeyman K, Cheng AJ, Ly T, Lloyd F, Fletcher S, Morgan JE, Partridge TA, Wilton SD (2001) Proc Natl Acad Sci USA 98:42–47
Manuvakhova M, Keeling K, Bedwell DM (2000) RNA 6:1044–1055
Matsumura K, Campbell KP (1994) Muscle Nerve 17:2–15
Mattis VB, Rai R, Wang J, Chang CW, Coady T, Lorson CL (2006) Hum Genet 120:589–601
Mazzon E, Britti D, De Sarro A, Caputi AP, Cuzzocrea S (2001) Eur J Pharmacol 424:75–83
McCaughan KK, Brown CM, Dalphin ME, Berry MJ, Tate WP (1995) Proc Natl Acad Sci USA 92:5431–5435
Mercatante DR, Kole R (2002) Biochim Biophys Acta 1587:126–132
Mercatante DR, Sazani P, Kole R (2001) Curr Cancer Drug Targets 1:211–230
Mingeot-Leclercq MP, Glupczynski Y, Tulkens PM (1999) Antimicrob Agents Chemother 43: 727–737
Mirabella M, Galluzzi G, Manfredi G, Bertini E, Ricci E, De Leo R, Tonali P, Servidei S (1998) Neurology 51:592–595
Mitkevich VA, Kononenko AV, Petrushanko IY, Yanvarev DV, Makarov AA, Kisselev LL (2006) Nucleic Acids Res 34:3947–3954
Moazed D, Noller HF (1986) Cell 47:985–994
Mori N, Funatsu Y, Hiruta K, Goto S (1985) Biochemistry 24:1231–1239
Muntoni F, Wells D (2007) Curr Opin Neurol 20:590–594
Nagai J, Takano M (2004) Drug Metab Pharmacokinet 19:159–170
Nakashima T, Teranishi M, Hibi T, Kobayashi M, Umemura M (2000) Acta Otolaryngol 120: 904–911
Namy O, Hatin I, Rousset JP (2001) EMBO Rep 2:787–793
Nudelman I, Rebibo-Sabbah A, Shallom-Shezifi D, Hainrichson M, Stahl I, Ben-Yosef T, Baasov T (2006) Bioorg Med Chem Lett 16:6310–6315

Olson TM, Alekseev AE, Liu XK, Park S., Zingman LV, Bienengraeber M, Sattiraju S, Ballew JD, Jahangir A, Terzic A (2006) Hum Mol Genet 15:2185–2191
Olsthoorn RC, Laurs M, Sohet F, Hilbers CW, Heus HA, Pleij CW (2004) RNA 10:1702–1703
Palmer E, Wilhelm JM, Sherman F (1979) Nature 277:148–150
Pinotti M, Rizzotto L, Pinton P, Ferraresi P, Chuansumrit A, Charoenkwan P, Marchetti G, Rizzuto R, Mariani G, Bernardi F (2006) J Thromb Haemost 4:1308–1314
Politano L, Nigro G, Nigro V, Piluso G, Papparella S, Paciello O, Comi LI (2003) Acta Myol 22:15–21
Purohit P, Stern S (1994) Nature 370:659–662
Rebibo-Sabbah A, Nudelman I, Ahmed ZM, Baasov T, Ben-Yosef T (2007) Hum Genet 122: 373–381
Recht MI, Douthwaite S, Puglisi JD (1999) EMBO J 18:3133–3138
Recht MI, Fourmy D, Blanchard SC, Dahlquist KD, Puglisi JD (1996) J Mol Biol 262:421–436
Renshaw J, Orr RM, Walton MI, Te Poele R, Williams RD, Wancewicz EV, Monia BP, Workman P, Pritchard-Jones K (2004) Mol Cancer Ther 3:1467–1484
Roberts RG, Coffey AJ, Bobrow M, Bentley DR (1993) Genomics 16:536–538
Sako Y, Usuki F, Suga H (2006) Nucleic Acids Symp Ser (Oxf.):239–240
Salas-Marco J, Bedwell DM (2004) Mol Cell Biol 24:7769–7778
Salas-Marco J, Bedwell DM (2005) J Mol Biol 348:801–815
Sandoval RM, Reilly JP, Running W, Campos SB, Santos JR, Phillips CL, Molitoris BA (2006) J Am Soc Nephrol 17:2697–2705
Sangkuhl K, Schulz A, Rompler H, Yun J, Wess J, Schoneberg T (2004) Hum Mol Genet 13: 893–903
Scaffidi P, Misteli T (2005) Nat Med 11:440–445
Schiffelers R, Storm G, Bakker-Woudenberg I (2001) J Antimicrob Chemother 48:333–344
Schroeder SJ, Blaha G, Moore PB (2007) Antimicrob Agents Chemother 51:4462–4465
Schroers A, Kley RA, Stachon A, Horvath R, Lochmuller H, Zange J, Vorgerd M (2006) Neurology 66:285–286
Seit-Nebi A, Frolova L, Justesen J, Kisselev L (2001) Nucleic Acids Res 29:3982–3987
Sener G, Sehirli AO, Altunbas HZ, Ersoy Y, Paskaloglu K, Arbak S, Ayanoglu-Dulger G (2002) J Pineal Res 32:231–236
Sierakowska H, Sambade MJ, Agrawal S, Kole R (1996) Proc Natl Acad Sci USA 93:12840–12844
Singh A, Ursic D, Davies J (1979) Nature 277:146–148
Singh G, Rebbapragada I, Lykke-Andersen J (2008) PLoS Biol 6: e111
Sleat DE, Sohar I, Gin RM, Lobel P (2001) Eur J Paediatr Neurol 5(Suppl A):57–62
Song H, Mugnier P, Das AK, Webb HM, Evans DR, Tuite MF, Hemmings BA, Barford D (2000) Cell 100:311–321
Sossi V, Giuli A, Vitali T, Tiziano F, Mirabella M, Antonelli A, Neri G, Brahe C (2001) Eur J Hum Genet 9:113–120
Stansfield I, Jones KM, Herbert P, Lewendon A, Shaw WV, Tuite MF (1998) J Mol Biol 282: 13–24
Tate WP, Poole ES, Horsfield JA, Mannering SA, Brown CM, Moffat JG, Dalphin ME, McCaughan KK, Major LL, Wilson DN (1995) Biochem Cell Biol 73:1095–1103
Temple GF, Dozy AM, Roy KL, Kan YW (1982) Nature 296:537–540
Thibault N, Grenier L, Simard M, Bergeron MG, Beauchamp D (1994) Antimicrob Agents Chemother 38:1027–1035
Thibault N, Grenier L, Simard M, Bergeron MG, Beauchamp D (1995) Life Sci 56: 1877–1887
Uehara Y, Hori M, Umezawa H (1976) Biochim Biophys Acta 442:251–262
Uehara Y, Kondo S, Umezawa H, Suzukake K, Hori M (1972) J Antibiot (Tokyo) 25:685–688
van Deutekom JC, Janson AA, Ginjaar IB, Frankhuizen WS, Aartsma-Rus A, Bremmer-Bout M, den Dunnen JT, Koop K, van der Kooi AJ, Goemans NM, et al. (2007) N Engl J Med 357: 2677–2686
Vicens Q, Westhof E (2001) Structure (Camb) 9:647–658

Vickers TA, Zhang H, Graham MJ, Lemonidis KM, Zhao C, Dean NM (2006) J Immunol 176:3652–3661
Wagner KR, Hamed S, Hadley DW, Gropman AL, Burstein AH, Escolar DM, Hoffman EP, Fischbeck KH (2001) Ann Neurol 49:706–711
Welch EM, Barton ER, Zhuo J, Tomizawa Y, Friesen WJ, Trifillis P, Paushkin S, Patel M, Trotta CR, Hwang S, et al. (2007) Nature 447:87–91
Williams T, Kole R (2006) Oligonucleotides 16:186–195
Wilschanski M, Famini C, Blau H, Rivlin J, AugartenA, Avital A, Kerem B, Kerem E (2000) Am J Respir Crit Care Med 161:860–865
Wilschanski M, Yahav Y, Yaacov Y, Blau H, Bentur L, Rivlin J, Aviram M, Bdolah-Abram T, Bebok Z, Shushi L, et al. (2003) N Engl J Med 349:1433–1441
Woolstencroft RN, Beilharz TH, Cook MA, Preiss T, Durocher D, Tyers M (2006) J Cell Sci 119:5178–5192
Xiong YQ, Kupferwasser LI, Zack PM, Bayer AS (1999) Antimicrob Agents Chemother 43: 1737–1742
Yang C, Feng J, Song W, Wang J, Tsai B, Zhang Y, Scaringe WA, Hill KA, Margaritis P, High KA, Sommer SS (2007) Proc Natl Acad Sci USA 104:15394–15399
Yoshizawa S, Fourmy D, Puglisi JD (1998) EMBO J 17:6437–6448
Zhouravleva G, Frolova L, Le Goff X, Le Guellec R, Inge-Vechtomov S, Kisselev L, Philippe M (1995) EMBO J 14:4065–4072
Zsembery A, Jessner W, Sitter G, Spirli C, Strazzabosco M, Graf J (2002) Hepatol 35:95–104

Part II
Frameshifting – Redirection of Linear Readout

Chapter 7
Pseudoknot-Dependent Programmed −1 Ribosomal Frameshifting: Structures, Mechanisms and Models

Ian Brierley, Robert J.C. Gilbert, and Simon Pennell

Abstract Programmed −1 ribosomal frameshifting is a translational recoding strategy that takes place during the elongation phase of protein biosynthesis. Frameshifting occurs in response to specific signals in the mRNA; a slippery sequence, where the ribosome changes frame, and a stimulatory RNA secondary structure, usually a pseudoknot, located immediately downstream. During the frameshift the ribosome slips backwards by a single nucleotide (in the 5′-wards/−1 direction) and continues translation in the new, overlapping reading frame, generating a fusion protein composed of the products of both the original and the −1 frame coding regions. In eukaryotes, frameshifting is largely a phenomenon of virus gene expression and associated predominantly with the expression of viral replicases. Research on frameshifting impacts upon diverse topics, including the ribosomal elongation cycle, RNA structure and function, tRNA modification, virus replication, antiviral intervention, evolution and bioinformatics. This chapter focuses on the structure and function of frameshift-stimulatory RNA pseudoknots and mechanistic aspects of ribosomal frameshifting. A variety of models of the frameshifting process are discussed in the light of recent advances in our understanding of ribosome structure and the elongation cycle.

Contents

I. Brierley (✉)
Division of Virology, Department of Pathology, University of Cambridge, Tennis Court Road, Cambridge CB2 1QP, UK
e-mail: ib103@mole.bio.cam.ac.uk

J.F. Atkins, R.F. Gesteland (eds.), *Recoding: Expansion of Decoding Rules Enriches Gene Expression*, Nucleic Acids and Molecular Biology 24,
DOI 10.1007/978-0-387-89382-2_7,

7.1 Introduction

The discovery of programmed −1 ribosomal frameshifting (hereafter frameshifting for brevity) is intimately entwined with studies of retrovirus gene expression. In the early 1980s, as the first complete retrovirus genomic sequences began to appear, it was clear that in many cases the open reading frames (ORFs) encoding the structural and enzymatic components of the virion, *gag* and *pol*, respectively, overlapped at the 3′ end of *gag* with *pol* in the −1 reading frame (Fig. 7.1). As most retroviruses were known to express Pol solely as a C-terminal extension of Gag, it was thought that RNA processing would produce an in-frame spliced mRNA capable of expressing the Gag-Pol polyprotein. However, in papers describing the complete sequence of Rous sarcoma virus (RSV; Schwartz et al., 1983) and a bovine leukemia virus provirus (Rice et al., 1985) an alternative suggestion was put forward that Gag and Pol could be coupled at the translational level, through a frameshifting mechanism. Shortly afterwards, Tyler Jacks and Harold Varmus published their landmark paper confirming expression of the RSV Gag-Pol polyprotein by frameshifting (Jacks and Varmus, 1985). Since this time, frameshift signals have been described in viruses of all kingdoms and in a growing number of cellular genes. Research on frameshifting impacts upon diverse topics, including the ribosomal elongation cycle, RNA structure and function, tRNA modification, virus replication, antiviral intervention, evolution and bioinformatics. This chapter will focus on the mRNA structures involved in eukaryotic frameshifting and the mechanistic insights that have been gained over the last 20 years. Subsequent chapters in this book address specifically the human immunodeficiency virus (HIV) frameshifting signal, and also frameshifting in prokaryotes (including bacteriophages and insertion sequence elements), yeast and plants and so will not be discussed in detail here. Interested readers may also like to look at other reviews on this fascinating topic (Farabaugh, 1996;

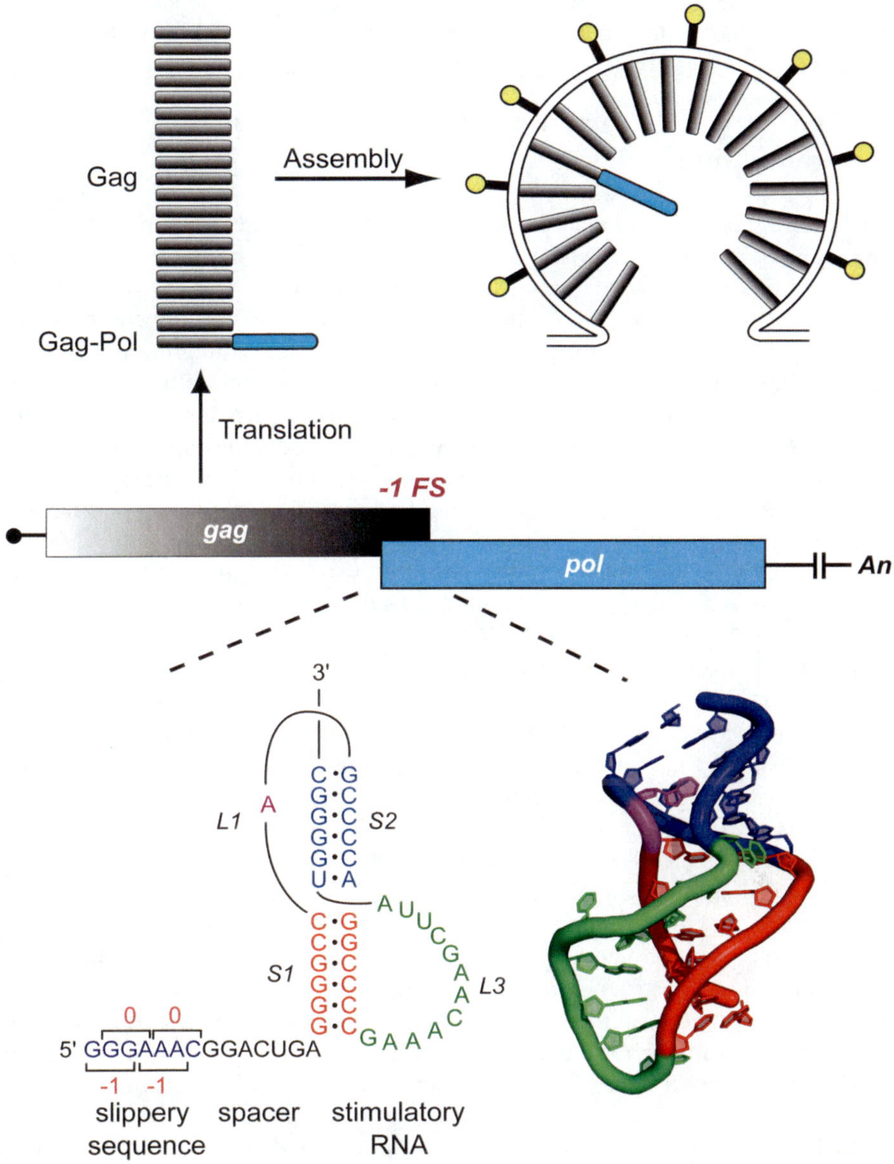

Fig. 7.1 Programmed −1 ribosomal frameshifting. In the centre, a schematic of a retrovirus genomic mRNA is shown with the 5′-proximal *gag* and overlapping *pol* coding sequences boxed. Translation of this mRNA (*upper panel*) yields mostly the Gag polyprotein, but occasional −1 frameshifting (-1FS) generates the Gag-Pol polyprotein. Assembly at the plasma membrane (*right*) includes the Pol domain in virions. The viral envelope glycoproteins are indicated as yellow and black spikes. Frameshifting (*lower panel*) is mediated by a signal composed of a slippery sequence (*purple*), a spacer region (*black*) and a downstream stimulatory RNA. The *gag* (0) and *pol* (−1) reading frames are indicated. This example shows the SRV-1 *gag/pro* stimulatory pseudoknot as a secondary structure (*left*) and as solved by NMR (*right*, Michiels et al., 2001; image prepared using PyMOL [www.pymol.org])

2000; Giedroc et al., 2000; Brierley and Pennell, 2001; Brierley and Dos Ramos, 2006; Dreher and Miller, 2006; Baranov et al., 2006; Giedroc and Cornish, 2008).

7.2 The Nature, Occurrence and Role of Ribosomal Frameshifting

Frameshifting is a translational recoding strategy that takes place during the elongation phase of protein biosynthesis. In response to signals in the mRNA, and at a certain frequency, ribosomes switch from the translation of one ORF (in the 0-frame) to an overlapping ORF (in the −1 reading frame) and continue translation, generating, in the majority of cases*, a fusion protein containing the products of both the upstream and the downstream ORFs (Fig. 7.1) (*frameshifting in the *dnaX* gene of *Escherichia coli* redirects ribosomes to a nearby termination codon in the −1 frame; in this case, and probably in some cellular examples, the frameshifting ribosomes produce a shorter product than that encoded by the 0-frame (reviewed in Chapter of Fayet and Prère, and see below]. Frameshifting in eukaryotes is largely a phenomenon of virus gene expression and associated predominantly with the expression of viral replicases. Thus in retroviruses, frameshifting at the overlap of *gag* and *pol* permits expression of the viral reverse transcriptase (as a component of the Gag-Pol polyprotein) and in other RNA viruses, frameshifting mediates expression of RNA-dependent RNA polymerases (Brierley et al., 1995; Dreher and Miller et al., 2006). As the expression of distal ORFs on viral mRNAs can be achieved in numerous ways (Pe'ery and Mathews, 2000), it is pertinent to question why a frameshift strategy has evolved. One explanation likely to be of general relevance is that frameshifting allows the generation of a defined ratio of protein products. In HIV-1, for example, modulation of the cytoplasmic Gag:Gag-Pol ratio is detrimental to replication (Shehu-Xhilaga et al., 2001; reviewed in Brierley and Dos Ramos, 2006) and the same is true for the retrovirus-like dsRNA virus of yeast, L-A (Dinman and Wickner, 1992). Similarly, in other RNA viruses, changing the stoichiometry of non-frameshifted and frameshifted products is also likely to be disadvantageous. Thus frameshifting has emerged as a potential target for antiviral therapeutics (Dinman et al., 1998). The frameshifting process also generates a fusion protein which itself may be biologically relevant, for example, in the incorporation of retroviral reverse transcriptase into virions (Fig. 7.1).

Examples of frameshifting in eukaryotic cellular genes have begun to emerge in the last few years. The signals present in the mouse embryonal carcinoma differentiation regulated (Edr, now known as PEG10) gene (Shigemoto et al., 2001; Manktelow et al., 2005) and the human paraneoplastic Ma3 gene (Wills et al., 2006) are clearly of retroviral origin, perhaps subsumed into the cellular genome from an endogenous retrovirus or retroviral relic. However, many candidates in the yeast genome that were identified by a bioinformatics approach appear to be genuinely unrelated to retroviruses, standing alone as the first true examples of novel cellular −1 ribosomal frameshifting signals (Jacobs et al., 2007). In contrast to viral frameshifting, the vast majority of these, and those predicted in other genomes

(Belew et al., 2008), direct the ribosome to a premature termination codon and may have a biological role in destabilising the mRNA (Plant et al., 2004).

7.3 The Structure of Ribosomal Frameshifting Signals

Eukaryotic ribosomal frameshift signals are composed of two essential elements: a "slippery" sequence, where the ribosome changes reading frame, and a stimulatory RNA secondary structure, often an RNA pseudoknot, located a few nucleotides (nt) downstream (Jacks et al., 1988; Brierley et al., 1989; ten Dam et al., 1990; see Fig. 7.1). A spacer region between the slippery sequence and the stimulatory RNA is also required and the precise length of this spacer must be maintained for maximal frameshifting efficiency (Brierley et al., 1989, 1992; Kollmus et al., 1994). The slippery sequence is a heptanucleotide stretch that contains two homopolymeric triplets and conforms in the vast majority of cases to the motif XXXYYYZ (for example, UUUAAAC in the coronavirus infectious bronchitis virus (IBV) *1a/1b* signal). Frameshifting at this sequence has been proposed to occur by "simultaneous-slippage" of the ribosome-bound peptidyl- and aminoacyl-tRNAs, which detach from the zero frame codons (XXY YYZ) and re-pair in the −1 phase (XXX YYY) (Jacks et al., 1988). The homopolymeric nature of the slippery sequence seems to be required to allow the tRNAs to remain base-paired to the mRNA in at least two out of three anticodon positions following the slip. Frameshift assays, largely carried out in vitro, have revealed that XXX can be represented by any homopolymeric triplet (A, C, G or U), but the Y triplet must be AAA or UUU (Jacks et al., 1988, Dinman et al., 1991; Brierley et al., 1992). In addition to these restrictions, slippery sequences ending in G (XXXAAAG or XXXUUUG) do not function efficiently in in vitro translation systems (Brierley et al., 1992), yeast (Dinman et al., 1991) or mammalian cells (Marczinke et al., 2000). At naturally occurring frameshift sites, of the possible codons which are decoded in the ribosomal A-site prior to tRNA slippage (XXXY<u>YYZ</u>), only five are represented in eukaryotes, AAC, AAU, UUA, UUC, UUU. Together with the in vitro data, it seems that the sequence restrictions observed are a manifestation of the need for the pre-slippage codon–anticodon complex in the A-site to be weak enough such that the tRNAs can detach from the codon during the process of frameshifting. G-C pairs are therefore avoided. Recent work suggests that the slippery sequence actually forms part of a somewhat larger motif, as certain bases immediately 5′ of the heptanucleotide stretch can influence the efficiency of frameshifting (Bekaert and Rousset, 2005; Léger et al., 2007). While this has not been examined systematically, these studies suggest an involvement of the ribosomal E-site in frameshifting (see below and the Chapter by Pech and colleagues).

7.4 Stimulatory RNA Structures

Slippery sequences alone can direct frameshifting (typically to 0.1–0.2% in mammalian cells; e.g. Marczinke et al., 2000; Dulude et al., 2002) to levels significantly higher than the estimated natural frameshift error rate (10^{-5} per codon for *E. coli* ribosomes; Kurland, 1992), indicating that these stretches are intrinsically frameshift-prone. However, to obtain the levels of frameshifting needed for their biological function (1–30% depending on the particular system), a downstream stimulatory RNA structure is also required (see Table 7.1). In eukaryotes, some stem-loop stimulatory RNAs have been described, such as those present at the *1a/1b* frameshift site of astroviruses (Marczinke et al. 1994) and the HIV-1 *gag/pol* overlap (discussed in more detail in the Chapter by Brakier-Gingras), but most commonly an H (hairpin)-type RNA pseudoknot is present (Brierley et al., 1989; ten Dam et al., 1990). These form when a single-stranded region in the loop of a hairpin base pairs with a stretch of complementary nucleotides elsewhere in the RNA chain (reviewed in Brierley et al., 2007; see Fig. 7.1). Such pseudoknots have two base-paired stem regions (S1 and S2) and, depending upon the number of loop bases that participate in the pseudoknotting interaction, two or three single-stranded loops (L1, L2 and L3). In almost all frameshift-promoting pseudoknots, L2 is absent (Table 7.1, Fig. 7.1) or very short (Fig. 7.2) and the base-paired stems stack to form a quasi-continuous helix. In these structures, L1 spans S2 and crosses the deep groove of the helix, whereas L3 spans S1 and crosses the shallow groove. While the precise folding of and, to some extent, the stability of pseudoknots has been shown to be critical to the frameshift process, few specific primary sequence requirements have been identified. This suggests that they are involved in few, if any, sequence-specific interactions, as evidenced by the failure to identify protein binding partners (see below). Where particular sequence requirements have been suspected, it may be that the bases concerned are involved in unanticipated tertiary interactions. The analysis of pseudoknot structure and function has involved a range of experimental approaches, including in vitro and in vivo structure–function assays, biophysical investigation of folding/unfolding and atomic resolution structure determination. Pseudoknots have been classified on the basis of secondary structural features (Brierley and Pennell, 2001) or by virus group (Giedroc and Cornish, 2008), but as our knowledge is incomplete, this is mostly for convenience of description. For the purposes of this review, examples of specific pseudoknot "classes" will be presented as they were described historically and linked, where possible, to related examples described more recently.

7.4.1 The IBV Pseudoknot and Relatives

A role for pseudoknots in frameshifting was first described for the coronavirus infectious bronchitis virus (IBV) *1a/1b* signal (Brierley et al., 1989; see Fig. 7.2). This pseudoknot is characterised by the possession of a long, stable S1, a coaxially stacked S2 and a capacity to direct high levels of frameshifting, up to ~60% in the rabbit reticulocyte lysate (RRL) in vitro translation system (Napthine et al., 1999). Although no atomic resolution structural model is available, several features important for function have been identified. The most striking of these is the importance of maintaining S1 length. Napthine and colleagues (1999) prepared a series of

Table 7.1 Details of established –1 ribosomal frameshift signals

Origin	System	Slippery site	DS	SP	S1	L1	S2	L2/ISE	L3	Comments
Stem-loop stimulatory RNA										
HAst-V1 orf1a/orf1b	Astrovirus	AAAAAAC	SL	6	7	10				A short stimulatory RNA
GLV p100/p190	Totivirus	CCCUUUA	SL	5	9	11				Rare slippery sequence
HTLV-I gag/pro	Deltaretrovirus	AAAAAAC	SL	7	10	3				HTLV-II gag/pro signal is almost identical
CfMV orf2a/orf2b	Sobemovirus	UUUAAAC	SL	7	11	4				Flanking sequences may stimulate frameshifting (~twofold)
SIV gag/pol	Lentivirus	UUUUUUA	SL	4 (9)*	16 (11)*	12				Single helix with ordered loop, NMR
HIV-1 gag/pol	Lentivirus	UUUUUUA	SL	0 (8)	19 (11)*	4				Two-stem helix with a purine-rich bulge, NMR
Pseudoknot stimulator with short S1										
PLRV orf2a/orf2b	Polerovirus	UUUAAAU	PK	5	4	2	3	1	9	PK like that of BWYV, X-ray
BWYV orf2/orf3	Polerovirus	GGGAAAC	PK	6	5	2	3	1	7	Compact, loop-helix, non-Watson-Crick, 1st X-ray
ScYLV P1/P2	Polerovirus	GGGAAAC	PK	6	5	2	3	1	9	PK like that of BWYV, NMR
PEMV P1/P2	Polerovirus	GGGAAAC	PK	7	5	2	3	1	8	PK like that of BWYV, NMR
EIAV gag/pol	Lentivirus	AAAAAAC	PK	9	5	4	5	0	9	Long SP, L3 may interact with S1
MMTV gag/pro	Betaretrovirus	AAAAAAC	PK	7	5	1	7	1	8	Kinked PK, 1st NMR
FIV gag/pol	Lentivirus	GGGAAAC	PK	8	5	2	5	1	12	From MS3D, likely to be kinked like MMTV PK
SRV-1 gag/pro	Betaretrovirus	GGGAAAC	PK	7	6	1	6	0	12	Coaxially stacked, no kink, NMR
IAP gag/pro	Endogenous RV	AAAAAAC	PK	8	6	1	8	0	9	PK not well characterised
HERV K-10 gag/pro	Endogenous RV	GGGAAAC	PK	7	6	1	6	0	9	NMR, L3 may interact with S1
VMV gag/pol	Lentivirus	GGGAAAC	PK	6	7	5	7	7	14	Contains unstructured ISE (cf. RSV, GAV)
Pseudoknot stimulator with long S1										
PEG10 (Edr) RF1/RF2	Cellular gene	GGGAAAC	PK	5	10	3	9	0	9	1st non-viral example of –1 frameshifting
Ma-3 orf1/orf2	Cellular gene	GGGAAAC	PK	6	11	3	5	1	10	Non-viral example identified by bioinformatics

Table 7.1 (continued)

Origin	System	Slippery site	DS	SP	S1	L1	S2	L2/ISE	L3	Comments
IBV orf1a/1b	Coronavirus	UUUAAAC	PK	6	11	2	6	0	32	First identified frameshift-promoting PK
MHV orf1a/1b	Coronavirus	UUUAAAC	PK	5	11	2	11	0	30	Long S2 predicted
SARS-CoV orf1a/1b	Coronavirus	UUUAAAC	PK	5	11	3	7	0	29	Substructure in L3
BEV orf1a/1b	Arterivirus	UUUAAAC	PK	5	11	4	5	0	69	Most likely PK is shown
EAV orf1a/1b	Arterivirus	GUUAAAC	PK	6	11	2	6	1	69	Unusual slippery sequence
HCV orf1a/1b	Coronavirus	UUUAAAC	PK	5	12	3	5	0	164	S2 formed by kissing interaction
L-A gag/pol	Totivirus	GGGUUUA	PK	5	13	8	6	2	11	Yeast dsRNA virus signal used as model system
HTLV-I pro/pol	Deltaretrovirus	UUUAAAC	PK	6	13	6	10	0	20	Girnary & Brierley, unpublished
TGEV orf1a/1b	Coronavirus	UUUAAAC	PK	3	14	8	5	3	163	S2 formed as for HCV, precise SP/S1 length unknown
RSV gag/pol	Alpharetrovirus	AAAUUUA	PK	1 (6)	14 (9)	11	8	53	17	Contains structured ISE
GAV orf1a/orf1b	Nidovirales	AAAUUUU	PK	8	22	3	7	64	41	A very large PK with substantial S1 and ISE
Pseudoknot stimulator with long–range interaction										
BYDV39K/60 K	Luteovirus	GGGUUUU	PK	5	8	57	6	0	3828	Unusual kissing PK, very long L3, see Fig. 7.2
RCNMV p27/p57	Dianthovirus	GGAUUUU	PK	5	12	46	6	0	2591	Similar to BYDV
Pseudoknot stimulator unclassified										
HIV gag/pol (group O)	Lentivirus	UUUUUUA	PK	8	8	2	8	0	20	PK features reminiscent of a viral PK that promotes stop codon suppression (Wills et al., 1991)

The table lists those established frameshift signals where details of the stimulatory RNA are available. Abbreviations used are DS, downstream stimulator; SP, spacer region (length in nt); S1, stem 1 (bp; unpaired residues within stems are not included in the total length); L1, loop 1 (nt); S2, stem 2 (bp); L2/ISE, loop 2/interstem element (nt); L3, loop 3 (nt); SL, stem-loop structure; PK, pseudoknot; NMR or X-ray, nuclear magnetic resonance or X-ray structure available; MS3D, three-dimensional mass spectrometric structural analysis; RV, retrovirus.

*Some or all of the SP may be incorporated into S1 (bracketed values show length of un-paired SP/S1).

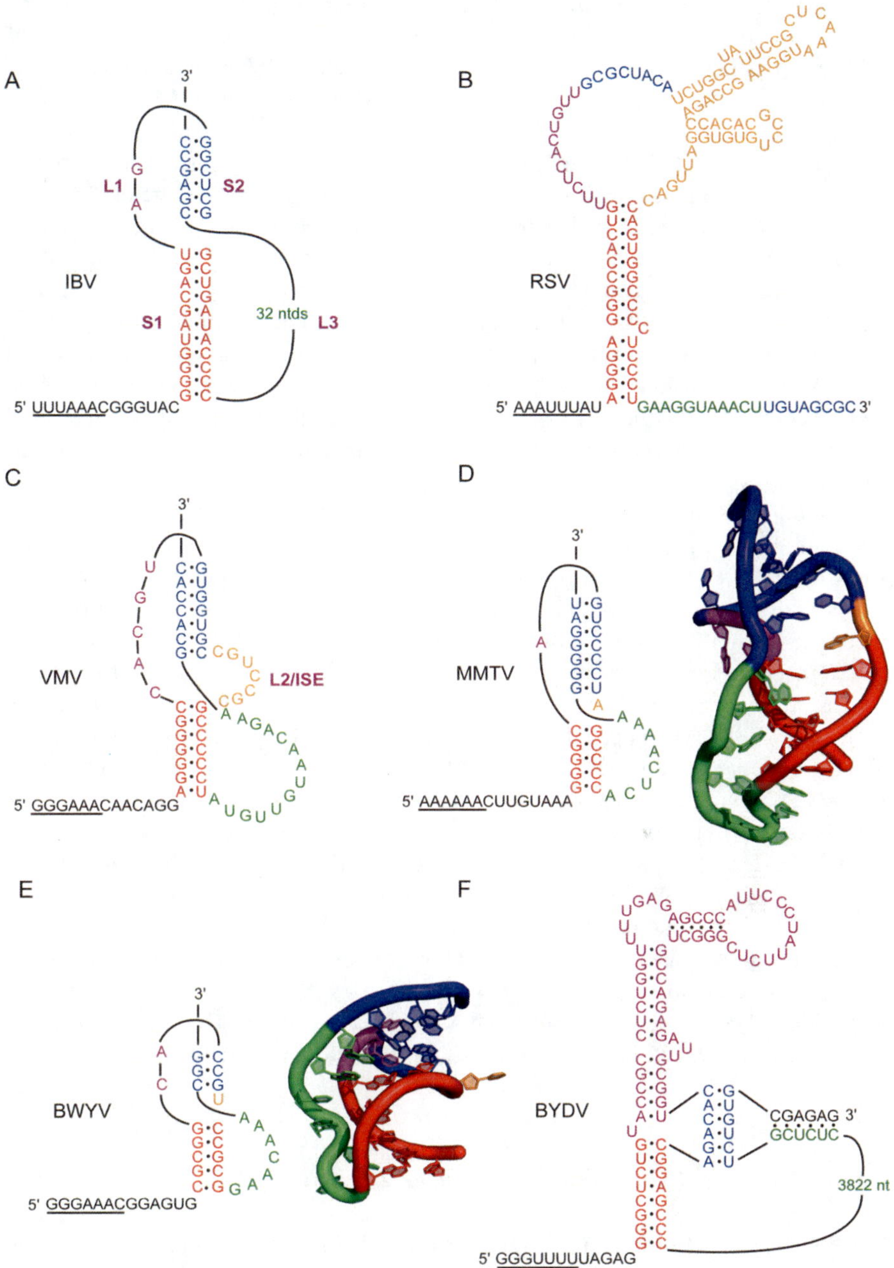

Fig. 7.2 Examples of RNA pseudoknot structures found at viral frameshift sites. Secondary structure models of a variety of pseudoknot stimulatory RNAs are shown, alongside three-dimensional models where available (S1, *red*; S2, *blue*; L1, *purple*; L2, *orange*; L3 *green*; details in Table 7.1). (**A**) Infectious bronchitis virus (IBV), (**B**) Rous sarcoma virus (RSV), (**C**) Visna-Maedi virus (VMV), (**D**) mouse mammary tumour virus *gag/pro* (MMTV), (**E**) beet western yellows virus (BWYV) and (**F**) barley yellow dwarf virus (BYDV). The slippery sequences are underlined. For simplicity of drawing, S2 of the RSV pseudoknot is not shown in base-paired representation

IBV-based pseudoknots in which this stem was varied from 4 to 13 base pairs (bp). Remarkably, frameshifting occurred very inefficiently until the wild-type length of 11 bp was reached, after which full function was retained. This phenomenon was unrelated to the thermodynamic stability of the stem, as the added bases were G-C pairs and the shorter stems of 7–10 bp were as stable as or more stable than the wild-type 11 bp stem. Whether there is any significance to the fact that 11 bp corresponds to a complete turn of an A-form RNA helix remains to be ascertained. A similar puzzle is the clear benefit of a G-rich stretch at the beginning of the 5′ arm of S1. We assume that these features are important in terms of how the pseudoknot is recognised and unwound by the ribosome during the frameshift. The 10 and 11 bp variants have been studied in mechanical unfolding experiments but have produced somewhat contradictory data; in one study the 11 bp construct required twofold greater unfolding force (Hansen et al., 2007) but in another, there was no obvious correlation between frameshift efficiency and the mechanical force required to unwind the pseudoknots (Green et al., 2008). The latter study revealed, however, that pseudoknots have unusual mechanical properties (discussed below). L3 of the IBV pseudoknot can be reduced in length to eight bases (from 32), close to the minimum number required to span S1 (Pleij et al., 1985), with no loss of function (Brierley et al., 1991). A lack of obvious primary sequence requirements within L3 argues against the formation of a L3–S1 interaction as is seen in some other pseudoknots (see below). The frameshift signal of the severe acquired respiratory syndrome (SARS) coronavirus is very similar to that of IBV although L3 is mostly folded into a stem-loop, which has alternatively been named SL1 or S3 (Baranov et al., 2005; Plant et al., 2005). Unexpectedly, although most of this loop can be deleted without effect, maintenance of an appropriate conformation of the wild-type L3 seems to be required for optimal frameshift efficiency (Baranov et al., 2005; Plant et al., 2005). Pseudoknots containing an SL1 motif can be predicted for all group 2 coronaviruses (Plant et al., 2005). In group 3 coronaviruses (e.g. IBV), however, the region equivalent to SL1 is probably a single-stranded loop (Brierley et al., 1989, 1991; Plant et al., 2005). The pseudoknots present at the *1a/1b* overlap of group 1 coronaviruses (e.g. human coronavirus 229E) appear to form a more "elaborated" pseudoknot that can be viewed as "kissing" stem-loops separated by a long (~150 nt) L3 (Herold and Siddell, 1993). The stimulatory RNA is still a pseudoknot but in this case, the downstream sequence that pairs with the loop is itself constrained within a hairpin.

7.4.2 The RSV Pseudoknot and Interstem Elements

The RSV *gag/pol* frameshifting signal has the distinction of being the first example of both −1 frameshifting and pseudoknot-dependent frameshifting (Jacks et al., 1988), although the pseudoknot was confirmed some years after the original report (Marczinke et al., 1998). The RSV stimulatory RNA is one of the most complex found to date and belongs to a family of frameshift-promoting pseudoknots that possess a L2 and thus have an element at the junction of the two stems (an interstem element, or ISE; Fig. 7.2). L2 of the RSV pseudoknot is 70 nt in length most of which is constrained within a substructure of two adjacent stem-loops (Fig. 7.2B). How the ISE folds within the pseudoknot is unknown. A similarly extensive ISE is predicted in the pseudoknot of gill-associated virus (GAV), a prawn-infecting member of the order *Nidovirales* (which includes coronaviruses; Cowley et al., 2000). The GAV pseudoknot is interesting both from the perspective of its sheer size (a total of ~170 nt, of which L1 and L3 contribute only 3 and 41 nt, respectively) and its resemblance to the RSV signal, with a huge S1 (22 bp), an extensive ISE and the use of an RSV-like slippery sequence (AAAUUUU; Table 7.1). The role of these large ISEs in frameshifting, if any, remains to be clarified, but they could add to pseudoknot stability or present additional challenges to a ribosome-associated helicase. In principle, they could also offer a binding site for translation components or regulatory proteins of viral (or cellular) origin.

The coaxial stacking of stems to generate a quasi-continuous helix has long been considered important for pseudoknot function (ten Dam et al., 1992) and the paucity of L2-containing pseudoknots is consistent with the hypothesis. However, recent structural studies (see below) indicate that the key issue is more likely to be the maintenance of an appropriate architecture at the helical junction. For the RSV and GAV stimulatory RNAs, we predict therefore that the ISEs do not compromise this important region of the pseudoknot. In support of this, we recently characterised a shorter, unstructured ISE (7 nt) in the pseudoknot of the lentivirus Visna-Maedi virus (VMV) (Pennell et al., 2008). The ISE of the VMV pseudoknot proved to be essential for frameshifting and is therefore a key component of the folded, active pseudoknot (Fig. 7.2C).

7.4.3 The "Kinked" Pseudoknots

The first frameshift-promoting pseudoknot to be described in three dimensions was derived from the mouse mammary tumour virus (MMTV) *gag/pro* overlap (Shen and Tinoco 1995; Fig. 7.2D). This pseudoknot is characterised by the presence of an unpaired, intercalated adenosine residue between two short stems of similar length. The presence of this adenosine (A14) forces the stems to bend relative to each other and this "kink" is thought to be important for function (Chen et al., 1996, Kang et al., 1996). However, the general importance of the kink was challenged with the release of the NMR structure of the closely related pseudoknot from simian retrovirus 1 (SRV-1) *gag/pro* (Michiels et al., 2001). This pseudoknot was predicted from its primary sequence to have an organisation similar to that of MMTV and was presumed to have a kink. However, it was shown that the stems were in fact coaxially stacked, with a base predicted to be in L1 being present at the stem–stem junction and base pairing to the "intercalated" adenosine (Fig. 7.1). As detailed further below, this observation adds to the general belief that for most, if not all, pseudoknots, the specific interactions and resultant architectures of the helical junction that are required for frameshifting are strongly context dependent (Cornish et al., 2006).

7.4.4 The Luteoviral Pseudoknots and Loop–Helix Interactions

Luteovirus frameshift-promoting pseudoknots are small, typically encompassing only ~25 nt, with an especially short S2 of three base pairs. Their small size and unexpected stability have made them amenable to structural studies and numerous three-dimensional models are available, including the only frameshift pseudoknot structures solved by X-ray crystallography (Fig. 7.2E). The structure of the pseudoknot of beet western yellows virus (BWYV) reveals a tightly folded motif with extensive non-Watson–Crick base pairing, explaining how these structures have a greater stability than would be predicted from their secondary structure (Su et al., 1999). The main stabilising feature is the presence of an extended triplex formed between the minor groove of S1 and L3, mediated predominantly by A-minor interactions. In addition, the helical junctions of these pseudoknots are bounded by base triple and base quadruple interactions which stabilise the fold and are important for frameshifting (Kim et al., 1999). Several luteoviral pseudoknot structures have been solved, including pea enation mosaic virus (Nixon et al., 2002), potato leaf roll virus (Pallan et al., 2005) and sugar cane yellow leaf virus (ScYLV; Cornish et al., 2005) and all are similar structures. The structures of mutants of the ScYLV pseudoknot have been especially informative (Cornish et al., 2006) revealing that, at least in this class of pseudoknot, the global "ground-state" structure is not strongly correlated with frameshifting efficiency. Giedroc and Cornish (2008) propose that the helical junction is mechanically stable and functions as a kinetic barrier to force-induced unfolding. Frameshift efficiency would therefore be affected by different kinetics of unfolding, diminishing in those pseudoknots that unfold at lower applied forces. Luteoviral frameshift-promoting pseudoknots (and other plant stimulatory RNAs) are discussed in more detail elsewhere in this volume (Chapter of Miller and Giedroc).

7.4.5 Long-Range Pseudoknots

Perhaps the most remarkable example of a non-canonical frameshift pseudoknot comes from the recoding signal of barley yellow dwarf virus (BYDV) (Paul et al., 2001). Initially thought to be a stem-loop stimulator, the BYDV signal was later found to be a "kissing loop" pseudoknot, with the sequence required for S2 formation located some 4 kb downstream (Fig. 7.2F). In addition to being the longest L3 sequence identified to date, the position of the interaction near the 3′ end of the genome directly links the control of translation to genomic replication (Barry and Miller 2002). This topic is discussed further in the Chapter by Miller and Giedroc.

7.5 Mechanistic Aspects of Ribosomal Frameshifting

The precise mechanism of programmed −1 ribosomal frameshifting remains unclear. On the surface it seems relatively straightforward; the ribosome encounters the stimulatory RNA structure while in the act of decoding the slippery sequence, something perturbs normal frame maintenance and a proportion of the ribosomes enter the −1 frame. However, the details are entwined with the natural movements of tRNAs, ribosomal subunits, ribosomal proteins and elongation factors, the dynamics of which we have only recently begun to understand at the molecular level. The frameshift process is most often considered within the framework of the simultaneous tRNA slippage model, specifically how the movement of tRNAs into the −1 frame occurs and how this can be influenced by the stimulatory RNA. Within this context, a role of the stimulatory RNA in ribosomal pausing or in the recruitment of *trans*-acting components has also been considered, along with the specific RNA structural features that could promote these events. In the following sections, these general themes are debated prior to an examination of the different models of frameshifting that have been proposed.

7.5.1 tRNA Slippage

In the ribosomal elongation cycle, tRNAs occupy a series of specific positions in the intersubunit space, known as the aminoacyl- (A), peptidyl- (P) and exit- (E) sites. Translocation of tRNAs between these sites, while still attached to the mRNA due to codon–anticodon recognition, combined with the close fit of the tRNAs to the intersubunit space, is the basis of ribosomal processivity along the mRNA and the maintenance of a single reading frame (Moazed and Noller, 1989; Rodnina et al., 1997). During a −1 frameshift, the anticodons of the tRNAs (or tRNA) must detach from the mRNA and re-associate in the −1 frame, and this raises several issues for the design of models. For example, at which stage in the ribosomal elongation cycle does the frameshift occur? What are the parameters involved in promoting dissociation of the tRNAs? Do the tRNAs truly slip simultaneously or does it occur sequentially? In the re-binding phase, do the anticodons sample the −1, zero or (even) the +1 frame and re-pair in the energetically most favourable phase, or are the tRNAs directed specifically to the −1 frame? Another relevant issue is how the mechanisms of frame maintenance inherent to the ribosome are overcome. In the discussion of putative models for −1 frameshifting detailed below, these issues will be addressed where possible. Many of these questions are also relevant to other recoding events and are debated extensively in other chapters of this book.

7.5.2 Ribosomal Pausing

Ribosomal pausing has been central to models of −1 frameshifting (and indeed, most recoding phenomena) where it may act to increase the time that ribosomes engage with the slippery sequence, promoting tRNA movements that would normally be unfavourable kinetically (Farabaugh, 2000). Polypeptide intermediates corresponding to ribosomes paused at RNA pseudoknots have been detected at the frameshift sites of RSV (unpublished data cited in Jacks et al., 1988), IBV (Somogyi et al., 1993) and L-A (Lopinski et al., 2000) and footprinting studies of elongating ribosomes have defined the site of pausing at the L-A (Tu et al., 1992; Lopinski et al., 2000), IBV, SRV-1 and pKA-A signals (Kontos et al., 2001). Pausing at the frameshift signal in the *E. coli dnaX* gene has also been witnessed, again from examination of translational intermediates (Tsuchihashi, 1991). Consistent with a role for pausing in frameshifting, the footprinting studies reveal that the ribosome is stalled over the slippery sequence while in contact with, and perhaps having unwound partially, the pseudoknot (Tu et al., 1992; Kontos et al., 2001). One of the virtues of the pausing model is that it can accommodate the variety of stimulatory RNAs present at −1 frameshifting signals. Notwithstanding the range of secondary and tertiary features presented to the ribosome, as long as pausing occurs, frameshifting results. Unfortunately, the idea that a pause alone is sufficient to induce frameshifting is highly questionable. Simple provision of a roadblock to ribosomes in the form of stable RNA hairpins (Brierley et al., 1991; Somogyi et al., 1993), a tRNA (Chen et al., 1995) or even different kinds of RNA pseudoknot (Napthine et al., 1999; Liphardt et al., 1999) is not sufficient to bring about frameshifting. Furthermore, non-frameshifting pseudoknots and stem-loops exist that can still pause ribosomes (Tu et al., 1992; Somogyi et al., 1993; Lopinski et al., 2000; Kontos et al., 2001). These experiments, of course, do not rule out a contribution of pausing to the mechanism of frameshifting since there are no documented examples where frameshifting has occurred in the absence of a detectable pause. Indeed, one cannot ignore the possibility that a precise "kinetic pause" is required for frameshifting (Bidou et al., 1997), which only certain stimulatory RNAs can generate. For example, during a −1 frameshift, two pauses could occur, one productive (in terms of frameshifting) upon initial encounter of the stimulatory RNA structure, and a second, non-productive pause, corresponding to a delay in unwinding of the structure *after* the crucial event in frameshifting has taken place. The magnitude of the initial pause would perhaps influence the extent of the frameshift, whereas the second pause, occurring during the time that the ribosomal "unwinding activity" (see below) deals with the secondary structure, would be irrelevant. The pausing assays employed currently cannot distinguish between two such pausing events and a detailed analysis of the kinetics of pausing will require further experimentation, including the analysis of translational elongation at the level of individual ribosomes (Wen et al., 2008).

7.5.3 The Stimulatory RNA Resists Unwinding by the Ribosome

As suggested several years ago (Brierley et al., 1991; Wills et al., 1991), the frameshift mechanism is likely to be linked to an unwinding activity of the ribosome, with the stimulatory RNA showing resistance to unwinding, perhaps by presenting an unusual topology (Yusupov et al., 2001; Yusupova et al., 2001; Plant et al., 2003, 2005; Takyar et al., 2005; Namy et al., 2006). Indeed, such resistance is likely to be responsible for the observed ribosomal pause. An examination of the path of the mRNA through the ribosome reveals a narrowing of the mRNA entry tunnel that would block entry of a folded mRNA (Yusupova et al., 2001). Recently, Takyar and colleagues (2005) demonstrated that the prokaryotic 70S ribosome can itself act as a helicase to unwind mRNA secondary structures before decoding, with the active

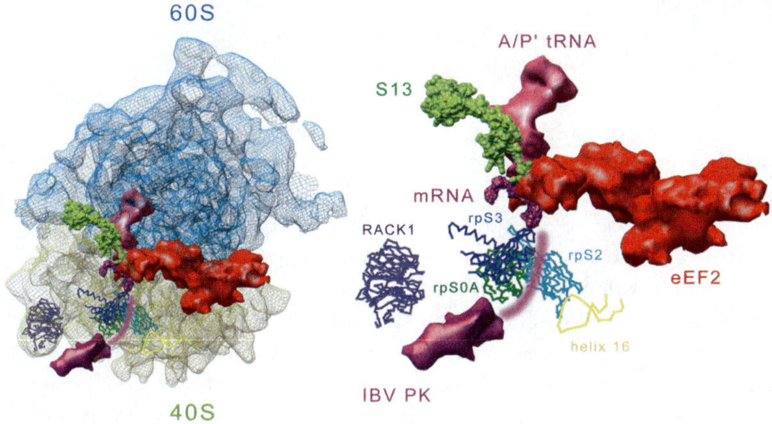

Fig. 7.3 Cryo-EM visualisation of 80S ribosomes stalled at the IBV pseudoknot. On the left is shown a representation of the 3D reconstruction of a pseudoknot-stalled rabbit 80S ribosome, with the large subunit coloured *blue* and the small *yellow*, bound with tRNA, mRNA and eEF2. On the right is shown an expanded view of these ligands, represented by either cryo-EM density or fitted atomic models. eEF2 is found between the subunits and coloured *red*, and the pseudoknot at the entry channel in the solvent face of the small subunit *purple*. The tRNA is distorted spring-like in an A/P' hybrid state and is shown in *pink*. The mRNA with a 45° kink found in the atomic structures for 70S prokaryotic ribosomes is shown in *purple*, linked by a sketched line in the same colour to the pseudoknot. The atomic structure of S13 (prokaryotic equivalent of rpS18) is shown in *light green*, its C-terminus interposed between the A- and P-sites. Also shown are the structures of RACK1, a ribosome-regulatory kinase, rpS0A and the helicase activity-associated proteins rpS3 (S3 in prokaryotes), and rpS2 (S5 in prokaryotes). The structure of the third helicase activity-associated protein rpS9 (S4 in prokaryotes) is not shown since it has not been identified as of yet in a eukaryotic reconstruction. However, it is known to be bound to 18S rRNA helix 16, which is shown. This image demonstrates how the pseudoknot is engaged with the proteins responsible for the helicase activity of the small subunit solvent face (Takyar et al., 2005)

site located between the head and the shoulder of the 30S subunit. Prokaryotic ribosomal proteins S3, S4 and S5 that line the mRNA entry tunnel are implicated in the helicase activity (Takyar et al., 2005), with the eukaryotic 80S counterparts rpS3, rpS9 and rpS2 thus likely to form important elements of an equivalent helicase. The cryo-EM structure of a stalled rabbit 80S ribosome–IBV pseudoknot complex (Namy et al., 2006; see below and Fig. 7.3) has features consistent with an interaction of this proposed helicase with the stimulatory pseudoknot. The specific structural features of the stimulatory RNAs that may affect helicase activity have not been determined. An examination of the confirmed and predicted stimulatory RNAs of Table 7.1 indicates that two basic pseudoknot groupings exist based on S1 length. In the first group, S1 is short (7 bp or less) and stabilised by interactions with a relatively short L3 ($\leq$14 nt). The conformation of the helical junction is crucial in these pseudoknots and is stabilised by non-Watson–Crick interactions. In the second group, S1 is longer ($\sim$11 bp or more), the length and nucleotide composition of L3 seem to be less important to function and there is little evidence for loop–helix interactions (although the lack of high-resolution structures limits such comparisons). Given these differences, it is unlikely that a single shared feature is responsible for function – more likely, each pseudoknot class has features that promote resistance to unwinding. For example, in those pseudoknots with an S1-L3 RNA triplex, the presence of the "third strand" may conceivably confound unwinding. The triplex may also provide unusual mechanical resistance to unwinding, as could a stable helical junction. In those pseudoknots with a longer S1, the length of the stem, in conjunction with the anchoring stem 2, may lead to torsional resistance as S1 begins to unwind (Plant and Dinman, 2005). Recently, mechanical unwinding studies have revealed that a functional IBV-based pseudoknot is a "brittle" structure, with a shallow dependence of the unfolding rate on applied force and a slower unfolding rate than component hairpin structures (Green et al., 2008). This greater mechanical stability and kinetic insensitivity to force is consistent with a role in resistance to unwinding. A caveat that must be mentioned here is that some viral stimulatory RNAs are stem-loops and have no obvious features that would suggest a specific capacity to resist helicase unwinding. In addition, it has been shown that simply annealing an RNA oligonucleotide at a suitable point downstream of a slippery sequence can in some circumstances promote efficient frameshifting, at least in vitro (Howard et al., 2004; Olsthoorn et al., 2004).

7.5.4 A Role for trans-Acting Protein Factors?

At present there is no evidence that frameshift-stimulatory RNAs are recognised in a specific functional manner by any cellular or viral proteins. This is surprising given the enormous number of known RNA-binding proteins (including pseudoknot-binding proteins) and the potential advantages that could accrue from a capacity to regulate frameshifting. Clearly, the extent to which researchers in the field have searched for frameshift-regulatory factors is hard to judge, given that a negative outcome of such research would likely remain unpublished. However, from our own perspective, we have been unable to identify specific binding partners of the IBV and MMTV pseudoknots in standard RNA-binding assays and in three-hybrid screens (Brierley, I. and colleagues, unpublished), and similar observations have been reported in at least one other laboratory (Dinman, J.D., unpublished). While this topic would benefit from a more extensive and rigorous analysis, the current information leads to the conclusion that the stimulatory RNA acts alone and is not targeted for regulatory purposes.

7.5.5 Conceivable Points for Frameshifting During the Elongation Cycle

The elongation cycle begins when a ternary complex of eEF1A-tRNA-GTP associates with a ribosome containing a P-site peptidyl-tRNA and an empty A-site. The newly delivered cognate tRNA is accommodated into the A-site and almost immediately, the peptidyl transfer reaction takes place, leaving a deacylated tRNA in the

P-site and a peptidyl-tRNA in the A-site (the pre-translocational, or PRE state). Subsequently, the acceptor ends of the A- and P-site tRNAs move, probably spontaneously, with respect to the large ribosomal subunit, but the anticodon ends remain in their original positions relative to the small ribosomal subunit, to yield hybrid state tRNA intermediates P/E and A/P. Complete translocation of the mRNA–tRNA complex into the E- and P-sites is catalyzed by a monomeric G protein, elongation factor 2 (eEF2), with associated GTP hydrolysis. Release of the eEF2–GDP complex and E-site tRNA results in the formation of the post-translocational complex (POST; peptidyl-tRNA in the P-site, empty A-site), competent for the next round of elongation. It is well established that the integrity of both homopolymeric triplets of the slippery sequence (X XXY YYZ) are required for efficient frameshifting, and this has led to the long-held view that the XXY and YYZ codons are present in the ribosomal P- and A-sites (respectively) prior to the frameshift (Jacks et al., 1988). Recent work also suggests the involvement of the 3 nt upstream of the slippery sequence, indicating that the key element is longer (denoted A BCX XXY YYZ) and occupies the E-site as well. There is currently a debate about whether the E-site tRNA anticodon forms an authentic complex with the mRNA but at present, the balance of evidence favours the view that such an interaction is likely. When considering tRNA slippage, it is also pertinent to ask how the tRNAs transit from A- to P- and P- to E-sites, as structural features have been noted that appear to define the boundary between the sites and may act as gates (Yusupov et al., 2001; Selmer et al., 2006; Giedroc and Cornish, 2008). For example, there is a ∼45° kink in the mRNA between A- and P-sites (Yusupov et al., 2001), stabilised by a critically placed magnesium ion (Selmer et al., 2006) and additionally, the C-terminus of S13 (eukaryotic homologue is S18) can also form a barrier between the two sites (Selmer et al., 2006; Moran et al., 2008). During tRNA transit, it is also relevant to question whether this movement is most likely to occur when the ribosome is dynamic, for example in the so-called "unlocked" conformation (Valle et al., 2003). Within the elongation cycle, frameshifting could conceivably take place at one of five steps (at each step, a specific model that has been proposed is given in brackets): (1) during accommodation of the A-site tRNA (the 9Å model), (2) subsequent to accommodation, but prior to peptidyl transfer (the simultaneous slippage model), (3) while tRNA hybrid-state intermediates are present (the dynamic model), (4) during the transition from hybrid state to POST state (the mechanical model) and (5) at the start of the next round of elongation (the three-tRNA model). Each has its merits and failings and aspects of each model are shared by others (Fig. 7.4).

7.6 Models of Frameshifting

7.6.1 The Integrated and 9Å Models

That frameshifting could occur during accommodation of the A-site tRNA has its origins in the "integrated" model of ribosomal frameshifting (Harger et al., 2002). Here, the authors argued convincingly, on the basis of the biochemical,

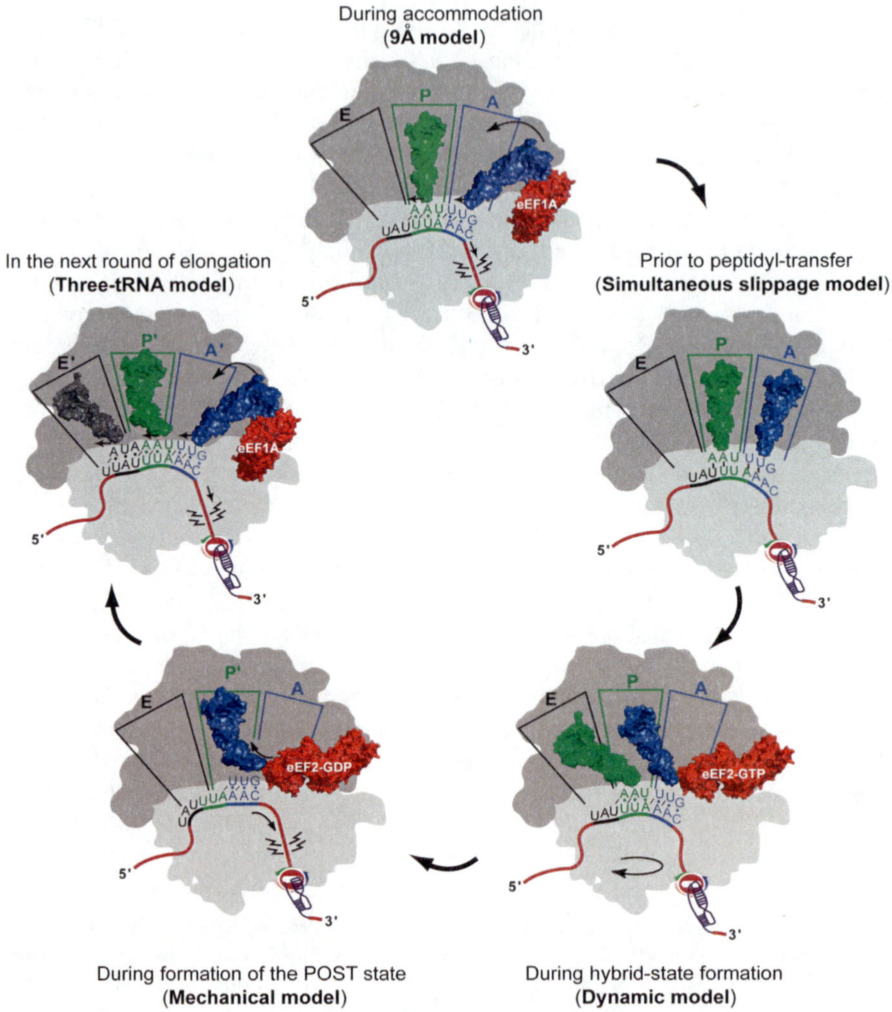

Fig. 7.4 Models of −1 ribosomal frameshifting. Cartoons of five models of frameshifting are shown, each occurring at a different stage of the elongation cycle. The 60S and 40S ribosomal subunits are in *dark grey* and *light grey*, the mRNA in *red*, the A-, P- and E-sites, codons and corresponding tRNAs in *blue*, *green* and *black*. The pseudoknot is in *purple*. The 40S helicase is shown as a *three-arrowed wheel* whose inhibition by the pseudoknot is indicated by a superimposed "no-entry" symbol. Initial codon–anticodon base pairs are shown as dots, post-slippage pairs as lines. All models show the pre-slippage pairing except for the simultaneous slippage model. For those models where mRNA tension has been explicitly implicated, this is indicated as a taut mRNA between helicase and tRNA anticodon, highlighted by jagged lines. The curved arrow in the 40S subunit of the dynamic model signifies the ratchet-like movements (see text). Short arrows close to the anticodons or mRNA show directions of movement or force

pharmacological and genetic data available at the time, that the −1 frameshift must occur prior to peptidyl transfer. The evidence was manifold and included the observation that several eEF1A, but no eEF2, mutants had been described that could modulate −1 frameshifting (Dinman and Kinzy, 1997), that anti-ribosomal antibiotics like sparsomycin that inhibit peptidyl transfer should thus increase the residence time of ribosomes paused over the slippery sequence stimulated −1 frameshifting (Dinman et al., 1997) and that mutant ribosomal proteins that result in slower rates of peptidyl transfer also increased −1 frameshifting (Meskauskas et al., 2003). Within the context of the integrated model, the 9Å model (Plant et al., 2003) explores a purely mechanistic theme and has its basis in modelling data (Noller et al., 2002) which suggests a 9Å movement of the anticodon of the A-site tRNA upon accommodation. Plant and colleagues (2003) proposed that resistance to unwinding of the stimulatory RNA would hold the mRNA in place such that the movement of the anticodon (and thus the associated mRNA) during accommodation would introduce local "tension" into the mRNA that would be relieved by disruption of the codon–anticodon interaction and re-pairing in the −1 frame. While the 9Å model satisfies many experimental observations, others indicate that it is not completely correct. The 9Å movement of the anticodon predicted by Noller and colleagues (2002) remains unconfirmed and some eEF2 mutants that affect programmed frameshifting have now been described (Ortiz et al., 2006). Also, it is not clear how the "local" tension placed on the A-site tRNA is transmitted to the P-site tRNA, which is also thought to slip, nor how the documented effects of mutations within eEF1A on frameshifting could or would equate to different local "tensions".

7.6.2 The Simultaneous Slippage Model

This original model of frameshifting (Jacks et al., 1988) postulated movement of P- and A-site tRNAs soon after accommodation and prior to peptidyl transfer. The model's strength lies in the fact that it rationalises the role of both slippery sequence codons in the process, but it is not clear what would drive tRNA movement. Ribosomal pausing alone does not appear to be sufficient to promote tRNA realignment into alternative reading frames and there are no large-scale global movements of ribosomal subunits at this point in the elongation cycle. In addition, it is hard to see how the tRNA could passively shift over the mRNA kink and S13 (S18) between the A- and P-sites. Furthermore, there is little time for the shift to take place between accommodation and peptidyl transfer.

7.6.3 The Dynamic Model

Variants of the simultaneous slippage model have been proposed in which the tRNAs slip during the formation of hybrid-state intermediates on the ribosome, or during translocation itself (Weiss et al., 1989; Farabaugh, 1996). These models seek ways to harness energy for tRNA dissociation from the molecular movements of the ribosome that occur during the elongation cycle. Recently, these ideas have been revisited and discussed in the light of the recent progress in understanding of ribosome dynamics (Giedroc and Cornish, 2008). In this review, Giedroc and Cornish highlight that the ribosome is most dynamic during the movement of the tRNA acceptor ends into the hybrid state conformation, a state in which −1 frameshifting has been shown to be promoted provided the translocating tRNA is deacylated (McGarry et al., 2005). Further, addition of EF-G.GDPNP to ribosomal complexes stabilises the P/E hybrid state tRNA, allowing for destabilisation of the P-site codon–anticodon hydrogen bonding interactions. Thus frameshifting could be promoted by the binding of eEF2-GTP to the hybrid state ribosome (tRNAs in P/E, A/P state), stabilising the P/E state and promoting frameshifting.

7.6.4 The Mechanical Model

This model was developed following cryo-EM studies of a RRL 80S ribosomal complex stalled at the IBV pseudoknot ($80S_{PK}$) and provided the first visualisation (at 16.2 Å) of a ribosome stalled at a frameshift-promoting pseudoknot (Namy et al., 2006; Moran et al., 2008). When compared with a reconstruction of an empty ribosome ($80S_{Apo}$; 14Å), the $80S_{PK}$ complexes were found to contain additional densities corresponding to a tRNA (in a hybrid state, see below), the translocase eEF2 and the pseudoknot located at the entrance to the mRNA channel. Overall, these features reveal that encounter of the pseudoknot by the ribosome leads to stalling of the elongation cycle during the translocation step (Fig. 7.3). In support of this, eEF2 has not been released and thus the POST state has not been reached. In addition, the anticodon of the tRNA is not fully in the P-site and thus the tRNA is present in a hybrid state: we refer to this state as A/P' and consider it to be different from the "canonical" A/P state(s) that occur(s) spontaneously following the peptidyl transferase reaction (see above; Moazed and Noller, 1989). The A/P' state does, however, appear to be a necessary intermediate in passage of tRNA from the A to the P-site, given the obstacles on its path. A surprising feature of the A/P' tRNA is its location and structure, with the T-loop pushed towards the intersubunit face of the large subunit and the anticodon arm bent markedly towards the A-site, while retaining contact with domain IV of eEF2. On the basis of this bending, and the presence of the pseudoknot at the mRNA entry channel in close association with components of the putative 80S helicase, a mechanical model of frameshifting was proposed in which the pseudoknot resists unwinding during eEF2-mediated translocation such that tension builds up in the mRNA, subsequently placing strain on the tRNA and resulting in the adoption of a bent conformation (Namy et al., 2006). The opposing actions of translocation, catalysed by eEF2, and pulling from the mRNA strand account for the movement of the elbow of the tRNA into the roof of the P-site and the bending of the tRNA, spring-like, in a (+) sense (3′ direction). These opposing forces place a strain on the codon–anticodon interaction that promotes breakage. Subsequent relaxation of the bent tRNA structure in a (−) sense direction

(5′) provides an opportunity for the tRNA to re-pair with the mRNA in the −1 position. At the present resolution, it is uncertain whether the tRNA is still bound to the mRNA in these complexes, but the close association of domain IV of eEF2 with the anticodon loop of the tRNA supports this view. In these complexes, the slippery sequence was omitted in an attempt to minimise conformational heterogeneity in the ribosome population, but the consequence is that the tRNA does not have an option to repair in the −1 (or +1) phase. The sequence that replaced the slippery sequence was more GC rich and would stabilise mRNA–tRNA contacts and so prevent breakage. The complexes observed, however, are not "dead-end" products since in ribosomal pausing assays, almost all pseudoknot-stalled ribosomes continue translation after the delay (Somogyi et al., 1993; Lopinski et al., 2000; Kontos et al., 2001) and with this particular combination of pseudoknot and translation system, about 50% having entered the −1 reading frame (Brierley et al., 1992).

That frameshifting could take place during translocation has been suggested previously (Weiss et al., 1989; Yelverton et al., 1994; Horsfield et al., 1995; Farabaugh et al., 1996; Kontos et al., 2001). The mechanical model offers a mechanism for disruption of the slippery sequence codon–anticodon interactions which is similar, in some respects, to the 9Å model (although the coupling with translocation would likely provide more force). However, it is difficult to tease apart the features of the stalled $80S_{PK}$ complex that are specific to frameshifting from those specific to translocation, as the molecular details of both processes are incompletely understood. Undoubtedly, the movement of the anticodon of the tRNA towards the A-site is suggestive of pseudoknot-induced mRNA tension, but the compression that places the anticodon above the plane of the mRNA and the movement of the elbow towards the top of the P-site is most likely a result of eEF2 action. It has been suggested recently that the $80S_{PK}$ complex could in fact be a previously undescribed natural intermediate of translocation that pseudoknot-induced stalling has allowed visualisation of (Moran et al., 2008). In this model, as the hybrid state tRNA is translocated from the (authentic) A/P state to the P/P-site proper, eEF2 action compresses the tRNA like a spring, with the anticodon sliding first up the face of eEF2 domain IV then over the "gate" between A and P-sites before the spring relaxes into the P-site, guided by the face of domain IV. This "molecular spring" model of natural translocation is not inconsistent with the mechanical model of frameshifting, although the precise point at which the pseudoknot acts to disrupt the codon–anticodon contacts is not clear. One possibility is that the disruption occurs as the tRNA is fully compressed at the "top" of the gate. At this point, eEF2 would guide the tRNA into the P-site, but as the anticodon is not engaged with the mRNA, the tRNA may be misplaced into the −1 frame. Alternatively, the tRNA may be placed accurately, but the mRNA may have slipped +1 (in the 3′ direction, effectively a −1 frameshift). This would be due to the pulling action of the pseudoknot as tension is released following detachment of the anticodon, or as has also been suggested, if a partially unwound pseudoknot at the entry channel were to refold, dragging the mRNA in a 3′ direction (Farabaugh, 1997). At present, however, there are too many uncertainties to favour a particular mechanism. For example, our complexes lack an E-site tRNA, a state that has been linked to frameshifting (Marquez et al., 2004; Sanders et al.,

2008), but we do not know whether this is a genuine absence. It may reflect the use of cycloheximide to stabilise complexes or structural changes in the L1 stalk, either of which could influence E-site occupancy (Pestova and Hellen, 2003). Further, the E-site tRNA may simply have been lost during purification.

While the features of the PK-stalled complexes are consistent with frameshifting during the translocation step, they could in principle represent ribosomes stalled during translocation that have already experienced the molecular events that lead to frameshifting, or as discussed in the final model below, have yet to frameshift. Further work, including the analysis of complexes containing a slippery sequence, and ribosomes stalled at alternative stimulatory RNAs would be of value in addressing some of these issues. A key question to answer is whether the presence of bound eEF2 is causally linked to frameshifting. A reconstruction of a complex comprising a ribosome stalled at a related stem-loop structure ($80S_{SL}$; 15.7Å) with reduced activity in frameshifting revealed a POST state ribosome lacking eEF2 and with a single tRNA in an authentic P/P state (Namy et al., 2006). No obvious density at the mRNA entry channel that could correspond to the stem-loop was observed, however, perhaps as a result of conformational flexibility. While this stem-loop has reduced activity in frameshifting, it is not negligible (~5- to 10-fold less efficient than the pseudoknot in RRL; Kontos et al., 2001). It may be that a sub-population of stalled ribosomes active in frameshifting are present that contain bound eEF2 but have evaded detection as such a species would represent only ~5% of the total ribosome population.

7.6.5 The Three-tRNA Model

Recent work by Léger and colleagues (2007) has provided evidence that the codon preceding the canonical slippery sequence can influence frameshifting. Mutations (underlined) within an "extended" HIV-1 frameshift signal (A BCX XXY YYZ; in HIV-1 U AAU UUU UUA) led to modest increases or decreases in frameshifting efficiency and this was also true for variant slippery sequences derived from other viruses. To account for these findings, the authors propose that the ribosomal P and A-sites are in fact occupied by BCX and XXY during encounter of the HIV-1 stimulatory RNA, with the YYZ codon (known to be crucial to frameshifting) becoming involved in the following cycle of elongation. In natural frameshift signals, nucleotides in the A BCX positions are not noticeably homopolymeric and could not form a stable interaction with the mRNA codon post-slippage (in most cases, only a single Watson–Crick pair can form). It may be that there are relaxed requirements for re-pairing in the E-site (if re-pairing occurs at all) and that the mutations act by changing the identity of tRNA (which in turn may also affect the capacity of the P-site tRNA to frameshift). The involvement of this extended frameshift signal led the authors to suggest a "three-tRNA" model of frameshifting which could also reconcile the conflicting arguments for frameshifting during translocation (see above) or accommodation (Farabaugh, 1997; Léger et al., 2004, Plant et al., 2005). In this model, the translocation of P- and A-site tRNAs decoding the first two triplets of the slippery sequence (BCX XXY) is blocked with the tRNAs in a position between the PRE and POST states (based on the intermediates identified in cryo-EM [Namy et al., 2006; see above] and kinetic studies [Pan et al., 2007]). Upon release of eEF2, the elongation cycle is completed, but the tRNAs remain out-of-register in transition intermediate sites (referred to as E*/E*, P*/P* by Léger and colleagues). Subsequently, the next tRNA is brought to the ribosome as a ternary complex with eEF1A and during accommodation, all ribosome-associated tRNAs frameshift −1 such that they occupy the standard binding sites (E/E, P/P, A/A). The authors suggest that the tRNAs would slip sequentially, first E-site then P-site then A-site (as discussed by Baranov et al., 2004) and that the driving force for the frameshift would likely be derived from the more stable association of the tRNAs with their authentic sites on the ribosome.

A key question that emerges from the three-tRNA model is whether the elongation cycle can complete while translocation is in an intermediate state. Under normal circumstances, the answer would be no, but in the presence of a stimulatory RNA inducing a strained conformation, it remains possible and might, for example, involve the occupation of another intermediate/hybrid state by the A-to-P translocating tRNA, distinct from canonical A/P and our A/P – which one might call A/P''. More discussion of this intriguing model can be found in the chapter by Brakier-Gingras later in this book.

7.7 Perspective

In the last 20 years, significant progress has been made in the field: the number of confirmed frameshift signals has expanded, examples from cellular genes have been identified and we have a better understanding of frameshift site organisation and structure as they will influence ribosome function. These studies have made important general contributions to our understanding of protein synthesis, translational regulation, RNA structure and function and virus gene expression. However, in general, the biology of frameshifting has lagged behind structural studies such that the role of frameshifting in many viruses (and cellular genes) remains incompletely understood. Nevertheless, frameshifting offers a potential target for antiviral intervention and this will help biological understanding. Also, bioinformatics analyses will be invaluable in assessing just how extensively frameshifting is used in cellular gene expression, and whether there are likely to be side effects associated with drugs targeting the frameshift process. Meanwhile, mechanistic studies of frameshifting will continue to benefit from high-resolution structural and biophysical analysis of stimulatory RNAs and ribosomal complexes. Given the recent progress in ribosomal crystallographic analysis, it is not implausible that the crystal structure of a frameshifting ribosomal complex could be obtained. Further mechanistic insights will be gained from the comparison of RNA oligonucleotide-mediated frameshifting (e.g. Howard et al., 2004) and from studies involving other recoding signals. This is important since frameshifting mediated by RNA oligonucleotides raises the intriguing possibility of *trans*-signalling in vivo where the stimulatory RNA might be synthesised elsewhere on the genome (e. g. as a miRNA) allowing frameshifting to be regulated temporally and spatially.

References

Baranov PV, Gesteland RF, Atkins JF (2004) RNA 10:221–230
Baranov PV, Henderson CM, Anderson CB, Gesteland RF, Atkins JF, Howard MT (2005) Virology 332:498–510
Baranov, PV, Fayet O, Hendrix RW, Atkins JF (2006) Trends Genet 22:174–181
Barry JK and Miller WA (2002) Proc Natl Acad Sci USA 99:11133–11138
Bekaert M, Rousset JP (2005) Mol Cell 17:61–68
Belew AT, Hepler NL, Jacobs JL, Dinman JD (2008) BMC Genomics 9:339
Bidou L, Stahl G, Grima B, Liu H, Cassan M, Rousset J-P (1997) RNA 3:1153–1158
Brierley I, Digard P, Inglis SC (1989) Cell 57: 537–547
Brierley I, Rolley NJ, Jenner AJ, Inglis SC (1991) J Mol Biol 220:889–902
Brierley I, Jenner AJ, Inglis SC (1992) J Mol Biol 20:463–479
Brierley I (1995) J. Gen. Virol. 76:1885–1892
Brierley I, Pennell S (2001) Cold Spring Harbor Symposium of Quantitative Biology LXV: 233–248
Brierley I, Dos Ramos FJ (2006) Virus Res 119:29–42
Brierley I, Pennell S, Gilbert RJC (2007) Nat Rev Microbiol 5:598–610
Chen X, Chamorro M, Lee SI, Shen LX, Hines JV, Tinoco I Jr, Varmus HE (1995) EMBO J. 14:842–852
Chen X, Kang H, Shen LX, Chamorro M, Varmus HE, Tinoco I Jr (1996) J Mol Biol 260:479–483
Cornish PV, Hennig M, Giedroc DP (2005) Proc Natl Acad Sci USA 102:12694–12699
Cornish PV, Stammler SN, Giedroc DP (2006) RNA 12:1959–1969
Cowley JA, Dimmock CM, Spann KM, Walker PJ (2000) J Gen Virol 81:1473–1484
Dinman JD, Icho T. and Wickner RB (1991) Proc Natl Acad Sci USA 88:174–178
Dinman JD, Wickner RB (1992) J Virol 66:3669–3676
Dinman JD, Kinzy TG (1997) RNA 3:870–881
Dinman JD, Ruiz-Echevarria MJ, Czaplinski K, Peltz SW (1997) Proc Natl Acad Sci USA 94:6606–6611
Dinman JD, Ruiz-Echevarria MJ, Peltz SW (1998) Trends Biotech 16:190–196
Dreher TW, Miller WA (2006) Virology 344:185–197
Dulude D, Baril M, Brakier-Gingras L (2002) Nucleic Acids Res 30:5094–5102
Farabaugh PJ (1996) Microb Rev 60:103–134
Farabaugh PJ (1997) Programmed Alternative Reading of the Genetic Code, pp. 69–102. Landes Bioscience, Austin, Texas and Springer, Heidelberg, Germany
Farabaugh PJ (2000) Prog Nucleic Acid Res Mol Biol 64:131–170
Giedroc DP, Theimer CA, Nixon PL (2000) J Mol Biol 298:67–185
Giedroc DP, Cornish PV (2008) Virus Res E pub
Green L, Kim CH, Bustamante C, Tinoco I Jr (2008) J Mol Biol 375:511–528
Hansen TM, Reihani SN, Oddershede LB, Sørensen MA (2007) Proc Natl Acad Sci USA 104:5830–5835
Harger JW, Meskauskas A, Dinman JD (2002) Trends Biochem Sci 27:448–454
Herold J, Siddell SG (1993) Nucleic Acids Res 21:5838–5842
Horsfield JA, Wilson DN, Mannering SA, Adamski FM, Tate WP (1995) Nucleic Acids Res 23:1487–1494
Howard MT, Gesteland RF, Atkins JF (2004) RNA 10:1653–1661
Jacks T, Varmus HE (1985) Science 230:1237–1242
Jacks T, Madhani HD, Masiarz FR, Varmus HE (1988) Cell 55:447–458
Jacobs JL, Belew AT, Rakauskaite R, Dinman JD (2007) Nucleic Acids Res 35:165–174
Kang HS, Hines JV, Tinoco I (1996) J Mol Biol 259:135–147

Kim YG, Su L, Maas S, O'Neill A, Rich A (1999) Proc Natl Acad Sci USA 96:14234–14239
Kollmus H, Honigman A, Panet A, Hauser H (1994) J Virol 68:6087–6091
Kontos H, Napthine S, Brierley I (2001) Mol Cell Biol 21:8657–8670
Kurland CG (1992) Ann Rev Genet 26:29–50
Léger M, Sidani S, Brakier-Gingras L (2004) RNA 10:1225–1235
Léger M, Dulude D, Steinberg SV, Brakier-Gingras L (2007) Nucleic Acids Res. 35:5581–5592
Liphardt J, Napthine S, Kontos H, Brierley I (1999) J Mol Biol 288: 321–335
Lopinski JD, Dinman JD, Bruenn JA (2000) Mol Cell Biol 20:1095–1103
Marczinke B, Bloys AJ, Brown TD K, Willcocks MM, Carter MJ, Brierley I (1994) J Virol 68:5588–5595
Marczinke B, Fisher R, Vidakovic M, Bloys AJ, Brierley I (1998) J Mol Biol 284:205–225
Marczinke B, Hagervall T, Brierley I (2000) J Mol Biol 295:179–191
Manktelow E, Shigemoto K, Brierley I (2005) Nucleic Acids Res 33:1553–1563
Marquez V, Wilson DN, Tate WP, Triana-Alonso F, Nierhaus KH (2004) Cell 118:45–55
McGarry KG, Walker SE, Wang H, Fredrick K (2005) Mol Cell 20:613–622
Meskauskas A, Harger JW, Jacobs KL, Dinman JD (2003) RNA 9:982–992
Michiels PJ, Versleijen AA, Verlaan PW, Pleij CW, Hilbers CW, Heus HA (2001) J Mol Biol 310:1109–1123
Moazed D, Noller HF (1989) Nature 342:142–148
Moran SJ, Flanagan JF IV, Namy O, Stuart DI Brierley I, Gilbert RJC (2008) Structure 16:664–672
Namy O, Moran SJ, Stuart DI, Gilbert RJC, Brierley I (2006) Nature 441:244–247
Napthine S, Liphardt J, Bloys A, Routledge S, Brierley I (1999) J Mol Biol 288:305–320
Nixon PL, Rangan A, Kim YG, Rich A, Hoffman DW, Hennig M, Giedroc DP (2002) J Mol Biol 322:621–633
Noller HF, Yusupov MM, Yusupova GZ, Baucom A, Cate JHD (2002) FEBS Lett 514:11–16
Olsthoorn RC, Laurs M, Sohet F, Hilbers CW, Heus HA, Pleij CW (2004) RNA 10:1702–1703
Ortiz PA, Ulloque R, Kihara GK, Zheng H, Kinzy TG (2006) J Biol Chem 281:32639–32648
Pallan PS, Marshall WS, Harp J, Jewett FC 3rd, Wawrzak Z, Brown BA 2nd, Rich A, Egli M (2005) Biochemistry 44:11315–11322
Pan D, Kirillov SV, Cooperman BS (2007) Mol Cell 25:519–529
Paul CP, Barry JK, Dinesh-Kumar SP, Brault V, Miller WA (2001) J Mol Biol 310:987–999
Pe'ery T, Mathews MB (2000) In Sonenberg N, Hershey J Mathews M (eds), Translational control of gene expression, 371pp. Cold Spring Harbor Laboratory Press, New York
Pennell S, Manktelow E, Flatt A, Kelly G, Smerdon SJ, Brierley I (2008) RNA 14:1366–1377
Pestova TV, Hellen CU (2003) Genes Dev 17:181–186
Plant EP, Jacobs KL, Harger J.W, Meskauskas A, Jacobs JL, Baxter JL, Petrov AN, Dinman JD (2003) RNA 9:168–174
Plant EP, Wang P, Jacobs JL, Dinman JD (2004) Nucleic Acids Res 32:784–790
Plant EP, Dinman JD (2005) Nucleic Acids Res 33:1825–1833
Plant EP, Perez-Alvarado GC, Jacobs JL, Mukhopadhyay B, Hennig M, Dinman JD (2005) PLoS Biology. 3:e172
Pleij CW A, Rietveld K, Bosch L (1985) Nucleic Acids Res 13:1717–1731
Rice NR, Stephens RM, Burny A, Gilden RV (1985) Virology 142:357–377
Rodnina MV, Savelsbergh A, Katunin VI, Wintermeyer W (1997) Nature 385:37–41
Sanders CL, Lohr KJ, Gambill HL, Curran RB, Curran JF (2008) RNA 14:1874–1881
Selmer M, Dunham CM, Murphy FV IV, Weixlbaumer A, Petry S, Kelley AC, Weir JR, Ramakrishnan V (2006) Science 313:1935–1942
Schwartz DE, Tizard R, Gilbert W (1983) Cell 32:853–869
Shehu-Xhilaga M, Crowe SM, Mak J (2001) J Virol 75:1834–1841
Shen LX, Tinoco I (1995) J Mol Biol 247:963–978
Shigemoto K, Brennan J, Walls E, Watson CJ, Stott D, Rigby PW, Reith AD (2001) Nucleic Acids Res 29:4079–4088
Somogyi P, Jenner AJ, Brierley I, Inglis SC (1993) Mol Cell Biol 13:6931–6940

Su L, Chen L, Egli M, Berger JM, Rich A (1999) Nat Struct Biol 6:285–292
Takyar S, Hickerson RP, Noller HF (2005) Cell 120:49–58
ten Dam E.B, Pleij CWA, Bosch L (1990) Virus Genes 4:121–136
ten Dam E, Pleij K, Draper D (1992) Biochemistry 31:11665–11676
Tsuchihashi Z (1991) Nucleic Acids Res 19:2457–2462
Tu C, Tzeng T-H, Bruenn JA (1992) Proc Natl Acad Sci USA 89:8636–8640
Valle M, Zavialov A, Li W, Stagg SM, Sengupta J, Nielsen RC, Nissen P, Harvey SC, Ehrenberg M, Frank J (2003) Nat Struct Biol 10:899–906
Weiss RB, Dunn DM, Shuh M, Atkins JF, Gesteland RF (1989) New Biol 1:159–169
Wen JD, Lancaster L, Hodges C, Zeri AC, Yoshimura SH, Noller HF, Bustamante C, Tinoco I (2008) Nature 452:598–603
Wills NM, Gesteland RF, Atkins JF (1991) Proc Natl Acad Sci USA 88:6991–6995
Wills NM, Moore B, Hammer A, Gesteland RF, Atkins JF (2006) J Biol Chem 281:7082–7088
Yelverton E, Lindsley D, Yamauchi P, Gallant JA (1994) Mol Microbiol 11:303–313
Yusupov MM, Yusupova GZ, Baucom A,. Lieberman K, Earnest TN, Cate JH, Noller HF (2001) Science 292:883–896
Yusupova GZ, Yusupov MM, Cate JH, Noller HF (2001) Cell 106:233–241

Chapter 8
Programmed –1 Ribosomal Frameshift in the Human Immunodeficiency Virus of Type 1

Léa Brakier-Gingras and Dominic Dulude

Abstract A programmed –1 ribosomal frameshift enables the human immunodeficiency virus of type 1 (HIV-1) to produce its enzymes in a precise proportion relative to its structural proteins, which is necessary to control viral assembly and maturation. Here, we critically review models that account for this phenomenon, focusing on the most recent model, which postulates that the frameshift is triggered by an incomplete translocation and involves the slippage of three tRNAs. The effect of changes in the rate of translation initiation and elongation and the possible involvement of cellular factors in frameshifting are briefly examined. Finally, we highlight recent efforts intended to interfere with this type of frameshift as a strategy to develop novel anti-HIV drugs.

Contents

8.1 Characteristics of the Slippery Sequence and the Frameshift Stimulatory Signal of HIV-1

Some 20 years ago, it was demonstrated for the first time that the human immunodeficiency virus of type 1 (HIV-1), which is responsible for AIDS, uses a

L. Brakier-Gingras (✉)
Département de Biochimie, Université de Montréal, Montréal, Québec, Canada, H3T 1J4
e-mail: lea.brakier.gingras@umontreal.ca

J.F. Atkins, R.F. Gesteland (eds.), *Recoding: Expansion of Decoding Rules Enriches Gene Expression*, Nucleic Acids and Molecular Biology 24,
DOI 10.1007/978-0-387-89382-2_8,

programmed –1 ribosomal frameshift to produce the Gag-Pol fusion protein, the precursor of its enzymes (the protease, reverse transcriptase and integrase). The frameshift occurs when ribosomes in the infected cells translate the full-length viral mRNA (Jacks et al., 1988b; Parkin et al., 1992). Translation of this full-length viral mRNA produces Gag, the precursor of viral structural proteins, according to the conventional rules. The ratio of Gag-Pol to Gag is critical for the virus replication (Park and Morrow, 1991; Karacostas et al., 1993; Hung et al., 1998; Shehu-Xhilaga et al., 2001; Biswas et al., 2004). Increasing or decreasing this ratio results in protein processing defects and in reduction of the stability of the dimeric RNA genome of the virus. Even small variations in this ratio impair the viral replication (Telenti et al., 2002; Dulude et al., 2006).

A programmed –1 frameshift occurs at a much higher frequency than an accidental frameshift and requires two *cis*-acting elements in the mRNA, the slippery sequence where the frameshift occurs and a downstream secondary structure, the frameshift stimulatory signal, which promotes the frameshift (reviewed by Brierley and Pennell, 2001; Brierley and Dos Ramos, 2006-see also Chapter 7 by Brierley et al.). In HIV-1 mRNA, the slippery sequence is U UUU UUA (where the 0 frame is indicated by spaces) and it was found to be invariant when comparing 1,000 HIV-1 sequences. Moreover, when this slippery sequence was replaced with other slippery sequences, the frameshift efficiency decreased and the viral replication was severely impaired (Biswas et al., 2004). For HIV-1 group M subtype B, the prevalent subtype in the Western world, mutagenesis studies first showed that the frameshift stimulatory signal is an 11-bp stem capped by an ACAA tetraloop. Further investigation, involving mutagenesis studies and probing with enzymatic attack, revealed that this stimulatory signal is more complex and consists of a long irregular stem-loop with two stems separated by a three-purine bulge (GGA) (Dulude et al., 2002). The upper stem-loop, called the classic signal, represents the previously defined 11-bp stem-loop while the lower stem results from the pairing of the spacer region following the slippery sequence and preceding the upper stem-loop with a complementary sequence downstream of this stem-loop. This structure was later substantiated by NMR studies (Gaudin et al., 2005; Staple and Butcher, 2005) (Fig. 8.1a). The other subtypes of HIV-1 group M share the same structure of the stimulatory signal as subtype B (Baril et al., 2003a). An independent in silico analysis also supported a two-stem structure for the stimulatory signal, with the exception that the GGA bulge is replaced with a double A bulge (Parisien and Major, 2008).

However, it is unclear how this structure exists dynamically during protein synthesis leading to the frameshift event. According to ribosome–mRNA–tRNA crystal structures (Yusupova et al., 2001), one can imagine that when the frameshift takes place, the lower stem of the stimulatory signal must be unwound to allow the ribosome access to the slippery sequence and the role of the effective stimulatory signal is played by the upper stem-loop. Thus, the lower stem may contribute to optimizing an initial interaction between the upper stem-loop and the ribosome (Léger et al., 2004; Staple and Butcher, 2005). This view is reinforced if one considers other related frameshift signals. In the simian immunodeficiency virus (SIV), which uses the same slippery sequence as HIV-1, the frameshift stimulatory signal is also a structured stem-loop (Marcheschi et al., 2007). This helix contains two stems that

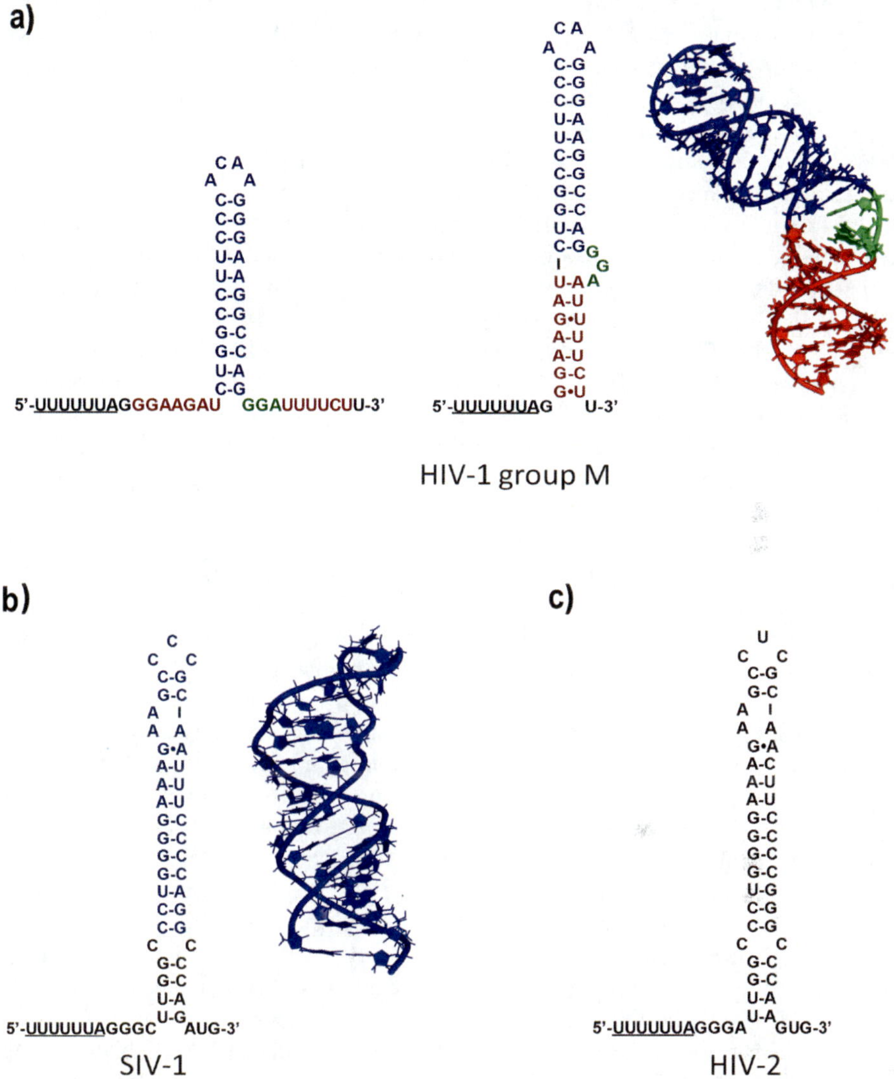

Fig. 8.1 Secondary and tertiary structures of frameshift stimulatory signals adopting a stem-loop conformation. (**a**) The HIV-1 group M frameshift stimulatory signal. (*Left*) The stem-loop structure defined by Jacks et al. (1988b). The structure consists of an 11-bp stem capped by an ACAA tetraloop separated from the slippery sequence (*underlined*) by an 8-nt spacer. (*Centre*) The two-stem helix determined by Dulude et al. (2002). (*Right*) The corresponding three-dimensional structure determined by NMR studies (Staple and Butcher, 2005; see also Gaudin et al., 2005) In this structure, the upper stem-loop (*blue*) and the lower stem (*red*) are separated by a purine bulge (*green*). (**b**) The SIV-1 frameshift stimulatory signal. (*Left*) The stem-loop structure consists of an irregular two-stem helix, in which the upper stem is capped by a complex loop separated from the lower stem by two bulged cytosines. (*Right*) The three-dimensional structure of the upper stem-loop as determined by NMR studies (Marcheschi et al., 2007). (**c**) The HIV-2 frameshift stimulatory signal. The stem-loop structure is similar to that found in SIV-1 (Marcheschi et al., 2007). The Data Bank accession codes of the atomic coordinates from the RCSB Protein Data Bank are 1Z2J (HIV-1 group M) and 2JTP (SIV-1)

are separated by two bulged cytosines. The loop capping the upper stem is more complex than that in HIV-1 and consists of a sheared G–A pair, a cross-strand adenine stacking, two G–C pairs and a novel cytidine triloop. A structure similar to that of SIV likely exists in the human immunodeficiency virus of type 2 (HIV-2) (Fig. 8.1b and c). As for HIV-1, one can predict that the lower stem of the SIV and HIV-2 frameshift stimulatory signal must be unwound when the frameshift takes place, while the upper stem-loop acts as the effective signal.

NMR studies of the HIV-1 group M frameshift region have revealed a 60° bend between the stems, reminiscent of the bend seen in a number of hairpin-type pseudoknots inducing a –1 frameshift, such as those in the mouse mammary tumour virus (MMTV) (Shen et al., 1995; Su et al., 1999), and in the feline immunodeficiency virus (FIV) (Yu et al., 2005). Interestingly, in HIV-1 group O (outlier), which is most common in and largely restricted to Cameroon and neighbouring countries, the frameshift stimulatory signal is a hairpin-type pseudoknot with two coaxially stacked 8-bp stems connected by loops of 2 and 20 nt (Baril et al., 2003b). This different signal supports previous phylogenetic analyses suggesting that HIV-1 group M and group O originate from different ape to human transmissions (Simon et al., 1998; Gao et al., 1999; Sharp et al., 2001). The pseudoknot found in HIV-1 group O is reminiscent of the simian retrovirus –1 (SRV-1) (Michiels et al., 2001) and the infectious bronchitis virus (IBV) frameshift stimulatory signals (Brierley et al., 1991) (Fig. 8.2a, b, and c). A pseudoknot structure has been proposed for the frameshift stimulatory signal of HIV-1 group M (Dinman et al., 2002), but this structure, which involves a pairing between the tetraloop capping the classic stem and a complementary sequence in the 3′ adjacent region, was not supported by NMR data.

Several sensitive assays have been developed in recent years, which have facilitated investigations of frameshifting, leading to specific models of the underlying mechanism. First, a dual-luciferase reporter system expressing the *Renilla* and firefly luciferase separated by the HIV-1 frameshift region was pioneered by Grentzmann et al. (1998). The insertion of the HIV-1 frameshift region is such that expression of the firefly luciferase requires a –1 frameshift in the HIV-1 frameshift region. In cultured mammalian cells, this efficiency has been found to be around 5–10%, when the complete frameshift region with the two-stem signal is used (Dulude et al., 2006; Plant and Dinman, 2006; Gendron et al., 2008). This value is consistently higher when measured in vitro (see Grentzmann et al., 1998), likely resulting from differences in the kinetics of translation and in the control of translational accuracy. Second, it has been shown that HIV-1 frameshift can be recapitulated in yeast (Harger and Dinman, 2003) and in bacteria (Léger et al., 2004), which therefore also constitute valuable systems to study HIV-1 frameshifting.

8.2 Models of –1 Ribosomal Frameshift in HIV-1

Several models have been proposed to explain the mechanism of the –1 ribosomal frameshift in HIV-1 and in other viruses and organisms that use a –1 frameshift. We summarize here the features of four specific models. As we indicate below, some

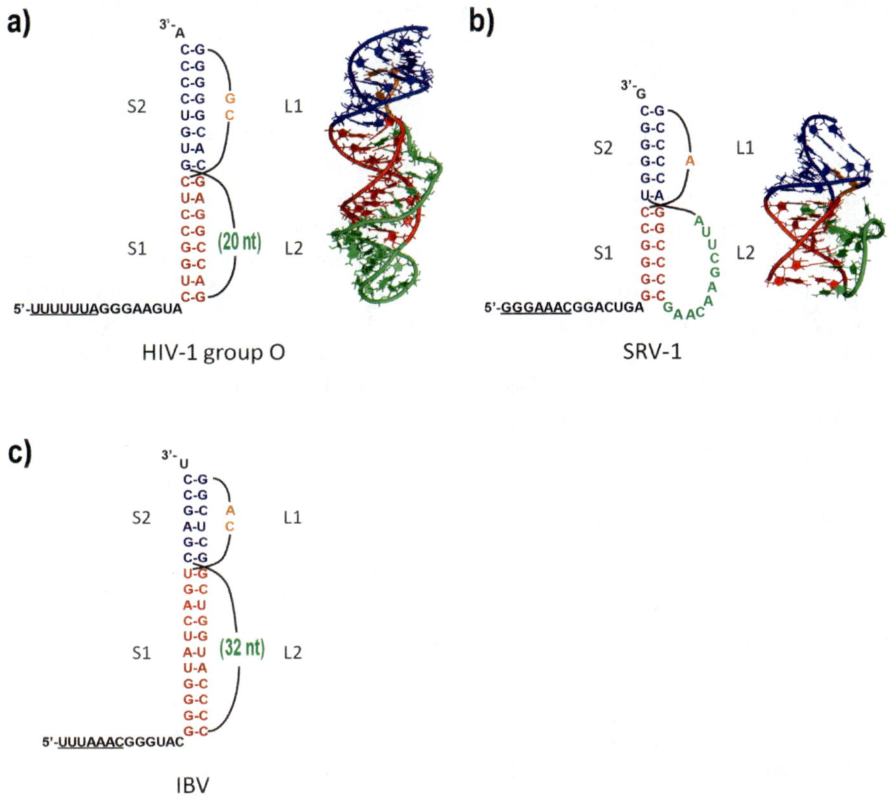

Fig. 8.2 Structural representation of frameshift stimulatory signals adopting a pseudoknot conformation. (**a**) The frameshift stimulatory signal of HIV-1 group O, as determined by Baril et al. (2003b). It consists of a pseudoknot containing an 8-nt stem 1 (S1), a 2-nt loop 1 (L1), an 8-nt stem 2 (S2) and a 20-nt loop 2 (L2). This structure is similar to that of the pseudoknots acting as frameshift stimulatory signals found in SRV-1 (Michiels et al., 2001) (**b**) and in IBV (Brierley et al., 1991) (**c**). For (**a**) and (**b**), the three-dimensional structure, which is determined, respectively, by modelling and NMR studies, is shown (*right*). S1 is *red*, S2 is *blue*, L1 is *orange* and L2 is *green*. The Data Bank accession code of the atomic coordinates from the RCSB Protein Data Bank is 1E95 for SRV-1. Coordinates for HIV-1 group O are available on request

features are shared among the four models. The precise mechanism of frameshifting remains, however, to be elucidated by future studies.

The initial model originated from the group of Harold Varmus (Jacks et al., 1988a). According to this model, the so-called two-tRNA simultaneous slippage, the frameshift occurs when the slippery sequence of the mRNA occupies the P and A sites of the ribosome. When encountering the frameshift stimulatory signal, i.e. the upper stem-loop of the signal, the ribosome pauses in translation. During this pause, unconventional events such as a –1 frameshift can occur with a fraction of ribosomes. That the ribosome pauses when encountering the frameshift stimulatory signal was demonstrated subsequently, although it was also shown that the pause is necessary but not sufficient and can be observed with structures that do not promote a frameshift (Tu et al., 1992; Somogyi et al., 1993). Also, it is not known whether the pausing that was detected is relevant for the frameshift (see Brierley and Pennell, 2001; Kontos et al., 2001; Howard et al., 2004). In the Varmus model, the codon–anticodon interactions between the mRNA and the peptidyl- and aminoacyl-tRNAs can be broken during the pause of the ribosome, the two tRNAs can then slip back by 1 nt, together with the ribosome, in the 5′ direction on the slippery sequence and, after the shift, re-pair to the mRNA in the new reading frame. After peptide bond formation, the ribosome continues translation in the new frame and unfolds the frameshift stimulatory signal which can re-form after its passage (see also Section 8.4 below). Another perspective is that, rather than a slippage of the two tRNAs and the ribosome, the mRNA could move by 1 nt in the 3′ direction relative to the tRNAs and the ribosome. With the HIV-1 slippery sequence, after the shift, the pairing to the mRNA is perfect for the peptidyl-tRNA, a peptidyl-$tRNA^{Phe}$, but there is a non-Watson–Crick U•U base pair at the third position for the aminoacyl-tRNA, Leu-$tRNA^{Leu}$ (Fig. 8.3a). However, the Varmus model raises a quandary: the peptide bond is formed almost immediately after the aminoacyl-tRNA occupies the A site, so that there is not much time for the shift to occur. Also, the model does not consider the nature of the driving force for the movement of the tRNAs and the ribosome relative to the mRNA. Plant et al. (2003) suggested that this driving force originates from a tension imposed in the mRNA when the aminoacyl-tRNA moves from the entry (A/T) site to the A site. The anticodon loop of the tRNA would be displaced by 9 Å, which would create a tension in the mRNA when the latter is pulled by the tRNA. This tension, which could be relieved by the displacement of the mRNA, would be caused by the resistance to unwinding of the stimulatory signal that is at the entrance tunnel of the ribosome and is too large to penetrate this tunnel if it remains folded.

A second model was proposed by the group of John Atkins, which addresses the problem raised by the lack of time for the frameshift to occur in the Varmus model. The Atkins model suggested that the frameshift occurs during translocation (Weiss et al., 1989). After peptide bond formation, the deacylated and the peptidyl-tRNAs occupy, respectively, the hybrid P/E and A/P sites on the ribosome. The next step in translation elongation consists of a movement of the anticodon stem-loops of the tRNAs to the E and P sites of the small ribosomal subunit, dragging along the mRNA. The latter movement is promoted by an exogenous translocase factor (called eEF2 in eukaryotes) and it completes the translocation (reviewed by Frank et al., 2007). Similar to the Varmus model, in the Atkins model, the resistance to unwinding of the frameshift stimulatory signal would create a tension in the mRNA that interferes with this movement, which would be relieved if the two tRNAs move in the 5′ direction with the ribosome, or if the mRNA moves in the 3′ direction, with respect to the ribosome (Fig. 8.3b). An involvement of the translocation step in the ribosomal frameshifting is supported by cryo-electron microscopy studies (Namy et al., 2006; Moran et al., 2008), showing that eEF2 is bound to a ribosome–mRNA complex stalled in the process of –1 frameshifting.

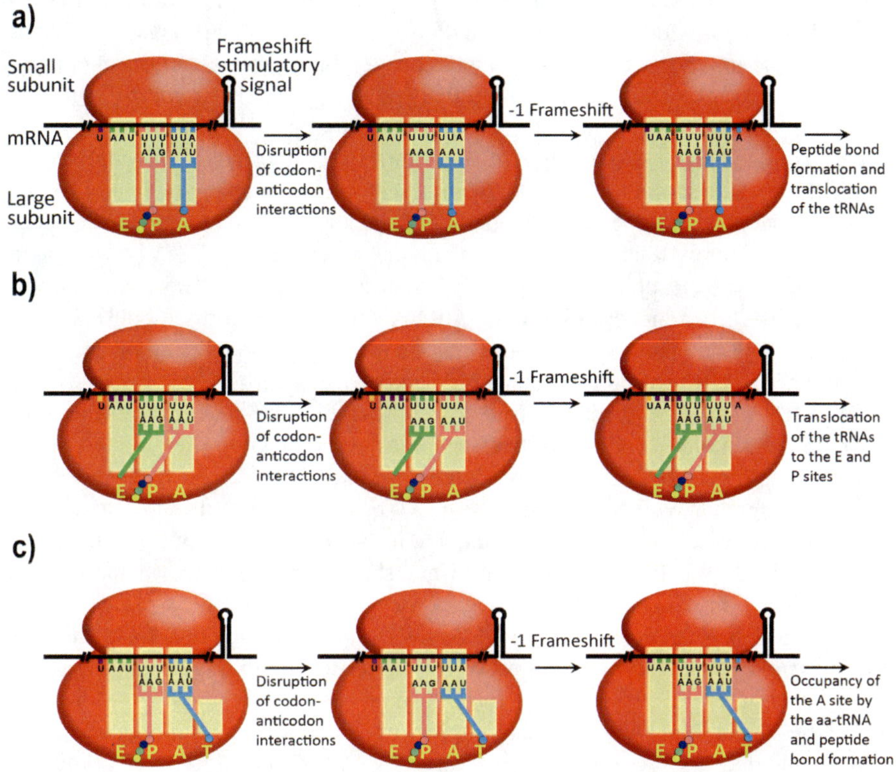

Fig. 8.3 Models that account for the programmed –1 ribosomal frameshift in HIV-1. The ribosomal subunits, the frameshift stimulatory signal (corresponding to the upper stem-loop of the signal – see the text), the slippery sequence (UUUUUUA) and the ribosomal sites (E, P, A and T sites) occupied by the corresponding tRNAs are indicated. The deacylated tRNA (deac-tRNA), the peptidyl-tRNA (pept-tRNA) and the aminoacyl-tRNA (aa-tRNA) are shown in *green*, *pink* and *blue*, respectively. *Coloured circles* represent amino acids attached to the 3′ end of tRNAs. Elongation factors are not shown. (**a**) In the model of simultaneous slippage of two tRNAs (Jacks et al., 1988a), the –1 frameshift occurs prior to peptide bond formation, when the pept-tRNA and the aa-tRNA occupy, respectively, the P and the A sites. The codon–anticodon interactions of the two adjacent tRNAs with the mRNA are broken, the two tRNAs slip back together with the ribosome by 1 nt and then re-pair with the mRNA in the new reading frame. (**b**) In the model proposed by Weiss et al. (1989), the –1 frameshift occurs during translocation. After peptide bond formation, the deac-tRNA and the pept-tRNA occupying, respectively, the hybrid P/E and A/P sites unpair from mRNA, move in the 5′direction with the ribosome and re-pair in the new reading frame. (**c**) In the model suggested by Farabaugh (1997), the –1 frameshift occurs when the pept-tRNA and the aa-tRNA occupy, respectively, the P and the entry (A/T) sites

A third model that proposes an alternative solution to the problem raised in the Varmus model was suggested by Phil Farabaugh (1997). This model proposes that the frameshift involves the peptidyl-tRNA in the P site and the aminoacyl-tRNA in the entry site, before it occupies the A site. Thus, the ribosome has much more time to shift before peptide formation occurs. We provided experimental support to this model by showing that ribosomal mutations that speed up the movement of the aminoacyl-tRNA from the entry site to the A site decrease the HIV-1 frameshift efficiency (Léger et al., 2004). However, the driving force for the shift is not explained and a possible involvement of the translocation step in the frameshift is not considered (Fig. 8.3c). It is interesting to note that sequencing of the frameshift product obtained with the HIV-1 frameshift region revealed that in 70% of the cases leucine was incorporated at the A site, while in 30% of the cases it was phenylalanine (Jacks et al., 1988b; Yelverton et al., 1994). This observation cannot be explained by the first two models describing the frameshift. Indeed, the ribosome controls the accuracy of translation by rejecting a non-cognate or a near-cognate tRNA before it can occupy the A site (Rodnina et al., 2005; Daviter et al., 2006). The occupancy of the A site by the aminoacyl-tRNA, Leu-tRNA^{Leu}, in the Varmus model, and the incorporation of leucine in the protein chain, in the Atkins model, take place before the shift occurs. However, if the frameshift occurs when Leu-tRNA^{Leu} is in the entry site, this tRNA could be rejected before it occupies the A site. Indeed, after the shift, it re-pairs with a UUU codon and thus becomes near-cognate. The rejected tRNA could then be replaced by Phe-tRNA^{Phe}, the cognate tRNA.

Finally, together with Sergey Steinberg, we proposed a fourth model that integrates the various experimental data related to the –1 ribosomal frameshift in HIV-1 (Léger et al., 2007). In this model, the frameshift is triggered by an incomplete translocation, which occurs in the elongation cycle preceding the cycle where the slippery sequence occupies the P and A sites. Such incomplete translocation is caused by the resistance of the stimulatory signal to unwinding when it reaches the entrance of the mRNA tunnel in the ribosome. As a result, the tRNAs drag the mRNA by 2, not 3 nt. After this incomplete translocation, the slippery sequence occupies the decoding region, but is displaced by 1 nt in the 3′ direction, compared to the position it would occupy after a complete translocation. As to the deacylated and the peptidyl-tRNAs, they occupy intermediate sites on the ribosome through which they transit (see Wen et al., 2008) when moving from the hybrid P/E and A/P sites to classic E and P sites. The frameshift occurs upon the arrival of the aminoacyl-tRNA, which also cannot occupy the standard entry site but an intermediate site. The frameshift involves the rupture of the codon–anticodon interactions for the three tRNAs, i.e. the deacylated, the peptidyl- and the aminoacyl-tRNA in the entry site, followed by the successive sliding of these tRNAs in the 5′ direction and their re-pairing to the mRNA in the new frame. The driving force for the shift is the tendency of the tRNAs to occupy the standard sites for which they have a greater affinity than for intermediate sites. In this model, the slippery sequence, which we name the extended slippery sequence, is thus not restricted to the codons occupying the P and A sites but also involves the E-site codon. Two pauses are associated with the shift in the fourth model. The first pause occurs after the incomplete translocation, while the second pause takes place after the binding of the aminoacyl-tRNA to the ribosome and before frameshifting takes place (Fig. 8.4). A conundrum with the last model is what happens to the codon–anticodon interaction for the E-site tRNA after the frameshift. Based on in vitro translation assays, Knud Nierhaus and co-workers provided evidence that there is a codon–anticodon interaction at the E site (reviewed in Nierhaus, 2006; see also Sanders and Curran, 2007 and Chapter 16

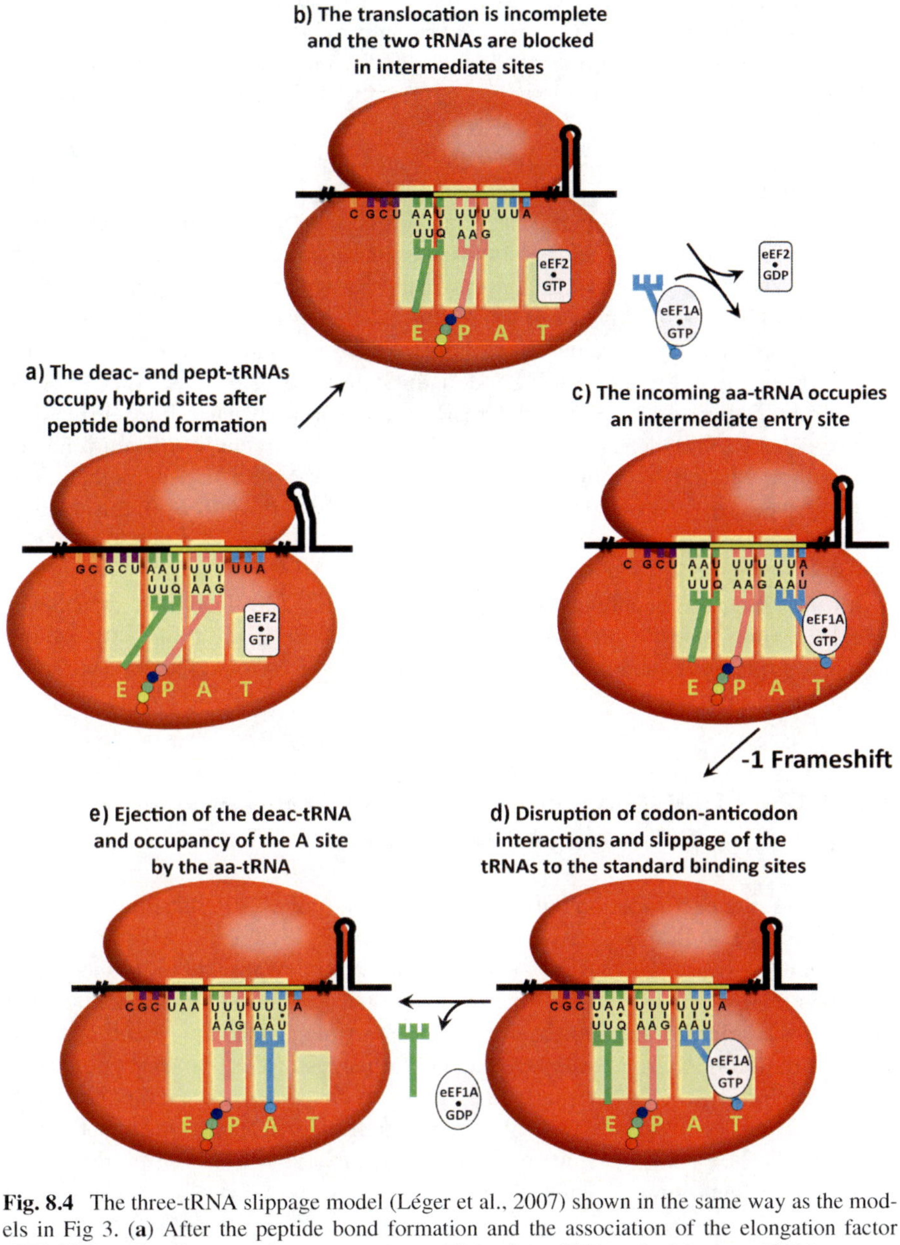

Fig. 8.4 The three-tRNA slippage model (Léger et al., 2007) shown in the same way as the models in Fig 3. (**a**) After the peptide bond formation and the association of the elongation factor eEF2·GTP to the ribosome, the acceptor stems of the deac-tRNA and of the pept-tRNA move to the E and P sites, respectively, so that the tRNAs occupy the P/E and A/P hybrid sites on the ribosome (Frank et al., 2007). (**b**) After GTP hydrolysis, the anticodon stem-loops of both tRNAs move together with the mRNA towards the E and P sites of the small ribosomal subunit. The resistance to unwinding of the frameshift stimulatory signal limits the displacement of the mRNA to 2 nt instead of 3, such that the deac-tRNA and the pept-tRNA occupy intermediate sites and not the classic E and P sites. (**c**) The ribosome selects a ternary complex containing the incoming aa-tRNA bound

by Pech et al.). Recently, X-ray crystallographic evidence revealed that the E-codon nucleotides can form Watson–Crick base pairs with the anticodon of the E-site tRNA (Jenner et al., 2007), supporting the conclusions of the Nierhaus group. However, with the HIV-1 extended slippery sequence, after the shift, only one base pair is possible for the E-site tRNA, which is also the case for most slippery sequences used in –1 ribosomal frameshifts. One can hypothesize that, in addition to the canonical Watson–Crick base pairs, the ribosome could tolerate a large variety of base pairs in the E site. It is also possible that weak codon–anticodon interactions trigger a rapid release of the E-site tRNA and, consequently, accelerate the occupancy of the A site by the shifted aminoacyl-tRNA, Leu-tRNA^{Leu}, which is near-cognate and thus risks rejection before it gets to the A site. If this tRNA is rejected, the cognate Phe-tRNA^{Phe} could be incorporated, as explained above. Another possibility could be that, after rejection of Leu-tRNA^{Leu}, codon–anticodon interactions between the deacylated and the peptidyl-tRNAs and the mRNA are disrupted. The ribosome bearing the two tRNAs could then slip forwards by 1 nt. This movement would contribute to decreasing the frameshift efficiency since the ribosome would be back in the initial reading frame, with the two tRNAs re-pairing perfectly to the mRNA in this reading frame. However, a rapid occupancy of the A site by Leu-tRNA^{Leu} would limit the probability of rejection of this tRNA (Fig. 8.5).

An additional point emphasized by Baranov et al. (2004) is that, whichever model accounts for the programmed –1 ribosomal frameshift, the shift of the tRNAs is probably sequential and not simultaneous, which would require a complex synchronization of the movement of the tRNAs.

8.3 Rates of Initiation and Elongation of Translation and Their Effect on the Frameshift

Altering the rate of translation initiation changes the frameshift efficiency, as shown with the yeast L-A double-stranded RNA virus (Lopinski et al., 2000) and with the barley yellow dwarf virus (BYDV) (Paul et al., 2001). We made the same observation with HIV-1, using a standard dual-luciferase system in cultured mammalian cells under conditions that decrease or increase the rate of the cap-dependent translation initiation (Gendron et al., 2008). These changes in the frameshift efficiency can be ascribed to changes in the spacing between ribosomes. When the frameshift region is translated, the stimulatory signal is unfolded. Depending upon the distance between the ribosomes, which in turn depends upon the rate of initiation, the signal can or cannot refold between the passages of two successive ribosomes. In

Fig. 8.4 (continued) to eEF1A·GTP. This complex binds to an intermediate entry site, which differs from the standard A/T site. (**d**) The tRNAs in the intermediate sites unpair from mRNA, successively shift to the standard E, P and A/T sites and re-pair to mRNA in the –1 reading frame. (**e**) The deac-tRNA is ejected from the ribosome upon the occupancy of the A site by the aa-tRNA

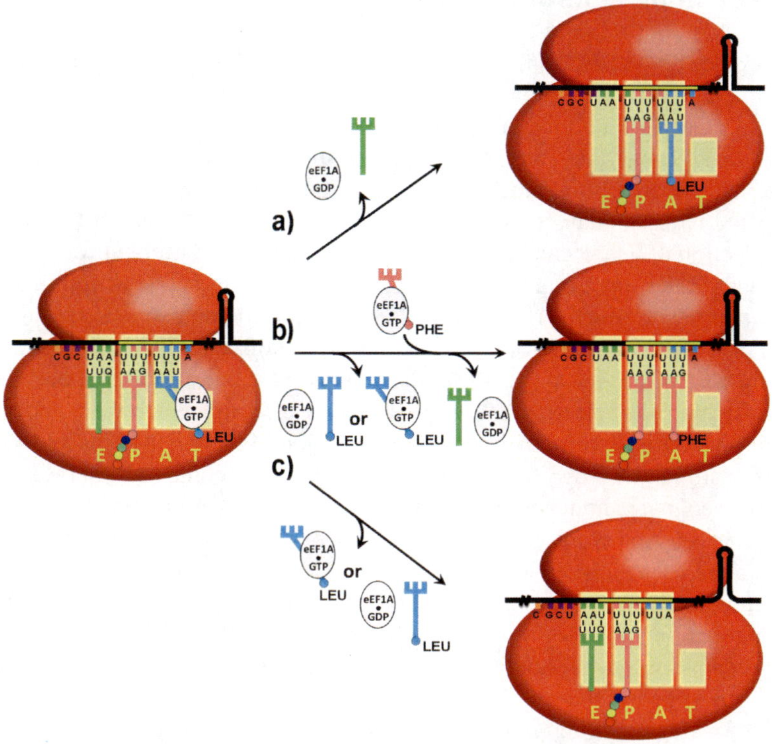

Fig. 8.5 Fate of the incoming aminoacyl-tRNA in the three-tRNA slippage model. All elements are shown as in Fig.3. (**a**) The incoming aminoacyl-tRNA, Leu-tRNA^{Leu}, occupies the A site and the deac-tRNA is ejected. (**b**) Leu-tRNA^{Leu} is rejected, before or after hydrolysis of the GTP bound to eEF1A, and replaced with the cognate tRNA, Phe-tRNA^{Phe}, which occupies the A site while the deac-tRNA is ejected. (**c**) Leu-tRNA^{Leu} is rejected, codon–anticodon interactions are broken and the ribosome and the two tRNA slip by 1 nt in the 3′ direction. The deac-tRNA and the pept-tRNA can then re-pair to the mRNA in the original frame

agreement with Lopinski et al. (2000), we showed that under basal conditions of translation initiation, every other ribosome encounters the folded stimulatory signal. A maximal two-fold increase in the frameshift efficiency can be reached when the initiation rate is decreased, corresponding to the situation where every ribosome encounters a folded stimulatory signal. In a similar way, changing the rate of translation elongation should also affect the spacing between the ribosomes. In addition, inhibitors of protein synthesis acting on the elongation step can also directly influence the frameshift process and thus its efficiency. For example, in the presence of sparsomycin, a well-known translation elongation inhibitor, it was observed that the frameshift efficiency increases with the yeast L-A virus in yeast cells (Dinman et al., 1997) and with a dual-luciferase construct containing the HIV-1 frameshift region in cultured mammalian cells (our own unpublished results).

8.4 Remaining Questions to Address

As mentioned above, the frameshift stimulatory signal could form a specific interaction with the ribosome, and it is now important to fully characterize this interaction. One can propose that this interaction decreases the activity of the ribosome-associated helicase (Takyar et al., 2005) that unwinds the secondary structure of mRNA. This decrease in activity would contribute to the resistance to unwinding of the stimulatory signal. If we consider our model of –1 frameshift (Léger et al., 2007) (Section 8.2), a decrease in the helicase activity could account for the incomplete translocation that triggers the frameshift. Another attractive hypothesis is that the interaction between the ribosome and the frameshift stimulatory signal alters the control of translation accuracy by the ribosome. As discussed above (Section 8.2), in our model, the aminoacyl-tRNA in the entry site becomes near-cognate after the frameshift and risks rejection before it occupies the A site, which would contribute to decrease the frameshift efficiency. A perturbation of the fidelity control when the frameshift stimulatory signal interacts with the ribosome could prevent the rejection of the near-cognate aminoacyl-tRNA. We are currently exploring this possibility.

Another question worthy of investigation is whether cellular proteins are able to modulate the frameshift efficiency of HIV-1. The HIV-1 frameshift can be recapitulated in bacteria (Léger et al., 2004) and in yeast (Harger and Dinman, 2003). It can also function, albeit less efficiently, in the *murine* leukaemia virus that normally uses a readthrough to produce its enzymes (Brunelle et al., 2003; Gendron et al., 2005). This suggests that the HIV-1 frameshift region contains the main elements required for frameshifting, but does not exclude the possibility that additional factors could modulate the frameshift efficiency. Screening a library of small hairpin or small interfering RNAs that target all human genes can be used to investigate whether there are cellular factors that participate in the HIV-1 frameshift. Recently, Brass et al. (2008) performed a large-scale screen with small interfering RNAs and identified over 250 cellular factors involved in HIV-1 replication. Preliminary results in our laboratory point to the involvement of two factors identified by Brass et al. in HIV-1 frameshifting, the release factor eRF1 and the rRNA dimethylase Dim1p. eRF1 is a translation termination factor that recognizes the termination codons in mRNA and triggers peptidyl-tRNA hydrolysis, a prerequisite to the release of the ribosomes. Dim1p is an evolutionary conserved modification enzyme that methylates two adjacent adenines in the terminal loop of the small subunit ribosomal RNA. Silencing either protein increased the HIV-1 frameshift efficiency by about 30%. Silencing Dim1p could affect the conformation of the small ribosomal subunit E site and also the ribosomal subunit association, since the rRNA loop it modifies is located in proximity of the E site (Jenner et al., 2007) and is part of a bridge involved in subunit association (Yusupov et al., 2001; reviewed in Liiv and O'Connor, 2006). One could hypothesize, considering our model of –1 frameshift, that changes in the E site could promote the sliding of the E-site tRNA, which is followed by the successive sliding of the two other tRNAs. Moreover, changes in subunit association could contribute to increase the proportion of ribosomes for which translocation is incomplete, by interfering with intersubunit movements required for completing translocation. The effect of silencing eRF1 could be indirect, reducing the number of ribosomes that initiate translation and, therefore, increasing the probability that these ribosomes encounter a folded frameshift stimulatory signal. It is likely that other cellular factors that influence HIV-1 frameshift will be discovered, thus making this frameshift a more complex event than presently assumed. Another interesting candidate for involvement in this frameshift (see also Brierley et al., 2007) is the ribosomal protein RACK1 (receptor for activated protein kinase C), which recruits the activated kinase C and links signal-transduction pathways to the ribosome (Nilsson et al., 2004). This protein is located in the head of the small ribosomal subunit (Sengupta et al., 2004) and could influence the interaction between the frameshift stimulatory signal and the ribosome.

Unmodified or modified oligonucleotides (e.g. 2′O-methyl oligonucleotides or morpholino-oligonucleotides) that pair to a region downstream of either the slippery sequence (Howard et al., 2004; Olsthoorn et al., 2004) or the classic frameshift stimulatory signal, i.e. the upper stem-loop (Vickers and Ecker, 1992), have been found to increase the HIV-1 frameshift efficiency. However, these results were obtained in vitro and Vickers and Ecker were unable to reproduce the observed effect in cultured cells. Also, so far, it is unknown whether the increase in –1 frameshift efficiency observed when a stem was created by adding complementary oligonucleotides downstream of the slippery sequence can be reproduced in cultured cells. Interestingly, Kollmus et al. (1996) replaced the HIV-1 upper stem-loop signal with the iron-response element (IRE) of ferritin mRNA. IRE is involved in the control of the iron metabolism and consists of a bulged helix. This structure increased the efficiency of frameshift to about the same level as the original signal in cultured cells. We investigated HIV-1 frameshift efficiency in bacteria, with a reporter whose expression depends upon HIV-1 –1 frameshift. Experiments were performed under conditions where the HIV-1 frameshift stimulatory signal was replaced with a variety of hairpins, such as the human IRE, the transactivation response element of the bovine immunodeficiency virus (BIV-TAR), and the *rev* response element of HIV-1 (HIV-1 RRE). In all these assays, the inserted hairpins were about as efficient as the HIV-1 stimulatory signal (Dulude et al., 2008). Taken altogether, these observations raise the question of whether a specific structure downstream of the slippery sequence is really required to promote the –1 frameshift. Could a large variety of hairpins stimulate HIV-1 –1 frameshift? We now have to figure out if and how these observations can be reconciled with the hypothesis that the frameshift stimulatory signal has a specific interaction with the ribosome and is not a simple roadblock.

The tRNAs are important players in the frameshift story and it has been shown that modifications of the bases in the anticodon loop can affect the frameshift efficiency (see Brierley et al., 1997; Carlson et al., 1999, 2001; Waas et al., 2007; Agris, 2008 and references herein). The rules that account for the role of tRNA modifications appear to be rather complex because these modifications can affect not only the codon–anticodon strength but also the conformational flexibility and the capacity of the tRNAs to slide from one site to another (see Chapter 7 by Brierley et al.). Thus, more studies are needed to characterize the role of the tRNAs in the frameshift.

8.5 The Frameshift Event as a Target for Novel Anti-HIV Drugs

The sensitivity of HIV-1 to small variations in the frameshift efficiency obviously makes this process an attractive target for the development of novel anti-HIV drugs. A large variety of compounds have already been assayed for their ability to change the frameshift efficiency. These compounds could target either the ribosome or the frameshift stimulatory signal. For instance, inhibitors of protein synthesis at concentrations that do not affect the translational machinery can still significantly alter the frameshift efficiency. However, the window of concentrations over which these drugs can be used to alter the frameshift efficiency without altering normal translation is rather narrow, and toxic side-effects can be observed at higher concentrations. High-throughput screening of chemical compounds is currently used in different laboratories to select those compounds that interfere with frameshifting in HIV-1. One of these compounds, RG501, increased HIV-1 frameshift efficiency about two-fold and interfered with viral propagation in cultured cells (Hung et al., 1998). However, this compound affected multiple frameshift signals and was toxic. Although the authors concluded that RG501 could bind to different signals, we would rather suggest that it exerts its effect via binding to the ribosome, which would readily account for the large spectrum of frameshift signals it affects. Also, bacteria were transformed with a plasmid encoding a fluorescent reporter whose expression depends upon HIV-1 frameshift and a second fluorescent reporter, which was expressed following the standard translation rules. The bacteria were next transformed with expression plasmids harbouring a library of 14-mer arginine-rich peptides and screened by flow cytometry (Dulude et al., 2008). Candidates were selected that were able to decrease the frameshift efficiency about two-fold. These peptides were also assayed with different stimulatory signals. Again, the effect was independent of the nature of the stimulatory signal, suggesting that the peptides target the ribosome. A fraction of these peptides were also active in a eukaryote system and are currently being assayed for their effect on HIV-1 propagation. Clearly, much remains to be done to identify compounds that selectively target frameshift efficiency but do not affect normal translation.

If the target of a compound is the HIV-1 frameshift stimulatory signal, one can hypothesize that this compound could stabilize the structure of the signal and interfere with its unwinding, thus increasing the frameshift efficiency. We mentioned above the results of Kollmus et al. (1996), showing that replacement of HIV-1 frameshift stimulatory signal with the human IRE stimulated frameshifting at the same level of efficiency. In addition, the frameshift efficiency was further stimulated upon binding of an iron-regulated protein to the IRE element, thus demonstrating the possibility to alter the frameshift efficiency with proteins that bind to the stimulatory signal. Interestingly, a small ligand binding to the HIV-1 frameshift stimulatory signal was identified from a resin-bound dynamic combinatorial library (McNaughton et al., 2007). A chemically modified aminoglycoside antibiotic, guanidinoneomycin B, was also found to bind to this stimulatory signal (Staple et al., 2008). However, it is possible that such drugs bind non-discriminatorily to many RNA sequences and motifs, and their effect on HIV-1 has not yet been tested.

8.6 Conclusions and Perspectives

The characterization of the programmed −1 ribosomal frameshift in HIV-1 has considerably progressed in the last few years, although the potential role of cellular factors in this process remains to be discovered. The discovery and evaluation of the role of cellular factors in frameshifting constitutes a priority and could lead to a deeper understanding of the regulation of the frameshift during virus infection. The possibility of targeting the frameshift as an HIV therapeutic strategy is a worthy goal. A major problem with the current treatments of patients infected with HIV-1 is the emergence of drug-resistant strains, which creates a compelling need to find novel targets. The development of drugs targeting the frameshift, alone or in combination with other drugs, for the treatment of individuals affected with HIV-1 constitutes an exciting avenue for future investigation.

Acknowledgments We thank Sergey Steinberg, Kevin Wilson and Steve Michnick for very stimulating discussions and for critical reading of this review. We are also grateful to Pascal Chartrand, Gerardo Ferbeyre, Nikolaus Heveker, Luis Rokeach and all the members of the Brakier-Gingras group for critical reading of this manuscript. Work from this laboratory that is cited herein was supported by the Canadian Institutes for Health Research.

References

Agris PF (2008) Bringing order to translation: the contributions of transfer RNA anticodon-domain modifications. EMBO Rep 9:629–635

Baranov PV, Gesteland RF, Atkins JF (2004) P-site tRNA is a crucial initiator of ribosomal frameshifting. RNA 10:221–230

Baril M, Dulude D, Gendron K, Lemay G, Brakier-Gingras L (2003a) Efficiency of a programmed −1 ribosomal frameshift in the different subtypes of the human immunodeficiency virus type 1 group M. RNA 9:1246–1253

Baril M, Dulude D, Steinberg SV, Brakier-Gingras L (2003b) The frameshift stimulatory signal of human immunodeficiency virus type 1 group O is a pseudoknot. J Mol Biol 331:571–583

Biswas P, Jiang X, Pacchia AL, Dougherty JP, Peltz SW (2004) The human immunodeficiency virus type 1 ribosomal frameshifting site is an invariant sequence determinant and an important target for antiviral therapy. J Virol 78:2082–2087

Brass AL, Dykxhoorn DM, Benita Y, Yan N, Engelman A, Xavier RJ, Lieberman J, Elledge SJ (2008) Identification of host proteins required for HIV infection through a functional genomic screen. Science 319:921–926

Brierley I, Dos Ramos FJ (2006) Programmed ribosomal frameshifting in HIV-1 and the SARS-CoV. Virus Res 119:29–42

Brierley I, Meredith MR, Bloys AJ, Hagervall TG (1997) Expression of a coronavirus ribosomal frameshift signal in *Escherichia coli*: influence of tRNA anticodon modification on frameshifting. J Mol Biol 270:360–373

Brierley I, Pennell S (2001) Structure and function of the stimulatory RNAs involved in programmed eukaryotic−1 ribosomal frameshifting. Cold Spring Harb Symp Quant Biol 66:233–248

Brierley I, Pennell S, Gilbert RJ (2007) Viral RNA pseudoknots: versatile motifs in gene expression and replication. Nature Rev Microbiol 5:598–610

Brierley I, Rolley NJ, Jenner AJ, Inglis SC (1991) Mutational analysis of the RNA pseudoknot component of a coronavirus ribosomal frameshifting signal. J Mol Biol 220:889–902

Brunelle MN, Brakier-Gingras L, Lemay G (2003) Replacement of murine leukemia virus readthrough mechanism by human immunodeficiency virus frameshift allows synthesis of viral proteins and virus replication. J Virol 77:3345–3350

Carlson BA, Kwon SY, Chamorro M, Oroszlan S, Hatfield DL, Lee BJ (1999) Transfer RNA modification status influences retroviral ribosomal frameshifting. Virology 255:2–8

Carlson BA, Mushinski JF, Henderson DW, Kwon SY, Crain PF, Lee BJ, Hatfield DL (2001) 1-Methylguanosine in place of Y base at position 37 in phenylalanine tRNA is responsible for its shiftiness in retroviral ribosomal frameshifting. Virology 279:130–135

Daviter T, Gromadski KB, Rodnina MV (2006) The ribosome's response to codon-anticodon mismatches. Biochimie 88:1001–1011

Dinman JD, Richter S., Plant EP, Taylor RC, Hammell AB, Rana TM (2002) The frameshift signal of HIV-1 involves a potential intramolecular triplex RNA structure. Proc Natl Acad Sci USA 99:5331–5336

Dinman JD, Ruiz-Echevarria MJ, Czaplinski K., Peltz SW (1997) Peptidyl-transferase inhibitors have antiviral properties by altering programmed –1 ribosomal frameshifting efficiencies: development of model systems. Proc Natl Acad Sci USA 94:6606–6611

Dulude D, Baril M, Brakier-Gingras L (2002) Characterization of the frameshift stimulatory signal controlling a programmed –1 ribosomal frameshift in the human immunodeficiency virus type 1. Nucleic Acids Res 30:5094–5102

Dulude D, Berchiche YA, Gendron K, Brakier-Gingras L, Heveker N (2006) Decreasing the frameshift efficiency translates into an equivalent reduction of the replication of the human immunodeficiency virus type 1. Virology 345:127–136

Dulude D, Theberge-Julien G, Brakier-Gingras L, Heveker N (2008) Selection of peptides interfering with a ribosomal frameshift in the human immunodeficiency virus type 1. RNA 14:981–991

Farabaugh PJ (1997) Programmed alternative reading of the genetic code: RG Landes Company. Austin, Texas, USA and Springer-Verlag, Heidelberg, Germany, pp 69–101

Frank J, Gao H, Sengupta J, Gao N, Taylor DJ (2007) The process of mRNA-tRNA translocation. Proc Natl Acad Sci USA 104:19671–19678

Gao F, Bailes E, Robertson DL, Chen Y, Rodenburg CM, Michael SF, Cummins LB, Arthur LO, Peeters M., Shaw GM, Sharp PM, Hahn BH (1999) Origin of HIV-1 in the chimpanzee *Pan troglodytes troglodytes*. Nature 397:436–441

Gaudin C, Mazauric MH, Traikia M, Guittet E, Yoshizawa S, Fourmy D (2005) Structure of the RNA signal essential for translational frameshifting in HIV-1. J Mol Biol 349: 1024–1035

Gendron K, Charbonneau J, Dulude D, Heveker N, Ferbeyre G, Brakier Gingras L (2008) The presence of the TAR RNA structure alters the programmed –1 ribosomal frameshift efficiency of the human immunodeficiency virus type 1 (HIV-1) by modifying the rate of translation initiation. Nucleic Acids Res 36:30–40

Gendron K, Dulude D, Lemay G, Ferbeyre G, Brakier-Gingras L (2005) The virion-associated Gag-Pol is decreased in chimeric Moloney murine leukemia viruses in which the readthrough region is replaced by the frameshift region of the human immunodeficiency virus type 1. Virology 334:342–352

Grentzmann G, Ingram JA, Kelly PJ, Gesteland RF, Atkins JF (1998) A dual-luciferase reporter system for studying recoding signals. RNA 4:479–486

Harger JW, Dinman JD (2003) An in vivo dual-luciferase assay system for studying translational recoding in the yeast *Saccharomyces cerevisiae*. RNA 9:1019–1024

Howard MT, Gesteland RF, Atkins JF (2004) Efficient stimulation of site-specific ribosome frameshifting by antisense oligonucleotides. RNA 10:1653–1661

Hung M, Patel P, Davis S, Green SR (1998) Importance of ribosomal frameshifting for human immunodeficiency virus type 1 particle assembly and replication. J Virol 72: 4819–4824

Jacks T, Madhani HD, Masiarz FR, Varmus HE (1988a) Signals for ribosomal frameshifting in the Rous sarcoma virus gag-pol region. Cell 55:447–458

Jacks T., Power MD, Masiarz FR, Luciw PA, Barr PJ, Varmus HE (1988b) Characterization of ribosomal frameshifting in HIV-1 gag-pol expression. Nature 331:280–283

Jenner L, Rees B, Yusupov M, Yusupova G (2007) Messenger RNA conformations in the ribosomal E site revealed by X-ray crystallography. EMBO Rep 8:846–850

Karacostas V, Wolffe EJ, Nagashima K, Gonda MA, Moss B (1993) Overexpression of the HIV-1 gag-pol polyprotein results in intracellular activation of HIV-1 protease and inhibition of assembly and budding of virus-like particles. Virology 193:661–671

Kollmus H, Hentze MW, Hauser H (1996) Regulated ribosomal frameshifting by an RNA-protein interaction. RNA 2:316–323

Kontos H, Napthine S, Brierley I (2001) Ribosomal pausing at a frameshifter RNA pseudoknot is sensitive to reading phase but shows little correlation with frameshift efficiency. Mol Cell Biol 21:8657–8670

Léger M, Dulude D, Steinberg SV, Brakier-Gingras L (2007) The three transfer RNAs occupying the A, P and E sites on the ribosome are involved in viral programmed –1 ribosomal frameshift. Nucleic Acids Res 35:5581–5592

Léger M, Sidani S, Brakier-Gingras L (2004) A reassessment of the response of the bacterial ribosome to the frameshift stimulatory signal of the human immunodeficiency virus type 1. RNA 10:1225–1235

Liiv A, O'Connor M (2006) Mutations in the intersubunit bridge regions of 23S rRNA. J Biol Chem 281:29850–29862

Lopinski JD, Dinman JD, Bruenn JA (2000) Kinetics of ribosomal pausing during programmed –1 translational frameshifting. Mol Cell Biol 20:1095–1103

Marcheschi RJ, Staple DW, Butcher SE (2007) Programmed ribosomal frameshifting in SIV is induced by a highly structured RNA stem-loop. J Mol Biol 373:652–663

McNaughton BR, Gareiss PC, Miller BL (2007) Identification of a selective small-molecule ligand for HIV-1 frameshift-inducing stem-loop RNA from an 11,325 member resin bound dynamic combinatorial library. J Am Chem Soc 129:11306–11307

Michiels PJ, Versleijen AA, Verlaan PW, Pleij CW, Hilbers CW, Heus HA (2001) Solution structure of the pseudoknot of SRV-1 RNA, involved in ribosomal frameshifting. J Mol Biol 310: 1109–1123

Moran SJ, Flanagan JF, Namy O, Stuart DI, Brierley I, Gilbert RJ (2008) The mechanics of translocation: a molecular "spring-and-ratchet" system. Structure 16:664–672

Namy O, Moran SJ, Stuart DI, Gilbert RJ, Brierley I (2006) A mechanical explanation of RNA pseudoknot function in programmed ribosomal frameshifting. Nature 441:244–247

Nierhaus KH (2006) Decoding errors and the involvement of the E-site. Biochimie 88: 1013–1019

Nilsson J, Sengupta J, Frank J, Nissen P (2004) Regulation of eukaryotic translation by the RACK1 protein: a platform for signalling molecules on the ribosome. EMBO Rep 5:1137–1141

Olsthoorn RC, Laurs M, Sohet F, Hilbers CW, Heus HA, Pleij CW (2004) Novel application of sRNA: stimulation of ribosomal frameshifting. RNA 10:1702–1703

Parisien M, Major F (2008) The MC-Fold and MC-Sym pipeline infers RNA structure from sequence data. Nature 452:51–55

Park J, Morrow CD (1991) Overexpression of the gag-pol precursor from human immunodeficiency virus type 1 proviral genomes results in efficient proteolytic processing in the absence of virion production. J Virol 65:5111–5117

Parkin NT, Chamorro M, Varmus HE (1992) Human immunodeficiency virus type 1 gag-pol frameshifting is dependent on downstream mRNA secondary structure: demonstration by expression in vivo. J Virol 66:5147–5151

Paul CP, Barry JK, Dinesh-Kumar SP, Brault V, Miller WA (2001) A sequence required for –1 ribosomal frameshifting located four kilobases downstream of the frameshift site. J Mol Biol 310:987–999

Plant EP, Dinman JD (2006) Comparative study of the effects of heptameric slippery site composition on –1 frameshifting among different eukaryotic systems. RNA 12: 666–673

Plant EP, Jacobs KL, Harger JW, Meskauskas A, Jacobs JL, Baxter JL, Petrov AN, Dinman JD (2003) The 9-A solution: how mRNA pseudoknots promote efficient programmed –1 ribosomal frameshifting. RNA 9:168–174

Rodnina MV, Gromadski KB, Kothe U, Wieden HJ (2005) Recognition and selection of tRNA in translation. FEBS Lett 579:938–942

Sanders CL, Curran JF (2007) Genetic analysis of the E site during RF2 programmed frameshifting. RNA 13:1483–1491

Sengupta J, Nilsson J, Gursky R, Spahn CM, Nissen P, Frank J (2004) Identification of the versatile scaffold protein RACK1 on the eukaryotic ribosome by cryo-EM. Nature Struct Mol Biol 11:957–962

Sharp PM, Bailes E, Chaudhuri RR, Rodenburg CM, Santiago MO, Hahn BH (2001) The origins of acquired immune deficiency syndrome viruses: where and when? Phil Trans Roy Soc (Lond.) 356:867–876

Shehu-Xhilaga M, Crowe SM, Mak J (2001) Maintenance of the Gag/Gag-Pol ratio is important for human immunodeficiency virus type 1 RNA dimerization and viral infectivity. J Virol 75: 1834–1841

Shen LX, Cai Z, Tinoco I Jr (1995) RNA structure at high resolution. FASEB J 9:1023–1033

Simon F, Mauclere P, Roques P, Loussert-Ajaka I, Muller-Trutwin MC, Saragosti S, Georges-Courbot MC, Barre-Sinoussi F, Brun-Vezinet F (1998) Identification of a new human immunodeficiency virus type 1 distinct from group M and group O. Nat Med 4:1032–1037

Somogyi P, Jenner AJ, Brierley I, Inglis SC (1993) Ribosomal pausing during translation of an RNA pseudoknot. Mol Cell Biol 13:6931–6940

Staple DW, Butcher SE (2005) Solution structure and thermodynamic investigation of the HIV-1 frameshift inducing element. J Mol Biol 349:1011–1023

Staple DW, Venditti V, Niccolai N, Elson-Schwab L, Tor Y, Butcher SE (2008) Guanidinoneomycin B recognition of an HIV-1 RNA helix. Chembiochem 9:93–102

Su L, Chen L, Egli M, Berger JM, Rich A (1999) Minor groove RNA triplex in the crystal structure of a ribosomal frameshifting viral pseudoknot. Nature Struct Biol 6:285–292

Takyar S, Hickerson RP, Noller HF (2005) mRNA helicase activity of the ribosome. Cell 120: 49–58

Telenti A, Martinez R, Munoz M, Bleiber G, Greub G, Sanglard D, Peters S (2002) Analysis of natural variants of the human immunodeficiency virus type 1 gag-pol frameshift stem-loop structure. J Virol 76:7868–7873

Tu C, Tzeng TH, Bruenn JA (1992) Ribosomal movement impeded at a pseudoknot required for frameshifting. Proc Natl Acad Sci USA 89:8636–8640

Vickers TA, Ecker DJ (1992) Enhancement of ribosomal frameshifting by oligonucleotides targeted to the HIV gag-pol region. Nucleic Acids Res 20:3945–3953

Waas WF, Druzina Z, Hanan M, Schimmel P (2007) Role of a tRNA base modification and its precursors in frameshifting in eukaryotes. J Biol Chem 282:26026–26034

Weiss RB, Dunn DM, Shuh M, Atkins JF, Gesteland RF (1989) *E. coli* ribosomes re-phase on retroviral frameshift signals at rates ranging from 2 to 50 percent. New Biol 1:159–169.

Wen JD, Lancaster L, Hodges C, Zeri AC, Yoshimura SH, Noller HF, Bustamante C, Tinoco I (2008) Following translation by single ribosomes one codon at a time. Nature 452: 598–603

Yelverton E, Lindsley D, Yamauchi P, Gallant JA (1994) The function of a ribosomal frameshifting signal from human immunodeficiency virus-1 in *Escherichia coli*. Mol Microbiol 11: 303–313

Yu ET, Zhang Q, Fabris D (2005) Untying the FIV frameshifting pseudoknot structure by MS3D. J Mol Biol 345:69–80

Yusupov MM, Yusupova GZ, Baucom A, Lieberman K, Earnest TN, Cate JH, Noller HF (2001) Crystal structure of the ribosome at 5.5 A resolution. Science 292:883–896

Yusupova GZ, Yusupov MM, Cate JH, Noller HF (2001) The path of messenger RNA through the ribosome. Cell 106:233–241

Chapter 9
Ribosomal Frameshifting in Decoding Plant Viral RNAs

W. Allen Miller and David P. Giedroc

Abstract Frameshifting provides an elegant mechanism by which viral RNA both encodes overlapping genes and controls expression levels of those genes. As in animal viruses, the −1 ribosomal frameshift site in the viral mRNA consists of a canonical shifty heptanucleotide followed by a highly structured frameshift stimulatory element, and the gene translated as a result of frameshifting usually encodes the RNA-dependent RNA polymerase. In plant viruses, the −1 frameshift stimulatory element consists of either (i) a small pseudoknot stabilized by many triple-stranded regions and a triple base pair containing a protonated cytidine at the helical junction, (ii) an unusual apical loop–internal loop interaction in which a stem-loop in the 3′ untranslated region 4 kb downstream base pairs to a bulged stem-loop at the frameshift site, or (iii) a potential simple stem-loop. Other less well-characterized changes in reading frame occur on plant viral RNAs, including a possible +1 frameshift, and net −1 reading frame changes that do not utilize canonical frameshift signals. All these studies reveal the remarkable ways in which plant viral RNAs interact with ribosomes to precisely control protein expression at the ratios needed to sustain virus replication.

Contents

W.A. Miller (✉)
Plant Pathology Department, and Biochemistry, Biophysics & Molecular Biology Departments, Iowa State University, 351 Bessey Hall, Ames, IA 50011, USA
e-mail: wamiller@iastate.edu

J.F. Atkins, R.F. Gesteland (eds.), *Recoding: Expansion of Decoding Rules Enriches Gene Expression*, Nucleic Acids and Molecular Biology 24,
DOI 10.1007/978-0-387-89382-2_9,

9.1 Introduction

9.1.1 Frameshifting Plant Viruses

Plant viruses known to employ minus one (−1) programmed ribosomal frameshifting include one genus (*Dianthovirus*) in the Tombusviridae family, all three genera (*Luteovirus, Polerovirus*, and *Enamovirus*) of the Luteoviridae, and the *Umbravirus* and *Sobemovirus* genera. The latter two genera bear some resemblance in genome organization and sequence to the *Luteovirus* and *Polerovirus* genera, respectively, but have not been assigned to a family (Hull and Fargette, 2005; Miller et al., 2002; Taliansky et al., 2005). It is likely that translation of the genome of an entirely unrelated genus, *Closterovirus*, undergoes a net +1 reading frame change to translate the viral RdRp coding region (Karasev et al., 1995). This would be the first known +1 frameshift in any plant viral RNA. A carlavirus may use −1 frameshifting by a novel "P-site only" mechanism (Gramstat et al., 1994). Recently a new ORF was discovered in the large Potyviridae family of plant viruses (Chung et al., 2008). The new ORF, called pipo, overlaps with the major ORF which encodes a large polyproteins. One likely mechanism for expression of the pipo ORF is −1 ribosomal frameshifting.

9.1.2 Why Frameshift?

RNA viruses have evolved to compress maximal protein-coding capacity and regulatory signals for RNA replication, localization, encapsidation, and gene expression into minimal sequence space (Belshaw et al., 2007; Holmes, 2003). Programmed ribosomal frameshifting allows the virus to control levels of synthesis of two proteins using one sequence that codes for a portion of both proteins. The result of the frameshift is two proteins that have the same amino terminal sequence up to the frameshift site, at which point the amino acid sequences diverge, with a small portion of the proteins having the −1 frame (frameshifted) sequence, while the majority of the proteins result from conventional unshifted translation of the initial frame (called the zero frame). Thus, only a small percentage of ribosomes change reading frames in plant viral frameshifting. In many viruses, such as those in the *Luteovirus, Dianthovirus*, and *Umbravirus* genera, the frameshift occurs immediately upstream of the zero frame stop codon, so the translated proteins consist of a short and long (C-terminally extended) version of the same protein (Fig. 9.1). In contrast, in the

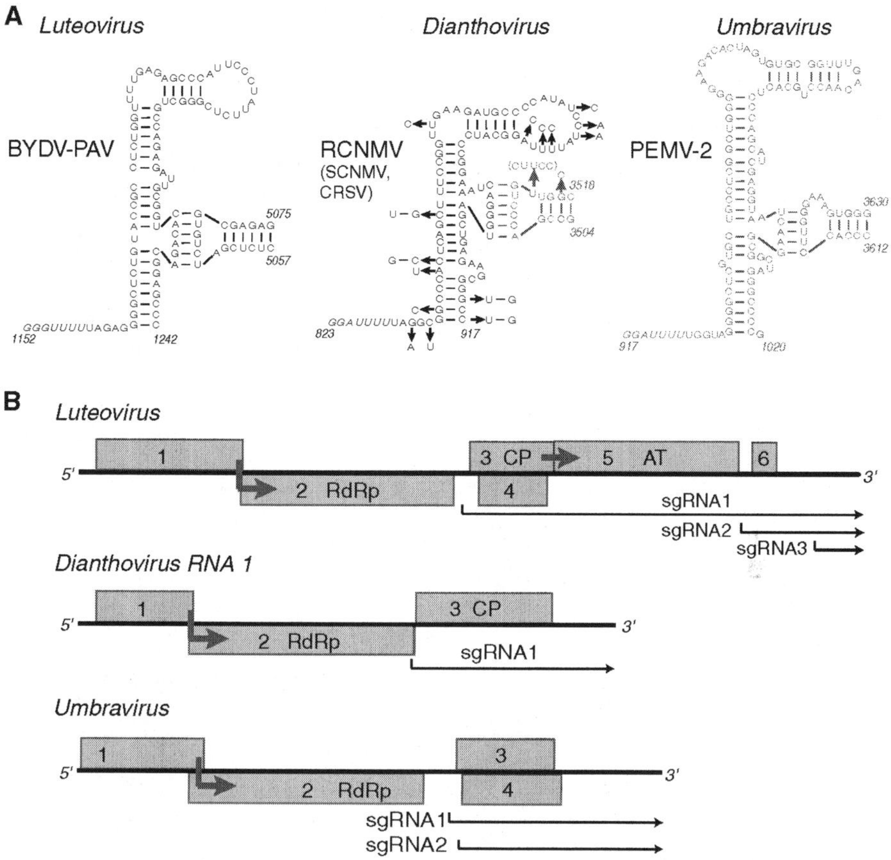

Fig. 9.1 Frameshifting in Luteo-, Diantho-, and Umbraviruses. (**A**) Representative secondary structures required for −1 frameshifting by each of the three genera known or thought to harbor such structures. The shifty heptanucleotide (SH) at which the ribosomes change reading frame is indicated in italics. The apical loop–internal loop frameshift structures comprising the adjacent downstream-bulged stem-loop (ADSL) base-paired to the long-distance frameshift element (LDFE) in the 3′ UTR are indicated for each virus. Base numbering indicates positions of the bases in the viral genome. These structures are predicted here for the first time in Diantho- and Umbraviruses. They are conserved in all genus *Luteovirus* (Barry and Miller, 2002; Salem et al., 2008) and *Umbravirus* RNAs but diverge significantly in sequence (not shown). The bases and base pairs that differ from RCNMV in the other two dianthoviruses, SCNMV and CRSV, are shown with *arrows* (as in (Kim and Lommel, 1998)). Known structures are in *black*; our predicted structures are in *gray* text. (**B**) Genome organizations of viruses in each of the three genera. *Gray square arrow* indicates frameshift site. ORF 2 encodes the RdRp and is translated via −1 frameshift from ORF 1. Luteovirus ORFs 3, 4, and 5 encode coat protein (CP), putative movement protein, and a 3′ extension of ORF 3 required for aphid transmission (AT), respectively. ORF 5 is translated via ribosomal readthrough of the leaky ORF 3 stop codon (gray straight arrow). Genomes in all three genera lack a 5′ cap or VPg and a poly(A) tail. The approximate positions of sgRNAs are indicated by the *black arrows*

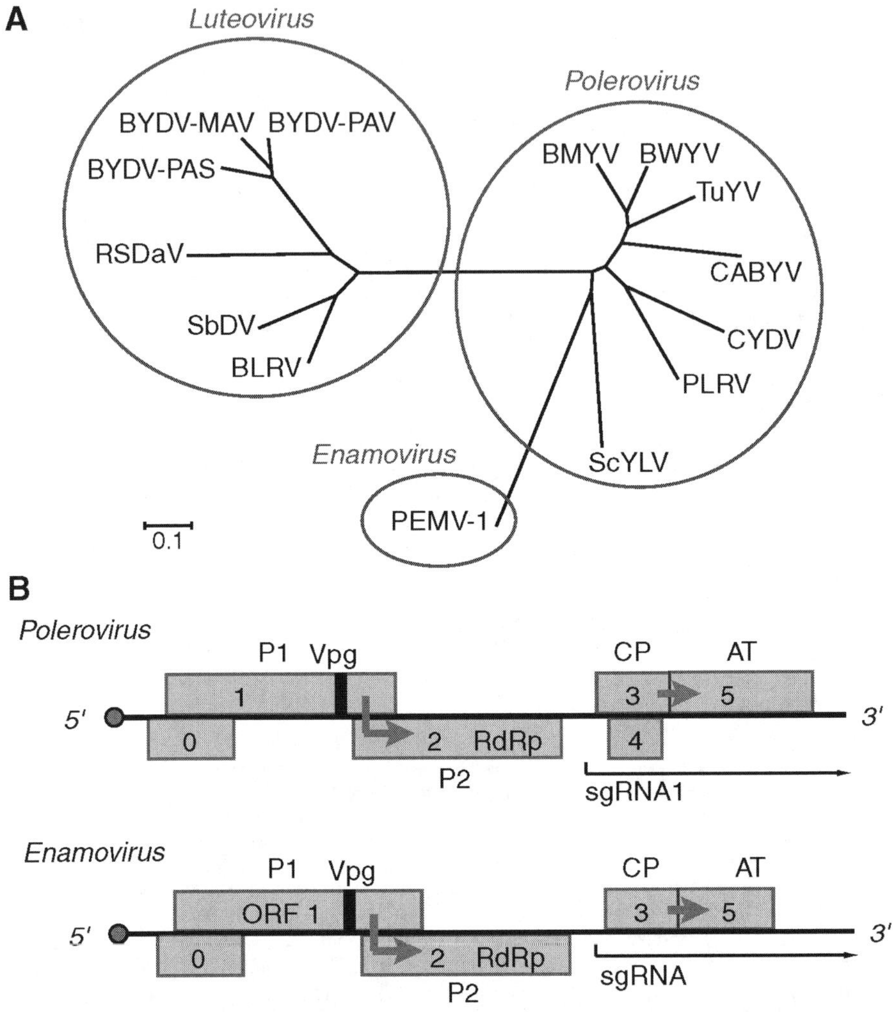

Fig. 9.2 Phylogenetic relationships among the *Luteoviridae* and schematic representations of *Polerovirus* and *Enamovirus* genomic RNAs. (**A**) Unrooted radial phylogenetic tree of selected *Luteoviridae* sequences. Members of the three genera are circled with the genus indicated in italics. (**B**) Genome organizations of polero- and enamoviruses, highlighting the P1–P2 frameshift sites *(gray square arrow)* in each case. ORF designations are from 0 to 5, with ORF 1 and ORF 2 encoding P1 and P2, respectively; CP, coat protein; AT, aphid transmission factor. *Gray circle*, VPg covalently linked to the 5′ end of the genomic RNA; *straight gray arrow,* termination codon between the CP and AT genes that is readthrough to make a CP–AT fusion protein. The approximate positions of sgRNAs are indicated by the *black arrows*

polero- and enamoviruses, significant translation continues in the zero frame beyond the frameshift site (Fig. 9.2). In all plant viruses, with the possible exception of the Potyviridae and carlaviruses, the protein translated via ribosomal frameshifting is the RNA-dependent RNA polymerase (RdRp).

The high ratio of a viral protein to a C-terminally extended version of the protein can also be achieved by in-frame readthrough of a stop codon (codon redefinition). In fact, different viruses in the same family (Tombusviridae) use either readthrough or frameshifting to express the RdRp (Lommel et al., 2005). The portion of the RdRp upstream of the leaky (recoded) stop codon in Tomato bushy stunt virus (TBSV, a tobamovirus), called p33, equivalent to the pre-shift portion of the RdRp in frameshifting viruses specifically binds viral RNA (Pogany et al., 2005). The full-length readthrough product ($p92^{pol}$) includes the p33 and downstream RdRp domains (Panavas et al., 2005). p33 then localizes the viral RNA and $p92^{pol}$ to membrane-bound vesicles derived from the peroxisomal membrane (McCartney et al., 2005). By analogy with Brome mosaic virus (Schwartz et al., 2002), these may serve as replication factories with many more copies of p33 than $p92^{pol}$. Thus infrequent in-frame readthrough of the stop codon is necessary to ensure a high ratio of p33 to RdRp in TBSV-infected cells. Based on sequence homologies (Koonin and Dolja, 1993) and the observation that the homologous protein (p27) of dianthovirus red clover necrotic mosaic virus (RCNMV) also localizes to membranes (Turner et al., 2004), it is highly likely that the pre-shift ORF of viruses in the *Dianthovirus, Luteovirus*, and *Umbravirus* genera play a role similar to that of p33. In these viruses, infrequent frameshifting, rather than readthrough, ensures a low level of RdRp expression. In addition to playing a role in RNA replication, the pre-shift part of ORF 2 of poleroviruses, enamovirus, and sobemoviruses also encodes the viral protease which is needed early in infection and the 5′ genome-linked protein (VPg) (van der Wilk et al., 1997) which is absent in the Tombusviridae and in genus *Luteovirus*.

Unlike in-frame stop codon readthrough, frameshifting requires overlapping genes which allow for compression of coding regions on the genome. Thus, at the slippery site and downstream frameshift-stimulating structure in the mRNA, one sequence tract includes a *cis*-acting frameshift signal and encodes portions of two proteins, exemplifying the remarkable multiple functions that evolution has bestowed on a single tract of viral sequence.

9.1.3 Proposed Mechanisms of −1 Frameshift Stimulation

Ribosomal frameshifting is programmed entirely by the mRNA. For a region of mRNA to promote programmed −1 frameshifting, two sequence elements are required at a minimum. These include a shifty heptanucleotide at which the ribosome pauses and shifts reading frame (below), followed six to eight bases downstream by a highly structured region, such as a pseudoknot, a stable stem-loop, or other structure that greatly stimulates the process (see Chapters 7 and 8). It is currently unknown exactly how a downstream structure, e.g., a pseudoknot, stimulates −1 frameshifting. However, several proposals have been made that attempt to identify the exact step in a translocation cycle where frameshifting occurs on the ribosome (Harger et al., 2002; Jacks et al., 1988; Leger et al., 2007; Plant et al., 2003) (for a review, see Giedroc and Cornish (2009)). For a pseudoknot, what is clear is that the relatively short spacer between the slip-site and the pseudoknot places the pseudoknot in direct contact with the mRNA entry channel of the translocating ribosome (Namy et al., 2006). In bacteria and probably also in eukaryotes, the mRNA channel is contained totally within the small (30S in bacteria) ribosomal subunit, between the head and body, lined with ribosomal proteins S3, S4, and S5 (Brodersen et al., 2002). Biochemical experiments reveal that the 70S ribosome has helicase activity (Takyar et al., 2005), which may function by passively trapping transiently unfolded secondary structure by dsRNA-binding protein S5; alternatively, the S3/S4/S5 proteins might function as a processivity clamp positioned at the entrance to the mRNA tunnel (Takyar et al., 2005).

Because error-free translation likely requires that the A- and P-site tRNAs maintain hydrogen bonding contact with the mRNA during both 60S and 40S translocation, at some point in an elongation cycle with the paused ribosome positioned over the slip-site, these interactions must be broken for frameshifting to occur. The sequence of the heptanucleotide slip-site in a −1 frameshift signal is such that during re-pairing of the P- and A-site tRNAs in the new −1 reading frame, only the wobble codon–anticodon interaction is changed ("0" frame: X $\underline{\text{XXY}}$ $\underline{\text{YYZ}}$ to "−1" frame: $\underline{\text{XXX}}$ $\underline{\text{YYY}}$ Z) (Brierley et al., 1992). As a result, the new −1 reading frame provides near cognate re-pairing partners for the bound tRNAs. This suggests that the total free energy of codon–anticodon formation of the −1 frame will be comparable to that of the reference frame (frame 0). Since frameshifting does in fact occur over such a sequence, but happens infrequently, a sizable transition state energy barrier to shifting reading frames must be present, for which there is now structural insight (Selmer et al., 2006). A downstream secondary structural element could therefore function by lowering this energy barrier, either by playing an active (mechanical) or passive role in this process.

A mechanical model of frameshift stimulation, using our understanding of prokaryotic ribosomes as a model, hypothesizes that the downstream pseudoknot functions as a kinetic barrier to normal translocation, and in so doing, lowers the energy barrier to tRNA–mRNA re-pairing. It can be imagined that this kinetic barrier would come into play either during spontaneous hybrid A/P P/E states formation (Cornish et al., 2008) (coincident with 50S translocation) and/or during the next step in the elongation cycle, EF-G (eEF2)•GTP-driven translocation (30S translocation) (Namy et al., 2006). In either case, movement of the ribosomal subunits may cause tension to build up in the mRNA strand due to the downstream structural element positioned in the mRNA entry channel. This tension could then be released via −1 frameshifting (Plant et al., 2003).

Recent support for a mechanical model of frameshift stimulation has emerged from cryo-electron microscopy images of mammalian 80S ribosomes (to a resolution of ≈16 Å) paused over the infectious bronchitis virus (IBV) pp1a/pp1b pseudoknot frameshift signal (Namy et al., 2006). As expected (Yusupova et al., 2001), the pseudoknot lies at the entrance to the mRNA channel apparently making direct contact with elements of the putative ribosomal helicase, including rpos3 (bacterial S3), 16S helix 16, rpS9 (S4), and rpS2 (S5), as well as the ribosome-regulatory protein RACK1. Furthermore, the clearly defined electron density observed for the pseudoknot strongly suggests that the pseudoknot is largely folded as the ribosome shifts reading frames. This structure is fully compatible with the idea that the ability of the downstream pseudoknot to actively lower the energy barrier for unpairing of the P-site codon–anticodon interaction will be more strongly correlated with frameshift stimulation. By extension, RNA motifs more capable of resisting the force of ribosomal helicase-mediated unwinding and eEF2(EF-G)-catalyzed translocation will thus be more efficient frameshift stimulators (Larsen et al., 1997).

In plant viral RNAs there appear to be three classes of RNA structure downstream of the slippery site that can facilitate −1 ribosomal frameshifting: (i) an "apical loop–internal loop" (ALIL) structure in which a bulged stem-loop, located 5–6 nt downstream of the slippery site, base pairs to a distant loop in the 3′ UTR; (ii) a very small, highly structured hairpin-type pseudoknot; or (iii) a stable, imperfect stem-loop. Other as-yet unidentified structures may facilitate frameshifting in other plant viruses such as the Potyviridae (see below). The following discussion provides an overview of these different structures, with an emphasis on the pseudoknots in the polero and enamoviruses which, due to their small size, are among the best characterized frameshift signals in eukaryotes (Cornish et al., 2005).

9.2 Plant Virus Frameshift Elements

9.2.1 Luteoviridae

The *Luteoviridae* are nonenveloped, icosahedral, aphid-transmissible, single-stranded positive(+)-sense RNA viruses. They are grouped in three genera: *Luteovirus*, *Polerovirus*, and *Enamovirus* (D'Arcy and Domier, 2005; Mayo and D'Arcy, 1999; Miller et al., 2002). The genomes of all *Luteoviridae* are 5.6–5.8 kb long and typically have six open reading frames (ORFs) which are divided into two clusters separated by a noncoding intergenic region (Figs. 9.1 and 9.2). ORFs 1 and 2 encode proteins P1 and P2 which form part of the viral replicase. In poleroviruses and enamovirus, P1 is a polyprotein precursor containing a putative helicase, a chymotrypsin-like (3C-like) proteinase, and VPg peptides (van der Wilk et al., 1997), while the exact biochemical function of P1 of genus *Luteovirus* P1 is ill-defined, as viruses in this genus have neither a VPg nor any known proteinase. In all Luteoviridae, P1 plays an essential role in RNA replication and P2 contains the RNA-dependent RNA polymerase active site (RdRp) (D'Arcy and Domier, 2005). The RdRp is translated as the C-terminal part of a P1–P2 fusion protein via −1 ribosomal frameshifting (Brault and Miller, 1992; Di et al., 1993; Garcia et al., 1993; Kujawa et al., 1993; Prüfer et al., 1992).

Within the Luteoviridae, the RNA structures that facilitate frameshifting differ markedly, depending on the genus. Viruses in genus *Luteovirus* are known or predicted to have a GGGUUUU shifty site. The structured region (Fig. 9.1A) that begins six to eight bases downstream of the shifty heptanucleotide consists of a large adjacent downstream-bulged stem-loop (ADSL) that forms a complex pseudoknot by base pairing of a bulge loop in the ADSL to a stem-loop in the long-distance frameshift element (LDFE) located four kilobases downstream in the 3′ UTR (Barry and Miller, 2002). In contrast, viruses in the other two genera of the Luteoviridae harbor a small, compact, highly structured pseudoknot adjacent to, and downstream of, the shifty site (see below). These striking differences between *Luteovirus* and *Polerovirus/Enamovirus* frameshift signals correlate with highly divergent sequences of the RdRp (ORF 2) and upstream (ORF 1) coding regions (D'Arcy and Domier, 2005). In fact, the RdRp of genus *Luteovirus* is more closely related to those of the Tombusviridae and Umbraviruses than to the other genera of the Luteoviridae (Miller, Liu and Beckett, 2002). In particular it is most closely related to the RdRps of viruses in the *Dianthovirus* genus (Tombusviridae) which, unlike other Tombusvirids, employ −1 frameshifting via an RNA structure very similar to that of genus *Luteovirus* (Fig. 9.1).

9.2.2 Frameshift Stimulators Involving Long-Distance Base Pairing

The RNA structures for frameshifting by Barley yellow dwarf virus (BYDV, genus *Luteovirus*) and RCNMV (genus *Dianthovirus*) RNAs have been investigated. Both consist of a shifty site (GGGUUUU in BYDV, GGAUUUU in RCNMV) followed by a large bulged stem-loop (the ADSL) (Fig. 9.1A). The BYDV frameshift structure requires an additional stem-loop, the LDFE, located 4 kb downstream in the genome (Paul et al., 2001). The loop of this stem-loop base pairs to a large single-stranded bulge in the ADSL (Barry and Miller, 2002). Based on our predictions, this structure, now called an apical loop-internal loop (ALIL) interaction (Mazauric et al., 2008), is conserved in all members of the *Luteovirus* (Barry and Miller, 2002; Salem et al., 2008) and *Dianthovirus* genera (WAM unpublished, Fig. 9.1). We predict here that members of genus *Umbravirus*, exemplified by pea enation mosaic virus RNA-2 (PEMV2), contain a similar ALIL structure to stimulate −1 frameshifting (Fig. 9.1A).

The requirements for −1 frameshifting on BYDV RNA were demonstrated using a bicistronic reporter system, consisting of a reporter ORF (encoding Renilla luciferase) in the zero frame fused at its 3′ end to the 3′ end of the viral ORF 1 in the frameshift region, which, in turn, is followed by the 5′ end of the −1 ORF (including the region of ORF overlap and the putative frameshift sequence) fused to a firefly luciferase ORF in the −1 frame, so that a −1 frameshift is required for expression of the firefly luciferase (Grentzmann et al., 1998). The ratio of firefly luciferase to Renilla luciferase produced in the frameshifted construct vs. a positive control with both luciferases in the same reading frame provides a measure of frameshift efficiency. Frameshift efficiency is about 1–2% for BYDV frameshift signals (Barry and Miller, 2002; Paul et al., 2001). Using this system Barry and Miller (2002) showed that mutations that disrupted the ADSL–LDFE base pairing reduced frameshifting to background levels in vitro and in plant cells. In the context of the full-length viral RNA genome, the disruptive mutations also blocked frameshifting in vitro and prevented virus replication in vivo (oat cells). Compensating mutations that restored ADSL–LDFE base pairing also restored frameshifting in vitro and allowed virus replication in oat cells (Barry and Miller, 2002).

In contrast to the requirement for an LDFE in BYDV RNA, Kim and Lommel (1998) reported that the shifty heptanucleotide and the adjacent bulged stem-loop (equivalent to the ADSL in BYDV) were necessary and sufficient for significant (~7%) frameshifting by RCNMV RNA and thus did not require downstream sequences including the stem-loop that we predict forms the LDFE (Fig. 9.1A). We suggest that these experiments may not have revealed all of the sequence requirements for frameshifting on the full-length viral RNA because the frameshifting constructs tested contained a very short (10 codon) first (zero frame) ORF from the start codon through the shifty site which is followed immediately by a stop codon. Indeed, in earlier studies of the BYDV frameshift sequences, Brault and Miller (1992) obtained similar results to those of Kim and Lommel (1998) with a similar construct containing a very short (nine codon) ORF upstream of the shifty site, along with the ADSL, but no LDFE. A peculiar finding was that, unlike in other well-studied −1 frameshift elements from retroviruses or coronaviruses, the stop codon immediately adjacent to the shifty site appeared to be required for the −1 frameshift (Brault and Miller, 1992). Thus, these early constructs in BYDV and more recent experiments with RCNMV, both characterized with a very short first (zero frame) ORF, may not behave "naturally." Certainly, a more native-like context is a full-length ORF encoding a functional protein upstream of the shifty site, as described more recently (Barry and Miller, 2002; Paul et al., 2001), employing either a dual luciferase reporter construct or the natural full-length viral genome. We speculate that ribosomes engaged in decoding a very short first ORF are in an initiation phase of the elongation process that may be "shiftier" than elongating ribosomes farther downstream from the start codon. This shiftiness could be aided by the nearby in-frame stop codon (Brault and Miller, 1992). Another possibility is that folding of the nascent protein may influence shiftiness. The nature of the polypeptide in the exit tunnel can influence ribosome elongation rate (Cruz-Vera et al., 2005; Lu and Deutsch, 2008) which, in turn, may influence frameshifting (Ivanov and Atkins, 2007). The amino terminus of the short ORF 1-encoded protein in the constructs (Brault and Miller, 1992; Kim and Lommel, 1998) would not have exited the polypeptide exit tunnel (~30 amino acids) when the ribosome encounters the shifty sequence and thus would be unable to fold. This is in contrast to the point at which the ribosome encounters the shifty site in the context of mRNAs encoding dual luciferase reporter or viral genome proteins. In this case, much or all of ORF1-encoded protein has been synthesized and likely folded outside of the ribosome, but it remains attached to the ribosome via the growing polypeptide chain near the carboxyl terminus. Folding of the nascent peptide after leaving the exit tunnel may affect elongation rate. Moreover, the dual luciferase reporter provides a more accurate measurement of frameshift rate because it allows measurement of the level of production of the first (zero frame) ORF as well as the post-shift (−1 frame) ORF. Finally, proximity of the frameshift site to the start codon has also been observed to affect +1 frameshifting in a misleading way. In this case, frameshifting was abolished when the +1 frameshift site was placed three codons downstream of the start codon (Belcourt and Farabaugh, 1990).

Additional strong evidence to support a requirement for the ALIL structure (ADSL–LDFE base pairing) is that it is phylogenetically conserved. A stem-loop in the 3′ UTR (putative LDFE, albeit sometimes with a very weak predicted stem) capable of forming five to six base pairs with the 3′ bulge in the ADSL is predicted in all members of the *Luteovirus* and *Dianthovirus* genera (Fig. 9.1), despite many differences in sequence. Interestingly, in the dianthoviruses and umbraviruses, the putative LDFE we predict is located upstream of the cap-independent translation element that is also located in the 3′ UTR (Mizumoto et al., 2003), whereas in the luteoviruses, the LDFE is downstream of the cap-independent translation element which is located in the 5′ end of the 3′ UTR (Paul et al., 2001). Phylogenetic conservation of base pairing is strong evidence of a biological function and given that the ADSL–LDFE base pairing is required for BYDV frameshifting, frameshifting is its likely role for the other viruses, including dianthoviruses.

The ALIL structure is not unique to plant viruses. A different type of ALIL in which the small stem-loop precedes the large bulged stem-loop, and base pairs to a bulge in the 5′ side rather than the 3′ side of the large bulged stem-loop, was shown recently to be required for frameshifting of bacterial transposable elements in the IS3 family (Mazauric et al., 2008). In that study, modeling predicted that the proximal (bottom) helix coaxially stacks with the bulge loop–stem-loop helix (ADSL–LDFE interaction) which forces a sharp bend at the junction of this helix with the upper helix in the ADSL. It is possible that a similar structure forms here and blocks the helicase activity of the ribosome.

The long-distance base pairing spanning up to 4 kb between the ADSL and the LDFE may serve as an RNA traffic control signal, allowing the virus to switch from translation to replication of the viral genome (Barry and Miller, 2002; Miller and White, 2006). All positive-strand RNA viruses must first translate the RdRp (replicase) prior to RNA replication. The replicase initiates complementary RNA synthesis at the 3′ end of the viral template. Early in infection this template must be the same genomic RNA that is being translated. However, ribosomes on the viral RNA would interfere with the replicase moving in the opposite direction toward the 5′ end (Gamarnik and Andino, 1998). Because frameshifting requires the long-distance interaction, the replicase could disrupt the LDFE in the 3′ UTR, and thus block frameshifting, before the replicase reaches the translated ORF. By the time the replicase would reach the RdRp-encoding ORF, that portion of the RNA would be ribosome-free. A similar process is predicted to occur via a cap-independent translation element, located near the LDFE, which must base pair to the 5′ UTR to initiate translation of both the pre-shift and post-shift (RdRp) ORFs (Guo et al., 2001). This interaction would also be disrupted by the replicase, shutting off translation initiation, freeing the entire viral genomic RNA of ribosomes, making it available for full replication (Barry and Miller, 2002). Future experimentation will determine the validity of this model.

9.2.3 Polerovirus and Enamovirus

The four viruses studied in these two genera are beet western yellows virus (BWYV), potato leaf roll virus (PLRV), sugarcane yellow leaf virus (ScYLV), and pea enation mosaic virus RNA1 (PEMV-1). BWYV, PLRV, and ScYLV are classified as poleroviruses, while PEMV-1 is the sole member of the *Enamovirus* genus (Fig. 9.2). PEMV is unique in that it harbors a second taxonomically distinct(+)-sense RNA, PEMV RNA2, classified in the *Umbravirus* genus (Demler et al., 1993). RNA1 and RNA2 can replicate independently but both are required for wild-type levels of infectivity (Demler et al., 1994).

The reported frameshifting efficiency of poleroviruses and enamoviruses ranges from 1−15% depending on the frameshift signal itself as well as the biological context within which the measurement was made; indeed, earlier measurements place this efficiency at ≈1–2% using standard in vitro eukaryotic translation systems (Garcia, van Duin and Pleij, 1993; Kujawa et al., 1993). More recent measurements using a dual luciferase assay place this value at 4–6% for BWYV, 9% for PEMV-1, and ≈15% for ScYLV −1 frameshifting P1–P2 wild-type signals, as measured by coupled transcription–translation assays with rabbit reticulocyte lysates (Cornish, Hennig and Giedroc, 2005; Kim et al., 2000, 1999; Nixon et al., 2002).

The slippery sequence at which the ribosomes change reading frame is GGGAAAC in PEMV-1 (Nixon et al., 2002), BWYV (Garcia et al., 1993), and ScYLV (Moonan et al., 2000; Smith et al., 2000), and a more efficient sequence UUUAAA(U/C) in PLRV (Garcia et al., 1993; Kujawa et al., 1993) (see Fig. 9.3 below). The second element is a downstream hairpin-type (H-type) RNA pseudoknot positioned 6–8 nt from the 3′ edge of the slip-site (see Figs. 9.3 and 9.4). The slip-site alone dramatically increases the intrinsic level of frameshift errors from 0.00005 to ≈0.005 per codon depending strongly on the sequence (Stahl et al., 2002), with the pseudoknot further stimulating this process from 5- to 30-fold more.

9.2.4 Frameshift Stimulators with Compact Pseudoknots

Although a diverse array of RNA motifs are capable of stimulating −1 frameshifting when placed ≈6–8 nt downstream of the slip-site (Giedroc and Cornish, 2009), the atomic (1.3−1.6 Å) resolution crystallographic structure of the 28-nt P1–P2 pseudoknot from BWYV clearly established how an RNA pseudoknot with a very short 3-bp pseudoknot-forming stem S2 could fold and stimulate −1 frameshifting (Egli et al., 2002; Su et al., 1999) (Fig. 9.3A). This RNA is compact and is largely triple-stranded RNA with the majority of the L1 and L2 loop nucleotides making

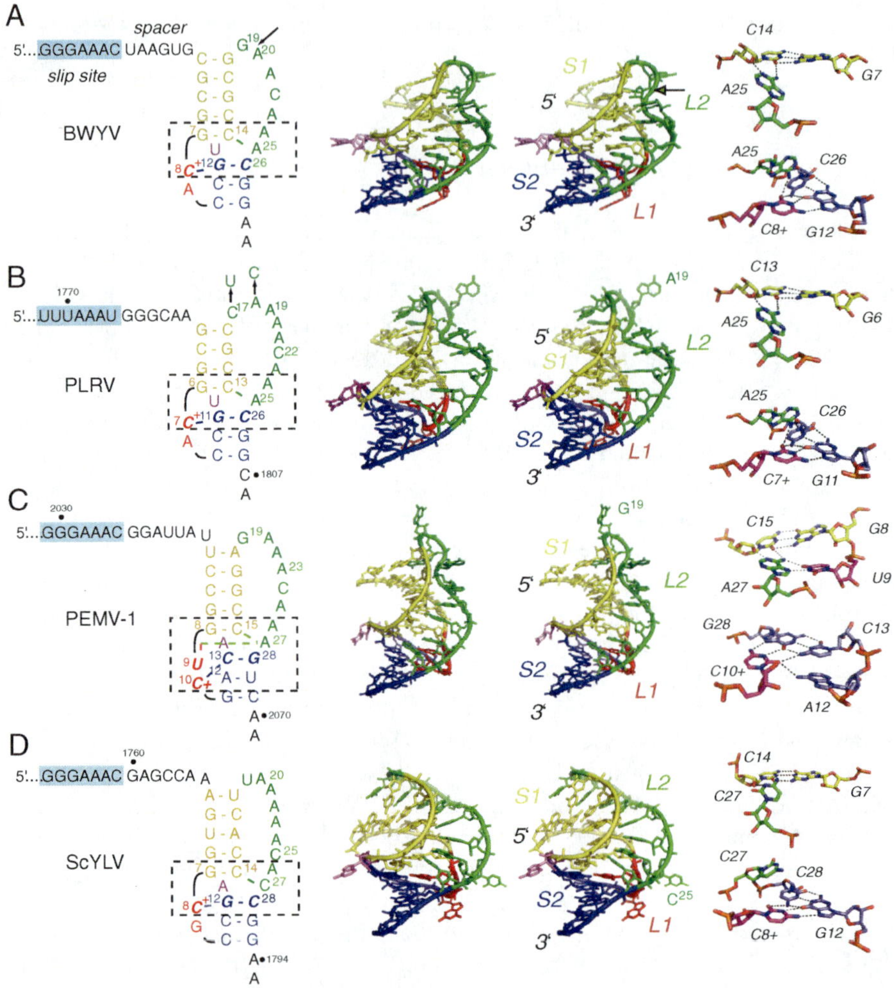

Fig. 9.3 Stereo views of the structures of four related luteoviral P1–P2 pseudoknots. (**A**) BWYV (PDB 1L2X solved to 1.25 Å resolution) (Egli et al., 2002); (**B**) PLRV (PDB 2A43 solved to 1.34 Å resolution) (Pallan et al., 2005); (**C**) PEMV-1 (PDB 2RP1) (Giedroc and Cornish, 2009); (**D**) ScYLV (PDB 1YG4) (Cornish, Hennig and Giedroc, 2005). Secondary structure models of the four frameshift signals are shown on the *far left*, with nucleotides in S1 (*yellow*), L1 (*red*), S2 (*blue*) and L2 (*green*) color-coded as in the stereo views of the structures (*middle*). Genomic RNA nucleotide numbers are also shown for the PLRV, PEMV-1, and ScYLV RNAs. Close-ups of the helical junction regions of all four RNAs are shown to the far *right* (these correspond to the regions *inside* the dashed box on the secondary structure diagrams, *far left*) with the L2–S1 minor groove base triple highlighted at the *top*, and the L1–S2 major groove base quadruple shown at the *bottom*. Stereo images and helical junctions are reprinted from Virus Research 139, Giedroc DP, Cornish PV, Frameshifting RNA pseudoknots: Structure and mechanism, pp 193–208, with permission from Elsevier

noncanonical base–base and base–sugar edge hydrogen bonds with base pairs of the two stems. A protonated C8+ from loop L1 forms a standard Hoogsteen-type base pair with an accepting G12 as part of a C8+•(G12–C26) base triple at the helical junction (Fig. 9.3A). The two helical stems are strongly overrotated relative to one another, with the S1 G7–C14 base pair actually stacked on the S2 C8+•G12 Hoogsteen pair; this overrotation is partly mediated by an unpaired nucleotide (U12 in BWYV) that is flipped out of the stack and may function as a "spacer" at the helical junction. Thermodynamic and structural studies of the three other luteoviral RNA pseudoknots confirm that the C+•(C–G) L1–S2 major groove triple base pair is a common feature of all P1–P2 RNA pseudoknots from poleroviruses and the enamovirus (Fig. 9.3) (Nixon et al., 2002; Nixon and Giedroc, 2000). Although protonation of this cytidine is strongly stabilizing, the degree to which protonation affects frameshifting efficiency or mechanical stability (Tinoco et al., 2006) remains unknown.

The other striking structural feature of the BWYV pseudoknot is a minor groove triplex, in which a run of three consecutive adenosines at the 3′ end of the loop L2 (A23–A24–A25) forms a series of Watson–Crick-sugar edge hydrogen bonding interactions with base pairs of the upper stem S1 closest to the helical junction (see Fig. 9.3A) (Egli et al., 2002; Su et al., 1999). While reminiscent of classical A-minor interactions found in other complex RNAs and the large ribosomal subunit (Doherty et al., 2001; Nissen et al., 2001), these interactions are characterized by direct adenosine N1–2′OH hydrogen bonding interactions, many of which could be detected directly in solution using NMR methods (Cornish et al., 2005, 2006; Giedroc et al., 2003; Nixon et al., 2002) (see below). Other distinct L2–S1 hydrogen bonding interactions are found further from the helical junction in all luteoviral RNAs but these tend to be unique to individual RNAs.

An extensive mutational analysis of BWYV pseudoknot revealed that the hydrogen bonding interactions that stabilize the distinct helical junction region and the minor groove triplex are critical for supporting high levels of frameshift stimulation in vitro (Kim et al., 1999). However, a remarkable finding not predicted by the structure was that a single nucleotide insertion between G19 (or U19) and A20 at the tip of loop L2 dramatically increased frameshift stimulation, by upward of 300% (Kim et al., 1999) (see Fig. 9.3A, *arrow*). The origin of this stimulation is unknown and it is unclear the degree to which these mutations alter hydrogen bonding interactions of A20 with the G4–C17 base pair (Su et al., 1999). It is interesting to note, however, that this region of the pseudoknot is likely in closest proximity, or in direct contact, with the ribosome during frameshifting. These findings seem to parallel those found for a single nucleotide *deletion* mutant in loop L2 in the ScYLV P1–P2 frameshifting pseudoknot, which stimulates frameshifting by 200% relative to the wild-type RNA (ΔC25; see Fig. 9.4D, below) (Cornish, Hennig and Giedroc, 2005), via an as-yet unknown mechanism. Along similar lines, substitution of the extrahelical "spacer" nucleotide at the helical junction (U13 in the BWYV RNA; U12 in PLRV; see purple base in Figs. 9.3A–B) with A or G gives rise to a detectable *decrease* in frameshift stimulation (ranging from 35 to 90% of WT levels in vitro) (Kim et al., 1999, 2000). Since the analogous spacer adenosine in the PEMV-1 and ScYLV pseudoknots is highly dynamic in solution (Nixon et al., 2002;

Cornish et al., 2005), the structural origin of this small effect is unknown; one possibility of course is that this nucleotide also makes physical contact with the ribosome mRNA entry tunnel given its close proximity to the "top" of stem S1.

Figure 9.3B–D compares the structures of three other luteoviral P1–P2 frameshifting pseudoknots to that from BWYV (Fig. 9.3A) (Egli et al., 2002; Su et al., 1999). These include a crystallographic structure of the PLRV pseudoknot (Pallan et al., 2005), an NMR solution structure of the P1–P2 signal from PEMV-1 (Giedroc and Cornish, 2009; Nixon et al., 2002), and a solution structure of the frameshifting motif from ScYLV (Cornish et al., 2005). The PEMV-1 RNA structure clearly reveals that the five most 3′ residues of loop L2 (5′-A23-C24-A25-A26-A27) adopt a conformation that is essentially identical to that found in the BWYV (Egli et al., 2002) and PLRV (Pallan et al., 2005) pseudoknots, each of which contains the same 5′-ACAAA 3′ L2 sequence (Fig. 9.3). This L2 structure is achieved by anchoring the Watson–Crick edges of C24 through A27 into the S1 minor groove where they are engaged in numerous hydrogen bonding interactions.

A critical and unique aspect of the PEMV-1 RNA structure is that both Watson–Crick and Hoogsteen edges of the 3′ nucleotide of loop L2, A27 are tied up in hydrogen bonding (Giedroc et al., 2003; Nixon et al., 2002). In addition to A27 (and A25), the N1/N6 face of A26 forms two hydrogen bonds with the N3–N2 edge of G7. The C24 N4 amino group is also close to the 2′ OH of G17. A23 is stacked on C24, with the Watson–Crick edge rotated out of the stack, leaving the N7 and N6 nitrogens within hydrogen bonding distance of the 2′-OH of G17. At this point the PEMV-1 structure diverges from the BWYV/PLRV structures, with A22 stacked on A23, and essentially extruded from the triple helix, with A21 inserted back into the S1 minor groove near the top two S1 base pairs. G20 is extrahelical. In fact, if one considers A22 an extrahelical insertion in the PEMV-1 loop L2, the entire 5′-A(A)ACAAA loop sequence adopts essentially identical conformations in all three pseudoknots (Fig. 9.4). As was found for the BWYV RNA, strong support for the functional importance of the key structural features was also obtained for the PEMV-1 RNA; for example, efforts to replace the unique PEMV-1 helical junction region with that of the BWYV RNA failed (Nixon et al., 2002).

The solution structure of the ScYLV P1-P2 pseudoknot is characterized by several unique features relative to the PEMV-1 and BWYV/PLRV RNAs (see Fig. 9.3D) (Cornish et al., 2005). Notably, all of loop L2 is very well ordered

Fig. 9.4 Structures of the BWYV (**A**), PLRV (**B**), PEMV-1 (**C**), and ScYLV (**D**). P1–P2 pseudoknots emphasizing the structure of the minor groove spanning L2 loop (Giedroc and Cornish, 2009). The complete nucleotide sequence (5′–3′) of loop L2 is indicated for each RNA, with extrahelical nucleotides in parentheses. Lower case nucleotides in the PLRV RNA represent substitutions of the wild-type RNA sequence (C17u; A18c) incorporated to facilitate crystal packing (Pallan et al., 2005). u17 in the PLRV RNA (B) is grayed out since it is not colored in the structure shown. Bound Mg^{2+} ions are also indicated as *stippled spheres* in the BWYV and PLRV pseudoknot structures. Reprinted from Virus Research 139, Giedroc DP, Cornish PV, Frameshifting RNA pseudoknots: Structure and mechanism, pp 193–208, with permission from Elsevier

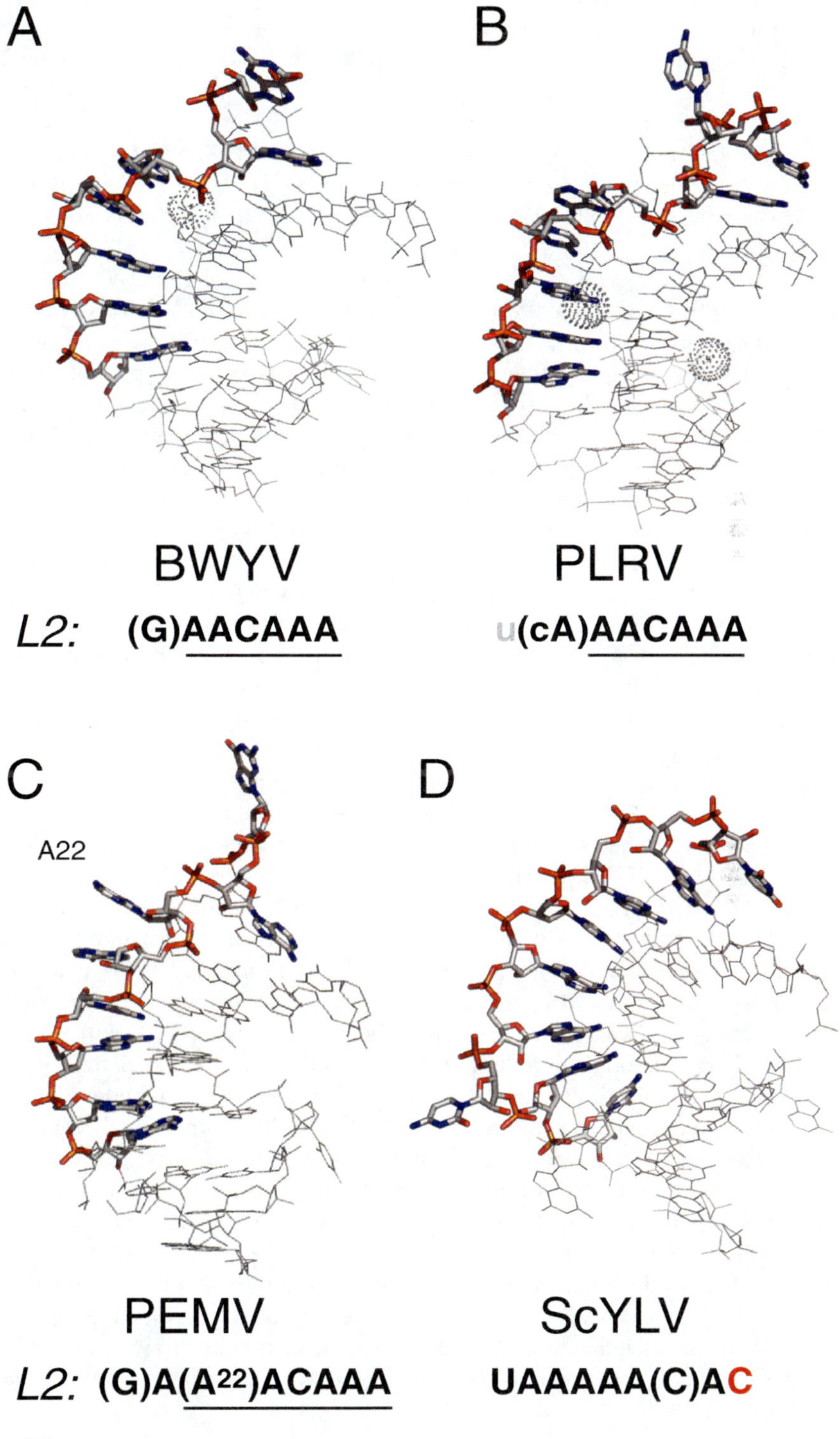
A
B
BWYV
PLRV
L2: (G)AACAAA
u(cA)AACAAA
C
D
A22
PEMV
ScYLV
L2: (G)A(A22)ACAAA
UAAAAA(C)AC

Fig. 9.4

and exhibits continuous stacking of A20 through C27 into the minor groove of S1, with C25 flipped out of the triple-stranded stack. Five consecutive triple base pairs flank the helical junction where the 3′ nucleotide of L2, *C27*, adopts a cytidine 27 N3–cytidine 14 2′-OH hydrogen bonding interaction with the C14–G7 base pair (Cornish et al., 2006). This interaction is isosteric with the adenosine N1–2′-OH interaction found in the BWYV (Su et al., 1999), PLRV (Pallan et al., 2005), and PEMV-1 RNAs (cf. Fig. 9.3, *right-most column*); however, the ScYLV and BWYV mRNA structures differ in their detailed L2–S1 hydrogen bonding and L2 stacking interactions.

Given the isosteric nature of the C27 N3•••C14 2′ OH (ScYLV) and A25 N1•••C14 2′ OH (BWYV) hydrogen bonds in these two RNAs by NMR (Cornish et al., 2006; Giedroc et al., 2003), the extent to which a BWYV-like adenosine in the ScYLV context could functionally substitute for the 3′ L2 cytidine (C27) was investigated using an in vitro coupled transcription–translation assay with rabbit reticulocyte lysates. The unexpected finding from these experiments is that the C27A ScYLV RNA is a very poor frameshift stimulator in vitro (Cornish et al., 2005). This surprising finding, in turn, made the prediction that substitution of A25 in the BWYV RNA with cytidine (A25C) would *increase* frameshifting in this context; this is exactly what was found. Thus, a 3′ L2 cytidine is a positive determinant for frameshift stimulation by luteoviral RNAs (Cornish et al., 2006). It would be of interest to determine the degree to which other *Luteoviridae* frameshifting pseudoknots with a cytidine in this position, e.g., cereal yellow dwarf virus (CYDV) and beet chlorosis virus (BCV) are also superior frameshift stimulators (Smith et al., 2000). The degree to which the C27A substitution (C1790A in the full-length genome) influences replication of ScYLV in plants is unknown; indeed, such an experiment is complicated by the fact that this substitution would result in a Thr-to-Asn substitution in the P1–P2 fusion protein (Moonan et al., 2000).

Strikingly, the global structure of the poorly functional C27A ScYLV RNA is nearly indistinguishable from the wild-type counterpart, despite the fact that the helical junction region is altered and incorporates the anticipated isostructural *A27*·(G7–C14) minor groove base triple (Cornish et al., 2006) in adopting a helical junction region that is basically indistinguishable from the BWYV and PLRV pseudoknots (Su et al., 1999), as expected. These results suggest that the lowest energy "ground-state" structure is *not* strongly correlated with frameshift stimulation; instead, these findings point to a reduced stability that derives from an altered helical junction architecture in the C27A ScYLV RNA as of critical importance, for which there is some evidence (Cornish and Giedroc, 2006; Cornish et al., 2005).

Studies in other frameshifting systems have uncovered a correlation between thermodynamic stability of the downstream element and frameshifting efficiency (Bidou et al., 1997; Larsen et al., 1997; ten Dam et al., 1995), although it is not obvious why this has to be the case (Cao and Chen, 2008; Theimer and Giedroc, 2000). In fact, as discussed above, insertion or deletion of an unpaired and extrahelical residue(s) in loop L2 can strongly *increase* frameshifting efficiency, a finding often attributed to a direct interaction with the ribosome (Kim et al., 1999, 2000). We have argued that distinct unfolding thermodynamics measured for closely related

RNAs, e.g., WT vs. C27A ScYLV RNAs, may in fact be reporting on different kinetics of pseudoknot unfolding, with increased rates of unfolding, i.e., unfolding at lower applied forces, negatively correlated with frameshift stimulation (Giedroc et al., 2000; Theimer and Giedroc, 2000). In other words, these studies suggest that the helical junction in luteoviral RNAs is mechanically stable and functions as a classic kinetic barrier to force-induced unfolding (Onoa et al., 2003), essentially placing the unfolding of the entire pseudoknot under kinetic control. This barrier is predicted to be altered in functionally compromised RNAs. Mechanical force-induced unfolding/refolding experiments will be required to obtain direct evidence for this proposal (Green et al., 2008; Hansen et al., 2007).

9.2.5 Sobemoviruses

Sobemoviruses are similar to the poleroviruses and enamovirus based on sequence homology of the RdRp and the presence of a VPg (Makinen et al., 2000; Skaf et al., 2000; van der Wilk et al., 1998). Despite homology at the genome level, sobemoviruses are not classified in the Luteoviridae because, unlike Luteovirids, sobemoviruses (i) have a significantly smaller genome (just over 4 kb), (ii) encode a larger coat protein that lacks the readthrough domain found in the Luteoviridae, (iii) are mechanically transmissible, (iv) are not persistently transmitted by aphids, and (v) are not limited to the phloem (Miller et al., 2002; Tamm and Truve, 2000). All sobemoviruses employ −1 frameshifting, but recent resequencing results have challenged previous conclusions on the nature of the ORF translated via the frameshift. The most recent (and now apparently incorrect, see below) ICTV classification (Hull and Fargette, 2005) divides sobemoviruses into two subgroups: those in which the frameshift event shifts the ribosome out of the RdRp frame into a much shorter reading frame (ORF 3) and those in which the ribosome is shifted into the RdRp ORF (ORF 2b) as is the case with all other viruses discussed previously in this chapter.

Some of the first sobemoviruses sequenced, including southern cowpea mosaic virus (SCPMV, Genbank accession no. NC_001625), southern bean mosaic virus (SBMV, NC_004060), Rice yellow mottle virus (RYMV, NC_001575), Ryegrass mosaic virus (RGMoV, NC_003747), and Lucerne transient streak virus (LTSV, NC_001696) were thought to encode a protease, VPg, and RdRp all in ORF 2, which would have been translated as one large polyprotein of about 100 kDa (Fig. 9.5) (Tamm and Truve, 2000). A ribosomal frameshift site identified in the middle of this ORF was proposed to allow translation of a short ORF 3 of unknown function (Fig. 9.5). Other sobemoviruses, the first being cocksfoot mottle virus (CfMV, (Makinen et al., 1995)), were reported to have a more polerovirus-like genome organization with the protease and VPg encoded in ORF 2a and the RdRp translated from the overlapping ORF2b via −1 frameshift (Fig. 9.5). The amino acid sequences just downstream of the frameshift site were shown to be homologous between the two groups of viruses, i.e., ORF 3 of the viruses with the SCPMV-like organization was

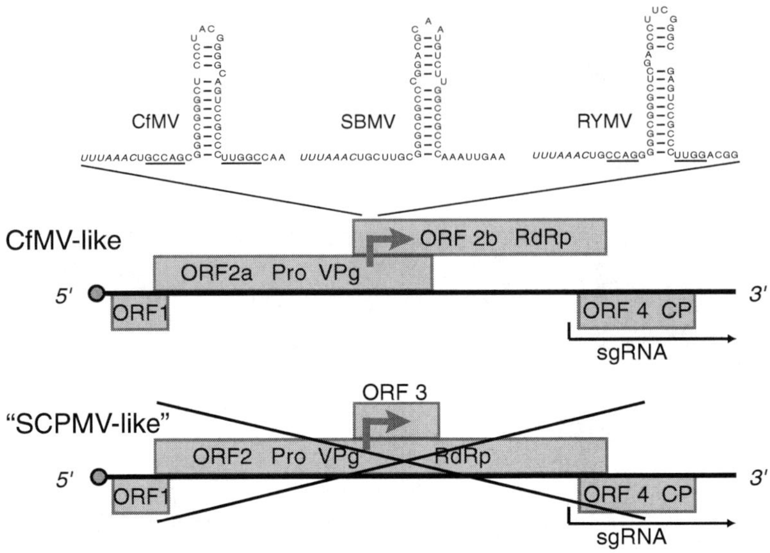

Fig. 9.5 Corrected genome organization of sobemoviruses and examples of −1 frameshift sequences. *Bottom*: Genome organizations of the two apparent subgroups of sobemoviruses are shown. Erroneous SCPMV-like organization is crossed out. All sobemoviruses have the CfMV-like organization (Meier and Truve, 2007). *Top*: Predicted secondary structures of the indicated sobemoviral RNAs at the −1 frameshift site (Makinen et al., 1995). Slippery site is in italics. Underlined bases extend the helix in selected viruses but this additional base pairing is not phylogenetically conserved. Frameshifting has been demonstrated only for CfMV and no data exist to support or refute the secondary structures

shown to be homologous to the portion of the 2b-encoded protein (RdRp) immediately downstream of the frameshift site in the CfMV-like viruses (Dwyer et al., 2003). Why would some viruses use frameshifting to produce a truncated version of the RdRp (those with ORF 3), while others use frameshifting to produce the entire RdRp (those with ORF 2b)? This puzzle was resolved recently by resequencing of all of the so-called SCPMV-like viral genomes which revealed that they all have CfMV-like genome organizations (Balke et al., 2007; Fargette et al., 2004; Lokesh et al., 2001; Meier and Truve, 2007). Meier and Truve (2007) showed that sequencing errors in the form of a single base insertion in the region of ORF 2–ORF 3 overlap led to the appearance that the RdRp coding region was in the same ORF as the upstream protease coding region. Thus, the sequences thought to be in "ORF 3" that were homologous to the 2b ORF are actually all encoded by ORF 2b. In short, resequencing revealed a paradigm shift on the role of the frameshift in sobemoviruses.

Frameshifting has been demonstrated for CfMV (Makinen et al., 1995). The portion of the genome that confers −1 ribosomal frameshifting was narrowed to a sequence that includes a simple imperfect stem-loop with one or two unpaired bases interrupting the helix (Lucchesi et al., 2000). No mutagenesis or structure probing has been performed to verify the structure, but it contains a highly stable, extremely GC-rich stem-loop predicted to be stabilized by a UNCG or GNRA tetraloop at the end (Fig. 9.5). It is conserved in all sobemoviruses, and no alternative structures such as a pseudoknot have been reported or predicted. This imperfect stem-loop superficially resembles the bipartite stem-loop that stimulates −1 frameshifting of HIV-1 and SIV RNAs (Gaudin et al., 2005; Marcheschi et al., 2007; Staple and Butcher, 2005).

Lucchesi et al. (2000) observed about a 10% frameshift rate by CfMV RNA in wheat germ extract. When a stop codon was placed downstream in the −1 ORF to mimic translation of the ORF 3 in what they thought was the situation in the SCPMV-like viruses, the authors found frameshift rates of about 50%. They proposed that this stop codon (which could be 53–462 nt in the −1 frame downstream of the shifty site) enhanced stalling of ribosomes on the shifty site. However, similarly placed stop codons downstream in the zero frame (ORF 2a frame) did not enhance frameshifting.

In yeast cells, using a β-galactosidase–firefly luciferase dual reporter assay designed so that expression of a luciferase gene required the CfMV −1 frameshift sequence, frameshifting efficiencies varied wildly. The small stem-loop alone gave 14% frameshifting, while a region that included about 750 nt of viral sequence spanning upstream and downstream, beyond the region of ORF overlap, gave a 26% frameshift rate (Makelainen and Makinen, 2005). Surprisingly, a subset of that ORF overlap sequence gave a whopping, apparent 68% frameshift rate in yeast. Using a dual luciferase reporter in wheat germ extract, the rates with the same inserts were 25%, 36%, and 28%, respectively (Makelainen and Makinen, 2005). Frameshift rates were less than 6% in bacterial cells but frameshifting was not tested in plant cells. The authors conclude that "reliable estimates for −1 programmed ribosomal frameshifting ... can be obtained only by using full-length viral sequences." Importantly, altering the UUUAAAC frameshift sequence to UUUAAGC or removal of the sequence containing the predicted adjacent stem-loop sequence totally abolished frameshifting in wheat germ extract (Lucchesi et al., 2000).

Unlike most studies, Makelainen and Makinen (2005) tested the effects of viral proteins on frameshifting efficiencies in yeast. They found that the P1 protein encoded by the pre-shift (zero frame) ORF, called p27, appeared to reduce frameshifting as measured by its negative effect on luciferase expression in yeast cells. This finding suggests a possible frameshift regulatory role for p27 in viral infection. Its significance would be improved if the results can be corroborated by studies in plant cells and with related viruses.

9.2.6 Closteroviruses

Viruses in the Closteroviridae family of positive sense RNA viruses have the largest genomic RNA of any plant virus, ranging from 12 to 20 kb. The genome organization and expression strategy resemble that of viruses in the Nidovirales order of animal viruses (see Chapter 7): The 5′ half to two-thirds of the genome encodes proteins involved in replication (Karasev, 2000). The 3′ portion of the genome contains several ORFs expressed from a nested set of 3′ co-terminal subgenomic RNAs. In closteroviruses, the protease and RNA replication genes, encoded by ORFs 1a and 1b, have similarities to those in the alphavirus-like supergroup, including putative helicase, methyltransferase, and two proteases encoded by ORF 1a followed by the RdRp encoded by ORF 1b (Karasev, 2000). Unlike the Nidovirales, expression of the RdRp (ORF1b) is predicted to require a +1, instead of a −1, reading frame change to be expressed (Agranovsky et al., 1994; Karasev et al., 1995).

It must be pointed out that frameshifting has yet to be experimentally demonstrated for closteroviruses. The +1 frameshift was proposed to occur either at the stop codon of the 1a ORF in beet yellow virus (BYV) (Agranovsky et al., 1994) or at a rare codon at the homologous position in citrus tristeza virus (CTV) (Karasev et al., 1995). Either codon would facilitate frameshifting by inducing ribosomes to pause, analogous to the mechanism of +1 frameshifting transposable elements in yeast (Belcourt and Farabaugh, 1990). Sequencing of additional closterovirids revealed no obvious conserved frameshift site. Similar to BYV, pineapple mealybug wilt-associated virus-1 (PMWaV-1) has a putative +1 frameshift site, GUUU*AGC*, in which underlined bases indicate the ORF 1a stop codon (UAA in BYV) and italics indicate first codon (asparagine) of the 1b ORF (Melzer et al., 2008). In contrast closteroviruses PMWaV-2 and grapevine leafroll-associated virus-3 (GLRaV-3) have an entirely different sequence at this location in the genome, UUUC*GAG* (proposed first codon of ORF 1b in italics), at which the +1 shift is proposed to occur (Melzer et al., 2008). We await experimental evidence to determine if either of these sites is the actual frameshift site.

Cevik et al. (2008) detected the RdRp of CTV in infected cells and by cell-free vitro translation. Enigmatically, western blots probed with antibodies against the 57-kDa RdRp domain (ORF 1b product) of CTV revealed only a ~50-kDa product, a ~10-kDa fragment but no 400-kDa product expected of the whole 1a–1b transframe protein. This led the authors to conclude that the 57-kDa RdRp domain is proteolytically cleaved immediately from the precursor 1a–1b fusion protein into ~50- and ~10-kDa fragments. The lack of detection of the expected full-length 1a–1b protein, or a larger precursor of the 57-kDa product, leaves the mechanism of expression of the RdRp domain in doubt. Even if the RdRp is produced as a cleavage product of the 1a–1b fusion, it is not known whether it is produced by a +1 frameshift or –2, +4 shift or a larger ribosome hopping mechanism.

9.2.7 Carlaviruses

Potato virus M in the genus *Carlavirus* has a 12-kDa extension added to the coat protein (CP) by a novel type of −1 frameshift. The 34-kDa CP ORF terminates with the sequence AAU AGA AAA UGA, with the A UG forming a potential start codon for the 12-kDa ORF. In reticulocyte lysates, Gramstat et al. (1994) showed that the 12-kDa protein can be synthesized by initiation at the AUG, but of greater interest is that the 12 kDa is also present fused to the C terminus of the CP via a −1 frameshift that does not require the canonical shifty heptanucleotide, nor does it require a downstream structured element. Instead, this frameshift requires only a homotetramer of AAAA or UUUU immediately upstream of an adjacent stop codon. The overlapping AUG in the 12-kDa ORF frame is not necessary for frameshifting. The authors propose a model in which ribosome slippage occurs with the tRNA:mRNA base-paired only at the P-site, rather than the canonical simultaneous A- and P-site slippage, because the A-site would be at the stop codon (Gramstat et al., 1994). Readers should be aware that the analysis of the frameshift signal was performed using only rabbit reticulocyte lysates which often have reduced translational fidelity relative to wheat germ extract (Dinesh-Kumar et al., 1992; Kozak, 1990) or, most importantly, compared to the environment in vivo. The evidence that this frameshift takes place in vivo is identification of a low-abundance protein of the expected size using antisera targeted to either the product of the 34-kDa CP ORF or the 12-kDa ORF. We are unaware of any subsequent research that confirms the proposed frameshift process in vivo.

9.2.8 Potyviruses

Viruses in the large, economically important Potyviridae family were thought to contain a single ORF encoding a ∼340–395-kDa polyprotein that is cleaved by viral proteases into the functional proteins, like viruses in the Picornavirales (Berger et al., 2005; Le Gall et al., 2008). However, a small overlapping ORF called pipo was discovered recently using bioinformatic methods (Chung et al., 2008). This ORF, which ranges in size from 60 to 115 codons depending on the virus, is in the +2 (or −1) frame relative to the polyprotein ORF and overlaps with the polyprotein ORF region coding for the P3 protein (Fig. 9.6). In turnip mosaic potyvirus (TuMV), the amino acid sequence (referred to in uppercase as PIPO) encoded by pipo has a molecular weight of 7 kDa, but antibodies against this protein detected only a ∼25-kDa protein in TuMV-infected plants (Chung et al., 2008). A protein of this size is consistent with a fusion product of the amino terminal portion of the P3 protein (after proteolytic cleavage of P3 from the polyprotein) fused to PIPO and terminating at the pipo stop codon (Fig. 9.6). Thus, PIPO does not appear to be translated independently.

A possible mechanism for expression of PIPO is by frameshifting of some ribosomes from the P3 coding region into the −1 (= +2) frame at the beginning of the pipo ORF. A likely frameshift site is the $G_{1-2}A_{6-7}$ motif which is the only well-conserved sequence in pipo and is predicted to be the start of the pipo ORF based on the presence of in-frame stop codons upstream in some potyviruses (Chung et al., 2008). GAA AAA A (gaps separate codons in the polyprotein ORF) does not fit the canonical X XXY YYZ shifty sequence at the −1 frameshift site used by other viruses, although it is conceivable that ribosomes could shift reading frame in the homopolymeric A tract. It is noteworthy that the potyvirus $G_{1-2}A_{6-7}$ motif resembles the AGA AAA UGA sequence at the reading frame change site in carlaviruses (above), suggesting that a possible common frameshift mechanism may be employed by these unrelated virus groups. No conserved secondary structures, such as those required for conventional −1 frameshifting, have been detected downstream of the $G_{1-2}A_{6-7}$ motif in potyviral RNAs (Chung et al., 2008). Thus, if −1 or +2 frameshifting or a type of ribosome hop that leads to a net change of +2 in reading frame takes place, it is likely to be via a mechanism different from that of "canonical" −1 frameshifting. Other nontranslational mechanisms can explain the production of a 25-kDa protein containing PIPO epitopes, so further experimentation is necessary to determine whether potyviruses employ frameshifting.

9.3 Summary

In most cases, the biological role and basic mechanism of −1 ribosome frameshifting are likely the same in plant viruses as in animal viruses. While some of the shifty heptanucleotide sites are the same in plant and animal viruses, the specific downstream structures that facilitate −1 frameshifting differ significantly between viruses of the two kingdoms. No ALIL-like structures that base pair with a stem-loop kilobases downstream are known outside of the plant viruses. At the other extreme, the polero- and enamoviruses have the smallest known frameshift pseudoknots, enabling the highest levels of structural characterization of any frameshift-inducing structure. As these structures facilitate frameshifting in rabbit reticulocyte lysates (Cornish et al., 2005; Kim et al., 1999, 2000; Nixon et al., 2002), the reason why they exist only in plant viruses is not likely due to special features unique to plant ribosomes. These plant virus frameshift signals add to the large and diverse pool of eukaryotic frameshift elements, which should perhaps help us solve the question of why some RNA structures, but not others of similar apparent stability, interact with the ribosome in a way that brings about a −1 frameshift. The apparent +1 frameshift of closteroviruses appears to be confined to the plant virus world as are the possible −1 frameshift signals in carlaviruses and potyviruses. Thus, we eagerly

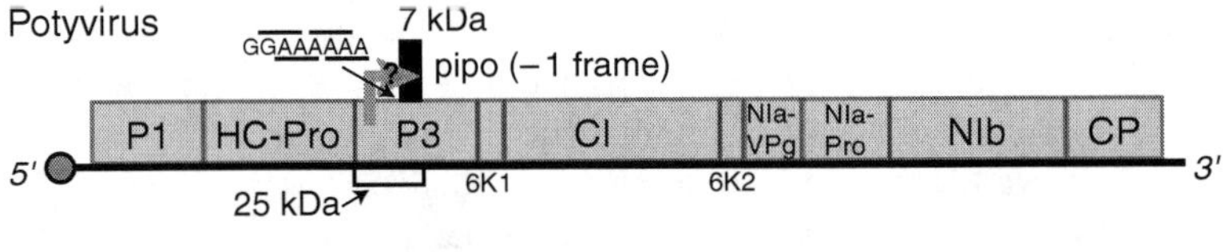

Fig. 9.6 Potyvirus genome organization. *Gray boxes* indicate individual proteins generated by proteolytic cleavage of the polyprotein. *Black box* above P3 represents the pipo ORF in the −1 frame. *Large gray arrow* with question mark indicates possible frameshift event. An example of the conserved $A_{1-2}G_{6-7}$ site is shown at the start of the pipo ORF with lines under the codons in the zero frame (polyprotein ORF) and lines over the codons in the −1 frame (pipo ORF). The 25-kDa protein that could be generated by a frameshift followed by cleavage at the HC-Pro/P3 cleavage site is indicated (Chung et al., 2008)

anticipate results of future research and structural analysis to determine how these diverse plant viral RNAs induce ribosomes to change reading frames by what may be novel mechanisms.

Acknowledgments The authors thank Andrew Firth, John Atkins, and Alex Karasev for valuable advice, and Nikki Krueger for constructing the phylogenetic tree in Fig. 9.2A. This work was funded by USDA National Research Initiative grant 2008-35319-19196 and NIH grant GM067104 to WAM, and NIH grants AI040187 and AI067416 to DPG.

References

Agranovsky AA, Koonin EV, Boyko VP, Maiss E, Frotschl R., Lunina NA, Atabekov JG (1994) Beet Yellows Closterovirus: Complete genome structure and identification of a leader papain-like Thiol protease. Virology 198:311–324

Balke I, Resevica G, Zeltins A (2007) The ryegrass mottle virus genome codes for a sobemovirus 3C-like serine protease and RNA-dependent RNA polymerase translated via −1 ribosomal frameshifting. Virus Genes 35:395–398

Barry JK, Miller WA (2002) A -1 ribosomal frameshift element that requires base pairing across four kilobases suggests a mechanism of regulating ribosome and replicase traffic on a viral RNA. Proc Natl Acad Sci USA 99:11133–11138

Belcourt MF, Farabaugh PJ (1990) Ribosomal frameshifting in the yeast retrotransposon Ty - Transfer-RNAs induce slippage on a 7-Nucleotide minimal site. Cell 62:339–352

Belshaw R, Pybus OG, Rambaut A (2007) The evolution of genome compression and genomic novelty in RNA viruses. Genome Res 17:1496–1504

Berger PH, Adams MJ, Barnett OW, Brunt AA, Hammond J, Hill JH, Jordan RL, Kashiwazaki S, Rybicki E, Spence N, Stenger DC, Ohki ST, Uyeda I, van Zaayen A, Valkonen J, Vetten HJ (2005) Family potyviridae. In: Fauquet CM, Mayo MA, Maniloff J, Desselberger U, Ball LA (eds) Virus taxonomy: Eighth report of the international committee on the taxonomy of viruses, pp. 819–841. Elsevier Academic Press, San Diego

Bidou L, Stahl G, Grima B, Liu H, Cassan M, Rousset JP (1997) In vivo HIV-1 frameshifting efficiency is directly related to the stability of the stem-loop stimulatory signal. RNA 3: 1153–1158

Brault V, Miller WA (1992) Translational frameshifting mediated by a viral sequence in plant cells. Proc Natl Acad Sci USA 89:2262–2266

Brierley I, Jenner AJ, Inglis SC (1992) Mutational analysis of the "slippery-sequence" component of a coronavirus ribosomal frameshifting signal. J Mol Biol 227:463–479

Brodersen DE, Clemons WM Jr, Carter AP Wimberly BT, Ramakrishnan V (2002) Crystal structure of the 30 S ribosomal subunit from Thermus thermophilus: Structure of the proteins and their interactions with 16 S RNA. J Mol Biol 316:725–768

Cao S, Chen SJ (2008) Predicting ribosomal frameshifting efficiency. Phys Biol 5:16002

Cevik B, Lee RF, Niblett CL (2008) In vivo and in vitro expression analysis of the RNA-dependent RNA polymerase of Citrus tristeza virus. Arch Virol 153:315–321

Chung BY-W, Miller WA, Atkins JF, Firth AE (2008) An overlapping essential gene in the Potyviridae. Proc Natl Acad Sci USA 105:5897–5902

Cornish PV, Ermolenko DN, Noller HF, Ha T (2008) Spontaneous intersubunit rotation in single ribosomes. Mol Cell 30:578–588

Cornish PV, Giedroc DP (2006) Pairwise coupling analysis of helical junction hydrogen bonding interactions in luteoviral RNA pseudoknots. Biochemistry 45:11162–11171

Cornish PV Giedroc DP, Hennig M (2006) Dissecting non-canonical interactions in frameshift-stimulating mRNA pseudoknots. J Biomol NMR 35:209–223

Cornish PV, Hennig M, Giedroc DP (2005) A loop 2 cytidine-stem 1 minor groove interaction as a positive determinant for pseudoknot-stimulated -1 ribosomal frameshifting. Proc Natl Acad Sci USA 102:12694–12699

Cornish PV, Stammler SN, Giedroc DP (2006) The global structures of a wild-type and poorly functional plant luteoviral mRNA pseudoknot are essentially identical. RNA 12: 1959–1969

Cruz-Vera LR, Rajagopal S, Squires C, Yanofsky C (2005) Features of ribosome-peptidyl-tRNA interactions essential for tryptophan induction of tna operon expression. Mol Cell 19: 333–343

D'Arcy CJ, Domier LL (2005) Luteoviridae. In: Fauquet CM, Mayo MA, Maniloff J, Desselberger U, Ball LA (eds) Virus taxonomy: Eighth report of the international committee on taxonomy of viruses, pp. 891–900. Elsevier, San Diego

Demler SA, Borkhsenious ON, Rucker DG, de Zoeten GA (1994) Assessment of the autonomy of replicative and structural functions encoded by the luteo-phase of pea enation mosaic virus. J Gen Virol 75:997–1007

Demler SA, Rucker DG, de Zoeten GA (1993) The chimeric nature of the genome of pea enation mosaic virus: The independent replication of RNA 2. J Gen Virol 74:1–14

Di R, Dinesh-Kumar SP, Miller WA (1993) Translational frameshifting by barley yellow dwarf virus RNA (PAV serotype) in *Escherichia coli* and in eukaryotic cell-free extracts. Molec Plant-Microbe Interact 6:444–452

Dinesh-Kumar SP, Brault V, Miller WA (1992) Precise mapping and in vitro translation of a trifunctional subgenomic RNA of barley yellow dwarf virus. Virology 187:711–722

Doherty EA, Batey RT, Masquida B, Doudna JA (2001) A universal mode of helix packing in RNA. Nat Struct Biol 8:339–343

Dwyer GI, Njeru R, Williamson S, Fosu-Nyarko J, Hopkins R, Jones RA, Waterhouse PM, Jones MG (2003) The complete nucleotide sequence of Subterranean clover mottle virus. Arch Virol 148:2237–2247

Egli M, Minasov G, Su L, Rich A (2002) Metal ions and flexibility in a viral RNA pseudoknot at atomic resolution. Proc Natl Acad Sci USA 99:4302–4307

Fargette D, Pinel A, Abubakar Z, Traore O, Brugidou C, Fatogoma S, Hebrard E, Choisy M, Sere Y, Fauquet C, Konate G (2004) Inferring the evolutionary history of rice yellow mottle virus from genomic, phylogenetic, and phylogeographic studies. J Virol 78:3252–3261

Gamarnik AV, Andino R (1998) Switch from translation to RNA replication in a positive-stranded RNA virus. Genes Dev 12:2293–2304

Garcia A, van Duin J, Pleij CW (1993) Differential response to frameshift signals in eukaryotic and prokaryotic translational systems. Nucleic Acids Res 21:401–406

Gaudin C, Mazauric MH, Traikia M, Guittet E, Yoshizawa S, Fourmy D (2005) Structure of the RNA signal essential for translational frameshifting in HIV-1. J Mol Biol 349: 1024–1035

Giedroc DP, Cornish PV (2009) Frameshifting RNA pseudoknots: Structure and mechanism. Virus Res 139:193–208.

Giedroc DP, Cornish PV, Hennig M (2003) Detection of scalar couplings involving 2′-hydroxyl protons across hydrogen bonds in a frameshifting mRNA pseudoknot. J Am Chem Soc 125:4676–4677

Giedroc DP, Theimer CA, Nixon PL (2000) Structure, stability and function of RNA pseudoknots involved in stimulating ribosomal frameshifting. J Mol Biol 298:167–185

Gramstat A, Prüfer D, Rohde W (1994) The nucleic acid-binding zinc finger protein of potato virus M is translated by internal initiation as well as by ribosomal frameshifting involving a shifty stop codon and a novel mechanism of P-site slippage. Nucleic Acids Res 22: 3911–3917

Green L, Kim CH, Bustamante C, Tinoco I Jr (2008) Characterization of the mechanical unfolding of RNA pseudoknots. J Mol Biol 375:511–528

Grentzmann G, Ingram JA, Kelly PJ, Gesteland RF, Atkins JF (1998) A dual-luciferase reporter system for studying recoding signals. RNA 4:479–486

Guo L, Allen E, Miller WA (2001) Base-pairing between untranslated regions facilitates translation of uncapped, nonpolyadenylated viral RNA. Mol Cell 7:1103–1109

Hansen T M, Reihani SN, Oddershede LB, Sorensen MA (2007) Correlation between mechanical strength of messenger RNA pseudoknots and ribosomal frameshifting. Proc Natl Acad Sci USA 104:5830–5835

Harger JW, Meskauskas A, Dinman JD (2002) An "integrated model" of programmed ribosomal frameshifting. Trends Biochem Sci 27:448–454

Holmes EC (2003) Error thresholds and the constraints to RNA virus evolution. Trends Microbiol 11:543–546

Hull R, Fargette D (2005) Sobemovirus. In: Fauquet CM, Mayo MA, Maniloff J, Desselberger U, Ball LA (eds) Virus taxonomy: Eighth report of the international committee on taxonomy of viruses, pp. 885–890. Elsevier, San Diego

Ivanov IP, Atkins JF (2007) Ribosomal frameshifting in decoding antizyme mRNAs from yeast and protists to humans: Close to 300 cases reveal remarkable diversity despite underlying conservation. Nucleic Acids Res 35:1842–1858

Jacks T, Madhani HD, Masiarz FR, Varmus HE (1988) Signals for ribosomal frameshifting in the *Rous Sarcoma* virus gag pol region. Cell 55:447–458

Karasev AV (2000) Genetic diversity and evolution of Closteroviruses. Annu Rev Phytopathol 38:293–324

Karasev AV, Boyko VP, Gowda S, Nikolaeva OV, Hilf ME, Koonin EV, Niblett CL, Cline K, Gumpf DJ, Lee RF, et al (1995) Complete sequence of the citrus tristeza virus RNA genome. Virology 208:511–520

Kim KH, Lommel SA (1998) Sequence element required for efficient −1 ribosomal frameshifting in red clover necrotic mosaic dianthovirus. Virology 250:50–59

Kim YG, Maas S, Wang SC, Rich A (2000) Mutational study reveals that tertiary interactions are conserved in ribosomal frameshifting pseudoknots of two luteoviruses. RNA 6: 1157–1165

Kim YG, Su L, Maas S, O'Neill A, Rich A (1999) Specific mutations in a viral RNA pseudoknot drastically change ribosomal frameshifting efficiency. Proc Natl Acad Sci USA 96:14234–14239

Koonin EV, Dolja VV (1993) Evolution and taxonomy of positive-strand RNA viruses: Implications of comparative analysis of amino acid sequences. Critic Rev Biochem Molec Biol 28:375–430

Kozak M (1990) Evaluation of the fidelity of initiation of translation in reticulocyte lysates from commercial sources. Nucleic Acids Res 18:2828–2828

Kujawa AB, Drugeon G, Hulanicka D, Haenni AL (1993) Structural requirements for efficient translational frameshifting in the synthesis of the putative viral RNA-dependent RNA polymerase of potato leafroll virus. Nucleic Acids Res 21:2165–2171

Larsen B, Gesteland RF, Atkins JF (1997) Structural probing and mutagenic analysis of the stem-loop required for *Escherichia coli* dnaX ribosomal frameshifting: Programmed efficiency of 50%. J Mol Biol 271:47–60

Le Gall O, Christian P, Fauquet CM, King AM, Knowles NJ, Nakashima N, Stanway G, Gorbalenya AE (2008) Picornavirales, a proposed order of positive-sense single-stranded RNA viruses with a pseudo-T = 3 virion architecture. Arch Virol 153:715–727

Leger M, Dulude D, Steinberg SV, Brakier-Gingras L (2007) The three transfer RNAs occupying the A, P and E sites on the ribosome are involved in viral programmed -1 ribosomal frameshift. Nucleic Acids Res 35:5581–5592

Lokesh GL, Gopinath K., Satheshkumar PS, Savithri HS (2001) Complete nucleotide sequence of Sesbania mosaic virus: A new virus species of the genus Sobemovirus. Arch Virol 146: 209–223

Lommel SA, Martelli GP, Rubino L., Russo M (2005) Tombusviridae. In: Fauquet CM, Mayo MA, Maniloff J, Desselberger U, Ball LA (eds) Virus taxonomy: Eighth report of the international committee on taxonomy of viruses, pp. 907–936. Elsevier, San Diego

Lu J, Deutsch C (2008) Electrostatics in the ribosomal tunnel modulate chain elongation rates. J Mol Biol 384:73–86

Lucchesi J, Makelainen K, Merits A, Tamm T, Makinen K (2000) Regulation of −1 ribosomal frameshifting directed by cocksfoot mottle sobemovirus genome. Eur J Biochem 267: 3523–3529

Makelainen K, Makinen K (2005) Factors affecting translation at the programmed −1 ribosomal frameshifting site of Cocksfoot mottle virus RNA in vivo. Nucleic Acids Res 33:2239–2247

Makinen K, Makelainen K, Arshava N, Tamm T, Merits A, Truve E, Zavriev S, Saarma M (2000) Characterization of VPg and the polyprotein processing of cocksfoot mottle virus (genus sobemovirus). J Gen Virol 81:2783–2789

Makinen K, Naess V, Tamm T, Truve E, Aaspollu A, Saarma M (1995) The putative replicase of cocksfoot mottle sobemovirus is translated as a part of the polyprotein by −1 ribosomal frameshift. Virology 207:566–571

Makinen K, Tamm T, Naess V, Truve E, Puurand ö, Munthe T, Saarma M (1995) Characterization of cocksfoot mottle sobemovirus genomic RNA and sequence comparison with related viruses. J Gen Virol 76:2817–2825

Marcheschi RJ, Staple DW, Butcher SE (2007) Programmed ribosomal frameshifting in SIV is induced by a highly structured RNA stem-loop. J Mol Biol 373:652–663

Mayo MA, D'Arcy CJ (1999) Family *Luteoviridae*: A reclassification of luteoviruses. In: Smith HG, Barker H (eds) The *Luteoviridae*. CABI Publishing, Wallingford, Oxon

Mazauric MH, Licznar P, Prere MF, Canal I, Fayet O (2008) Apical loop-internal loop RNA pseudoknots: A new type of stimulator of −1 translational frameshifting in bacteria. J Biol Chem 283:20421–20432

McCartney AW, Greenwood JS, Fabian MR, White KA, Mullen RT (2005) Localization of the tomato bushy stunt virus replication protein p33 reveals a peroxisome-to-endoplasmic reticulum sorting pathway. Plant Cell 17:3513–3531

Meier M, Truve E (2007) Sobemoviruses possess a common CfMV-like genomic organization. Arch Virol 152:635–640

Melzer MJ, Sether DM, Karasev AV, Borth W, Hu JS (2008) Complete nucleotide sequence and genome organization of pineapple mealybug wilt-associated virus-1. Arch Virol 153:707–714

Miller WA, Liu S, Beckett R (2002) Barley yellow dwarf virus: *Luteoviridae* or *Tombusviridae*? Mol Plant Pathol 3:177–183

Miller WA, White KA (2006) Long distance RNA-RNA interactions in plant virus gene expression and replication. Ann Rev Phytopathol 44: 447–467

Mizumoto H, Tatsuta M, Kaido M, Mise K, Okuno T (2003) Cap-independent translational enhancement by the 3′ untranslated region of red clover necrotic mosaic virus RNA1. J Virol 77:12113–12121

Moonan F, Molina J, Mirkov TE (2000) Sugarcane yellow leaf virus: An emerging virus that has evolved by recombination between luteoviral and poleroviral ancestors. Virology 269:156–171

Namy O, Moran SJ, Stuart DI, Gilbert RJ, Brierley I (2006) A mechanical explanation of RNA pseudoknot function in programmed ribosomal frameshifting. Nature 441:244–247

Nissen P, Ippolito JA, Ban N, Moore PB, Steitz TA (2001) RNA tertiary interactions in the large ribosomal subunit: The A-minor motif. Proc Natl Acad Sci USA 98:4899–4903

Nixon PL, Cornish PV, Suram SV, Giedroc DP (2002) Thermodynamic analysis of conserved loop-stem interactions in P1-P2 frameshifting RNA pseudoknots from plant Luteoviridae. Biochemistry 41:10665–10674

Nixon PL, Giedroc DP (2000) Energetics of a strongly pH dependent RNA tertiary structure in a frameshifting pseudoknot. J Mol Biol 296:659–671

Nixon PL, Rangan A, Kim YG, Rich A, Hoffman DW, Hennig M., Giedroc DP (2002) Solution structure of a luteoviral P1–P2 frameshifting mRNA pseudoknot. J Mol Biol 322:621–633

Onoa B, Dumont S, Liphardt J, Smith SB, Tinoco I Jr, Bustamante C (2003) Identifying kinetic barriers to mechanical unfolding of the T. thermophila ribozyme. Science 299:1892–1895

Pallan PS, Marshall WS, Harp J, Jewett FC, 3rd, Wawrzak Z, Brown BA 2nd, Rich A, Egli M (2005) Crystal structure of a luteoviral RNA pseudoknot and model for a minimal ribosomal frameshifting motif. Biochemistry 44:11315–11322

Panavas T, Hawkins CM, Panaviene Z, Nagy PD (2005) The role of the p33:p33/p92 interaction domain in RNA replication and intracellular localization of p33 and p92 proteins of Cucumber necrosis tombusvirus. Virology 338:81–95

Paul CP, Barry JK, Dinesh-Kumar SP, Brault V, Miller WA (2001) A sequence required for −1 ribosomal frameshifting located four kilobases downstream of the frameshift site. J Mol Biol 310:987–999

Plant EP, Jacobs KL, Harger JW, Meskauskas A, Jacobs JL, Baxter JL, Petrov AN, Dinman JD (2003) The 9-Å solution: How mRNA pseudoknots promote efficient programmed -1 ribosomal frameshifting. RNA 9:168–174

Pogany J, White KA, Nagy PD (2005) Specific binding of tombusvirus replication protein p33 to an internal replication element in the viral RNA is essential for replication. J Virol 79: 4859–4869

Prüfer D, Tacke E, Schmitz J, Kull B, Kaufmann A, Rohde W (1992) Ribosomal frameshifting in plants: A novel signal directs the −1 frameshift in the synthesis of the putative viral replicase of potato leafroll luteovirus. EMBO J 11:1111–1117

Salem NM, Miller WA, Rowhani AK, Golino DA, Moyne A.-L, Falk BW (2008) Rose spring dwarf-associated virus has RNA structural and gene-expression features like those of *Barley yellow dwarf virus*. Virology 375:354–360

Schwartz M, Chen J, Janda M, Sullivan M, den Boo J, Ahlquist P (2002) A positive-strand RNA virus replication complex parallels form and function of retrovirus capsids. Mol Cell 9: 505–514

Selmer M, Dunham CM, Murphy FVt, Weixlbaumer A, Petry S, Kelley AC, Weir JR, Ramakrishnan V (2006) Structure of the 70S ribosome complexed with mRNA and tRNA. Science 313:1935–1942

Skaf JS, Schultz MH, Hirata H, de Zoeten GA (2000) Mutational evidence that the VPg is involved in the replication and not the movement of Pea enation mosaic virus-1. J Gen Virol 81: 1103–1109

Smith GR, Borg Z, Lockhart BE, Braithwaite KS, Gibbs MJ (2000) Sugarcane yellow leaf virus: A novel member of the Luteoviridae that probably arose by inter-species recombination. J Gen Virol 81:1865–1869

Stahl G, McCarty GP, Farabaugh PJ (2002) Ribosome structure: Revisiting the connection between translational accuracy and unconventional decoding. Trends Biochem Sci 27:178–183

Staple DW, Butcher SE (2005) Solution structure and thermodynamic investigation of the HIV-1 frameshift inducing element. J Mol Biol 349:1011–1023

Su L, Chen L, Egli M, Berger JM, Rich A (1999) Minor groove RNA triplex in the crystal structure of a ribosomal frameshifting viral pseudoknot. Nat Struct Biol 6:285–292

Takyar S, Hickerson RP, Noller HF (2005) mRNA helicase activity of the ribosome. Cell 120: 49–58

Taliansky ME, Robinson DJ, Waterhouse PM, Murant AF, De Zoeten GA, Falk BW, Gibbs MJ (2005) Umbravirus. In: Fauquet CM, Mayo MA, Maniloff J, Desselberger U, Ball LA (eds), Virus taxonomy: Eighth report of the international committee on taxonomy of viruses, pp. 901–906. Elsevier, San Diego

Tamm T, Truve E (2000) Sobemoviruses. J Virol 74:6231–6241

ten Dam EB, Verlaan PW, Pleij CW (1995) Analysis of the role of the pseudoknot component in the SRV-1 gag-pro ribosomal frameshift signal: Loop lengths and stability of the stem regions. RNA 1:146–154

Theimer CA, Giedroc DP (2000) Contribution of the intercalated adenosine at the helical junction to the stability of the gag-pro frameshifting pseudoknot from mouse mammary tumor virus. RNA 6:409–421

Tinoco I Jr, Li PT, Bustamante C (2006) Determination of thermodynamics and kinetics of RNA reactions by force. Q Rev Biophys 39:325–360

Turner KA, Sit TL, Callaway AS, Allen NS, Lommel SA (2004) Red clover necrotic mosaic virus replication proteins accumulate at the endoplasmic reticulum. Virology 320:276–290 van der Wilk F, Verbeek M, Dullemans A, van den Heuvel J (1998) The genome-linked protein (VPg) of southern bean mosaic virus is encoded by the ORF2. Virus Genes 17:21–24

van der Wilk F, Verbeek M, Dullemans AM, van den Heuvel JF (1997) The genome-linked protein of potato leafroll virus is located downstream of the putative protease domain of the ORF1 product. Virology 234:300–303

Yusupova GZ, Yusupov MM, Cate JH, Noller HF (2001) The path of messenger RNA through the ribosome. Cell 106:233–241

Chapter 10
Programmed Frameshifting in Budding Yeast

Philip J. Farabaugh

Abstract The budding yeast, *Saccharomyces cerevisiae*, provided two of the earliest analyzed examples of programmed translational frameshifting. The Ty family of retrotransposons uses +1 programmed frameshifting in the expression of their pol homologue, resulting in the production of a gag–pol fusion protein. The pol gene product encodes the enzymatic activities necessary for reverse transcription of the Ty mRNA. Similarly, the endogenous L-A virus uses −1 frameshifting in expression of the protein responsible for catalyzing replication of L-A, a double-stranded RNA virus. Subsequently, three chromosomal genes were shown to use +1 frameshifting in their expression: ABP140, EST3, and OAZ1. Their frameshifting strongly resembles the Ty event. Bioinformatic analysis suggests that many genes may employ −1 frameshifting although the function of the frameshift in those genes remains controversial. This review will discuss the mechanisms of these frameshift events, the evolution of programmed frameshifting in budding yeast, and the lessons to be learned about programmed alternative decoding events based on frameshifting in yeast.

Contents

P.J. Farabaugh (✉)
Department of Biological Sciences and Program in Molecular and Cell Biology, University of Maryland Baltimore County, Baltimore, MD 21250, USA
e-mail: farabaug@umbc.edu

J.F. Atkins, R.F. Gesteland (eds.), *Recoding: Expansion of Decoding Rules Enriches Gene Expression*, Nucleic Acids and Molecular Biology 24,
DOI 10.1007/978-0-387-89382-2_10,

10.1 Introduction

Programmed translational frameshifting is widely dispersed across all species. The predominant type of frameshifting overall arguably is programmed −1 frameshifting, which is commonly found in a wide variety of viruses and chromosomal elements. The first frameshift site identified in budding yeast *Saccharomyces cerevisiae* was a +1 programmed frameshift that occurs between the structural (*gag*) and enzymatic (*pol*) genes of the retrotransposon Ty1 (Clare et al. 1988; Belcourt and Farabaugh 1990). Ty1 frameshifting occurs by shifting +1 after decoding the first codon of a 7-nt shift site: CUU-AGG-C. Subsequently, +1 programmed frameshift sites were identified in the Ty3 retrotransposon (Farabaugh et al. 1993) and in three structural genes, *EST3* encoding a protein subunit of telomerase (Morris and Lundblad 1997), *ABP140* encoding a actin-binding protein and putative S-adenosylmethionine-dependent methyltransferase (Asakura et al. 1998), and *OAZ1* encoding ornithine decarboxylase antizyme (Palanimurugan et al. 2004).

S. cerevisiae genes also employ −1 frameshifting. The endogenous L-A double-stranded RNA virus employs −1 simultaneous slippage frameshifting (Dinman et al. 1991); as with the Ty elements, frameshifting occurs between the structural and enzymatic genes of this virus. Bioinformatic analysis has identified many potential −1 programmed frameshift sites in the *S. cerevisiae* genome (Hammell et al. 1999; Bekaert et al. 2005); some promote frameshifting in vivo (Jacobs et al. 2007). The frameshift sites sensitize the genes to nonsense-mediated decay, a possible genome-wide genetic control system (Plant et al. 2004). Even finding a phenotypically significant role in gene expression for a portion of these sites will show that yeast is more similar to other eukaryotes than had previously been thought.

This chapter will discuss the phenomenology and mechanism of programmed frameshifting in budding yeast. That discussion will focus on analysis of *cis*-acting and some *trans*-acting factors that are responsible for promoting +1 and −1 frameshifting. Chapter 15 will provide a fuller discussion of the genetics of frameshift-regulatory *trans*-acting factors.

10.2 Programmed +1 Frameshifting in S. cerevisiae

Programmed frameshifting in the yeast *S. cerevisiae* was among the first examples of the use of this type of expression strategy in any organism. Previous to identification of the +1 frameshift site in the Ty1 retrotransposon (Clare and Farabaugh 1985) the only known examples of the evolution of a frameshift site in a structural gene were gene *10* of T7 bacteriophage (Dunn and Studier 1983), the gene for peptide release factor 2 (Craigen et al. 1985; Craigen and Caskey 1986), and the Rous sarcoma virus (Jacks and Varmus 1985). Spontaneous, low-efficiency frameshifting at mutant sites had been known at that point for over a decade (see Atkins et al. 1972) but the concept of the sequence of a gene inducing the ribosome to frequently change reading frame was novel. Work on the Ty1 +1 programmed frameshift site defined the phenomenology of that type of event and was the first to provide insight into the mechanism of programmed +1 frameshifting. Subsequent work on frameshifting in the distantly related Ty3 retrotransposon and on the three chromosomal genes that employ frameshifting has expanded our understanding of +1 frameshifting.

10.2.1 Ty Transposons in S. cerevisiae

The Ty retrotransposons of *S. cerevisiae* – Ty1, Ty2, Ty3, Ty4, and Ty5 – are a family of retroviral-like transposable elements that transpose through an RNA intermediate using a reverse transcriptase encoded by the Ty *pol* gene to convert the mRNA into a cDNA copy (Boeke et al. 1985). The reaction occurs inside a virus-like particle with protein components encoded by the Ty *gag* gene. [The *gag* and *pol* genes of Ty1 were initially named *TYA* and *TYB* (Clare and Farabaugh 1985) but this chapter will use the retroviral nomenclature for clarity.] Ty elements encode a single full-length mRNA with two open reading frames, the 5′ *gag* and 3′ *pol* genes; the 3′ end of the *gag* ORF overlaps the 5′ end of the *pol* ORF, as diagrammed in Fig. 10.1A. In addition, the first AUG initiation codon is over 900 nt into the *pol* ORF and it is the 44th AUG in the Ty1 mRNA. Eukaryotic initiation usually occurs at the first AUG in an mRNA, which in Ty1 is the initiator of the *gag* gene. Clearly, expression of the Ty1 *pol* gene had to be "unusual" (Clare and Farabaugh 1985) and a simple explanation involved the ribosome shifting spontaneously into the *pol* reading frame and continuing translation. Clare and Farabaugh (1985) showed that the Ty1 *pol* translation product was a *gag–pol* fusion as predicted by the model.

10.2.1.1 Evidence for +1 Frameshifting in Ty1 Elements

The putative frameshift had to occur within the 38 nt *gag–pol* overlap region, which included a 14 nt sequence, conserved between the closely related Ty1 and Ty2 retrotransposons, that was necessary and sufficient to support +1 frameshifting (Clare, Belcourt and Farabaugh 1988). Subsequent mutagenesis identified a minimal 7 nt frameshift sequence =: CUU-AGG-C, shown in codons of the *gag* gene, which is also the site of the programmed +1 frameshift event decoding the CUU as

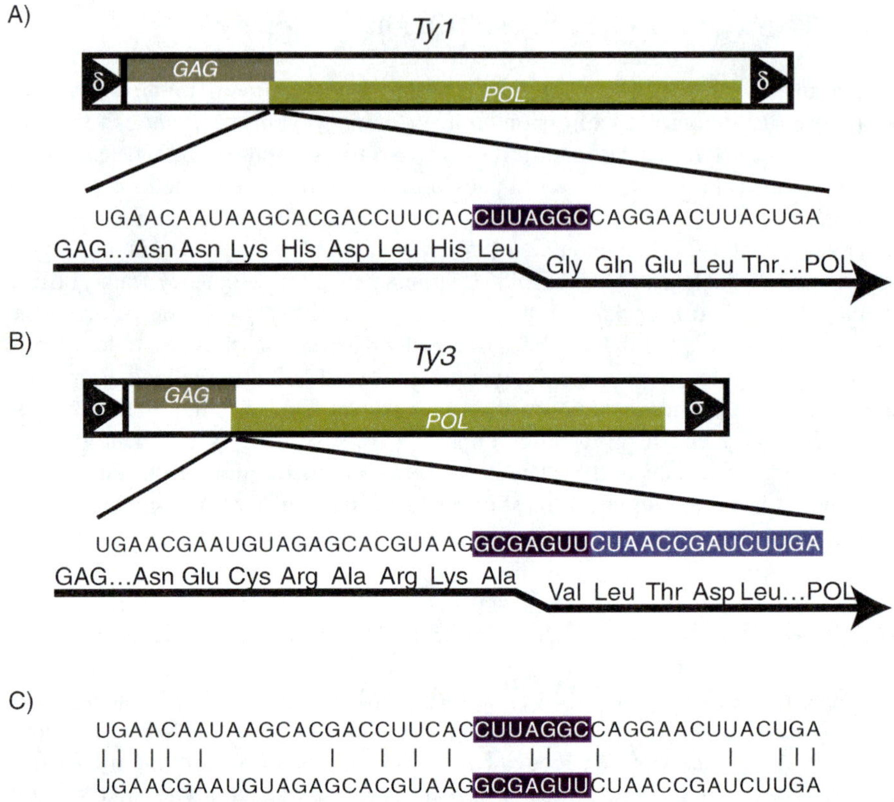

Fig. 10.1 Gene organization for Ty1 and Ty3 retrotransposons of *S. cerevisiae*. (**A**) The Ty1 element cartooned to show the 330 bp delta (d) terminal repeats and the two encoded genes (*GAG* and *POL*). The sequence of the overlap region between the two genes is given below with the CUU-AGG-C shifty heptamer highlighted in the *purple rectangle*. The sequence of the primary translation product in the zero (left) and +1 reading frames. (**B**) The organization of the Ty3 element is shown as in (A) but with sigma (s) terminal repeats. The sequence of the Ty3 stimulator is shown highlighted by *blue*. (**C**) A comparison of the primary sequences of the Ty1 and Ty3 *GAG–POL* overlaps indicating the low level of sequence similarity

leucine then GGC as glycine (Belcourt and Farabaugh 1990). Frameshifting is 40% efficient.

The frameshift signal consists of two codons in the normal or zero reading frame and a third overlapping codon in the shifted frame. The frameshift occurs when the CUU codon occupies the ribosomal P site with the ribosome selecting a tRNA recognizing the +1 frame GGC codon rather than the zero frame AGG codon. Peptidyl-tRNA reading CUU would slip +1 onto the overlapping UUA codon allowing reading of the then in-frame GGC codon by a glycyl-tRNA. Belcourt and Farabaugh (1990) showed that overexpression of the AGG-decoding $tRNA^{Arg}_{CCU}$ drastically reduced frameshifting indicating that decoding of AGG competes with frameshifting. The major CUU-decoding $tRNA^{Leu}_{UAG}$ is unusual in that its wobble uridine is unmodified; in other yeast tRNAs, wobble uridines are modified to restrict reading to A or to A and G (see Johansson et al. 2008 and references therein). As a result, $tRNA^{Leu}_{UAG}$ can decode all six Leu codons (Weissenbach et al. 1977; Randerath et al. 1979). The weak U•U pair formed on CUU or the ability of the tRNA to recognize the overlapping UUA codon could each explain the tendency toward frameshifting. As predicted, a novel tRNA making a normal Watson–Crick interaction with CUU drastically reduced frameshifting, and replacement of the CUU codon with CUC, CUA, or CUG, weakening pairing after slippage, strongly reduced frameshifting (Belcourt and Farabaugh 1990). These data are consistent with a model of +1 frameshifting in which slow decoding by $tRNA^{Arg}_{CCU}$ induces a translational pause during which peptidyl-$tRNA^{Leu}_{UAG}$ slips from CUU to UUA allowing in-frame recognition of the next codon, GGC, by $tRNA^{Gly}_{CCG}$ (Fig. 10.2).

10.2.1.2 The Purpose of +1 Frameshifting for Ty1 Propagation

Genes use programmed frameshifts for a variety of purposes. Frameshifting produces two primary translation products, one form by normal translation and a second less abundant form through frameshifting. The second form can either cause premature termination at a termination codon in the shifted frame or translation can continue in the shifted frame creating a translational fusion of the zero and shifted frame ORFs.

Frameshifting can provide translation products with distinct enzymatic function, as has been hypothesized for the *Escherichia coli dnaX* gene (Blinkowa and Walker 1990; Flower and McHenry 1990; Tsuchihashi and Kornberg 1990). The gene encodes the τ subunit of DNA polymerase III by normal translation of a truncated γ subunit by frameshifting followed immediately by termination; the γ subunit appears to be associated with distributive synthesis on the lagging strand while the full-length τ subunit provides the extreme processivity required for the leading strand (Blinkowa and Walker 1990; Flower and McHenry 1990; Tsuchihashi and Kornberg 1990).

Frameshifting can provide for autogenous control, for example, in the *prfB* gene of *E. coli*, which encodes peptide release factor 2 (RF2) (Craigen and Caskey 1986). Frameshifting occurs when the ribosome encounters a UGA termination codon, recognized by RF2. When RF2 is in limiting supply, the ribosome pauses at the stop codon inducing +1 frameshifting and expression of full-length product; when RF2 is not limiting, translation terminates at the UGA. This autogenous control mechanism insures that of sufficient factor is continuously present.

Like the retroviruses, the Ty1 frameshift accomplishes a morphogenetic purpose: the insertion of the *pol*-encoded enzymatic functions into the virus-like particle (VLP) formed by the *gag*-encoded structural proteins (Farabaugh 1995). Changing the ratio of Gag to Gag–Pol products by either reducing (Xu and Boeke 1990) or increasing frameshifting (Kawakami et al. 1993) severely reduces transposition frequency. The topology of the Gag–Pol fusion protein places the enzymatic activities within the VLP while enforcing a 50:1 Gag to Gag–Pol stoichiometry.

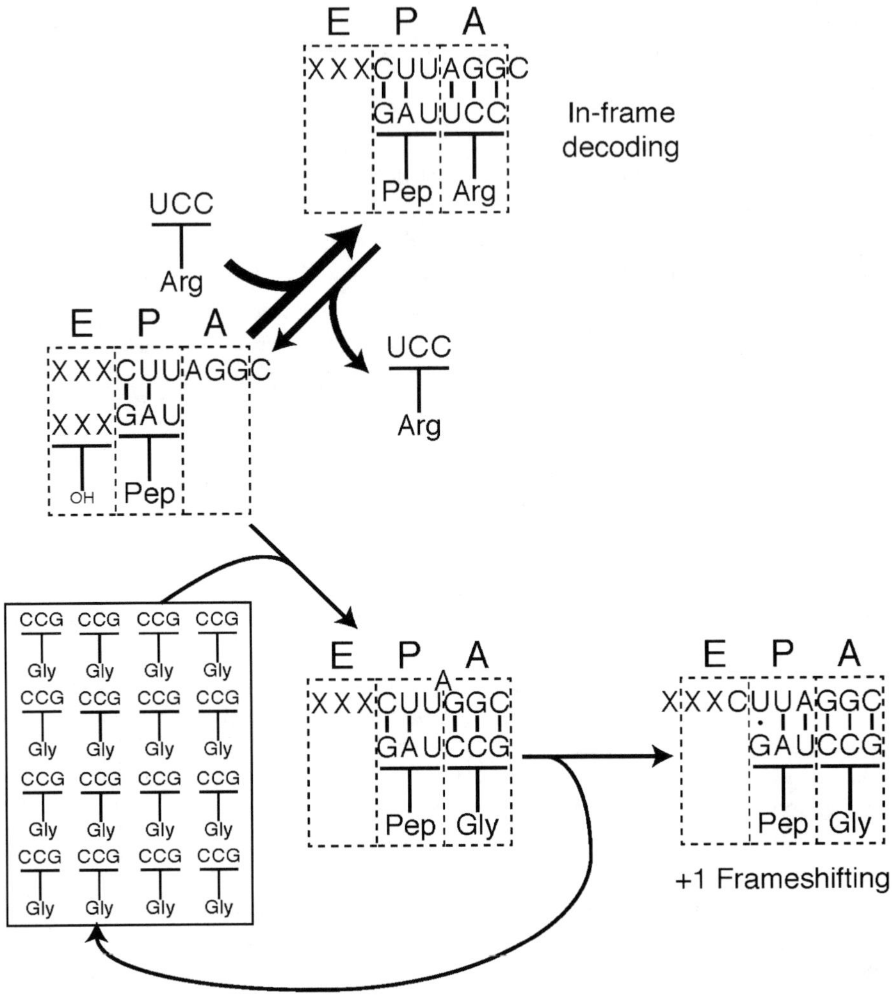

Fig. 10.2 Model of Ty1 frameshift mechanism. The three ribosomal tRNA-binding sites (E, P, and A) are diagrammed as dotted rectangles binding tRNAs cartooned as Ts; the anticodon of each tRNA appears above the tRNA running 5′–3′ from right to left. *On the left*, the P site is shown occupied by peptidyl-$tRNA^{Leu}_{UAG}$; the identity of the tRNA in the E site does not influence frameshifting and it is shown with XXX as anticodon. Two alternatives exist for the next step of elongation. Above, $tRNA^{Arg}_{CCU}$ is shown occupying the A site, leading to in-frame decoding; this reaction is shown as reversible because wobble mispairing in the P site appears to block cognate acceptance. Below, $tRNA^{Gly}_{GCC}$ is shown occupying the A site, also reversibly; its binding can lead to +1 frameshifting. Binding of this tRNA to the A site is shown requiring the middle A of the shifty heptamer to be excluded from the A site to allow the GGC anticodon to bind there, facilitating out-of-frame binding of $tRNA^{Gly}_{GCC}$. The number of tRNA shown corresponds to their relative concentration in the cell; $tRNA^{Gly}_{GCC}$ is present at approximately 16-times the concentration of $tRNA^{Arg}_{CCU}$

10.2.1.3 Ty1-Type Programmed Frameshifting Occurs in Multiple-Related Ty-Family Elements

The simple heptameric Ty1 frameshift site and its location in the *gag–pol* overlap simplified testing if other Ty retrotransposons use the same mechanism. Ty2 (Clare et al. 1988) and Ty4 (Janetzky and Lehle 1992; Stucka et al. 1992) include the same heptameric sequence in roughly the same location within the overlap allowing production of a Gag–Pol fusion protein by frameshifting. Ty5 does not code for separate *gag* and *pol* genes; instead the element is spanned by a single ORF that includes homology to both *gag* and *pol* (Voytas and Boeke 1992). Ty3 has a structure similar to Ty1, Ty2, and Ty4 but the sequence of the *gag–pol* overlap includes no sequence similar to their heptameric frameshift site (Hansen, Chalker and Sandmeyer 1988). Further analysis of the Ty3 frameshift region was required to understand the mechanism of expression of its *pol* product.

10.2.1.4 Evidence for +1 Frameshifting in Ty3 Elements

A ribosome entering the Ty3 overlap in the zero frame could shift within the overlap and continue into the +1 frame *pol* gene to generate the expected Gag–Pol product. Detailed mutagenesis of the region identified a minimal 21 nt region that supported maximal 15% frameshifting (Farabaugh et al. 1993). N-terminal protein sequencing identified an essential 7 nt sequence GCG-AGU-U, shown in codons of the *gag* gene, which corresponds to the first 7 nt of the 21 nt region (Fig.10.1B). A 14 nt downstream sequence, termed the Ty3 stimulator, stimulates frameshifting about 7.5-fold (Farabaugh et al. 1993; Li et al. 2001). The mechanism of action of *cis*-acting stimulator sequences will be addressed below.

Ty3 frameshifting resembles Ty1 in that the second codon of the frameshift heptamer, AGU, is slowly recognized in vivo; overexpressing this tRNA strongly reduces frameshift efficiency (Farabaugh et al. 1993). Farabaugh et al. (1993) concluded that unlike Ty1 frameshifting, Ty3 frameshifting does not occur by peptidyl-tRNA slippage. In Ty3 frameshifting, the $tRNA^{Ala}$ that reads the GCG codon could make no more than a wobble base pair with the +1 shifted codon, CGA. They hypothesized frameshifting by out-of-frame binding of incoming aminoacyl-tRNA. A recent paper by Hansen et al. (2003) suggested the possibility of such weak pairing in the shifted frame. This paper may have misinterpreted what Farabaugh et al. (1993) meant by out-of-frame binding since it rejected a model in which the incoming aminoacyl-tRNA binds to a shifted site overlapping the ribosomal A site. Stahl et al. (2002) suggested that out-of-frame binding would involve a tRNA correctly positioned in the A site interacting with a +1 shifted mRNA codon, which appears to be consistent with the arguments of Hansen et al. (2003). Frameshifting at the Ty3 site is much less efficient than in Ty1; without the stimulator, frameshifting is only about 2% efficient, while in Ty1 it is 40%. Clearly, this site very inefficiently induces frameshifting, regardless of whether it does so by out-of-frame recognition or peptidyl-tRNA slippage.

10.2.1.5 Translational Frameshifting Induced by Near-Cognate Peptidyl-tRNAs

Initially, it was very unclear why GCG as opposed to any other codon should allow +1 frameshifting. The predicted GCG-decoding tRNA, $tRNA^{Ala}_{CGC}$, by forming three G-C base pairs would be among the least likely to slip on the mRNA. After shifting +1 it would require purine–purine or pyrimidine–pyrimidine pairs in at least two positions, which available data suggested would block tRNA slippage (for example, see Brierley et al. 1992). It was therefore mystifying that the sequence GCG-AGU-U should induce +1 frameshifting. Presumably, the structure of $tRNA^{Ala}_{CGC}$ would explain this propensity. Efforts to identify a gene encoding this tRNA using genetic selections were unsuccessful (P. Farabaugh, unpublished data).

A solution to this conundrum came from computational analysis identifying 274 tRNA genes from 42 anticodon classes but lacking a gene for the putative $tRNA^{Ala}_{CGC}$ (Percudani et al. 1997; Hani and Feldmann 1998). The four Ala codons would be decoded by two tRNAs, $tRNA^{Ala}_{UGC}$ for GCA/GCG and $tRNA^{Ala}_{GGC}$ for GCU/GCC. The wobble base of $tRNA^{Ala}_{UGC}$ is 5-carbamoylmethyluridine (ncm^5U), which recognizes both A- and G-ending codons (Johansson et al. 2008). Yeast tRNA families, however, usually also include a tRNA with C in the wobble position dedicated to recognizing the G-ending codon, suggesting that ncm^5U might be somewhat inefficient. The fact that the Ala family lacks such a C-wobble tRNA might be relevant to frameshifting at GCG.

A total of 11 codons support +1 frameshifting significantly above background (Vimaladithan and Farabaugh 1994). Oddly, overexpression of the encoded cognate tRNA for seven of these codons actually *reduced* frameshifting (Sundararajan et al. 1999). The other four codons lack a predicted cognate decoder: GCG, CCG, CUG, and CGA; the missing tRNAs would have a wobble C or, for CGA, I. Expressing synthetic tRNAs with these wobble nucleotides *reduced* frameshifting (Sundararajan et al. 1999). Furthermore, eliminating the cognate tRNAs for AGG, GGG, and GUG strongly *increased* frameshifting (Sundararajan et al. 1999). Finally, frameshifting was increased by overexpressing near-cognate tRNAs for CCU/CCC/CCG ($tRNA^{Pro}_{UGG}$), GGG ($tRNA^{Gly}_{UCC}$ or $tRNA^{Gly}_{GCC}$) of GUG ($tRNA^{Gly}_{UAC}$). These data suggested that frameshifting required that the shift codon be decoded by a particular near-cognate isoaccepting tRNA. All 11 examples of +1 frameshifting require non-canonical wobble pyrimidine•pyrimidine or purine•purine wobble pairs (Sundararajan et al. 1999).

Surprisingly, there was no correlation between stability of the frameshift-stimulating tRNA in the +1 frame (a presumed measure of slipperiness) and ability to induce frameshifting (Vimaladithan and Farabaugh 1994). Some codons with the least predicted propensity for slippage are among the most active in stimulating frameshifting. This seems to contradict the proposal by Hansen et al. (2003) that frameshifting always involves slippage. However, whether slippage or out-of-frame binding is the correct mechanism remains hypothetical, that is, it cannot be directly observed and must be inferred from other data. Thus, the mechanism used cannot be definitely proven.

10.2.2 EST3 *Gene in S. cerevisiae*

EST3 is an essential gene encoding a protein subunit of telomerase (Lendvay et al. 1996) and immunoprecipitation analysis shows it is part of telomerase in vivo (Hughes et al. 2000) but the protein is dispensable for in vitro activity (Lingner et al. 1997). The *EST3* gene included no ORF of sufficient length to be annotated as a putative gene in yeast genome database. Telomerase includes *TLC1* RNA, which is the template in telomere elongation (Singer and Gottschling 1994). Morris and Lundblad (1997) suspected that *EST3* encoded a second RNA component. In fact, it does encode a protein product by frameshift translation of two short overlapping ORFs (Fig. 10.3A). Morris and Lundblad (1997) found that termination codons in

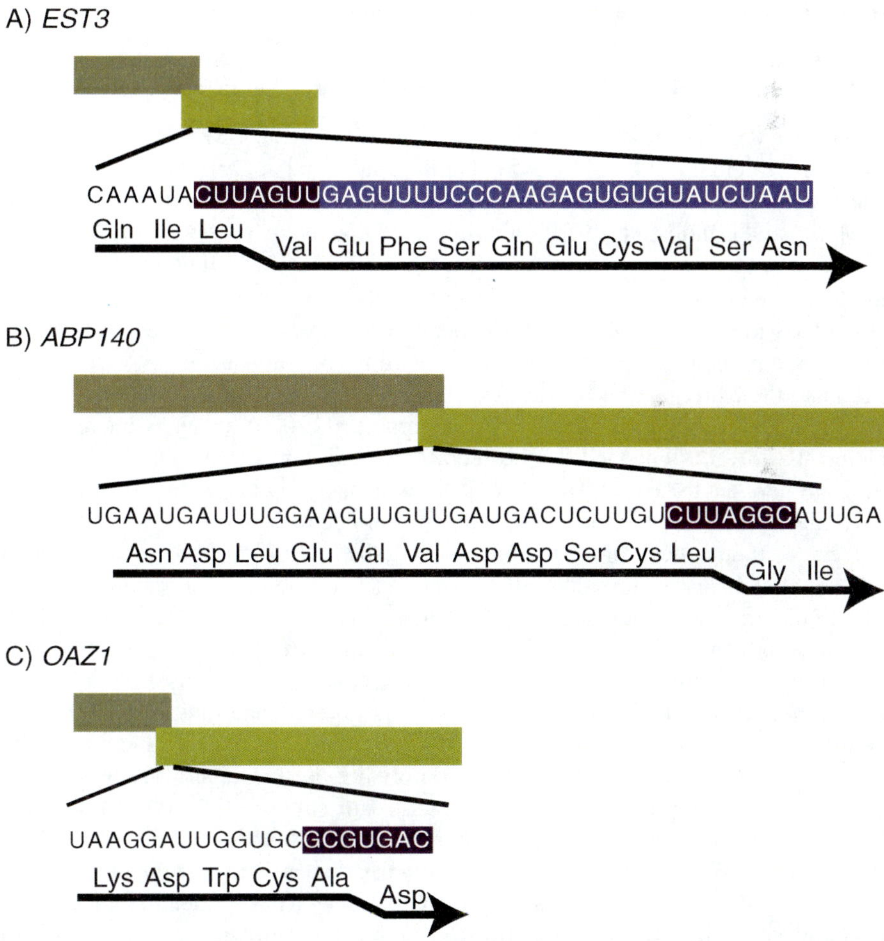

Fig. 10.3 Gene organization of (**A**) *EST3*, (**B**) *ABP140*, and (**C**) *OAZ1* shown as in Fig. 10.1

each of the ORFs caused a null phenotype and each could be suppressed by a nonsense suppressor tRNA, demonstrating that both ORFs participate in encoding the protein product.

The overlap between the two *EST3* ORFs includes a sequence that had previously been shown to stimulate frameshifting, CUU-AGU-U (Vimaladithan and Farabaugh 1994). Mutating this signal to replace either the CUU or AGU with synonymous codons (UUG and UCU) produces the null phenotype (Morris and Lundblad 1997). The primary translation products of the gene are two proteins, one consistent with termination just after the frameshift heptamer and a second the size predicted for a frameshift product (Morris and Lundblad 1997). The ratio of the two products predicts a frameshift efficiency of 75–90%, although the stability of the shorter product was not tested. Interestingly, a mutation that fuses the two reading frames into a single ORF, and therefore produces 100% full-length product, had no observable phenotype (Morris and Lundblad 1997). This raises the question of whether the frameshift is necessary for the function of the gene.

10.2.2.1 Identification of the Downstream Stimulator Sequence

When tested in the absence of any other sequences, the *EST3* heptamer has a frameshift efficiency of only 8% (Vimaladithan and Farabaugh 1994). This low efficiency contrasts with the apparent high efficiency of the *EST3* frameshift reported by Morris and Lundblad (1997). They suggested that overexpression of *EST3* might have titrated the low-abundance AGU-decoding $tRNA^{Ser}_{GCU}$, stimulating frameshifting. The explanation is unconvincing for several reasons. First, $tRNA^{Ser}_{GCU}$ is not particularly low abundance. The tRNA abundance is proportional to gene copy number (Percudani et al. 1997); the median gene copy for tRNA genes in *S. cerevisiae* is 5.5 and there are genes 4 for $tRNA^{Ser}_{GCU}$. More importantly, there are over 90,000 AGU codons in the yeast genome (Nakamura et al. 2000) and it is unclear how adding the mRNAs encoded by the reporter would compete against such a high level of endogenous usage. The alternative explanation is that some feature of the *EST3* mRNA increases frameshift efficiency significantly.

Detailed mutagenesis of the regions flanking the *EST3* frameshift signal revealed a 27 nt downstream stimulator that increases frameshifting approximately 10-fold (Taliaferro and Farabaugh 2007). There are many examples of secondary structures inducing −1 simultaneous slippage frameshifting and termination codon readthrough (reviewed by Farabaugh 1996; Gesteland and Atkins 1996; Brierley and Pennell 2001; Namy et al. 2004). There is no evidence that the Ty3 context forms a secondary structure (hairpin loop or pseudoknot). Taliaferro and Farabaugh (2007) showed that an oligonucleotide comprising the heptameric frameshift site with the regions 30 nt upstream and downstream supported 40% frameshifting, suggesting that all or nearly all of the information required for maximal frameshifting lies within that region. The upstream 30 nt is dispensable, but the downstream 30 nt stimulated frameshifting about 10-fold. A set of nested deletions identified a minimal region 27 nt downstream of the frameshift site and suggested it might be modular because frameshift efficiency declined at three points by twofold steps.

Saturation nucleotide substitution mutagenesis was inconsistent with any model in which the stimulator forms Watson–Crick base pairs with an RNA partner but was consistent with models that do not require such base pairing. Below we will discuss a general model for how this type of downstream stimulatory sequences might increase +1 frameshifting.

10.2.3 ABP140 Gene in S. cerevisiae

ABP140, a second chromosomal gene employing programmed +1 translational frameshifting, encodes an actin filament-binding protein with homology to an S-adenosylmethionine-dependent methyltransferase (Asakura et al. 1998; Katz et al. 2003). The gene is inessential (Niewmierzycka and Clarke 1999) so the frameshift must itself be inessential, at least during growth in laboratory cultures.

The gene was identified by searching for peptides from a novel filamentous actin (F-actin)-binding protein in the *S. cerevisiae* genome (Asakura et al. 1998). The peptides identified a region with two overlapping ORFs. One peptide spanned the overlap and was encoded partly in each ORF, suggesting that the protein was expressed by joining the two ORFs (Fig. 10.3B). The junction between the two ORFs occurs at the sequence CUU-AGG-C (in codons of the upstream frame), the Ty1 frameshift heptamer. Frameshifting by the Ty1 mechanism would produce the observed sequence of the Abp140 peptide. It is not known whether the gene includes any frameshift-stimulating context sequences.

10.2.4 OAZ1 Gene in S. cerevisiae

OAZ1, a third chromosomal +1 frameshifting gene, is a member of a large family of ornithine decarboxylase (ODCase) antizyme genes distributed throughout eukaryotes (Ivanov and Atkins 2007). Palanimurugan et al. (2004) identified the yeast homologue, *OAZ1*, by searching the yeast genome for sequences encoding proteins that resembled other known fungal antizymes. The *OAZ1* gene consists of two ORFs joined by a +1 shifty stop frameshift site (Fig. 10.3C). Most of the similarity to other members of the antizyme family are found in the downstream 225-codon ORF but the similarity extends across the sequence encoded at a putative frameshift signal, GCG-UGA-C (shown as codons of the upstream ORF), and the sequence of about 20 nt surrounding the frameshift site is highly similar to other fungal and metazoan antizymes at the DNA level (Palanimurugan et al. 2004). This frameshift site is used in budding yeasts most closely related to *S. cerevisiae*; more distantly related yeasts including *S. kluyveri*, *Kluyveromyces thermotolerans,* and *K. waltii* use the heptamer CCG-UGA-C and *Yarrowia lipolytica* uses UCC-UGA-C (Ivanov, Gesteland and Atkins 2006). Mass spectrometry of the encoded protein demonstrated that Oaz1 is encoded by +1 frameshifting that reads the shifty heptamer as Ala-Asp (Ivanov, Gesteland and Atkins 2006). The Oaz1 protein, like other ODCase antizymes, binds to and directs the proteasomal degradation of ODCase in the presence of excess spermidine, an eventual downstream product of ODCase.

Excess polyamines also regulate expression of mammalian ODCase antizymes by stimulating increased +1 frameshifting. Palanimurugan et al. (2004) showed that increasing spermidine in *S. cerevisiae* induced increasing expression of the antizyme protein; at the same time, a smaller protein corresponding to the first ORF declined with increasing spermidine. This is consistent with the frameshifting model but proof of a frameshift mode of expression requires direct confirmation.

The *S. cerevisiae OAZ1* frameshift signal involves a putative P site codon, GCG, also used in the Ty3 programmed frameshift (Farabaugh et al. 1993) and a poorly recognized UGA-C tetranucleotide known to be competent to stimulate 37% frameshifting (Pande et al. 1995). The high efficiency of the site in the absence of any flanking context sequences suggests that the *OAZ1* site may be as simple as that of the Ty1 retrotransposon and may consist of only the 7 nt frameshift signal. The variant frameshift sites CCG-UGA-C that appears in distantly related budding yeast involves a CCG P site codon. A cognate tRNA for this codon is lacking in one yeast using this site, *S. kluyveri*, but is present in low copy in another, *K. waltii* (Ivanov et al. 2006). The alternative site used in *Y. lipolytica*, UCC-UGA-C, also uses P site codon recognized by a low copy tRNA (Ivanov et al. 2006). These sites are consistent with the near-cognate decoding model since all of the frameshifts could involve recognition by a higher copy isoaccepting species, though a direct test of that hypothesis has not been done. It is also unknown whether similar near-cognate decoding stimulates frameshifting in non-budding yeast antizyme genes.

10.2.5 Bioinformatic Analysis Identifies Novel +1 Frameshift Sites in S. cerevisiae

Analysis of known programmed frameshift sites identified a small number of potential frameshift sites involving the P site codons CUU and GCG. Mutagenesis of those sites identified a set of 11 codons that stimulate frameshifting significantly above background (Vimaladithan and Farabaugh 1994). These were identified in combination with pausing at the AGG-C A site signal from the Ty1 frameshift site in a background lacking the cognate tRNA for AGG, $tRNA^{Arg}_{CCU}$. The lack of this tRNA presumably stimulates an extended translational pause; frameshifting at the 11 codons is much less efficient in a wild-type background. It is unclear that these studies identified all possible frameshift sites since there might be unknown interactions between A and P site codons that would enable untested sequences to stimulate significant frameshifting.

A computational study identified the least common heptameric sequences in the *S. cerevisiae* genome and among those were many of the known programmed +1 frameshift sites (Shah et al. 2002). The low abundance of heptamers could have many causes but among them is the need for genes to avoid random frameshift signals. Several of the identified heptamers were tested for frameshift ability. Those that are known programmed frameshift sites supported high level frameshifting as did sites that differed from known signals only at the sixth or seventh base (that is, CUU-AGU-A or CUU-AGC-U instead of CUU-AGU-U); these divergent sites supported less frameshifting than the canonical frameshift site. Unexpectedly, a sequence supporting significant frameshifting, GGU-CAG-A, is unrelated to any known programmed frameshift site in *S. cerevisiae*, suggesting that the catalog of possible +1 frameshift sites may be larger than expected. The mechanism underlying frameshifting at this novel site has not been determined.

10.2.6 Factors That Stimulate Programmed +1 Frameshifting in S. cerevisiae

10.2.6.1 Near-Cognate Peptidyl-tRNAs Stimulate +1 Frameshifting

As described above, all +1 programmed frameshift events in *S. cerevisiae* occur when the ribosomal P site contains a near-cognate peptidyl-tRNA, one that fails to form a legal wobble base pairing interaction. Stahl et al. (2002) hypothesized that a non-canonical interaction in the P site would interfere with proper function of the A site based on fact that the ribosome contacts the codon•anticodon complexes in both the P and the A site simultaneously (Yusupov et al. 2001). By interacting with the previous complex while selecting the successive tRNA, the ribosome might improve selectivity.

Introduction of tRNA genes in which the anticodon has been mutated to that predicted for a cognate decoder of frameshift-inducing codons reduces the frequency of frameshifting, often by over a factor of 10 (Belcourt and Farabaugh 1990; Sundararajan et al. 1999). Since generating a cognate isoacceptor can be as easy as making a single change in the anticodon, it is significant that *S. cerevisiae* has not evolved genes encoding the missing cognate tRNAs. Of course, if frameshifting depends on the lack of efficient cognate decoding and if that frameshifting were required for cell viability, one would expect the mutations leading to poor decoding to be conserved over evolutionary time. I will return to this issue below when I discuss the evolution of programmed +1 frameshifting among budding yeasts.

10.2.6.2 Frameshifting Stimulated by tRNA Competition

Programmed +1 frameshifting also depends on poor tRNA decoding inducing a translational pause while the near-cognate peptidyl-tRNA occupies the P site. In every case except for the *OAZ1* gene, frameshifting occurs in competition with cognate decoding by a rare tRNA in the zero frame (Belcourt and Farabaugh 1990; Farabaugh et al. 1993; Morris and Lundblad 1997). In *OAZ1* frameshifting requires slow recognition of the termination sequence UGA-C by yeast peptide release factor 1 (eRF1) and eRF3 (Brown et al. 1990; Bonetti et al. 1995).

On the other hand, rapid cognate recognition of the first +1 frame codon stimulates frameshift efficiency requiring that the tRNA decoding the +1 frame must be abundant (Pande et al. 1995). The +1 and 0 frame codons overlap each other in the ribosomal A site during the translational pause leading to frameshifting. The two tRNAs can be understood to compete with each other for the A site. Significantly, there is little overlap between the codons that allow frequent near-cognate decoding and those that cause translational pausing. For example, the sense codons AGG, AGU, and UGG all can induce sufficient translational pausing (Pande et al. 1995) but not near-cognate decoding (Vimaladithan and Farabaugh 1994). UGG has no near-cognate tRNA. For AGG, the mcm^{5}U•G pair formed by its near cognate may be unable to stimulate frameshifting. AGU might induce frameshifting if recognized by tRNA$^{Ser}_{IGA}$ but preferential decoding by tRNA$^{Ser}_{UGA}$ may block frameshifting. These three codons could be among the least rapidly decoded codons, they may not allow frameshift-stimulating near-cognate decoding because of their inability to form an appropriate near-cognate pair. Other codons may be more rapidly decoded but allow frameshift-stimulating near-cognate decoding.

10.2.6.3 Cis-Acting RNA Sequences That Stimulate +1 Programmed Frameshifting

For most programmed frameshift sites, a nearby sequence increases frameshift efficiency (Larsen et al. 1995; Brierley and Pennell 2001; Plant et al. 2003). In −1 simultaneous slippage programmed frameshifting, typical of retroviruses, a pseudoknot downstream of the frameshift site increases frameshifting. Originally, the model for pseudoknot stimulation involved it causing the ribosome to pause while the shifty heptamer frameshift site occupied the ribosomal A and P sites (Jacks et al. 1987). More recent experiments brought this model into question (Kontos et al. 2001). I will discuss this issue in more depth below. Mammalian antizyme genes usually include context sequences, many involving pseudoknots (Ivanov and Atkins 2007), but the context requirements for the *S. cerevisiae OAZ1* gene to my knowledge have yet to be determined.

For two *S. cerevisiae* programmed +1 frameshift sites – Ty3 and *EST3* – downstream stimulator sequences increase frameshifting ~10-fold (Li et al. 2001; Guarraia et al. 2007; Taliaferro and Farabaugh 2007). The two stimulators share no significant sequence similarity and neither can form a stem loop or a pseudoknot. For Ty3, the primary RNA structure was shown to stimulate frameshifting and not its protein product (Li et al. 2001). Both stimulators significantly increase frameshifting only at sense pause codons and have no effect at a termination pause codon (Li et al. 2001; Taliaferro and Farabaugh 2007). The fact that the Ty3 stimulator included a region complementary to the loop of helix 18 (h18) of the small ribosomal subunit suggested that the stimulator might block this structure's monitoring cognate decoding in the A site (Li et al. 2001). Such complementary is not, however, true for the *EST3* stimulator.

Recent saturation mutagenesis studies of the two stimulators strongly suggest that the h18 complementarity model of Li et al. (2001) is invalid (Guarraia et al. 2007; Taliaferro and Farabaugh 2007). There is no correlation among Ty3 stimulator single base substitution mutants between their predicted effect on h18 complementarity and frameshift stimulation. Furthermore, in neither case does the mutant data suggest that either stimulator actually engages in Watson–Crick base pairing with any other RNA molecule. The data are consistent with the single stranded stimulators interacting with one or more partners in the ribosome or with ribosome-associated factors. The position of the stimulators immediately downstream of the A site predict that their first 9 nt might interact with the mRNA entrance tunnel (Yusupova et al. 2001). Part of this tunnel is composed of elements of three ribosomal proteins rpS3, rpS4, and rpS5 that form the ribosomal helicase, responsible for removing secondary structures in advance of the ribosome (Takyar et al. 2005). The two stimulators might act by blocking helicase activity, although how that might affect frameshifting is unclear. The Ty3 and *EST3* stimulators, respectively, include 5 and 21 nt past those proposed to occupy the tunnel; these regions could interact with surface exposed elements of rpS3, rpS4, or rpS5 or other nearby ribosomal structures. The length of the *EST3* region increases the list of possible targets to a large region of the 40S solvent surface or any ribosome-associated proteins that would be located in that region. How these interactions might affect frameshifting remains unclear.

10.3 Evolution of +1 Programmed Frameshift Systems in Budding Yeasts

Despite gross phenotypic similarities, budding yeasts are an ancient lineage. The divergence of all budding yeast has been dated to at least 150 million years ago (Mya) or to about the time of the divergence of monotremes from other mammals and of the breaking up of Pangea (reviewed by Weil 2001). A whole genome duplication event occurred much more recently, about 100 Mya, in the lineage leading to *S. cerevisiae* (Wolfe and Shields 1997). The *Saccharomyces sensu stricto* is a closely related group of species that includes *S. cerevisiae*; they diverged since 20 Mya.

10.3.1 Phylogeny of Retrotransposons in Budding Yeast

It was unclear if programmed frameshifting evolved rapidly, appearing in recent evolutionary history and perhaps disappearing in some lineages as rapidly, or if it was more deeply rooted in the evolution of budding yeasts. One way to approach this question is to study the evolution of retrotransposons in budding yeasts. Two such studies have reached slightly different conclusions. Liti et al. (2005) suggest that Ty1 and Ty5 retrotransposons both appeared in budding yeast relatively recently, during the divergence of *Saccharomyces sensu stricto*, suggesting that programmed frameshifting in this group is relatively transitory. Neuvéglise et al. (2002) using a more flexible definition of the elements assigned the appearance of Ty1 before the genome duplication event and of Ty5 at nearly the beginning of the divergence of the budding yeast. Importantly, Neuvéglise et al. (2002) find that all of the members of their Ty1-family appear to employ +1 frameshifting at the same CUU-AGG-C site, suggesting that frameshifting must be over 100 million years old.

10.3.2 Phylogeny of Frameshifting in the Three Chromosomal Genes

The *ABP140* gene is distributed widely across all eukaryotes and is present in all fungi metazoans tested and in some plants (Farabaugh et al. 2006) but frameshifting is present only in budding yeasts more closely related to *S. cerevisiae* than are *Candida albicans*, *C. tropicalis*, *Debaryomyces hansenii*, and *Y. lipolytica*. This puts the origin of frameshifting at approximately the time suggested by Neuvéglise et al. (2002), about 150 Mya.

ABP140 genes have diverged slowly enough that they can be identified widely throughout eukaryotes. By contrast, no homologue of *EST3* could be identified outside budding yeast (Farabaugh et al. 2006). However, again frameshifting is present in all species more closely related to *S. cerevisiae* than *C. albicans*, *D. hansenii*, and *K. lactis*. These data again place the origin of programmed +1 frameshifting at about 150 Mya during the divergence of budding yeast.

The *OAZ1* gene and its expression by shifty stop programmed +1 frameshifting is deeply rooted in eukaryotes; the evolution of these genes will be discussed in Chapter 13.

10.3.3 Phylogeny of tRNAs Important to Programmed +1 Frameshifting

The presence or absence of certain tRNAs are essential for maximal +1 frameshift efficiency. Therefore, if programmed frameshifting is as old as suggested by phylogenetics, we would expect the distribution of tRNA isoacceptors to be similarly old. A comparison of tRNA genes encoded across budding yeasts showed that this expectation is correct; the distribution of essential tRNAs also arose approximately 150 Mya (Farabaugh et al. 2006). The distribution of tRNAs has evolved by anticodon mutation. For example, a presumed non-frameshift-competent $tRNA^{Leu}_{IAG}$ in *D. hansenii* is most closely related to a clade of $tRNA^{Leu}_{GAG}$ present in frameshift-competent budding yeast. This relationship suggests that change between GAG and AAG in the anticodon is responsible for the lack of frameshifting in *D. hansenii*. Two other examples of such a switch were found that appear to have occurred approximately at the origin of programmed +1 frameshifting in these genes.

10.3.4 What Persistence Over Evolutionary Time Implies About the Function of Programmed Frameshift

The retention of programmed frameshifting for over 150 million years is significant since very little would have been necessary to eliminate it. Frameshifting per se is not necessary for budding yeast; in taxa distantly related to *S. cerevisiae* there is no evidence for +1 programmed frameshifting and the tRNAs necessary to support it are missing (Farabaugh et al. 2006).

If frameshifting had no phenotypic effect, loss of frameshifting should have been frequent. However, among 12 species examined, there is evidence of the loss of only one frameshift site from the *ABP140*, *EST3*, and *OAZ1* genes (Farabaugh et al. 2006). This strongly suggests that frameshifting is essential. This conclusion challenges genetic results suggesting that *ABP140* and *EST3* are inessential (Morris and Lundblad 1997; Asakura et al. 1998). The contradiction between the phylogenetics and genetics suggests that the role of frameshifting in these genes may be subtle, outside the ability of researchers working in defined media to ascertain. Presumably, the role of the genes and the frameshift evidences itself only in the more complex natural environment in which budding yeasts evolved.

The failure to express the full-length *EST3* frameshift product results in cell death due to loss of telomeres (Lendvay et al. 1996). Regulation of the protein by modulating frameshifting may be essential to cell viability under some conditions. This could be a homeostatic mechanism, for example, allowing cells to modulate the amount of protein by altering frameshifting. For example, carbon limitation causes cells to reduce growth rate and along with that reduction cells both reduce the rate of protein synthesis and destroy large proportions of the translational apparatus. A feedback regulatory loop might link telomere length to decreasing cell growth rate through *EST3* frameshifting. It is also conceivable that telomere loss regulated by *EST3* frameshifting could be used to control cellular lifetime. I know of no attempt to test these hypotheses.

Conservation of the *ABP140* programmed frameshift (Farabaugh et al. 2006) argues for its functional significance. Frameshifting often provides a morphogenetic function, as in retroviruses, other viruses, and Ty elements. It may be that frameshifting in *ABP140* allows the synthesis of alternative forms of the protein that achieve a similar morphogenetic purpose. The putative dual function of the protein, as an F-actin-binding protein and methyltransferase, resembles the multi-functional *gag–pol* proteins of retroviruses and retrotransposons in that it unites a structural and an enzymatic function into one protein using frameshifting. However, Abp140 has no similarity to other known actin-binding proteins and the region required for that function has not been identified (Asakura et al. 1998), so we do not know whether the two functions are encoded by the two separate ORFs.

10.4 Programmed −1 Frameshifting in S. cerevisiae

Programmed −1 frameshifting sites are common phylogenetically. The first such sites were identified in metazoan retroviruses in Rous sarcoma virus (Jacks and Varmus 1985) and human immunodeficiency virus-1, HIV-1 (Jacks et al. 1988b), and have been found in a wide variety of viruses and transposons in all three domains of life (Brierley 1995; Dinman 1995; Cobucci-Ponzano et al. 2005; Dreher and Miller 2006). The vast majority of these sites conform to a characteristic structure. The frameshift occurs at a slippery heptameric sequence of the form X-XXY-YYZ (shown in codons of the upstream gene) where XXX can be a run or any nucleotide (UUU, AAA, CCC, or GGG), YYY is either UUU or AAA, and Z is usually not G

(Jacks et al. 1988a). Frameshifting on these sites is thought to occur by simultaneous slippage of two tRNAs, bound to the XXY and YYZ codons, in the −1 direction onto the overlapping XXX and YYY codons (Jacks et al. 1988a). Based on this mechanism, these sites are referred to as simultaneous slippage −1 frameshift sites.

10.5 The L-A virus

The frameshift site in the yeast virus L-A was one of the first identified examples of simultaneous slippage −1 frameshifting (Icho and Wickner 1989; Dinman et al. 1991). L-A is a double-stranded endogenous virus of *S. cerevisiae* that, like retroviruses, includes genes encoding the structural and enzymatic proteins required for virus function (Icho and Wickner 1989). Frameshifting in the overlap between the upstream *orf1* gene, an analog of the retroviral *gag* gene that encodes a structural protein, and *orf2*, an analog of retroviral *pol* that encodes the viral RNA-dependent RNA polymerase, occurs at a typical simultaneous slippage −1 frameshift site (Icho and Wickner 1989; Dinman et al. 1991). The site includes a heptamer that conforms to a canonical simultaneous slippage site: G-GGU-UUA (Fig. 10.4). Although no protein sequencing of the primary translation product has demonstrated it, genetic

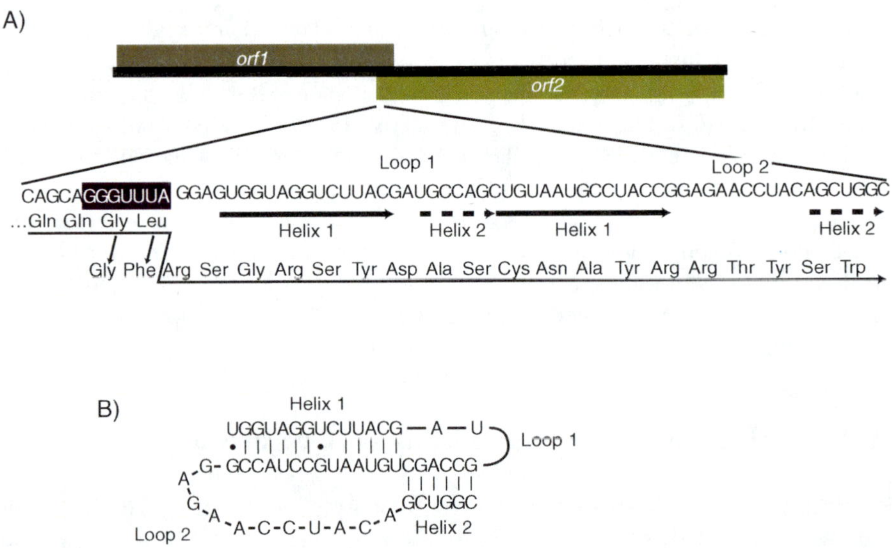

Fig. 10.4 Gene organization of the L-A double-stranded virus of *S. cerevisiae.* (**A**) Diagrams the two encoded genes, *orf1* and *orf2*, and the sequence of the portion of the overlap region essential for −1 frameshifting. The *two arrows* below the shifty heptamer (*highlighted in purple*) indicate the slippage of the two bound tRNAs during frameshifting. The downstream sequences involved in forming the frameshift-stimulating pseudoknot are indicated, including the sequences involved in the two helices and intervening loops. (**B**) The predicted structure of the pseudoknot is shown; in the structure Helix 1 and Helix 2 would be coaxial but are shown unstacked for clarity

analysis and similarity to retroviral frameshift sites suggest that the frameshift occurs in this heptamer (Dinman et al. 1991). The required region downstream of the heptamer can fold into a pseudoknot that resembles those found in retroviral frameshift sites (Brierley 1995; Dinman 1995; Giedroc et al. 2000).

Mutations of the L-A slippery heptamer that disrupt the runs of identical bases (e.g., GGG to AGG or UUU to CUU) strongly reduced frameshift efficiency consistent with frameshifting requiring slippage of the tRNAs (Dinman et al. 1991). Furthermore, replacement of the P site codon, GGG, with AAA, UUU, or CCC actually increased frameshifting; changing the A site codon, UUU, to AAA increased frameshifting while mutation to CCC strongly reduced it (Dinman et al. 1991). These data show that frameshifting in L-A follows the pattern established for metazoan viruses (Jacks et al. 1988a; Brierley et al. 1992). Similarly, mutations of the putative pseudoknot that disrupt its helical regions strongly reduced frameshifting but a combination of compensatory mutations restored it to normal levels, arguing that the pseudoknot stimulates maximal frameshifting efficiency (Dinman et al. 1991; Dinman and Wickner 1992).

The mechanism by which the pseudoknot stimulates frameshifting remains controversial. An early idea was that it would arrest the forward progress of the ribosome, causing a translational pause that would allow time for the two ribosome-bound tRNAs to shift reading frames −1 (Jacks et al. 1987). Chen et al. (1995) proposed an alternative model that the ribosome binds a protein factor required to promote the change in reading frames. Work in a variety of laboratories has been unsuccessful in identifying the hypothetical pseudoknot-binding factor, but work in yeast (Tu et al. 1992; Lopinski et al. 2000) and in mammalian cell extracts (Somogyi et al. 1993) has demonstrated that the pseudoknot does induce a protracted translational pause. A variety of models have been proposed to explain the pausing effect including several that propose some kind of reversible alteration in its tertiary structure (Dinman 1995; Farabaugh 1996; Plant et al. 2003). These models are difficult to reconcile with observations made in mammalian cell extracts showing no correlation between translational pausing and frameshift efficiency (Kontos et al. 2001). Recent cryo-electron microscopy results show that a frameshift pseudoknot blocks completion of the translocation step (Namy et al. 2006), which suggests that the pseudoknot may actually not simply pause the ribosome but trap it in a usually transient intermediate conformation in which the ribosome fails to sufficiently stabilize base pairing and thus allows their repositioning on the mRNA. Pausing may, therefore, be a collateral consequence of the presence of the pseudoknot rather than the precipitating event leading to frameshifting. I will return to this issue below.

10.5.1 Role of the Frameshift in Virus Function

The frameshifts of both the Ty elements and the L-A virus are required for their replicative lifecycle. Mutations that alter frameshifting efficiency block both retrotransposition of Ty elements (Kawakami et al. 1993) and replication of the L-A virus (Dinman and Wickner 1992). Changes to frameshifting of as little as twofold, either increasing or decreasing, can drastically reduce replication (Dinman and Wickner 1992). Dinman and Wickner (1992) hypothesized that forming virus particle depends on precise stoichiometry of *gag* to *gag–pol* proteins and that an imbalance in those proteins would result in production of partially completed, non-functional particles. For Ty elements, changing stoichiometry does interfere with virus biogenesis but the effect is to block proteolytic processing of a Ty-encoded polyprotein. A Ty-encoded protease (p23) processes the *gag* product from a p58 full-length product to a p54 probable capsid protein (Adams et al. 1987; Muller et al. 1987) and the *gag–pol* protein to produce p54 and three other activities, the protease itself, integrase (p90), and reverse transcriptase (p60) (Eichinger and Boeke 1988; Garfinkel et al. 1991). Increasing frameshift efficiency blocked processing of the protease, integrase, and reverse transcriptase, producing an enzymatically inactive particle.

It is likely that the correct *gag* to *gag–pol* stoichiometry allows formation of a structure essential to correct proteolysis rather than blocking formation of VLPs per se. It is significant that a VLP with an excess of *gag–pol* protein is still proteolytically processed to release the capsid protein so the particle is not completely insensitive to protease (Kawakami et al. 1993). Presumably, the structure formed in the presence of excess *gag–pol* protein adopts a conformation making processing of the *pol* protein inaccessible to the protease, perhaps because of multimerization of the unprocessed protein.

Because frameshifting efficiency can vary with physiological changes, the fact that the formation of functional virus particles depends on a precise stoichiometry is counterintuitive. Rather than facilitating transposition of the element, the use of frameshifting for the expression of *gag–pol* protein means that transposition can be easily compromised. This fact provides a useful target for anti-retroviral gene therapy, as has been noted (Harford 1995; Dinman et al. 1998; Brierley and Dos Ramos 2006) but, from the perspective of the virus, the choice of frameshifting is problematic on this basis. The use of frameshifting to encode the extended polyprotein rather than other choices is often explained by reference to genome compression (c.f. Belshaw et al. 2007), although the amount of genome space saved by using frameshifting rather than alternative splicing is relatively small. Another common argument is that the fact that the *gag–pol* mRNA is identical to the template for genome replication suggests that the virus should avoid splicing because it would result in viruses carrying deletions of the intron (Icho and Wickner 1989). A more compelling reason is that the Gag domain must anchor the Pol domain in the structure of the virus particle insuring the deposition of the enzymatic activities inside the virus (Dinman et al. 1991). The fact that frameshifting imposes a cost in terms of generating non-functional virus particles simply underscores the impression, derived from many studies, that evolution rarely produces an optimal solution. The other steps in production of active virus are arguably much less efficient. For example, only about 1% of HIV-1 viruses are infective (Dimitrov et al. 1993); most defective viruses carry inactivating mutations caused by an extremely error-prone reverse transcriptase (Preston et al. 1988). In that context, it is clear that virus assembly need not be maximally efficient, as it indeed appears not to be for HIV-1 (Marozsan et al. 2004).

10.5.2 Evidence for other −1 Programmed Frameshift Sites in S. cerevisiae

Until recently, no chromosomal gene in *S. cerevisiae* was known to use programmed −1 frameshifting. This should not be considered a definitive statement since the programmed +1 frameshift in the *OAZ1* gene remained unrecognized for several years. One signal for such a −1 frameshift mechanism would be the presence of overlapping ORFs with the second ORF in the −1 frame with respect to the first and with sequence motifs typical of the encoded protein product appearing in both reading frames. There appear to be no examples of this type in the *S. cerevisiae* genome (Jacobs et al. 2007).

Jacobs et al. (2007) have argued that the requirement for frameshifting resulting in a C-terminally extended product is perhaps too strict. Programmed frameshifts need only produce an alternative protein product, and that product could be a shortened rather than an extended form, as with the *dnaX* gene of *E. coli*, discussed above. Such frameshifts are more difficult to identify by inspection but not more difficult to identify by motif searching algorithms. By searching for a motif juxtaposing a slippery heptamer (X-XXY-YYZ) immediately upstream of a predicted pseudoknot Jacobs et al. (2007) found putative sites in a large minority of genes in *S. cerevisiae*. The over 3000 putative signals identified a priori might seem too large a number but several candidates showing particularly high scores do promote −1 frameshifting in a reporter assay. Sites derived from the *CTS1*, *EST2*, and *PPR1* genes induced very high-efficiency frameshifting, from 43 to 64% and a site in the *TBF1* stimulated 5.2% frameshifting. Four other sites, *NUP82*, *BUB3*, *FLR1*, *SPR6*, stimulated only 0.4–0.9% frameshifting and an *FKS1* site stimulated <0.01% frameshifting. These data show the existence of −1 frameshifting sites in chromosomal genes of *S. cerevisiae* but that their efficiency varies widely.

Another search for simultaneous slippage frameshift sites in *S. cerevisiae* found almost an order of magnitude fewer potential sites (Theis, Reeder and Giegerich 2008). The authors of that study argue that the Jacobs et al. (2007) study identified sites that included most stable secondary structures that were not pseudoknots. They claim that many of the hits in the Jacobs et al. (2007) study actually involve stem loops or more complex non-pseudoknot structures. Still, the number of sites found by both searches, totaling 74 sites, is far more than had previously been thought to exist in the yeast genome.

No evidence was presented that the predicted sites are required for an observable phenotype at the cellular level (Jacobs et al. 2007). Given the widespread presence of these sites across the genome, however, Plant et al. (2004) suggested that the frameshifts might function to insure the instability of the encoded transcript by making it sensitive to nonsense-mediated decay or NMD (reviewed by Conti and Izaurralde 2005). They showed that bona fide −1 frameshift signals do induce NMD, probably because some proportion of the translating ribosomes read through the frameshift and terminate prematurely. Given the high activity of the *CTS1*, *EST2*, *PPR1*, and *TBF1* frameshifts in reporter studies, it seems likely that some of the identified sites might subject their mRNAs to NMD. A large proportion of the identified mRNAs are stabilized by mutations blocking NMD or the unrelated degradation mechanism, no-go decay (Doma and Parker 2006). These results need to be validated by experiments showing that the presence of the frameshift signal does induce mRNA decay in vivo of the endogenous mRNA.

10.6 General Lessons from Analysis of Programmed Frameshifting in Budding Yeast

The two forms of frameshifting in budding yeast have distinctly different mechanisms. They differ in the sign of the frameshift (+1 and −1), in the number of tRNAs that are required to realign on the mRNA, and in the steps of the translational elongation cycle during which they occur. These differences lead to many differences in the phenomenology of frameshifting including the nature of the genetic interactions between the programmed frameshift events and the host, the details of which are covered in Chapter 15. However, they are similar in that they both manipulate the process of translational elongation to facilitate an event that normally occurs with extremely low efficiency. The ways that the two types of frameshifts undermine continued maintenance of frame are different in detail but show interesting similarities.

10.6.1 How Do +1 Frameshift Signals Manipulate the Translation Machinery?

The basic mechanism of both +1 and −1 frameshifting requires that the frameshift signals interfere with the orderly progression of the elongation process to enable an alternative outcome to compete effectively with the normal codon-by-codon progress along the mRNA. The way that frameshift sites are able to block the continuation of this process helps to identify the aspects of elongation that are most critical to maintenance of reading frame. In both cases, the frameshift site imposes a kinetic block that stops the ribosome from continuing rapidly to the next cognate, in-frame decoding event and allowing a substitute event to occur that results in decoding shifting into an alternative reading frame. The way this is accomplished is different for the two mechanisms.

For +1 programmed frameshifting in *S. cerevisiae*, the decision to shift occurs when the ribosomal A site is empty. This is shown by the fact that frameshifting competes with normal recognition of the next zero frame codon (Belcourt and Farabaugh 1990; Farabaugh et al. 1993; Vimaladithan and Farabaugh 1994; Pande et al. 1995). This resembles +1 frameshifting in the bacterial *prfB* gene where frameshifting competes with recognition of the zero frame UGA termination codon (Adamski et al. 1993). However, pausing is not sufficient to allow frameshifting since mutations that alter the codon immediately upstream of the pause-inducing codon (the "P site codon") strongly reduce frameshifting (Belcourt and Farabaugh 1990; Farabaugh et al. 1993; Vimaladithan and Farabaugh 1994; Sundararajan et al. 1999). The crucial requirement for frameshifting is that the tRNA in the P site during the translational pause be a near-cognate isoacceptor that is mispaired in the wobble position (Sundararajan et al. 1999). Forcing near-cognate decoding of the P site codon invariably resulted in strongly increased frameshift efficiency and forcing cognate decoding strongly reduced it.

The requirement for near-cognate P site decoding identifies a critical feature of reading frame maintenance. Stahl et al. (2002) proposed that the ribosome constrains decoding in the A site to the nucleotide immediately following the P site codon by insuring that there is base pairing at the P site wobble position. The inverse of this statement is that the absence of stable wobble pairing in the P site reduces the constraint on accurate decoding in the A site resulting in frameshifting. Recent in vitro work has suggested that mispairing in the P site also induces missense errors as well (H. Zaher and R. Green, personal communication; Zaher and Green, 2009), suggesting that the P site is used to maintain A site accuracy in general.

During normal translation elongation, each successive in-frame cognate tRNA has a kinetic advantage over any incorrect tRNA, either in or out of frame. In vitro studies suggest that cognate aminoacyl-tRNAs induce a structural change in the ribosome that traps them in the A site and that near or non-cognate aminoacyl-tRNAs cannot induce this change (Gromadski and Rodnina 2004; Rodnina et al. 2005; Daviter et al. 2006). This structural transition increases activation of the EF-Tu GTPase approximately 50-fold; the change in GTPase activation provides much of the selection in favor of cognate and against near-cognate decoding. The fact that mismatching in the P site wobble position drastically increases errors strongly suggests that the lack of proper pairing interferes with some aspect of the structural transition and thus with GTPase activation. The inability to recognize the cognate tRNA as different in kind from the other available tRNAs would lead directly to increased recognition of errant tRNAs, including cognate tRNAs that bind out of frame. This is, of course, a hypothetical description of the frameshifting process and therefore is subject to invalidation, but no available evidence explicitly invalidates the model and much evidence is consistent with it.

Under this model, stimulation of +1 frameshifting results from a drastic change in the kinetics of tRNA discrimination. Whereas cognate aminoacyl-tRNAs normally have a large kinetic advantage over any other aminoacyl-tRNA, the lack of normal pairing in the P site wobble position eliminates the advantage, allowing other tRNAs to compete more effectively. The ratio of frameshifting to continued in-frame decoding is rather low, suggesting that even without this kinetic advantage the tRNA decoding in the zero frame is usually much more likely to be accepted. However, the kinetic change is sufficient to allow appreciable frameshifted recognition and the contexts around the frameshift sites in several cases have evolved sequences that further facilitate shifted decoding, though the mechanism of the context effects remains unclear.

10.6.2 −1 Frameshift Signals May Also Favor a Kinetically Unfavorable Event

The early model for programmed −1 frameshifting, involving pausing at a downstream pseudoknot stimulating frameshifting, has been strongly questioned (Kontos et al. 2001), as described above. It appears that a simple kinetic blockade model is inconsistent with the available data. More recent data from Namy et al. (2006) showed that a frameshift-inducing pseudoknot could cause a mammalian ribosome to become trapped in a translocation complex involving the two mRNA-bound tRNAs and EF2. In this complex, the presumed peptidyl-tRNA was observed to have moved part way from the A to P site of the 40S ribosomal subunit and to have adopted an unusual bent conformation. The apparent cause of this unusual complex is the fact that the downstream pseudoknot is tightly bound into the ribosomal helicase center, comprised of the mammalian counterparts of the *E. coli* rpS3, rpS4, and rpS5 (Takyar et al. 2005). The implication of the structure is that in the paused structure the helicase cannot unwind the pseudoknot, trapping EF2-promoted translocation in an intermediate complex because movement of the mRNA is insufficient to accommodate the 3 nt translocation step. Namy et al. (2006), in agreement with a previous proposal by Plant et al. (2003) suggest that a partial resolution of the resulting strained complex might be provided by slippage of the mRNA-bound tRNAs, which would move them in the −1 direction.

This model does not require any kinetic disruption, in accord with the lack of correlation between frameshifting and pseudoknot-induced pausing (Kontos et al. 2001). On the other hand, the failure of a simple kinetic blockade model to explain the data does not imply that −1 frameshifts do not manipulate a kinetic feature of the ribosome. The translocation step is normally extremely rapid (Rodnina et al. 1999) so a mid-translocation complex should be quite transient. Moreover, during translocation the ribosome undergoes a structural rearrangement, termed “unlocking,” that immediately precedes the transfer of the two bound tRNAs from the A to the P and from the P to the E sites of the small subunit (Savelsbergh et al. 2003). It may be that the kinetic blockage relevant to pseudoknot stimulation involves blocking the resolution of this “unlocked” structure and that the pseudoknot stimulates frameshifting by prolonging the time during which the codon•anticodon complexes are least stabilized by interactions with the ribosome. Translocation normally follows unlocking too rapidly to be observed (Savelsbergh et al. 2003), as might be expected if the ribosome needed to minimize the chances of loss of frame maintenance by remaining in the unlocked conformation for as little time as possible. The similarity between +1 and −1 frameshifting, therefore, may be that the programmed frameshift sites have identified two Achilles’ heels of the ribosome – the P site wobble interaction and the transient unlocking of the decoding center – and have exploited them to drastically increase the frequency of frameshifting.

References

Adams SE, Mellor J, Gull K, Sim RB, Tuite MF, Kingsman SM, Kingsman AJ (1987) Cell 49: 111–119
Adamski FM, Donly BC, Tate WP (1993) Nucleic Acids Res 21:5074–5078
Asakura T, Sasaki T, Nagano F, Satoh A, Obaishi H, Nishioka H, Imamura H, Hotta K, Tanaka K, Nakanishi H, Takai Y (1998) Oncogene 16:121–130
Atkins JF, Elseviers D, Gorini L (1972) Proc Natl Acad Sci USA 69:1192–1195
Bekaert M, Richard H, Prum B, Rousset JP (2005) Genome Res 15:1411–1420
Belcourt MF, Farabaugh PJ (1990) Cell 62:339–352.
Belshaw R, Pybus OG, Rambaut A (2007) Genome Res 17:1496–1504
Blinkowa AL, Walker JR (1990) Nucleic Acids Res 18:1725–1729
Boeke JD, Garfinkel DJ, Styles CA, Fink GR (1985) Cell 40:491–500
Bonetti B, Fu L, Moon J, Bedwell DM (1995) J Mol Biol 251:334–345
Brierley I (1995) J Gen Virol 76:1885–1892
Brierley I, Dos Ramos FJ (2006) Virus Res 119:29–42
Brierley I, Jenner AJ, Inglis SC (1992) J Mol Biol 227:463–479
Brierley I, Pennell S (2001) Cold Spring Harbor Symp Quant Biol 66:233–248
Brown CM, Stockwell PA, Trotman CAN, Tate WP (1990) Nucleic Acids Res 18:6339–6345
Chen X, Chamorro M, Lee SI, Shen LX, Hines JV, Tinoco I Jr, Varmus HE (1995) EMBO J 14:842–852
Clare J, Belcourt M, Farabaugh P (1988) Proc Natl Acad Sci USA 85:6816–6820
Clare J, Farabaugh PJ (1985) Proc Natl Acad Sci USA 82:2829–2833
Cobucci-Ponzano B, Rossi M, Moracci M (2005) Mol Microbiol 55:339–348.
Conti E, Izaurralde E (2005) Curr Opin Cell Biol 17:316–325
Craigen WJ, Caskey CT (1986) Nature 322:273–275
Craigen WJ, Cook RG, Tate WP, Caskey, CT (1985) Proc Natl Acad Sci USA 82:3616–3620
Daviter T, Gromadski KB, Rodnina MV (2006) Biochimie 88:1001–1011
Dimitrov DS, Willey RL, Sato H., Chang LJ, Blumenthal R, Martin MA (1993) J Virol 67: 2182–2190
Dinman JD (1995) Yeast 11:1115–1127
Dinman JD, Icho T, Wickner RB (1991) Proc Natl Acad Sci USA 88:174–178
Dinman JD, Ruiz-Echevarria MJ, Peltz SW (1998) Trends Biotechnol 16:190–196
Dinman JD, Wickner RB (1992) J Virol 66:3669–3676
Doma MK, Parker R (2006) Nature 440:561–564
Dreher TW, Miller WA (2006) Virology 344:185–197
Dunn JJ, Studier FW (1983) J Mol Biol 166:477–535
Eichinger DJ, Boeke JD (1988) Cell 54:955–966
Farabaugh P (1995) J Biol Chem 270:10361–10364
Farabaugh P (1996) Microbiol Rev 60:103–134
Farabaugh P, Zhao H, Vimaladithan A (1993) Cell 74:93–103
Farabaugh PJ, Kramer E, Vallabhaneni H, Raman A (2006) J Mol Evol 63:545–561
Flower AM, McHenry CS (1990) Proc Natl Acad Sci USA 87:3713–3717
Garfinkel DJ, Hedge AM, Youngren SD, Copel TD (1991) J Virol 65:4573–4581
Gesteland R, Atkins J (1996) Annu Rev Biochem 65:741–768
Giedroc DP, Theimer CA, Nixon PL (2000) J Mol Biol 298:167–185
Gromadski KB, Rodnina MV (2004) Mol Cell 13:191–200
Guarraia C, Norris L, Raman A, Farabaugh PJ (2007) RNA 13:1940–1947
Hammell AB, Taylor RC, Peltz SW, Dinman JD (1999) Genome Res 9:417–427
Hani J, Feldmann H (1998) Nucleic Acids Res 26:689–696
Hansen L, Chalker D, Sandmeyer S (1988) Mol Cell Biol 8:5245–5256
Hansen TM, Baranov PV, Ivanov IP, Gesteland RF, Atkins JF (2003) EMBO Rep 4:499–504
Harford JB (1995) Gene Expr 4:357–367

Hughes TR, Evans SK, Weilbaecher RG, Lundblad V (2000) Curr Biol 10:809–812
Icho T, Wickner RB (1989) J Biol Chem 264:6716–6723
Ivanov IP, Atkins JF (2007) Nucleic Acids Res 35:1842–1858
Ivanov IP, Gesteland RF, Atkins JF (2006) RNA 12:332–337
Jacks T, Madhani HD, Masiarz FR, Varmus HE (1988a) Cell 55:447–458
Jacks T, Power MD, Masiarz FR, Luciw PA, Barr PJ, Varmus HE (1988b) Nature 331:280–283
Jacks T, Townsley K, Varmus HE, Majors J (1987) Proc Natl Acad Sci USA 84:4298–4302
Jacks T, Varmus HE (1985) Science 230:1237–1242
Jacobs JL, Belew AT, Rakauskaite R, Dinman JD (2007) Nucleic Acids Res 35:165–174
Janetzky B, Lehle L (1992) J Biol Chem 267:19798–19805
Johansson MJ, Esberg A, Huang B, Bjork GR, Bystrom AS (2008) Mol Cell Biol 28:3301–3312
Katz JE, Dlakic M, Clarke S (2003) Mol Cell Proteomics 2:525–540
Kawakami K, Pande S, Faiola B, Moore D, Boeke J, Farabaugh P, Strathern J, Nakamura Y, Garfinkel D (1993) Genetics 135:309–320
Kontos H, Napthine S, Brierley I (2001) Mol Cell Biol 21:8657–8670
Larsen B, Peden J, Matsufuji S, Matsufuji T, Brady K, Maldonado R, Wills NM, Fayet O., Atkins JF, Gesteland RF (1995) Biochem Cell Biol 73:1123–1129
Lendvay TS, Morris DK, Sah J, Balasubramanian B, Lundblad V (1996) Genetics 144:1399–1412
Li Z, Stahl G, Farabaugh PJ (2001) RNA 7:275–284
Lingner J, Cech TR, Hughes TR, Lundblad V (1997) Proc Natl Acad Sci USA 94:11190–11195
Liti G, Peruffo A, James SA, Roberts IN, Louis EJ (2005) Yeast 22:177–192
Lopinski JD, Dinman JD, Bruenn JA (2000) Mol Cell Biol 20:1095–1103
Marozsan AJ, Fraundorf E, Abraha A, Baird H, Moore D, Troyer R, Nankja I, Arts EJ (2004) J Virol 78:11130–11141
Morris DK, Lundblad V (1997) Curr Biol 7:969–976
Muller F, Bruhl KH, Freidel K, Kowallik KV, Ciriacy M (1987) Mol Gen Genet 207:421–429
Nakamura Y, Gojobori T, Ikemura T (2000) Nucleic Acids Res 28:292
Namy O, Moran SJ, Stuart DI, Gilbert RJ, Brierley I (2006) Nature 441:244–247
Namy O, Rousset JP, Napthine S, Brierley I (2004) Mol Cell 13:157–168
Neuvéglise C, Feldmann H, Bon E, Gaillardin C, Casaregola S (2002) Genome Res 12:930–943
Niewmierzycka A, Clarke S (1999) J Biol Chem 274:814–824
Palanimurugan R, Scheel H, Hofmann K, Dohmen RJ (2004) EMBO J 23:4857–4867
Pande S, Vimaladithan A, Zhao H, Farabaugh P (1995) Mol Cell Biol 15:298–304
Percudani R, Pavesi A, Ottonello S (1997) J Mol Biol 268:322–330
Plant EP, Jacobs KL, Harger JW, Meskauskas A, Jacobs JL, Baxter JL, Petrov AN, Dinman JD (2003) RNA 9:168–174
Plant EP, Wang P, Jacobs JL, Dinman JD (2004) Nucleic Acids Res 32:784–790
Preston BD, Poiesz BJ, Loeb LA (1988) Science 242:1168–1171
Randerath E, Gupta RC, Chia LSY, Chang SH, Randerath K (1979) Eur J Biochem 93:79–94
Rodnina MV, Gromadski KB, Kothe U, Wieden HJ (2005) FEBS Lett 579:938–942
Rodnina MV, Savelsbergh A, Wintermeyer W (1999) FEMS Microbiol Rev 23:317–333
Savelsbergh A, Katunin VI, Mohr D, Peske F, Rodnina MV, Wintermeyer W (2003) Mol Cell 11:1517–1523
Shah AA, Giddings MC, Parvaz JB, Gesteland RF, Atkins JF, Ivanov IP (2002) Bioinformatics 18:1046–1053
Singer MS, Gottschling DE (1994) Science 266:404–409
Somogyi P, Jenner AJ, Brierley I, Inglis SC (1993) Mol Cell Biol 13:6931–6940
Stahl G, McCarty GP, Farabaugh PJ (2002) Trends Biochem Sci 27:178–183
Stucka R, Schwarzlose C, Lochmuller H, Hacker U, Feldmann H (1992) Gene 122:119–128
Sundararajan A, Michaud WA, Qian Q, Stahl G, Farabaugh PJ (1999) Mol Cell 4:1005–1015
Takyar S, Hickerson RP, Noller HF (2005) Cell 120:49–58
Taliaferro D, Farabaugh PJ (2007) RNA 13:606–613
Theis C, Reeder J, Giegerich R (2008) Nucleic Acids Res 36:6013–6020

Tsuchihashi Z, Kornberg A (1990) Proc Natl Acad Sci USA 87:2516–2520
Tu C, Tzeng TH, Bruenn JA (1992) Proc Natl Acad Sci USA 89:8636–8640
Vimaladithan A, Farabaugh P (1994) Mol Cell Biol 14:8107–8116
Voytas DF, Boeke JD (1992) Nature 358:717
Weil A (2001) Nature 409:28–29, 31
Weissenbach J, Dirheimer G, Falcoff R, Sanceau J, Falcoff E (1977) FEBS Lett 82:71–76
Wolfe KH, Shields DC (1997) Nature 387:708–713
Xu H, Boeke JD (1990) Proc Natl Acad Sci USA 87:8360–8364
Yusupov MM, Yusupova GZ, Baucom A, Lieberman K, Earnest TN, Cate JH, Noller HF (2001) Science 29:29
Yusupova GZ, Yusupov MM, Cate JH, Noller HF (2001) Cell 106:233–241
Zaher HS, Green R (2009) Nature 457:161–166.

Chapter 11
Recoding in Bacteriophages

Roger W. Hendrix

Abstract There are two classes of translational recoding, both frameshifts, known in the dsDNA tailed phages. The first is an inefficient frameshift between two overlapping tail genes, and both the shifted and unshifted products have essential roles as chaperones of tail assembly. This class is remarkable for the widespread conservation of a frameshift mechanism in the absence of conservation of the direction or magnitude of the shift. The second class of frameshifts adds an Ig-like domain to the C-terminus of one of the major structural proteins of the virion. In addition to the cases using a frameshift, some major structural proteins have a C-terminal Ig-like domain encoded directly in their gene, and some are missing such a domain. Among the non-tailed phages, some of the ssRNA phages have an essential termination codon readthrough event at the end of their coat protein gene.

Contents

Bacteriophages are very likely the most numerous biological organisms on the planet, with 5–10 phages present for every bacterial or archaeal cell in environmental surveys, and an estimated total of 10^{31} individual phages in the global population. The population apparently turns over every few days, implying about 10^{24} productive infections per second to maintain the population, and each of these infections is an opportunity for genetic mischief to generate diversity in the population (Wommack and Colwell, 2000; Wilhelm et al., 2002; Hendrix, 2003). There

R.W. Hendrix (✉)
Pittsburgh Bacteriophage Institute and Department of Biological Sciences, University of Pittsburgh, Pittsburgh, PA 15260, USA
e-mail: rhx@pitt.edu

J.F. Atkins, R.F. Gesteland (eds.), *Recoding: Expansion of Decoding Rules Enriches Gene Expression*, Nucleic Acids and Molecular Biology 24,
DOI 10.1007/978-0-387-89382-2_11,

is accumulating evidence that suggests phages have been around and evolving since the three current domains of cellular life split from each other about 3.5 billion years ago, and probably considerably earlier than that (Hendrix, 2004). Contemporary phages clearly have a profound influence on their hosts' evolution as well, serving inter alia as vectors of horizontal gene exchange and as agents of natural selection, and this has probably been going on since the early days of cellular life. In short, phages and their hosts have been engaged in a global orgy of mutation and gene swapping on a scale and speed that can scarcely be imagined, and they have been at it for billions of years.

Given this scenario, it may not be too far off the mark to suppose that in the realm of mechanisms of gene expression there is essentially nothing that phages have not tried, with natural selection imposing the only limitation on what has survived to be seen by us. We should therefore not be surprised to discover that phages, like their host cells, indulge in frameshifting and other forms of translational recoding. I describe here the main examples of recoding detected to date. Perhaps what is most surprising is that there are not more examples of this sort of non-canonical gene expression in phages, and at the end of the chapter I suggest that there are likely more examples yet to be discovered.

11.1 Tailed dsDNA Phages: Frameshift in the Tail Genes

One of the first examples of a programmed frameshift in a dsDNA tailed bacteriophage was found in the tail gene region of phage lambda (Levin et al., 1993). This occurs between two overlapping open reading frames called genes *G* and *T*, located immediately downstream from the gene encoding the major structural subunit of the tail (gene *V* in lambda) and immediately upstream from the gene (gene *H*) encoding the "tape measure protein," a protein whose size determines the length of the tail (Fig. 11.1). The frameshift is a −1 shift that takes place at a canonical −1 slippery site, G GGA AAG, located 6 and 7 codons upstream from the termination codon of gene *G*. The frequency of the shift is a relatively feeble 3.5%. The two proteins produced, called gpG and gpGT (gp = gene product), have masses of 16 and 32 kD, respectively.

It was not clear initially what biological significance, if any, this example of frameshifting might have, but the answer to this question has become clearer as the

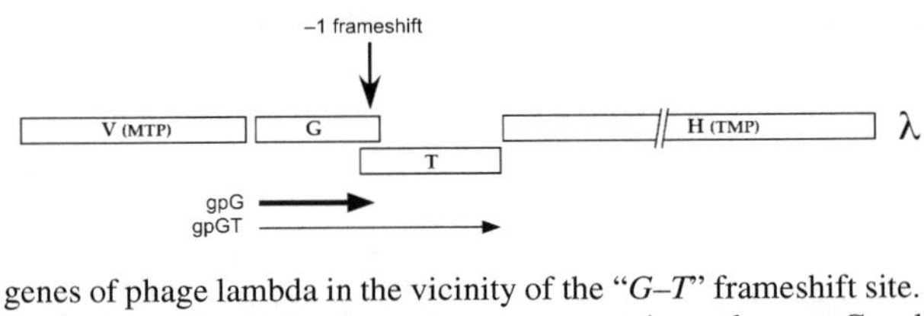

Fig. 11.1 Tail genes of phage lambda in the vicinity of the "*G–T*" frameshift site. Gene *V* encodes the *m*ajor *t*ail *p*rotein, gene *H* encodes the *t*ape *m*easure *p*rotein, and genes *G* and *T* are related by a −1 frameshift to give the two products indicated

result of two subsequent developments. First, it is now apparent that the majority of phages with long tails have an analogous pair of overlapping genes that appear from analysis of the sequence or by experimental test to be expressed by a frameshift mechanism. Known examples include phages with non-contractile tails, like lambda and many others, as well as contractile-tailed phages, like P2 (Christie et al., 2002) and Mu (Xu et al., 2004). Second, it is becoming clear that the two products of the frameshifting genes – gpG and gpGT in phage lambda – have a central role in phage tail assembly. In lambda both proteins are shown to be essential for tail formation, and efficient tail assembly requires that they be present in the appropriate ratio. They appear to act as assembly chaperones, mediating the assembly of the tail shaft around the tail length tape measure protein (Xu, 2001). As such they are integral components of the process of tail length determination. They are absent from the mature tail (Levin et al., 1993).

Analogues of the lambda *G–T* frameshifting gene pair are easily detected in the genome sequences of about two-thirds of long-tailed phages. In addition to the presence of two overlapping open reading frames and an appropriately situated slippery sequence, the feature most diagnostic of the presence of a "*G–T*" frameshift is the position of the candidate sequence in the highly conserved gene order in this section of the tail genes. The "*G–T*" gene pair is always found upstream from the tail length tape measure gene and downstream from the gene encoding the major tail shaft subunit, or the equivalent tail tube subunit in contractile-tailed phages. In most phages, including the prototype lambda, the only genes between the major tail subunit gene and the tape measure protein gene are the two genes that participate in frameshifting. In some, however, additional genes can be inserted within the otherwise conserved cluster. For example, in phage HK97 there is a small gene between the "*G–T*" pair and the tape measure gene; this interloper is flanked by a transcription promoter and terminator, and analysis of its sequence argues that it is a recent insertion into the genome (Juhala et al., 2000). In phage Twort, the tail tube subunit gene has three introns and is followed by an apparent homing nuclease and four small genes of unknown function (Stewart et al., 2009). Whatever the reason for the conserved order of the four tail genes under consideration here, these examples argue that they need not be contiguous to function efficiently.

Despite the conserved position of the "*G–T*" frameshifting genes among the other tail genes, there is great divergence in the primary sequences of the encoded proteins, to the extent that there is typically no detectable sequence similarity in a comparison of these sequences between two phages. Perhaps more surprising is that there is also diversity in the mechanism of frameshifting. In most of the examples of this class of frameshifts detected to date, the shift is into the −1 frame, and there is typically a canonical −1 slippery sequence, compatible with slippage while both the P and A sites of the ribosome are occupied. In the case of phage Mu, however, there is a −2 frameshift. There are not enough examples of such a shift to derive a convincing consensus, but the sequence of the phage Mu site suggests slippage with both ribosome sites occupied, and the sequence of the protein produced by the frameshift confirms this supposition (Xu et al., 2004). In phage SPO1 there is the potential for a +1 frameshift at the expected place in the gene order. This would occur with only the P site occupied, so the signal in the sequence suggesting a frameshift is only four bases, and there is no biochemical confirmation of the frameshift. However, the case for the SPO1 frameshift is strengthened considerably by the observation that five other phages of Gram-positive hosts have a pair of overlapping open reading frames in the corresponding position in their tail genes, and all five have potential +1 shifty sequences at comparable positions in the overlap (Stewart et al., 2009). A feature of nearly all the known or suspected "*G–T*" frameshift sites, or perhaps one should say a non-feature, is the lack of obvious frameshift-promoting elements in the sequence such as peudoknots or strategically placed Shine–Dalgarno sequences. This may simply reflect the fact that the frequency of frameshifting in these cases, typically 3–5%, is lower than observed in cases of frameshifting where such sequence elements are more apparent.

The conservation of the frameshifting mechanism of gene expression at the "*G–T*" position in the tail genes, in the face of variation in the detailed form of the frameshift (i.e., −1, −2, +1), argues that what is biologically important in this process is the protein products produced – two proteins with a particular sequence relationship made in a particular ratio – and not the details of the mechanism by which they are produced. It remains true that for a substantial minority of the long-tailed phages it has not been possible to identify a plausible candidate for a "*G–T*" pair of overlapping open reading frames. Whether this means that in these phages we are unable to identify the frameshift sequence because the gene organization in these phages is too different from the examples we know or that these phages accomplish tail assembly without the aid of a frameshift mechanism is a worthy topic for future investigations. Interestingly, no case has yet been identified in which the equivalent of gpG and gpGT is produced by a stop codon readthrough mechanism rather than by a frameshift. Nevertheless, a −1 "*G–T*" frameshift site from a mycobacteriophage was found to promote readthrough of its UGA termination codon at a level comparable to the level of frameshifting, albeit in the non-native context of the *Escherichia coli* cytoplasm (Xu et al., 2004).

11.2 Tailed dsDNA Phages: C-Terminal Domains

It has become clear recently that a number of structural proteins of tailed phage virions have a C-terminal immunoglobin-like domain (Fraser et al., 2006, 2007). Sequence analysis argues that these Ig-like domains have moved among their positions in different phages by horizontal exchange, because two Ig-like domains with very similar sequences are often found at the C-termini of two structural protein sequences that are not detectably similar outside of those C-terminal domains. There are also examples of pairs of homologous proteins for which the homologue in one phage has an Ig-like C-terminal domain and the homologue in the second phage lacks a corresponding C-terminal extension of any kind. The widespread occurrence of these C-terminal Ig-like domains suggests that they confer a selective benefit on the phage that carries them, but their absence in some phages argues that they are not essential. What function they do provide is not known, but the leading suggestion (Fraser et al., 2006, 2007) is that they bind to carbohydrates on the cell surface and thereby carry out an accessory role in adsorption of the phage to the cell.

The interest in these C-terminal domains in the present context is that some of them – about a dozen examples are known – are not encoded directly in the structural protein gene but are encoded in a separate overlapping open reading frame and added by a programmed frameshift. The first of these to be studied in detail, and in fact the first case of a frameshift of any sort to be detected in the tailed phages, occurs at the end of the major capsid protein of coliphage T7 (Dunn and Studier, 1983; Condron et al., 1991). This is a −1 frameshift that occurs with about 10% efficiency, adding 52 amino acids from the −1 frame to the end of the capsid protein. Both forms of the capsid protein are found in the assembled capsid, in about the same ratio as they are synthesized. This ratio roughly matches the ratio between the number of subunits in hexamers and in pentamers in the $T = 7$ icosahedral shell of this phage, and the suggestion was made in the case of another phage with similar numbers that the longer form of the capsid protein might be found exclusively in the pentamers, at the "corners" of the icosahedral shell (Zimmer et al., 2003), reminiscent of the case of phages like T4 which have a separate protein, homologous to the major capsid protein, occupying the pentamers. However, this attractive idea can be ruled out for T7 because the C-terminal domain is not visible in icosahedrally averaged cryo-EM images of the capsid (Agirrezabala et al., 2007). (Only icosahedrally symmetric features of the capsid, like the pentamer subunits, survive icosahedral averaging.) The long form of the T7 capsid protein is not essential for viability of the phage, as a mutant that makes only the short form is nearly fully functional in laboratory growth (Condron et al., 1991).

The T7 capsid protein frameshift is atypical of the cases under consideration in that the C-terminal domain added by the frameshift is not an Ig-like domain, nor a member of any recognized family (Fraser et al., 2007). Phage T3 is a close relative of T7 with a capsid protein of similar sequence; at the end of the T3 capsid protein gene there is a −1 frameshift that leads to the addition of an Ig-like domain with no detectable similarity to the C-terminal domain added in the T7 case (Condreay et al., 1989). This observation strengthens the impression that these C-terminal domains are horizontally mobile on a relatively short evolutionary timescale.

While the phage structural proteins with Ig-like domains encoded directly in the gene are found in a variety of different structural proteins, all those identified to date that are added by a frameshift are added to either the major capsid protein or the major tail protein (Fraser et al., 2007). Several have been characterized, and I cite some illustrative examples. *Listeria* phage PSA adds an Ig-like domain to the C-terminus of both its major capsid protein and its major tail protein, in both cases by a +1 frameshift. In the case of the major capsid protein frameshift, there is a potential pseudoknot in the mRNA that is thought to potentiate frameshifting (Zimmer et al., 2003). Similar potential pseudoknots are seen in some of the other examples as well, including the T7 and T3 cases. A second example of a phage that adds C-terminal domains to both the major capsid protein and the major tail protein by means of a frameshift is *Lactobacillus* phage A2. Both frameshifts are into the −1 frame

(Garcia et al., 2004; Rodriguez et al., 2005). A particularly interesting feature of the phage A2 major capsid protein frameshift is that, in contrast to the results cited above for T7, mutational analysis implies that neither form of the capsid protein alone is sufficient to support phage viability (Garcia et al., 2004). A third example comes from *Lactococcus* phage Q54. Here a +1 frameshift near the end of the major tail protein gene attaches an Ig-like domain to about 25% of the tail subunits. The Ig-like domain is annotated as the "receptor-binding protein" based on sequence matches to parts of structural proteins of other phages implicated in receptor binding (Fortier et al., 2006). This result supports the hypothesis that the role of the Ig-like domains on the ends of structural proteins of the virion is to assist in attaching the virion to the cell prior to infection. Phage Q54 also appears to have an un-annotated "*G–T*"-like −1 frameshift site in the next pair of genes downstream from the major tail protein gene and its Ig-like +1 frameshift partner. An analogous juxtaposition of two pairs of frameshift-related genes has been documented for phage PSA (Zimmer et al., 2003) (Fig. 11.2).

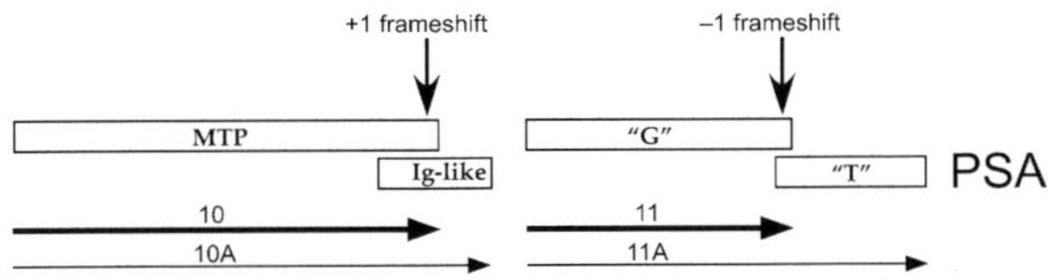

Fig. 11.2 Tail genes of phage PSA, which display the two main classes of frameshifts discussed here, in two adjacent pairs of overlapping genes. *Arrows* below the genes represent the protein products, labeled with the corresponding PSA gene names

The Ig-like domains on the ends of virion structural proteins present a rather confusing picture: a given protein in one phage will have an Ig-like domain and the homologous protein in another phage will lack the domain; one phage will have an Ig-like domain both on its major head proteins and on its major tail proteins and another will have it on only the tail proteins; and most phages encode the sequence of the Ig-like domain as part of the sequence of the structural protein gene, but some attach it to only a minority of the structural proteins through a frameshift. The widespread occurrence of these domains on virion structural proteins argues strongly that they confer a selective benefit on the phages, but if so, why don't the structural proteins always have them? Perhaps a useful way to think about these issues is to remember that phage evolution is a dynamic process. The Ig-like domains are evidently moving in and out of phage genomes rapidly on an evolutionary timescale, and if as we suspect they enter the genome by an essentially random process of non-homologous recombination, then most of them that land near the end of a structural protein gene will not be fused in-frame to that gene. Those that are fused in-frame will tend to persist because of the selective benefit they provide. Those that are not will either be lost or acquire mutations that allow their expression, by in-frame fusion or by frameshift. In this view, the difference between genes that express their Ig-like domain by frameshift and those that have it fused in-frame is a difference of history rather than function. Whether the frameshifting cases might have a selective advantage over the in-frame cases for some phages or some environmental situations is not clear at this point.

11.3 ssRNA Phages

The ssRNA phages, typified by coliphages MS2 and Qβ, have a positive sense ssRNA genome that functions as a polycistronic mRNA in the infected cell. These phages have four genes, somewhat differently arranged in the two subtypes (Fig. 11.3). Despite the apparent simplicity of having only a single type of mRNA, these phages mount a marvelously complex program of gene expression and genome replication. The genes are translated into proteins in a temporal order and at different levels appropriate to the proteins' roles in the phage life cycle. In addition, replication of the RNA is blocked when translation is taking place, and translation is blocked during replication. All of this is achieved by the secondary structure of the RNA as it interacts with ribosomes and with the phage-coded proteins. To cite just one aspect of this regulation – one of the classic examples – the coat protein gene is the only phage gene to be expressed initially because its ribosome-binding site (RBS) is the only one accessible in the folded RNA as it enters the cell. However, as the ribosome passes down the coat gene it disrupts secondary structure in the RNA that is sequestering the replicase gene RBS, allowing other ribosomes to enter and start translating the replicase. Later, when enough replicase has been made and the concentration of coat protein has built up, a coat protein dimer binds and stabilizes an element of RNA secondary structure that once again sequesters the replicase RBS. The remarkable molecular dance the phage RNA performs together with its ribosome and protein partners has been described in considerable detail (van Duin and Tsareva, 2006) and will not be repeated here. Most of this dance cannot properly be called recoding, but because it shares with true recoding events unusual and informative interactions between ribosomes and mRNA, it is similarly revealing of some of the unusual feats ribosomes are capable of.

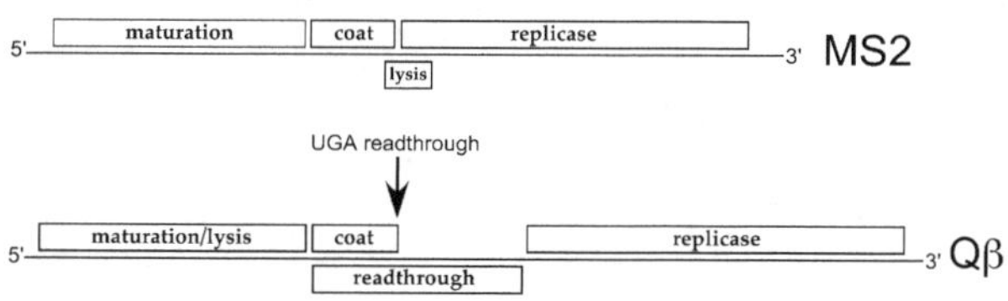

Fig. 11.3 Genomes of ssRNA phages MS2 and Qβ

The only canonical recoding event in the ssRNA phages with unequivocal importance for biological function is the UGA termination codon readthrough at the end of the coat protein gene of Qβ and its relatives. This occurs with about 5% efficiency, producing the coat readthrough protein, which is incorporated into the virion and is essential for phage infectivity (Weiner and Weber, 1971; Hofstetter et al., 1974). A frameshift during translation of the MS2 coat gene was originally proposed to lead to early out-of-frame termination, thereby positioning ribosomes to initiate translation of the overlapping lysis gene (Kastelein et al., 1982). However, this was later shown not to be correct; instead the ribosomes terminate normally at the end of the coat gene and drift back toward the 5′ end of the RNA until they encounter the beginning of the lysis gene and start translation (Berkhout et al., 1987; Adhin and van Duin, 1990). In Qβ the lysis function is not encoded in a separate gene as in MS2 but instead is incorporated into the maturation protein. However, a fortuitous cloning artifact has led to interesting speculation about the possibility that Qβ might also have a separate lysis gene overlapping the replicase gene as in MS2 that could be accessed by a frameshift, or more plausibly that it might have had such a lysis gene in its evolutionary past (Nishihara et al., 2004). There is evidence for frameshifting at the ends of both the coat and replicase genes of MS2 based on experiments with an in vitro translation system (Atkins et al., 1979) and with a cloned construct (de Smit et al., 1994). There is no evidence for any of these as to whether they have biological significance.

11.4 Undiscovered Examples of Phage Recoding?

The frameshift linking the *G* and *T* tail genes in phage lambda was described 15 years ago, but it is still rather common for newly published phage genome sequences not to have the corresponding frameshift site correctly identified and annotated. This observation highlights the fact that the standard gene calling programs do not detect the potential for this sort of translational recoding, and if the person doing the annotation does not have specific knowledge of this class of frameshift sites they will go un-annotated. Such an oversight is easily corrected because the biochemical work has been done in lambda and a few other phages to establish the reality of the frameshifts, and sequence analysis shows the widespread occurrence of the frameshift mechanism among these tail genes. On the other hand, any recoding events that have not yet been identified and characterized experimentally will escape the attention of both the gene calling programs and our current knowledge of phage biology.

What might such undiscovered recoding events be? In the absence of evidence to the contrary anything is possible, of course, but I would suggest that one promising place to look for new recoding events in phages is among the tRNAs that some phages encode. Less than half of the sequenced phage genomes have any detected tRNA genes; of those that do have some, the number of tRNA genes found in the genome varies between 1 and 30. In at least a few cases the phage tRNAs have been shown capable of participating in translation. A popular idea about the function of the phage tRNAs is that they modify the host translation system in a quantitative sense to make it more suitable to the phage's needs. For example, the phage tRNAs could optimize the translation system for reading highly expressed phage genes by supplementing tRNA pools corresponding to codons that occur frequently in those phage genes. There is some evidence to support this notion, and it may well explain the presence of at least some of the phage tRNAs.

However, there are three components of the translation system in addition to tRNAs that are found in phage genomes and whose presence suggests that the phage's participation in translation goes beyond simply adjusting tRNA pools. These are tmRNAs, which have been recognized in half a dozen phage genomes, Release Factor 1 homologues, found in five different mycobacteriophage genomes, and an apparent aminoacyl tRNA synthetase, found in the 500 kbp genome sequence of *Bacillus* phage G (our unpublished results). The presence of a tRNA synthetase brings to mind, for example, the case of pyrrolysine, which has a dedicated tRNA and tRNA synthetase, leading to recoding at certain UAG termination codons. Whether the presence of translational components in phages' genomes actually implies anything about translational recoding will of course require experimental test, but it would not be the first example of phages playing fast and loose with the "rules" of molecular biology.

References

Adhin MR, van Duin J (1990) Scanning model for translational reinitiation in eubacteria. J Mol Biol 213:811–818

Agirrezabala X, Velazquez-Muriel JA, Gomez-Puertas P, Scheres SH, Carazo JM, Carrascosa JL (2007) Quasi-atomic model of bacteriophage T7 procapsid shell: insights into the structure and evolution of a basic fold. Structure 15:461–472

Atkins JF, Gesteland RF, Reid BR, Anderson CW (1979) Normal tRNAs promote ribosomal frameshifting. Cell 18:1119–1131

Berkhout B, Schmidt BF, van Strien A, van Boom J, van Westrenen J, van Duin J (1987) Lysis gene of bacteriophage MS2 is activated by translation termination at the overlapping coat gene. J Mol Biol 195:517–524

Christie GE, Temple LM, Bartlett BA, Goodwin TS (2002) Programmed translational frameshift in the bacteriophage P2 FETUD tail gene operon. J Bacteriol 184:6522–6531

Condreay JP, Wright SE, Molineux IJ (1989) Nucleotide sequence and complementation studies of the gene 10 region of bacteriophage T3. J Mol Biol 207:555–561

Condron BG, Atkins JF, Gesteland RF (1991) Frameshifting in gene 10 of bacteriophage T7. J Bacteriol 173:6998–7003

de Smit MH, van Duin J., van Knippenberg PH, van Eijk HG (1994) CCC.UGA: a new site of ribosomal frameshifting in *Escherichia coli*. Gene 143:43–47

Dunn JJ, Studier FW (1983) Complete nucleotide sequence of bacteriophage T7 DNA and the locations of T7 genetic elements. J Mol Biol 166:477–535

Fortier LC, Bransi A, Moineau S (2006) Genome sequence and global gene expression of Q54, a new phage species linking the 936 and c2 phage species of *Lactococcus lactis*. J Bacteriol 188:6101–6114

Fraser JS, Maxwell KL, Davidson AR (2007) Immunoglobulin-like domains on bacteriophage: weapons of modest damage? Curr Opin Microbiol 10:382–387

Fraser JS, Yu Z, Maxwell KL, Davidson AR (2006) Ig-like domains on bacteriophages: a tale of promiscuity and deceit. J Mol Biol 359:496–507

Garcia P, Rodriguez I, Suarez JE (2004) A -1 ribosomal frameshift in the transcript that encodes the major head protein of bacteriophage A2 mediates biosynthesis of a second essential component of the capsid. J Bacteriol 186:1714–1719

Hendrix RW (2003) Bacteriophage genomics. Curr Opin Microbiol 6:506–511

Hendrix RW (2004) Hot new virus, deep connections. Proc Natl Acad Sci USA 101:7495–7496

Hofstetter H, Monstein HJ, Weissmann, C (1974) The readthrough protein A1 is essential for the formation of viable Q beta particles. Biochim Biophys Acta 374:238–251

Juhala RJ, Ford ME, Duda RL, Youlton A., Hatfull GF, Hendrix RW (2000) Genomic sequences of bacteriophages HK97 and HK022: pervasive genetic mosaicism in the lambdoid bacteriophages. J Mol Biol 299:27–51

Kastelein RA, Remaut E, Fiers W, van Duin J (1982) Lysis gene expression of RNA phage MS2 depends on a frameshift during translation of the overlapping coat protein gene. Nature 295: 35–41

Levin ME, Hendrix RW, Casjens SR (1993) A programmed translational frameshift is required for the synthesis of a bacteriophage lambda tail assembly protein. J Mol Biol 234:124–139

Nishihara T, Morisawa H, Ohta N, Atkins JF, Nishimura, Y (2004) A cryptic lysis gene near the start of the Qbeta replicase gene in the +1 frame. Genes Cells 9:877–889

Rodriguez I, Garcia P, Suarez JE (2005) A second case of -1 ribosomal frameshifting affecting a major virion protein of the Lactobacillus bacteriophage A2. J Bacteriol 187:8201–8204

Stewart CR, Casjens SR, Cresawn SG, Houtz JM, Smith AL, Ford ME, Peebles CL, Hatfull GF, Hendrix RW, Huang WM, Pedulla ML (2009) The genome of *Bacillus subtilis* bacteriophage SPO1. J Mol Biol 388:48–70

van Duin J, Tsareva N (2006) Single-stranded RNA phages. In: Calendar R (ed) The bacteriophages. Oxford University Press, New York, pp 175–196

Weiner AM, Weber K (1971) Natural read-through at the UGA termination signal of Q-beta coat protein cistron. Nat New Biol 234:206–209

Wilhelm SW, Brigden SM, Suttle CA (2002) A dilution technique for the direct measurement of viral production: a comparison in stratified and tidally mixed coastal waters. Microb Ecol 43:168–173

Wommack KE, Colwell RR (2000) Virioplankton: viruses in aquatic ecosystems. Microbiol Mol Biol Rev 64:69–114

Xu J (2001) A conserved frameshift strategy in dsDNA long tailed bacteriophages. PhD Thesis: University of Pittsburgh

Xu J, Hendrix RW, Duda RL (2004) Conserved translational frameshift in dsDNA bacteriophage tail assembly genes. Mol Cell 16:11–21

Zimmer M, Sattelberger E, Inman RB, Calendar R, Loessner MJ (2003) Genome and proteome of *Listeria monocytogenes* phage PSA: an unusual case for programmed + 1 translational frameshifting in structural protein synthesis. Mol Microbiol 50:303–317

Chapter 12
Programmed Ribosomal −1 Frameshifting as a Tradition: The Bacterial Transposable Elements of the IS*3* Family

Olivier Fayet and Marie-Françoise Prère

Abstract Insertion sequences (ISs) are small ubiquitous DNA transposable elements coding for one or two proteins that are found in the genome of most bacteria where they play an important role in genetic plasticity. Based on protein similarity, the ISs were grouped in 19 families, the largest being the IS3 family. Interestingly, most of its 418 members possess two overlapping genes and very likely use programmed ribosomal −1 frameshifting (PRF-1) to generate their transposase, the protein required for transposition, as was experimentally demonstrated for a few (e.g., IS3, IS150, IS911, IS3411). A systematic comparison of the IS3 family members was carried out to reveal the main features of the frameshift-programming signals present in their mRNA. The mandatory component is a short sequence where the shift from frame 0 to frame −1 occurs (Z-ZZN or more frequently X-XXZ-ZZN, the 0 frame codons are underlined). In the IS, there is a clear preference for the A-AA[A/G] and U-UU[U/C] tetramers (20%), and for the A-AAA-AA[A/G] heptamers (55%). The slippery motif is accompanied in 87% of the cases by one or two stimulatory elements. Like in eukaryotic viruses, it can be a structure formed by folding of the mRNA downstream of the motif. This is either a stem loop (60%) or a pseudoknot (13%). However, it can also be an upstream Shine–Dalgarno-like sequence (SD) that acts through pairing with 16S ribosomal RNA (in 56% of the IS). The two types of stimulators are both present in 42% of the IS and are both absent in 13% of them. Several lessons can be drawn from this comparative analysis and from more detailed analyses of frameshift signals of a few IS: (i) PRF-1 is a 2 (and perhaps 3) tRNA story and if ISs use a restricted set of frameshift motifs it is because prokaryotic ribosomes are less tolerant to near-cognate tRNA pairing than eukaryotic ribosomes. (ii) ISs have more flexibility in the design of their frameshift regions (use of 0, 1, or 2 stimulators) than eukaryotic viruses. (iii) The nucleotides immediately 3′ to the slippery motif modulate frameshifting and thus must play a role in frame maintenance possibly through yet to identify interactions with the ribosome.

O. Fayet (✉)
Université de Toulouse, Laboratoire de Microbiologie et Génétique Moléculaires, F-31000 Toulouse, France
e-mail: olivier.fayet@ibcg.biotoul.fr

J.F. Atkins, R.F. Gesteland (eds.), *Recoding: Expansion of Decoding Rules Enriches Gene Expression*, Nucleic Acids and Molecular Biology 24,
DOI 10.1007/978-0-387-89382-2_12,

Contents

Insertions sequences (ISs) are small (800–2700 bp long) ubiquitous bacterial transposable elements (Chandler and Mahillon 2002; for the ISFinder database see Siguier et al. 2006). They are found, sometime in many copies, in plasmids, bacteriophages, and chromosomes of most eubacteria where they play a major role in genome plasticity. Sequence comparison and functional analysis lead to the sorting of the 1,519 known insertion sequences (as of June 2008) into 19 different families. Members from four of these families appear to use programmed ribosomal −1 frameshifting (PRF-1) to express their transposase, the enzyme required for their mobility. As outlined in Fig. 12.1, programmed −1 frameshifting results from the encounter of three elements: (i) the ribosome, an exquisitely complex but necessarily imperfect machine (Kurland 1992); (ii) tRNAs, most of which are intrinsically promiscuous (Crick 1966; Agris et al. 2007); and (iii) signals embedded in the coding region of an mRNA (Gesteland and Atkins 1996). In the following, we will focus on the last constituent of the triumvirate and will lay emphasis on the specificity of frameshift signals of bacterial origin and on what in their design allows them to exploit tRNA promiscuity and ribosome permissiveness. The issue of the mechanism and timing of programmed −1 frameshifting during the elongation cycle has been addressed in former reviews and recent analyses (Farabaugh 1997; Harger et al. 2002; Plant et al. 2003; Baranov et al. 2004; Namy et al. 2006; Léger et al. 2007; Chapters 7 and 8). It is still a controversial topic because of a lack of entirely compelling experimental evidence. It should also be kept in mind that there is perhaps more than one way to do it: for example, programmed and non-programmed frameshifting may be mechanistically different and the eukaryotes' and prokaryotes' ribosomes may have diverged in their response to programmed −1 frameshifting inducing signals. This issue will not be resolved here, but relevant information from recent IS studies will be emphasized when appropriate.

Frameshifting was first identified in IS*1* where it occurs at a rather low frequency (below 1%) because of a rather inefficient slippery motif and because of a lack of a

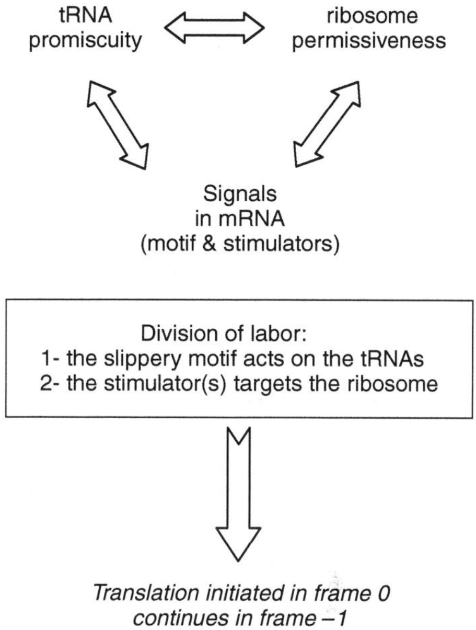

Fig. 12.1 The three pillars of −1 frameshifting. In the mRNA of certain genes, signals evolved to exploit two features of the translation machinery. The first is the capacity of tRNA to recognize more than one codon (here called promiscuity) and the second is ribosome flexibility (permissiveness) which allows it to deal with the unexpected, thus revealing an increased level of sophistication in its functioning

proper stimulator (Sekine and Ohtsubo 1992, 1989; Escoubas et al. 1991). In a few members of the IS*1*, IS*5*, and IS*630* families, use of programmed −1 frameshifting is suspected whereas in the large and widespread IS*3* family (27% of the known insertion sequences) such frameshifting appears to be the general rule. Occurrence of frameshifting was recognized, and demonstrated, at about the same time in several insertion sequences related to IS*3* of *Escherichia coli* (Prère et al. 1990; Polard et al. 1991; Vögele et al. 1991; Chandler and Fayet 1993; Sekine et al. 1994). Since then, members of that group were found in more than 200 bacterial species from all branches of the bacterial evolutionary tree: the IS database contains to date 418 entries for the IS*3* family (Table 12.1). Considering that about 742 complete bacterial genome sequences are now available (plus more than 900 genomes in progress) and that many of them were not examined for their IS content, the present number of IS*3*-related elements is a clear underestimate. The vast majority of the members of the IS*3* family possess two consecutive and overlapping genes, with the second, *orfB*, being in frame −1 relative to the first, *orfA* (Fig. 12.2A). Strikingly, nearly all of these appear to contain a potential PRF-1 signal in the *orfA–orfB* overlap region (Chandler and Fayet 1993). Direct evidence that frameshifting does indeed generate a OrfA::OrfB hybrid protein, the OrfAB transposase, has been obtained for a few insertion sequences. Figure 12.2B depicts a transposition pathway which, while established for IS*911*, likely applies for most members of the family. The OrfAB protein catalyses the excision of the IS, generating an IS circle that is subsequently re-inserted at a new location through the combined action of the OrfA and OrfAB proteins (Ton-Hoang et al. 1999; Loot et al. 2002); note that no function was

Table 12.1 Phylogenetic distribution of the IS from the IS3 family

Phylum	IS2 Group	IS*3* Group	IS*51* Group	IS*150* Group	IS*407* Group	Total	SEQUENCED Genomes
Chloroflexi					1	1	7
Deinococci		2		2		4	4
Spirochetes		4		1		5	14
Bacteroidetes		4			2	6	25
Planctomycetes					1	1	1
Chlamydia	1					1	13
Fusobacteria				1		1	1
Cyanobacteria		1				1	33
Gram+							
Gram+ Low GC (Firmicutes and Tenericutes)		17		38		55	151
Gram+ High GC (Actinobacteria)		5	16	5	3	29	54
Gram-							
Proteobacteria class α	6	8	12	4	28	58	90
Proteobacteria class ε	7	19	9	12	11	58	60
Proteobacteria class γ	5	55	20	9	26	115	176
Proteobacteria class δ	2	11	1		4	18	21
Proteobacteria class ε				1	1	2	20
Total	21	126	58	73	77	355	670

found for the OrfB protein, though it is indeed synthesized in the case of IS*911*. The purpose of an IS is to maintain and propagate itself, whereas its bacterial host would rather eliminate it since it does not fulfill an immediate housekeeping need as suggested by the frequent presence of mutated or deleted insertion sequences in bacterial genomes. To do so the IS uses transposition to keep constant its number of active copies and to have a chance to colonize other hosts through bacteriophages and plasmids, but it should not transpose too frequently. A more subtle way to ensure IS maintenance is that transposition-associated re-arrangements (e.g., inversions, deletions, gene fusions, and modifications of expression) provide genetic plasticity and thus may have an adaptative value for the host. The important issue with any mobile element is therefore control of transposition at a level ensuring survival and propagation of the IS while not being detrimental, and sometimes being beneficial, to the host.

To conclude, the IS*3* family constitutes a remarkable source, in terms of number and diversity, of bacterial frameshift signals and thus of strategies of transposition control.

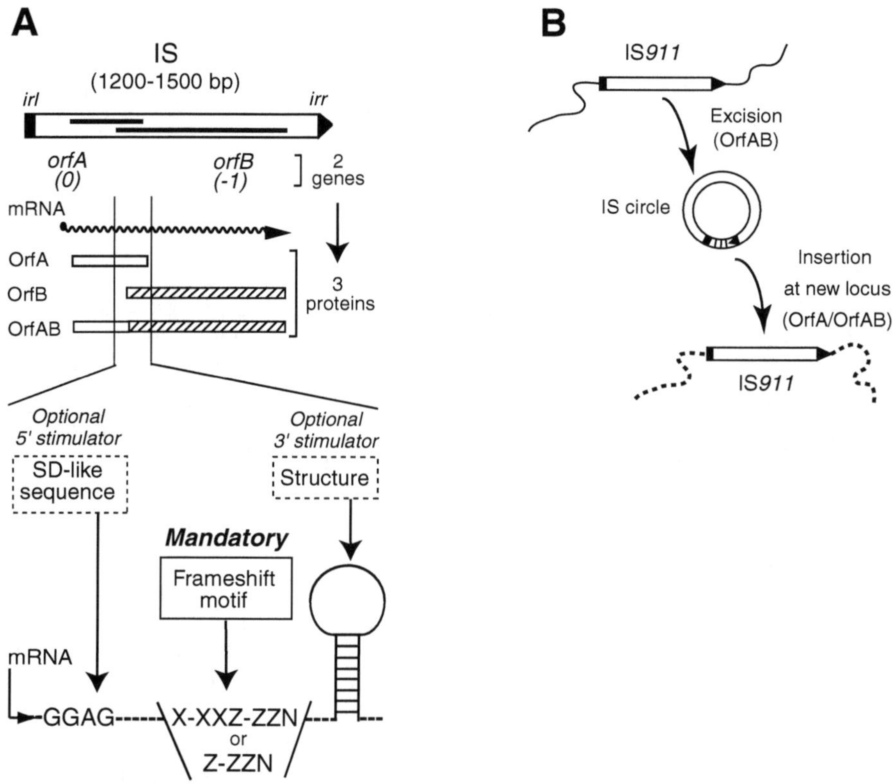

Fig. 12.2 Frameshifting in the transposable elements of the IS*3* family. (**A**) Prevalent genetic organization among the IS*3* family members and main features of their programmed −1 frameshifting region. The only mandatory signal is the frameshift motif (either an X-XXZ-ZZN heptamer or a Z-ZZN tetramer), but in most insertion sequences the motif is accompanied by one or two stimulators. (**B**) Scheme of the IS*911* transposition pathway. The frameshift product, the OrfAB transposase, is required for the excision of the IS and for its insertion at a new place

12.1 Comparative Analysis of IS*3* Family Members (ISFinder Database, June 2008)

After elimination of the obviously redundant or incomplete elements, 355 remaining IS3 members were individually examined for potential programmed −1 frameshift signals. In nearly all of them we could identify such a signal which was analyzed and compared in terms of frameshift motifs and potential upstream and downstream stimulators. Table 12.1 presents the phylogenetic distribution of the selected transposable elements of the IS*3* family. On the basis of similarity of the OrfB protein, the family has been subdivided into five groups named after one typical member, i.e., the IS*2*, IS*3*, IS*51*, IS*150*, and IS*407* groups (Chandler and Mahillon 2002). While these IS elements are predominant in proteobacteria, perhaps because these microorganisms are prevalent in nature, they are also frequently found in Gram-positive bacteria but with a notable group-specific pattern. Low GC Gram-positive bacteria only contain members of the IS*3* and IS*150* groups whereas high GC Gram-positive bacteria carry members of four groups. This suggests that IS*3* ancestors were present before separation of the Gram-positive and Gram-negative phyla.

12.1.1 The Frameshift Motifs

The mandatory component of −1 frameshift regions is a short sequence where the ribosome slips backward on the mRNA. It can either be a heptamer of the form X-XXZ-ZZN (where X, Z, and N can be different or identical nucleotides and where XXZ and ZZN are the zero frame codons) as initially established by the pioneering studies on animal viruses (Jacks and Varmus 1985; Jacks et al. 1988; Chapter 7; Brierley et al. 1987) or, as demonstrated later, a Z-ZZN tetramer (Sekine et al. 1994). (The letters X, Z, and N are used here instead of X, Y, and Z so as to avoid confusion with the standard IUPAC nucleotide nomenclature where Y designates pyrimidines.) The intrinsic shiftiness of these two types of sequences is due to the fact that they allow cognate or near-cognate pairing of one or two tRNA in the −1 frame. However, the capacity of the 16 possible Z-ZZN sequences and 64 X-XXZ-ZZN sequences to promote frameshifting is very variable (Brierley et al. 1992; Bekaert et al. 2003). Figure 12.3 gives the results from two studies where most of the heptamers were compared. In panel A, 44 heptamers were cloned in the infectious bronchitis virus (IBV) programmed −1 frameshifting region (Brierley et al. 1992) and in panel B the same set of heptamers was introduced into the frameshift region of IS*911*, an IS*3* family member (M.F. Prère, and O. Fayet, unpublished). The IBV constructs were assayed for their effect on eukaryotes ribosomes (rabbit reticulocytes lysate) whereas the IS*911* set was tested in *E. coli*. The points we would like to note are (i) the important differences in heptamer relative efficiencies in each series and (ii) the rather high proportion of "reasonably" efficient motifs for IBV (19 out of 44 display a PRF-1 frequency ≥ 0.25) opposed to the low number of such motifs in *E. coli*. Only five heptamers are markedly efficient in this bacterium (relative −1 frameshifting frequency ≥ 0.25). In addition, among the 24 motifs not tested in the IBV context, only 2, A-AAG-GGG and A-AAG-GGA, were shown to lead to a relative −1 frameshifting frequency above 0.25 in *E. coli*. Many of the "good" IBV heptamers have a heterogenous purine–pyrimidine composition – i.e., nucleotides X and Z (or Z and N) are, respectively, purines and pyrimidines, or pyrimidines and purines – without an adverse effect on frameshifting (U-UUA-AAC is the best IBV motif). In clear contrast, in *E. coli*, save for one exception (C-CCA-AAG is the best heptamer), there is a higher efficiency for motifs with a homogenous composition; this is also true for Z-ZZN motifs. The consequence of heterogeneity is that tRNAs that have read the XXY and ZZN codons will have to re-pair in the −1 frame on the XXX and ZZZ codons thus requiring a near-cognate interaction. In principle this makes re-pairing to the −1 frame thermodynamically less favorable than the initial 0 frame interaction. *E. coli* ribosomes, and presumably most bacterial ribosomes, must be less tolerant to near-cognate tRNA–codon interactions in the −1 frame than their mammalian counterparts. This suggests that when re-pairing happens,

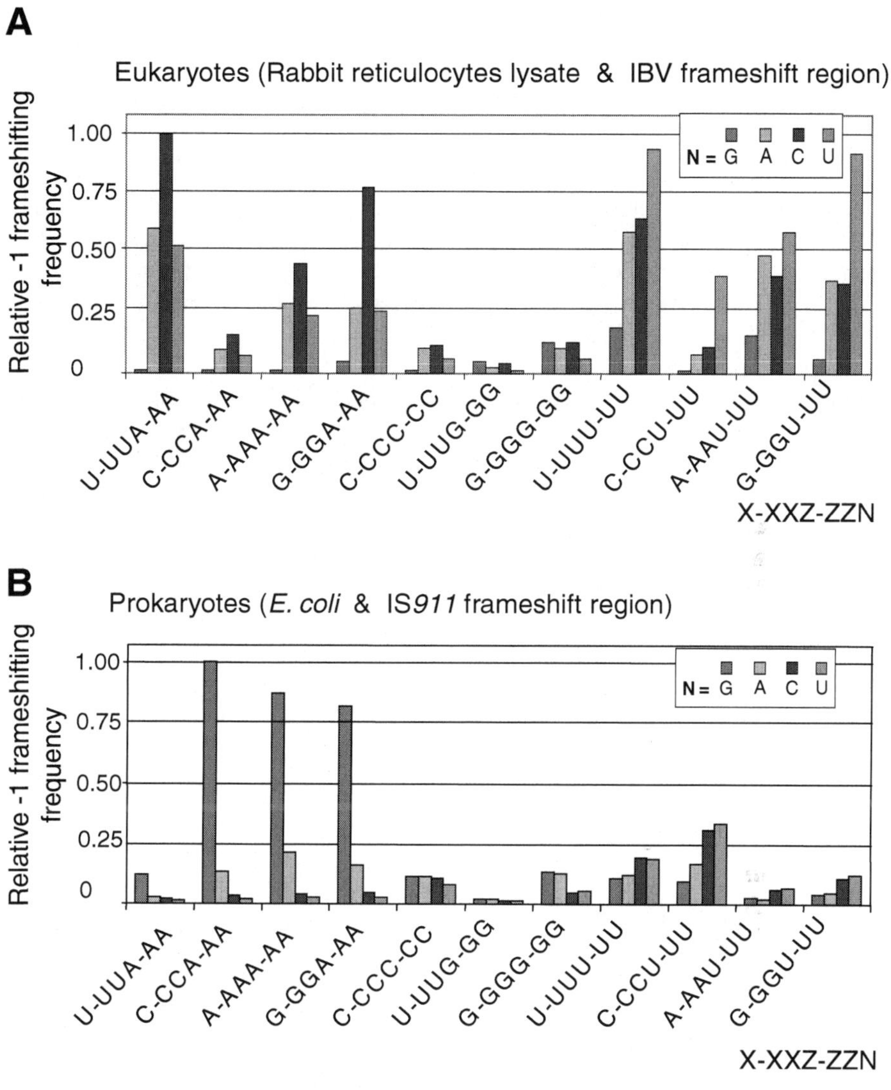

Fig. 12.3 Frameshift efficiency of a subset of 44 X-XXZ-ZZN heptamers. (**A**) Eukaryotic frameshifting: compilation of the data from Brierley et al. (1992) obtained in rabbit reticulocytes lysate. The various heptamers were cloned into the frameshift region of infectious bronchitis virus (IBV). The maximal programmed −1 frameshifting frequency of 1 (motif U-UUA-AAC) corresponds to an absolute frequency of 40%. (**B**) Bacterial frameshifting: the 44 heptamers were cloned in the context of the IS*911* frameshift region cloned in a 5′ fusion with *lacZ* and assayed in vivo in *E. coli*. The maximal frequency (C-CCA-AAG motif) is equivalent to an absolute frameshifting frequency of 2.7% (M.F. Prère and O. Fayet, unpublished)

bacterial ribosomes still have the time and capacity to sense the abnormal situation and react either (i) by forcing back the tRNAs to the 0 frame or (ii) by ejecting the improperly paired tRNA in the A site since this is where error correction can be implemented before peptidyl transfer (Rodnina and Wintermayer 2001). However, the demonstrated occurrence of −1 frameshifting on non-homogenous heptamers (Weiss et al. 1989) indicates that the *E. coli* ribosome sometimes fails to correct the tRNA-pairing anomaly. Instead it corrects the improper alignment with the A and P functional sites, caused by tRNA re-pairing, of the codon–anticodon complexes. By definition −1 frameshifting requires a 1 nt change in the respective positions of the mRNA and the ribosome: the ribosome has to move backward by 1 nt on the message. This is likely the limiting step for −1 frameshifting in both eukaryotes and prokaryotes. As will be detailed later, frameshifting efficiency is boosted to varying extents by stimulatory elements adjacent to the slippery motif; the most efficient stimulators bring the frameshifting frequency close to 80%, i.e., 8 ribosomes out of 10 are forced to change frame (Tsuchihashi and Brown 1992; Larsen et al. 1994). The role of these stimulators is probably to force this readjustment of the mRNA within the ribosome so that the −1 frame codons, XXX and ZZZ, are placed in the P and A sites. Once this repositioning has occurred there is probably no turning back and the change in frame becomes irrevocable even if the A site tRNA is rejected.

The large number of IS*3* family members should in principle reveal which slippery motifs are preferred in bacteria. However, there is a possible caveat. Comparative studies on all the IS*3* family elements and experimental analyses carried out on the IS*911* element (Haren et al. 2000) revealed that the frameshift motif lies within or very near the coding sequence for a potential leucine zipper motif which is likely involved in the homo- and hetero-dimerization of the OrfA and OrfAB proteins. Thus, the sequence of the frameshift motif is likely to be constrained in the IS*3* family, but the degree of stringency of this constraint may vary and remains to be determined. As shown in Fig. 12.4, only a limited repertoire of

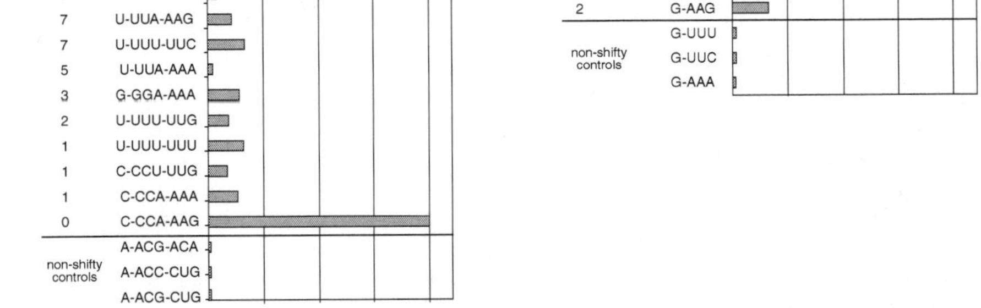

Fig. 12.4 Repartition of the heptameric (**A**) and tetrameric (**B**) frameshift motifs in the IS*3* family. On the right are shown the relative efficiencies of each motif (assayed as indicated in the legend of Fig. 12.3)

the 64 X-XXZ-ZZN motifs or 16 Z-ZZN motifs is used in insertion sequences. Heptamers are by far predominant over tetramers with 273 occurrences versus 70, and within the heptamers A-AAA-AAR sequences are the most frequent motifs (220 out of 273). The position of the motif relative to the leucine zipper appears to be well conserved within the IS*3* and IS*150* groups and in about half of the members of the IS*51* group: the first A generally belongs to a UUA or CUA codon (sometime to a AUA or GCA codon) corresponding to the third leucine of the zipper. In the IS*2* and IS*407* groups, the motif is located a few codons after the fourth leucine. Among tetramers, only A-AA[A/G] and U-UU[U/C] are used by insertion sequences, with a significant preference for the former pair over the latter (40 versus 28). Interestingly, all the U-UU[U/C] motifs are found among related elements all belonging to the IS*51* group. In each of these insertion sequences, the UUY codon of the motif codes for a phenylalanine which replaces the fourth leucine of the zipper. The A-AAR tetramers are more widespread since cases are found in all five groups; their position in relation with the leucine zipper is typical of the group.

The four tetramers used in the IS*3* family are the most shift-prone sequences of this type at least in *E. coli*. This is partly true also for the heptamers since about 50% of insertion sequences possess the highly efficient A-AAA-AAG and G-GGA-AAG motifs (see Figs. 12.3 and 12.4A). The total absence of the third highly frameshift-prone sequence, C-CCA-AAG, suggests that it must have been selected against, perhaps because it would introduce a proline right after the third leucine of the zipper. Among the "not so good" motifs, A-AAA-AAA is remarkably frequent. This implies that $tRNA^{Lys}$ is preferred perhaps because of structural idiosyncrasies that make it good at frameshifting (Murphy et al. 2004) and/or because there is a selection for protein sequence conservation. In many insertion sequences with an A-AAA-AAR heptamer reading in the 0 frame of the motif and flanking codons leads to the —ILKKA— aminoacid sequence, where L is the third leucine of the Leucine zipper motif (Haren et al. 2000); however, it has not yet been determined if any of the four other residues is functionally important.

The 10 other motifs range from passable to poor in terms of efficiency; some like G-GGA-AA[A/C], U-UUA-AAA, or A-AAA-AAC are slightly above the level of frameshifting of non-shifty sequences (Fig. 12.4); the same four heptamers are among the most efficient in eukaryotes (Fig. 12.3A). The use of motifs of moderate or low efficiency indicates that an IS does not necessarily need to achieve a high frameshift frequency to exist and survive.

12.1.2 Upstream Stimulators, Shine–Dalgarno-Like Sequences

A search of potential stimulators was carried out on the 355 selected insertion sequences. Table 12.2 presents a summary of this investigation. Eukaryotic viral frameshift signals contain a stimulator which is either a stem loop (SL) or more often a pseudoknot (PK) formed by folding on itself of the mRNA segment starting a few nucleotides downstream of the shifty motif (see Chapter 7). While 72.5% of the IS*3* family members possess similar downstream structures, 55% of them carry in their

Table 12.2 Frameshift stimulatory elements in the IS of the IS3 family∗

Stimulator Type	IS*2* Group	IS*3* Group	IS*51* Group	IS*150* Group	IS*407* Group	IS*3* Family
SD only	4	24	4	13	4	49 (13.8%)
SL only	8	21	11	22	13	75 (21%)
H-PK only	0	6	1	1	0	8 (2.2%)
A-PK	0	0	29	0	0	29 (8.2%)
SD+SL	5	47	8	20	59	139 (39%)
SD+H-PK	0	8	1	0	0	9 (2.5%)
SD+A-PK	0	0	1	0	0	1 (0.3%)
No stimulator	4	20	3	17	1	45 (13%)
Total	21	126	58	73	77	355
SD (± 3′ structure)	9	79	14	33	63	198 (56%)
3′ structure (± 5′ SD)	13	82	51	43	72	261 (73.5%)
Any stimulator	17	106	55	56	76	310 (87%)

*SD, Shine–Dalgarno-like sequence; SL, stem loop, H-PK, H-type pseudoknot; A-PK, ALIL pseudoknot.

mRNA a potential stimulatory sequence of a different nature upstream of the slippery motif: a Shine–Dalgarno-like sequence (SD), i.e., a sequence complementary to all or part of the 3′ end of 16S rRNA (the 3′-AUUCCUCC anti-Shine–Dalgarno sequence; see Fig. 12.5A). That an SD can act as a frameshift stimulator when located upstream of a −1 slippery motif was first demonstrated in the case of the chromosomal *dnaX* gene (Larsen et al. 1994) and in the case of IS*911* (Fig. 4.6A and B). Both of them have a GGAGM sequence that is located 10 (*dnaX*, with M=C) or 11 (IS*911,* with M=A) nucleotides from the P site when the 0 frame AAA and AAG codons of the motif, respectively, are in the P and A sites of the ribosome. Note that 10 and 11 nt correspond to the "corrected SD to P site spacing" as defined by Ringquist et al. (1992). For an SD involved in initiation it is the number of residues between the nucleotide expected to base-pair with C_{1535} of 16S-RNA and a start codon. For an SD stimulating −1 frameshifting it is the distance between the base expected to pair with C_{1535} and the XXZ codon of a shifty heptamer (or the nnZ codon in case of a Z-ZZN motif) as illustrated in Fig. 12.5B. Demonstration that the *dnaX* and IS*911* internal SD-like sequences operate via pairing with 16S RNA was obtained by a genetic approach (Larsen et al. 1994; M.F. Prère and O. Fayet, unpublished). The important implication is that an SD–anti SD helix can also be formed during the elongation phase. This is now substantiated by elegant single molecule analyses where individual *E. coli* ribosomes are followed when translating different types of mRNA, in particular one with an internal SD (Wen et al. 2008). One clear consequence of this interaction is transient ribosome stalling. The range of action of a stimulatory SD was experimentally determined for *dnaX* frameshifting (Larsen et al. 1994). When the stem loop is present the SD can be positioned between 9 and

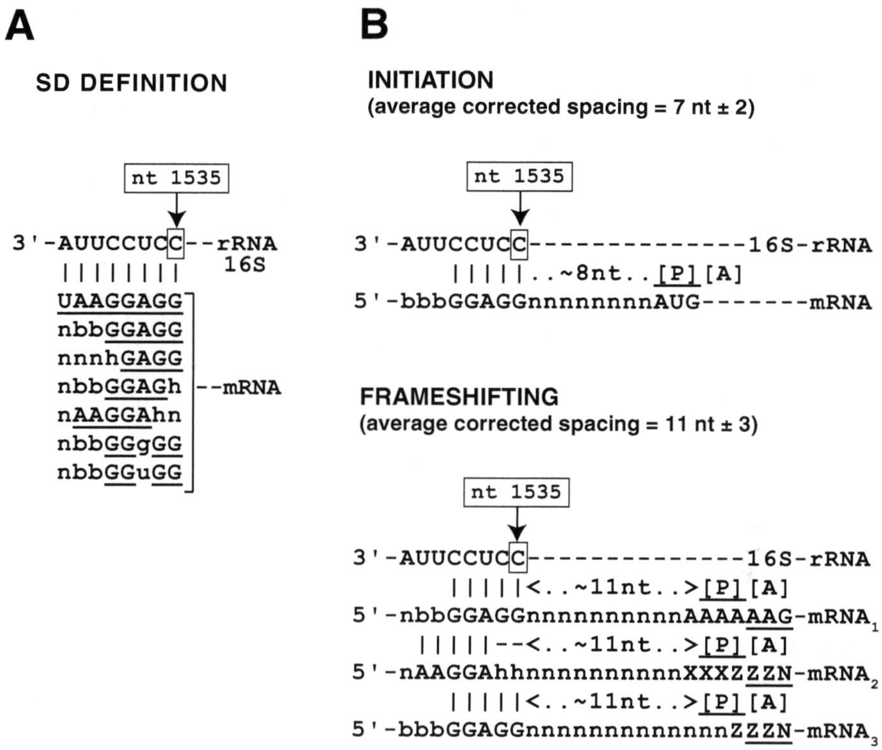

Fig. 12.5 Definition of the different types of Shine–Dalgarno sequences (SD) that were searched for in the frameshift regions of the IS*3* family members (**A**) and comparison of the average corrected spacing (according to Ringquist et al. 1992) in initiation and frameshifting (see text for details).

16 nt 5′ of the P site codon and still have the same stimulatory effect; at greater or lesser distances, the stimulation decreases rapidly.

All the IS*3* members were searched for SD-like sequences corresponding to either one of the seven shown in Fig. 12.5A and situated from 3 to 20 nt upstream of their slippery motif. About half do have such a sequence (Table 12.2) which is located at an average distance of 11 nt. To compare that average corrected spacing with that of SD involved in translation initiation, the same type of search was carried upstream of the start codon of the predicted ORFs of *E. coli* K12 MG1655. It turns out that the average corrected spacing is of 7 nt in that case (2665 ORFs have an SD as defined in Fig. 12.5A). Thus, on average when the ZZN codon of an heptamer or of a tetramer is in the A site, the frameshift-stimulatory SD is located 4 nt further away from the P site than one implicated in initiation is when the start codon is in the P site. This could displace the SD–aSD helix, an anomaly the ribosome tries to correct by pushing back the message by 1 nt. This scheme would explain why an SD-like sequence stimulates frameshifting, but it raises a question: if the SD is too far by 4 nt, why is the compensatory mRNA displacement limited to1 nt? There is no evidence in the literature that under influence of a −1 programmed frameshifting signal the position of the mRNA can be readjusted by 2, 3, or 4 nt more frequently than by 1 nt; on the contrary we found that −2 frameshifting occurs at a 25-fold lower frequency than −1 slippage even if the two tRNAs can re-pair in a cognate manner (M.F. Prère and O. Fayet, unpublished). One possibility is that the movement of the displaced SD–anti SD helix toward the P site is structurally limited. Recent structural analyses indeed suggest that the SD–anti SD helix sits within a rather crowded environment in the *Thermus thermophilus* ribosome, though there is room enough to allow a substantial rotational movement (Yusupova et al. 2006). The alternative is that the backward sliding of the mRNA in the ribosome is restricted to 1 nt.

12.1.3 Downstream Stimulatory Structures

12.1.3.1 Role in Frameshifting

There are two types of stimulatory structures that can enhance −1 frameshifting when located at the "right" distance downstream of a slippery heptamer or tetramer, namely stem loops and pseudoknots. The latter type can be divided into two classes, the H-type pseudoknots and the apical-loop–internal-loop pseudoknots (ALIL-PK). Comparative and experimental analyses have shown that the "right" distance for structures is generally of 5–7 nt (Brierley et al. 1992; Kollmus et al. 1994; Larsen et al. 1997). Table 12.3 shows that it is also the case for the IS*3* family members with an overall preferred distance of 6 nt except for the IS*51* group of elements where the favored spacer length is 5nt. The effect on ribosome functioning of downstream structures is probably mechanical. In contrast with an upstream SD which "pushes" the message into the −1 frame, a downstream structure has to "pull" on it. It probably exerts its action by sitting transiently at the entrance of the mRNA tunnel. The ribosome is forced to pause, while the XXZ and ZZN codons are in the

Table 12.3 Variation of the distance between the 3′-end of the frameshift motif and the 5′-end of the stimulatory structure

Spacing (nt)	IS2 GROUP	Is*3* Group	IS*51* Group	IS*150* Group	IS*407* Group	Total
3	–	–	3	1	–	4
4	–	2	4	1	–	7
5	–	34	26	9	1	70
6	8	36	14	25	50	133
7	4	6	1	7	15	33
8	1	4	3	–	2	10
9	–	–	–	–	4	4

P and A sites, before the helicase activity of the ribosome can melt the structure (Takyar et al. 2005). Pausing of eukaryotic ribosomes has been amply demonstrated by using in vitro translation systems, but until recently there was no clear evidence about prokaryotic ribosome behavior when encountering a stimulatory structure. The already mentioned in vitro single molecule analysis has shown that the *E. coli* ribosome is indeed pause-prone when it meets an RNA hairpin (Wen et al. 2008). Thus a ribosome faced with a frameshift region possessing an upstream SD and a downstream structure has two good reasons to pause.

A first model to explain the effect of a structure is that a pulling action may result from the mechanical tension imposed by a too short mRNA segment between the structure blocked at the entrance of the tunnel and the A site when the ZZN codon is being read. Earlier proposals were that this would cause disruption of the codon–anticodon interaction of the XXZ and ZZN tRNA and readjustment of the mRNA on the ribosome, followed by tRNA re-pairing in the −1 frame; this could occur before or during translocation (Jacks et al. 1988; Weiss et al. 1989; Farabaugh 1997). According to Plant et al. (2003), the tension would build during accommodation of the ZZN-decoding tRNA and be released by backward sliding of the ribosome on the message and concomitant re-pairing of the 2 tRNA in frame −1. In a second model, the pulling effect would occur earlier, during translocation of the NNX- and XXZ-decoding tRNAs (Namy et al. 2006), thus leading to misalignment by 1 nt of the XXZ and ZZN codons in the P and A sites. Slippage of the tRNA would occur after acceptance of the cognate ZZN-decoding tRNA, before its accommodation in the A site and perhaps while the E site is still occupied (Léger et al. 2007).

12.1.3.2 Stem Loops

From the summary presented in Table 12.2, stem loops are the most frequent type of stimulator in the IS*3* family (60% of the IS). As already shown for *dnaX* and IS*911* (Fig. 12.6), these stem loops tends to be associated with an upstream SD

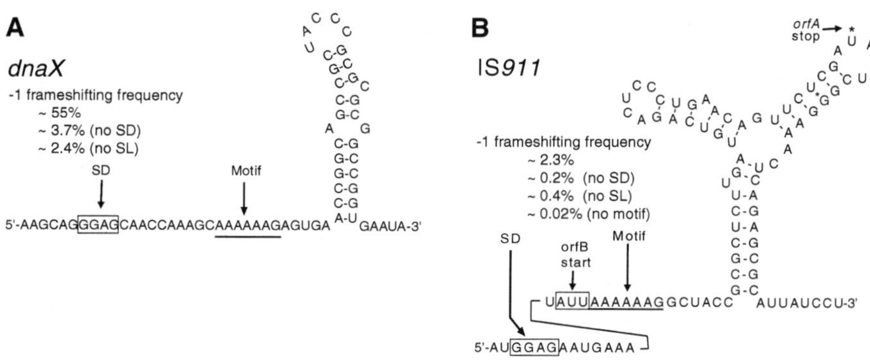

Fig. 12.6 Frameshift regions of the *dnaX* gene (**A**) and of the IS*911* transposable element (**B**). Frameshift frequencies, obtained using the same *LacZ*-based reporter as in Fig. 12.3, are indicated for mutants lacking the SD (no SD), or the stem loop (no SL), or the motif (no motif)

(39%). The end result can be very variable in terms of frameshifting efficiency as exemplified by analyses carried out on *dnaX* (Larsen et al. 1994, 1997) and IS*911* (Rettberg et al. 1999): the *dnaX* site is highly shift-prone whereas its IS*911* counterpart is significantly less so. The main reason for this lies in the design of the stem loop. A 10 bp stem, imperfect but all G/C pairs, capped by a small loop in *dnaX* (not counting the doubtful A-U pair at its base) versus a 3-stem structure for IS*911* where the 9 bp basal stem is also imperfect but contains fewer G/C pairs (6 out of 10). Both structures were validated by genetic and biochemical approaches. Probing in solution showed that the two arms of the Y-shaped IS*911* stem loop are not tightly structured and consequently the ensemble is probably more easily melted (spontaneously and/or by the ribosome) than the compact *dnaX* structure. In agreement with this proposal, removal of one or the other or the two arms of the Y transforms the IS*911* signal into one with a *dnaX*-like efficiency (M.F. Prère and O. Fayet, unpublished). These two examples suggest that, as a general rule, a single G-C-rich stem topped with a small loop should be a better stimulator for programmed −1 frameshifting than a multi-stem structure. Among the 214 insertion sequences with a stem loop, *dnaX*-like structures are indeed found (and sometime with a stronger looking stem than that of *dnaX*, e.g., in IS*1389* and IS*Aca1*) but in a majority of cases the stem loop is more convoluted, e.g., branched structures or successive stems separated by loops or bulges (the prize of extravagance being won by IS*Nha4* which tree-like structure comprises 270 nt, most of which are paired). Thus the general impression is one of high diversity in the size and design of the stem loop which makes it difficult to sort and categorize them. One exception exists within the IS*150* group. As exemplified by IS*150* itself (Vogële et al. 1991) and IS*1221* (Zheng and McIntosh 1995), a distinctive type of stem loop is found: a perfect 7–15 bp stem capped by a rather large loop (21 nt in IS*1221*, 28 nt in IS*150*, and up to 44 nt in IS*Bce15*). In spite of a large a priori unstructured loop, these stem loops are efficient frameshift stimulators: the IS*150* signal for example was reported to give a −1 frameshifting frequency of 60% (Vogële et al. 1991).

12.1.3.3 H-Type Pseudoknots

H-type pseudoknots typical of viral frameshift regions (Chapter 7; Giedroc et al. 2000; Giedroc and Cornish 2009; Brierley and Pennell 2001) are found in only 4.7% of insertion sequences, most of which belong to the IS*3* group (Table 12.2). In addition, about half of them are associated with an upstream SD and most are located 5 nt from a slippery tetramer (A-AAR). IS*3* itself possesses an H-type pseudoknot typical of the family (Sekine et al. 1994; Fig. 12.7A). It is associated with an A-AAG tetramer to give a signal of moderate frameshifting efficiency, i.e., 2–10%, depending on the reporter system and the method used (Sekine et al. 1994; Licznar et al. 2003; M.F. Prère and O. Fayet, unpublished). However, replacing its shift site motif by A-AAA-AAG boosts the efficiency to about 60% (Licznar et al. 2003), an overestimated value we subsequently re-assessed at 25%. Thus this is one of the most efficient natural frameshift stimulator when associated with a "good" motif, even if not quite on par with *dnaX*. Unfortunately, this pseudoknot has not been

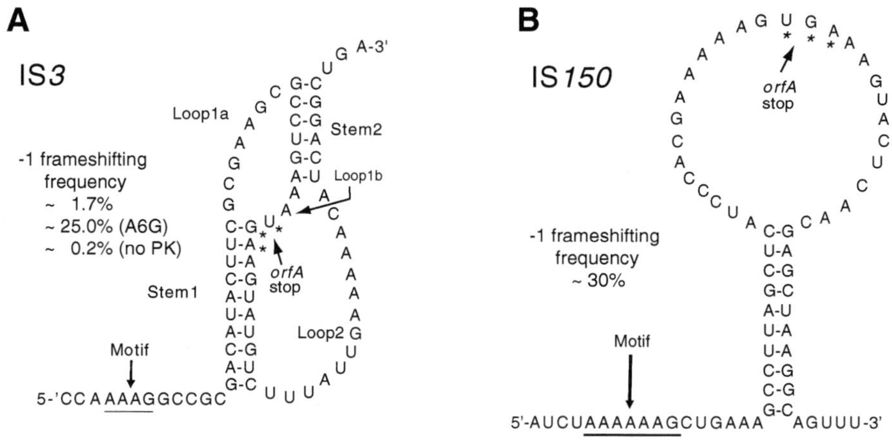

Fig. 12.7 Frameshift regions of the IS*3* (**A**) and of the IS*150* (**B**) transposable elements. The programmed −1 frameshifting frequencies (see Fig. 12.6 legend) were also measured in the case of IS*3* for mutants with an A-AAA-AAG motif (A6G) or without stem 2 (no PK)

thoroughly analyzed from a structural point of view, so it is not possible to make a full comparison with its well-known viral counterparts. The predicted 2D structure of the IS*3* pseudoknot, and of the other H-pseudoknots of the family, suggests that (i) stem 1 has to be about 10 bp long; (ii) stem 2 in general comprises 6 bp; (iii) loop 1a contains 5–10 nt and loop 1b, generally present, is less than 4 nt-long; and (iv) the size of loop 2 varies from 10 to 20 nt. These features are somewhat different from those of the simian retrovirus (SRV-1pk103 variant; Michiels et al. 2001), the mouse mammary tumor virus (MMTVvpk variant; Chen et al. 1995), and the beet western yellow virus (BWYV; Su et al. 1999) pseudoknots. They are characterized by a smaller stem 1 (and stem 2 for BWYV), a shorter loop 1a (1 or 2 nt), no loop 1b (except in MMTV where it is limited to 1 nt), and a smaller loop 2 (7–9 nt). The IS*3* family H-pseudoknots appear closer to the IBV pseudoknot (the 3D structure of which is not known; Brierley et al. 1992) for the size of stem 1 (11 bp), stem 2 (6 bp), and loop 2 (32 nt), but not for the size of loop 1a (2 nt) or loop 1b (absent). To compare the functional efficiency the IS*3* pseudoknot with the eukaryotic counterparts, the SRV-1pk103, MMTVvpk, and BWYV pseudoknots were cloned in the same bacterial reporter system as the IS*3* pseudoknot, downstream of an A-AAA-AAG motif (M.F. Prère and O. Fayet unpublished). The frameshift efficiencies observed in *E. coli* were 7% for SRV-1 (versus 16% in a eukaryotic reporter system), 0.5% for MMTV (versus 12% in a eukaryotic system), and 2.2% for BWYV (versus 4% in a eukaryotic system). These values are much smaller than the 25% in response to the A-AAA-AAG-associated IS*3* signal (Fig. 12.7A). This suggests that the rules of architecture of H-pseudknots stimulating programmed −1 frameshifting are not entirely the same for bacteria and eukaryotes presumably because of structural and functional differences in their target, the ribosome.

12.1.3.4 ALIL-PK

One interesting set of 30-related elements came out of a comparative analysis of the IS*51* group (Mazauric et al. 2008). All possess a U-UUY motif (in two the motif is extended to a U-UUU-UUY heptamer) which is not found in other groups of the IS*3* family. More interestingly, structure prediction suggested the presence of a new type of stimulator starting 5 nt downstream from the shift motif. It is a pseudoknot called an ALIL-PK because it results from pairing between an apical loop and an internal loop carried by two stem loops separated by a 45–65 nt spacer. A model is presented in Fig. 12.8. A related structure has also been proposed to be involved in barley yellow dwarf virus −1 frameshifting (Barry and Miller 2002; Chapter 9): notably, the stem loop with the apical loop is located 4 kb away from its pairing partner. All these suggest that the apical-loop/internal-loop interaction may indeed constitute a recurrent type of RNA interaction that allows formation of tight complexes (with Kd in the nM range; Aldaz-Carroll et al. 2002; Mazauric et al. 2008) between two segments of the same molecule or between segments of two separate molecules.

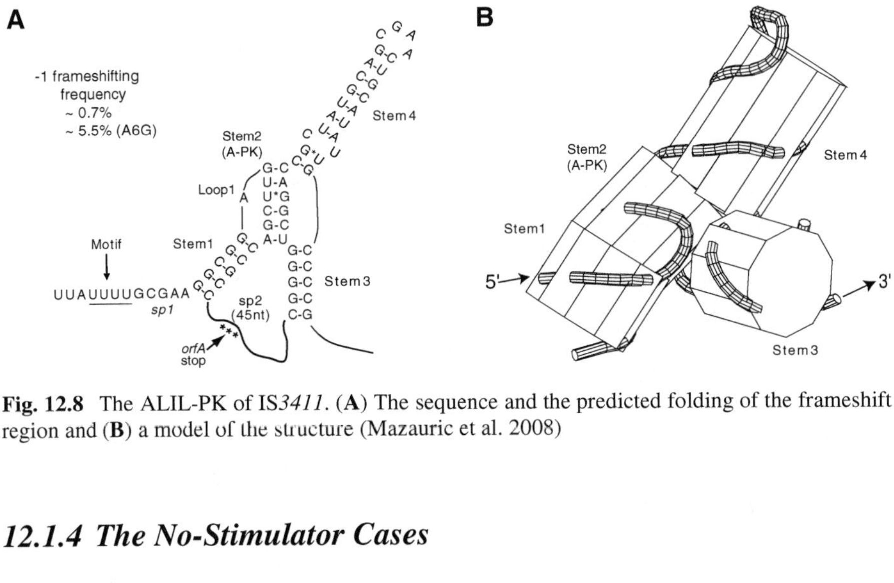

Fig. 12.8 The ALIL-PK of IS*3411*. (**A**) The sequence and the predicted folding of the frameshift region and (**B**) a model of the structure (Mazauric et al. 2008)

12.1.4 The No-Stimulator Cases

This is a minor and heterogenous category comprising 45 IS elements (Table 12.2). There are only three ISs that have no obvious motif. Like most insertion sequences from the other families, they have only one ORF and therefore have evolved a different strategy to control their transposition. Among the remaining 42 elements, five insertion sequences have an eight or nine A motif (a feature also found in two stimulator-containing IS). Such a stretch of identical nucleotides was shown to cause transcriptional slippage (see Chapter 19; Larsen et al. 2000; Baranov et al. 2005), thus leading to the same result as −1 frameshifting in terms of protein products. In the last 37 IS, 4 have an A-AAR tetramer and the rest have a shifty heptamer (mostly A-AAA-AAR). This significant number suggests that life without stimulator is possible for these transposable elements even though they probably have a low level of frameshifting.

12.2 Functional Analysis of Typical Cases

The basic features of the few IS frameshift regions that were analyzed in a more detailed manner have been described in the previous section and are shown in Figs. 12.7 and 12.8. In the following, we will highlight the contribution of these in-depth studies to the knowledge of programmed −1 frameshifting in prokaryotes.

*12.2.1 IS*911

An oddity of IS*911* is the use an AUU codon, located in the −1 frame just before the shift site (**A-UU**A-AAA-AAG), as initiator for expression of its second gene, *orfB*, at a low but significant level (Polard et al. 1991). Both frameshifting and AUU initiation rely on the upstream SD sequence, but they are independent processes. If a function for the OrfB protein is still lacking, the role of recoding is clearly to provide the OrfAB transposase required for mobility of the element (Fig. 4.2B; Prère et al. 1990; Polard et al. 1991; Ton-Hoang et al. 1999). Besides the structure/function analysis of the recoding signal, it was possible to assess the effect on transposition of mutations affecting frameshifting (Rettberg et al. 1999). Mutations neutral for OrfAB but destabilizing stem 1 reduce both frameshifting and transposition several fold. Hence, the level of frameshifting determines the frequency of mobility. Recoding is therefore a means to control the biological activity of the IS.

The IS*911* programmed −1 frameshifting region served as a basis to investigate the effect on frameshifting of the immediate context of an A-AAA-AAG motif in *E. coli* (Bertrand et al. 2002). This issue had only been partially explored in bacteria and eukaryotes using other motifs and stimulators (Horsfield et al. 1995; Kim et al. 2001). The 6 nt 3′ of A-AAA-AAG connect the motif to the stem loop. They are in the mRNA entry tunnel when the motif is in the P and A sites of the ribosome (Yusupova et al. 2001) and they can therefore interact with 16S-rRNA and/or with proteins S3, S4, S5, and S12. Surprisingly, they were shown to modulate frameshifting over a wide range since among 158 random mutants, a 14-fold difference in recoding frequency was observed (the same effect was observed with the three other X-XXA-AAG motifs). The key part is played by the first 1 nt, with generally U>C>A>G, in terms of decreasing frameshifting efficiency, but the other nucleotides are also implicated. This suggests a role of the nucleotide adjacent to the motif in the stabilization of the codon–anticodon interaction in the A site, at least when this codon is AAG. However, it was not possible to deduce further rules concerning the influence of the five other 3′ nt.

While many studies have focused on the shifty motifs and stimulatory structures, this work brings to light the important modulatory role played by the few nucleotides that link these two elements. It therefore raises the more general question of the role in frame maintenance of the mRNA segment situated between the A site and the entry of the message tunnel in the 30S subunit of the ribosome.

12.2.2 IS3

Tetrameric Z-ZZN motifs should in principle allow re-pairing of only one tRNA. If, as with heptamers, frameshifting occurs when the ZZN codon is in the A site, then the P site tRNA has to make room for its neighbor. But are all P site tRNA equally obliging? Only limited analyses had been carried out to answer this, the most relevant, though not complete, being that on the programmed −1 frameshifting site of the bacterial *cdd* gene (Mejlhede et al. 1999). The effect of the 3 nt preceding the A-AAG motif (i.e., N_1-N_2N_3A-AAG) on frameshifting was therefore analyzed in a systematic manner within the context of the IS*3* recoding signal [Fig. 12.7B; Licznar et al. 2003; see also Napthine et al. (2003) for a similar analysis in eukaryotes]. (i) Nucleotides N_2N_3 have a strong modulatory role which is correlated with the wobble properties of the tRNA reading each N_2N_3A codon; some tRNAs make room more readily than others. (ii) The N1 nucleotide constitutes another level of modulation: in general the frequency of programmed −1 frameshifting increases when N_1 and N_2 are identical. (iii) As observed with the X-XXA-AAG heptamers, the nucleotides between the motif and the structure also modulate frameshifting. (iv) Importantly, there is no specific effect of a rare codon (AGA or AGG) or of a stop codon positioned 3′ adjacent to the AAG codon of the motif.

The mechanistic implication of point (iv) is that the N_2N_3A and AAG codons, respectively, are in the P and A sites when frameshifting occurs. In contrast with shifting on A-AAA-AAG, re-pairing of the P site tRNA in the −1 frame does not appear essential, but at the price of a substantial reduction in efficiency. This is also clearly different from cases of non-programmed low-frequency frameshifting (i.e., non-stimulated −1 shift on low-efficiency tetrameric or heptameric motifs) where an adjacent stop or rare codon has a marked effect that strongly suggests P site tRNA shifting, possibly with a coordinated movement of the E site tRNA, while the A site remains empty because of the stop/rare codon (Weiss et al. 1987; Barak et al. 1996; Horsfield et al. 1995).

Another aspect revealed by the IS*3* study is the possible role of nucleotide modifications in the anticodon loop of the P site tRNA in determining its propensity to liberate the third base of the codon so that the A site tRNA can re-pair in the −1 frame (for a recent review on tRNA anticodon modifications see Agris 2008). The tRNAs that promote frameshifting are those with a xo^5-type modification of U_{34} (or that have inosine as base 34), the nucleotide that pairs with the third base of the codon. These tRNAs can generally read three codons (extended wobble). The tRNAs that reduce programmed −1 frameshifting are those with modifications of the xm^5 type on base U_{34} (or have k^2C as base 34). They are known to have a restricted wobble (tRNA reading only two codons). This provides experimental evidence of the importance for frameshifting of P site codon–anticodon interaction (Baranov et al. 2004) and reveals an unexpected level of its subtlety.

12.2.3 *IS*3411

The ALIL-PK stimulator typical of the IS*51* group has already been presented. The demonstration of its existence and functional importance was revealed using IS*3411* (Fig. 12.8; Chen and Hu 2006; Mazauric et al. 2008). Some features of its architecture were disclosed, for example, the necessity of all four stems, the requirement for a single nucleotide in loop 1, and the similarity of stems 1 and 2 with the equivalent stems of the SRV-1 pseudoknot (Fig. 12.8; Michiels et al. 2001). The importance of its integrity for transposition at a normal level was demonstrated. Furthermore, it was shown that the ALIL complex could be formed *in trans* in vivo also and could then lead to frameshift stimulation. This finding expands to bacteria the possibility of influencing the framing of elongating ribosomes via small antisense oligonucleotides (Howard et al. 2004; Olsthoorn et al. 2004; Plant and Dinman 2005) and perhaps through small non-coding RNA.

12.3 Lessons from IS*3* Family Studies

12.3.1 *A Summary of How Motifs and Stimulators Combine in the IS*3 *Family*

From the overview of the IS*3* family it is possible to formulate minimal rules (with many exceptions) concerning the way the motifs and stimulators associate. These rules probably reflect general preferences among prokaryotes in terms of programmed −1 frameshifting motifs and stimulatory elements.

(i) A heptameric motif can be alone (42 cases out of 355), or more often be associated with an usptream SD, or with a downstream stem loop, or with both.
(ii) Tetrameric motifs are nearly always accompanied by a stimulatory structure (5 exceptions out of 68 cases).
(iii) H-pseudoknots are generally preceded by an A-AAR motif and are half of the time associated with an SD.
(iv) ALIL-pseudoknots are associated with an U-UUY motif (or in some cases with a U-UUU-UUY motif) and are very rarely combined with an upstream SD.

12.3.2 *The Three Ways of Translational −1 Frameshifting in Bacteria*

Ribosomal −1 slippage can probably occur at least in three different ways in *E. coli*.

(1) By P site tRNA slippage (with perhaps a concomitant E site tRNA movement to liberate the base 5′ of the P site codon) on “low-”efficiency shifty tetramer or

heptamer without nearby stimulator and probably also on any codon but at an even lower frequency.

(2) By simultaneous slippage of P and A sites tRNA on the "high-"efficiency heptamers with a stimulator and probably without stimulator also, as suggested by the significant number of IS possessing a heptamer without associated stimulators (where the line between "low-" and "high-"efficiency slippery sequences is not known yet and may vary with the species).

(3) By A site tRNA slippage assisted by P site tRNA movement, to liberate the nucleotide 5′ of the A site codon, on the A-AAG tetramer. This may also happen on A-AAA and U-UUY motifs, but whatever the motif, our guess would be that a stimulator (preferably a pseudoknot) must be present.

12.4 Concluding Remarks

As judged from insertion sequences, bacteria utilize a more limited repertoire of shifty heptamers than eukaryotic viruses. This is probably because the bacterial ribosome is much less tolerant to near-cognate tRNA pairing that its eukaryote counterpart. However, while the known eukaryotic programmed −1 frameshifting signals always appear to require a downstream stimulator, the IS frameshift regions do not have such an absolute requirement. When they do have a downstream stimulator, the preference is for stem loops over pseudoknots or upstream SD-like sequence stimulators which can be associated with a stem loop or a pseudoknot and thus lead to more flexibility in overall design. The fact that 45 insertion sequences possess only a shifty heptamer suggests that frameshifting can occur without the help of stimulators. The inverse is not true, i.e., frameshifting in a region where there are stimulators but no shifty motif. This is a reminder that the shifty motif, hence the possibility for one or two tRNAs to re-pair in the −1 frame is the only mandatory requirement for −1 slippage. Another lesson from studying IS elements is that there is probably a structural restriction that limits the amplitude of the mRNA/ribosome relative movement to 1 nt. An unexpected finding is that the 6 nt on the 3′ side of the motif influence the efficiency of −1 frameshifting. This part of the mRNA which lies in a dedicated tunnel plays a part in frame maintenance through yet unknown interactions with the ribosome. Hopefully, the progresses in sophisticated structural analyses of prokaryotic ribosomes either frozen in different states with various partners (mRNA, tRNA, and protein factors; Berk and Cate, 2007; Frank et al. 2007; Bashan and Yonath 2008; Ramakrishnan 2008) or in solution while they are functioning (Lee et al. 2007; Wen et al. 2008; Stapulionis et al. 2008) will soon facilitate an understanding of the molecular mechanisms whereby the ribosome can be so efficiently driven out of its normal path.

Acknowledgments The help of Mick Chandler and Patricia Siguier with the IS database has been greatly appreciated. This work was funded by the Centre National de la Recherche Scientifique (CNRS), the University of Toulouse, and by a grant to OF from the Agence National pour la Recherche (#ANR-05-BLAN-0048-01)

References

Agris PF (2008) EMBO Rep 9:629–635
Agris PF, Vendeix FA, Graham WD (2007) J Mol Biol 366:1–13
Aldaz-Carroll L, Tallet B, Dausse E, Yurchenko L, Toulme JJ (2002) Biochemistry 41:5883–5893
Bashan A, Yonath A (2008) Trends Microbiol 16:326–335
Barak Z, Lindsley D, Gallant J (1996) J Mol Biol 256:676–684
Baranov PV, Gesteland RF, Atkins JF (2004) RNA 10:221–230
Baranov PV, Hammer AW, Zhou J, Gesteland RF, Atkins JF (2005) Genome Biol 6:R25
Barry JK, Miller WA (2002) Proc Natl Acad Sci USA 99:11133–11138
Bekaert M, Bidou L, Denise A, Duchateau-Nguyen G, Forest JP, Froidevaux C, Hatin I, Rousset JP, Termier M (2003) Bioinformatics 19:327–335
Berk V, Cate JH (2007) Curr Opin Struct Biol 17:302–309
Bertrand C, Prère MF, Gesteland RF, Atkins JF, Fayet O (2002) RNA 8:16–28
Brierley I, Boursnell ME, Binns MM, Bilimoria B, Blok VC, Brown TD, Inglis SC (1987) EMBO J 6:3779–3785
Brierley I, Jenner AJ, Inglis SC (1992) J Mol Biol 227:463–479
Brierley I, Pennell S (2001) Cold Spring Harbor Symp. Quant Biol 66:233–248
Chandler M, Fayet O (1993) Mol Microbiol 7:497–503
Chandler M, Mahillon J (2002) Insertion Sequences revisited. In Craig NL, Craigie R, Gellert M, Lambowitz AM (eds) Mobile DNA II, American Society for Microbiology, Washington DC, –pp 305–366
Crick FH (1966) J Mol Biol 19:548–555
Chen X, Chamorro M, Lee SI, Shen LX, Hines JV, Tinoco I Jr, Varmus HE (1995) EMBO J 14:842–852
Chen CC, Hu ST (2006) J Biol Chem 281:21617–21628
Escoubas JM, Prère MF, Fayet O, Salvignol I, Galas D, Zerbib D, Chandler M (1991) EMBO J 10:705–712
Farabaugh PJ (1997) Programmed Alternative Reading of the Genetic Code. Landes Bioscience, Austin, Texas and Springer, Heidelberg, Germany, pp 69–102
Frank J, Gao H, Sengupta J, Gao N, Taylor DJ (2007) Proc Natl Acad Sci USA 104:19671–19678
Gesteland RF, Atkins JF (1996) Annu Rev Biochem 65:741–68
Giedroc DP, Cornish PV (2009) Virus Res 139:193–208
Giedroc DP, Theimer CA, Nixon PL (2000) J Mol Biol 298:167–185
Haren L, Normand C, Polard P, Alazard R, Chandler M (2000) J Mol Biol 296:757–768
Harger JW, Meskauskas A, Dinman JD (2002) Trends Biochem Sci 27:448–454
Horsfield JA, Wilson DN, Mannering SA, Adamski FM, Tate WP (1995) Nucleic Acids Res 23:1487–1494
Howard MT, Gesteland RF, Atkins JF (2004) RNA 10:1653–1661
Jacks T, Varmus HE (1985) Science 230:1237–1242
Jacks T, Madhani HD, Masiarz FR, Varmus HE (1988) Cell 55:447–458
Kim YG, Maas S, Rich A (2001) Nucleic Acids Res 29:1125–1131
Kollmus H, Honigman A, Panet A, Hauser H (1994) J Virol 68:6087–6091
Kurland CG (1992) Annu Rev Genet 26:29–50
Larsen B, Wills NM, Gesteland RF, Atkins JF (1994) J Bacteriol 176:6842–6851
Larsen B, Gesteland RF, Atkins JF (1997) J Mol Biol 271:47–60
Larsen B, Wills NM, Nelson C, Atkins JF, Gesteland RF (2000) Proc Natl Acad Sci USA 97: 1683–1688
Lee TH, Blanchard SC, Kim HD, Puglisi JD, Chu S (2007) Proc Natl Acad Sci USA 104: 13661–13665
Léger M, Dulude D, Steinberg SV, Brakier-Gingras L (2007) Nucleic Acids Res 35:5581–5592
Licznar P, Mejlhede N, Prère MF, Wills N, Gesteland RF, Atkins JF, Fayet O (2003) EMBO J 22:4770–4778

Loot C, Turlan C, Rousseau P, Ton-Hoang B, Chandler M (2002) EMBO J 21:4172–4182
Mazauric MH, Licznar P, Prère MF, Canal I, Fayet O (2008) J Biol Chem 2008 283:20421–20432
Mejlhede N, Atkins JF, Neuhard J (1999) J Bacteriol 181:2930–2937
Murphy FV 4th, Ramakrishnan V, Malkiewicz A, Agris PF (2004) Nat Struct Mol Biol 11: 1186–1191
Michiels PJ, Versleijen AA, Verlaan PW, Pleij CW, Hilbers CW, Heus HA(2001) J Mol Biol 310:1109–1112
Namy O, Moran SJ, Stuart DI, Gilbert RJ, Brierley I (2006) Nature 441:244–247
Olsthoorn RC, Laurs, Sohet F, Hilbers CW, Heus HA, Pleij CW (2004) RNA 10:1702–1703
Napthine S, Vidakovic M, Girnary R, Namy O, Brierley I (2003) EMBO J 22:3941–3950
Plant EP, Dinman JD(2005) Nucleic Acids Res 33:1825–1833
Plant EP, Jacobs KL, Harger JW, Meskauskas A, Jacobs JL, Baxter JL, Petrov AN, Dinman JD (2003) RNA 9:168–174
Polard P, Prère MF, Chandler M, Fayet O (1991) J Mol Biol 222:465–477
Prère MF, Chandler M, Fayet O (1990) J Bacteriol 172:4090–4099
Ramakrishnan V (2008) Biochem Soc Trans 36:567–574
Rettberg CC, Prère MF, Gesteland RF, Atkins JF, Fayet O (1999) J Mol Biol 286:1365–1378
Ringquist S, Shinedling S, Barrick D, Green L, Binkley J, Stormo GD, Gold L (1992) Mol Microbiol 6:1219–1229
Rodnina MV, Wintermeyer W (2001) Annu Rev Biochem 70:415–435
Sekine Y, Ohtsubo E (1989) Proc Natl Acad Sci USA 86:4609–4613
Sekine Y, Ohtsubo E (1992) Mol Gen Genet 235:325–332
Sekine Y, Eisaki N, Ohtsubo E (1994) J Mol Biol 235:1406–1420
Siguier P, Perochon J, Lestrade L, Mahillon J, Chandler M (2006) Nucleic Acids Res 34(Database issue):D32–36
Stapulionis R, Wang Y, Dempsey GT, Khudaravalli R, Nielsen KM, Cooperman BS, Goldman YE, Knudsen CR (2008) Biol Chem 389:1239–1249
Su L, Chen L, Egli M, Berger JM, Rich A (1999) Nat Struct Biol 6:285–292
Takyar S, Hickerson RP, Noller HF(2005) Cell 120:49–58
Ton-Hoang B, Polard P, Haren L, Turlan C, Chandler M (1999) Mol Microbiol 32:617–627
Tsuchihashi Z, Brown PO(1992) Genes Dev 6:511–519
Yusupova GZ, Yusupov M, Cate JH, Noller HF (2001) Cell 106:233–241
Yusupova G, Jenner L, Rees B, Moras D, Yusupov M (2006) Nature 444:391–394
Vögele K, Schwartz E, Welz C, Schiltz E, Rak B (1991) Nucleic Acids Res 19:4377–4385
Weiss RB, Dunn DM, Atkins JF, Gesteland RF (1987) Cold Spring Harbor Symp Quant Biol 52:687–693
Weiss RB, Dunn DM, Shuh M, Atkins JF, Gesteland RF(1989) New Biol 1:159–169
Wen JD, Lancaster L, Hodges C, Zeri AC, Yoshimura SH, Noller HF, Bustamante C, Tinoco I (2008) Nature 452:598–603
Zheng J, McIntosh MA (1995) Mol Microbiol 16:669–685

Chapter 13
Autoregulatory Frameshifting in Antizyme Gene Expression Governs Polyamine Levels from Yeast to Mammals

Ivaylo P. Ivanov and Senya Matsufuji

Abstract Polyamines play an important role in a variety of biological processes. Antizyme is a key regulator of polyamine levels in eukaryotes through its inhibitory role on polyamine biosynthesis and transport. Antizyme mRNA expression requires +1 ribosomal frameshifting linking two partially overlapping open reading frames. The frameshifting is enhanced in the presence of elevated levels of polyamines thus closing an autoregulatory loop. The ribosomal frameshift site, which is present at the end of the first open reading frame, is usually surrounded by *cis*-acting stimulators of frameshifting. The extraordinary antiquity of the antizyme gene has enabled it to acquire large number of diverse *cis*-acting stimulators in different evolutionary branches. Our current understanding of the frameshifting mechanism suggests that termination of translation at the end of the first open reading frame is a crucial component of the polyamine sensor.

Contents

Too much of good thing can be a disaster. Though all cells need polyamines, they have to carefully control polyamine levels, and eukaryotic cells employ the protein antizyme to do so. Overaccumulation of polyamines in cells may lead either to

---

I.P. Ivanov (✉)
BioSciences Institute, University College Cork, Cork, Ireland and Department of Human Genetics, University of Utah, Salt Lake City, Utah, USA
e-mail: ivaylopi@hotmail.com

J.F. Atkins, R.F. Gesteland (eds.), *Recoding: Expansion of Decoding Rules Enriches Gene Expression*, Nucleic Acids and Molecular Biology 24,
DOI 10.1007/978-0-387-89382-2_13,

malignant transformation (Auvinen et al., 1992) or to cell death (Poulin et al., 1993, Seiler and Raul, 2005). In some sense polyamines are Mg^{2+}-like ions with periodically distributed charges that facilitate binding to nucleic acids. Apart from crucial roles in charge neutralization and the proximity of nucleic acid strands, polyamines perform other roles such as a posttranslational constituent of translation elongation factor eIF5A which is an essential cellular component (Schnier et al., 1991; Park et al., 1997; Sami et al., 2009) and regulates nerve process extension (Huang et al., 2007). Another example is in inward rectification of K(+), Kir, channels for endocochlear potential relevant to hearing and balance functions of the inner ear (Wang et al., 2009).

A reflection of the widespread functions and investigations is that Pubmed currently has over 75,000 listings for polyamines since 1947. Long predating Pubmed is the discovery of spermidine and spermine in seminal fluid. In 1678 van Leeuwenhoek reported the slow crystallization of a substance which was much later shown to be spermine phosphate; spermine was later identified in many tissues in addition to seminal fluid (Rosenheim, 1924). Also putrescene was identified a long time ago in putrefying meat (rats bury their dead after bacterial action has liberated polyamines – they also bury bits of wood sprinkled with polyamines!). The first issue of the *Journal of Biological Chemistry* reported decarboxylation of ornithine to putrescene (Dakin, 1906) from which spermidine and spermine are derived (Review, Cohen, 1998). However, the enzyme that catalyzes the first step of the pathway, ornithine decarboxylase (ODC), is rate limiting and has been the focus of intensive research.

Polyamine biosynthesis is a prerequisite for cell growth and is increased in many tumors. In particular, ornithine decarboxylase (ODC) which catalyzes the first and rate-limiting step is dramatically augmented in rapidly growing cells. ODC inhibitors prevent cell growth and have been intensively studied. One such ODC inhibitor is α-difluoromethylornithine (DFMO, also known as eflornithine), which is being tested on a mouse squamous cell carcinoma model (Burns et al., 2009), neuroblastoma (Hogarty et al., 2008), prostate cancer (Simoneau et al., 2008), and in phase III trials of colorectal adenomas (Gerner and Meyskens, 2009). Eflornithine under the trade name Vaniqua has long been approved for aesthetic (cosmetic) facial hair reduction, where it is a useful adjuvant for hormonal therapy for hirsutism (Blume-Peytavi and Hahn, 2008). There are significant differences in polyamine metabolism between mammalian cells and protozoan parasites such as trypanosomes. Trypanosomes require polyamines not only for supporting their proliferation but also as a component of a protective reducing molecule, trypanothionine (Fairlamb et al., 1985). The half-life of ODC in mammalian cells is about 20 minutes, while in *Trypanosoma brucei*, the causative agent of the African sleeping sickness, the half-life is long, about 18 hours. The parasite is susceptible to DFMO which is an important drug for this disease (Balasegaram et al., 2008; Sanderson et al., 2008).

Antizyme, a natural inhibitor of ODC in mammalian cells, was postulated by E.S. Canllakis and coworkers and was named "ODC antizyme," a derivative of "anti-enzyme" (Fong et al., 1976; Heller et al., 1976), because they expected that negative regulation of an enzyme by a protein whose synthesis is induced by a product of the enzyme might not be restricted to ODC. Since then, however, enzyme regulatory mechanisms of this type have not been found except for ODC antizyme, which is now simply called antizyme. Subsequently, antizyme has turned out to be exceptional in other two facets: it represses ODC by triggering ubiquitin-independent degradation by the 26S proteasomes (Murakami et al., 1992; Takeuchi et al., 2008) and it requires a recoding event for its expression and for induction of synthesis by polyamines studied (Matsufuji et al., 1990, 1995; Miyazaki et al.,1992; Review, Coffino, 2001) as shown in Fig. 13.1 and discussed in detail later. Antizyme also controls cellular polyamine levels through cellular uptake or excretion of polyamines (Suzuki et al., 1994; Mitchell et al., 1994; Sakata et al., 2000 but see Porat et al., 2008). Since de novo synthesis and cellular uptake are the only supply source of cellular polyamines, antizyme can exert a complete block on the polyamine supply.

A wide range of eukaryotes, including protists, fungi, and animals, have antizyme (see below). In mammals, the budding yeast, *Saccharomyces cerevisiae*, and the fission yeast *Schizosaccharomyces pombe* it has been definitively demonstrated that antizyme regulates the cellular polyamine levels (Murakami et al., 1994; Ivanov et al., 2000; Palanimurugan et al., 2004). Interestingly, one of the yeast prions, [PSI^+], a non-functional form of eukaryotic release factor 3 (eRF3), induces antizyme synthesis by stimulating translational frameshifting, leading to a decrease in cellular polyamine levels (Namy et al., 2008). This

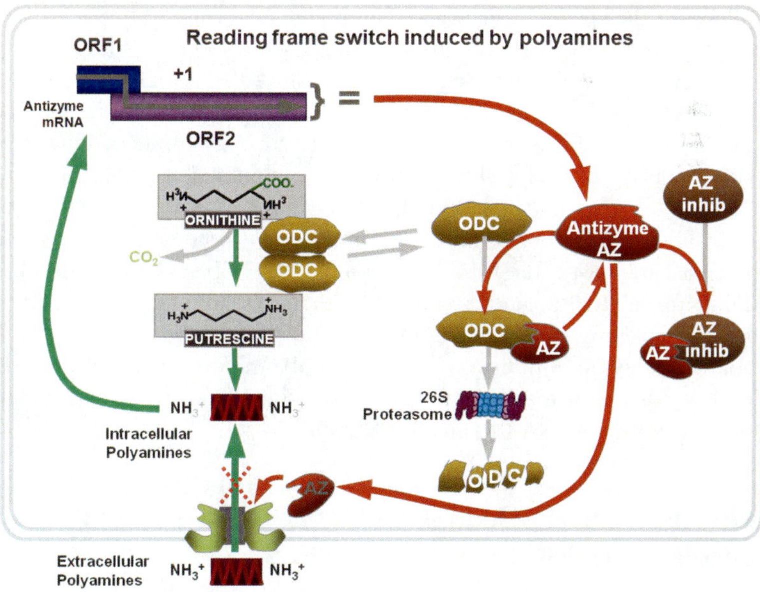

Fig. 13.1 Schematic representation of the interaction between antizyme, ODC, and polyamines

explains half of the phenotypes epigenetically associated with [PSI^+]. Mammals have three paralogs of antizyme (also see below). Antizyme 2, like antizyme 1, is being widely expressed in different tissues, but is more highly conserved and is expressed at lower levels than antizyme 1 (Ivanov et al., 1998a; Zhu et al., 1999). Unlike Antizyme 1, it is predominately nuclear (Murai et al., 2009). Before its identity as an antizyme was known it was found to be overexpressed in the brains of mice with drug-induced seizures (Kajiwara et al., 1996). Homozygous knockout of the antizyme 1 gene in mice results in partial embryonic death, the extent of which depends on the genetic background and a double knockout of antizyme 1 and 2 genes resulted in complete lethality. This indicates that polyamine regulation by systemic type antizymes (antizyme 1 and 2) is essential in mammals (Matsufuji et al., unpublished). Antizyme 3 is only expressed in mid- to late spermatogenesis (Ivanov et al., 2000a; Tosaka et al., 2000a; Ike et al., 2004). Its expression is quite different and is described in a separate following section.

Antizyme is a part of the negative regulatory circuit of polyamines, but there are some reports that antizyme may regulate proteins other than ODC or polyamine transporter. In these cases, antizyme mediates a bifurcated response to polyamine levels. It has been reported that antizyme binds to and promotes degradation of cyclin D1 (Newman et al., 2004), Smad1 (Gruendler et al., 2001), and Aurora-A (Lim and Gopalan, 2007). In addition, binding of antizyme 3 to testicular germ cell-specific protein GGN1 has been reported (Zhang et al., 2005). Antizyme 2 also has specific binding partners (Murai and Matsufuji, unpublished). Although the consequence of binding is unknown in all of these cases, the interactions involved could provide clues for the specific functions of antizyme isoforms.

In mammals there are two proteins with similarities to ODC which lack decarboxylase activity that bind to and sequester antizyme – these are known as antizyme inhibitor 1 and 2 (ODCp) (Fujita et al., 1982; Murakami et al., 1996; López-Contreras et al., 2006). Both antizyme inhibitor 1 and 2 are implicated in the regulation of polyamines and other cellular functions (Mangold et al., 2008; Snapir et al., 2008), and is cell cycle regulated (Murakami et al., 2009). Antizyme inhibitor 1 has wide tissue distribution, whereas antizyme inhibitor 2 is predominately expressed in the brain, testis and mast cells (Murakami et al., 1996; López-Contreras et al., 2006; 2009; Mäkitie et al., 2009; Kanerva et al. 2009). Analysis of knockout mice revealed that most pups lacking antizyme inhibitor 1 die perinatally (Tang et al., 2009). Regulation of antizyme inhibitor synthesis by polyamines is not via frameshifting but at least in part, is at the initiation of translation at the AUU start codon of a uORF in its encoding mRNA (Ivanov et al., 2008).

13.1 Alternative Start Codons and Possible Regulation Through the 3′ UTR

The +1 translational frameshifting is not the only posttranscriptional regulation of antizyme genes. Vertebrate antizyme 1 ORF1 has two AUG codons, the positions of which are highly conserved. Initiation at the first encodes a conserved peptide sequence that is missing if the downstream AUG is utilized instead. Initiation at the first AUG results in a protein that is targeted for mitochondrial localization. In vitro and in vivo experiments show that both conserved AUG ORF1 codons are utilized for translation initiation of vertebrate antizyme 1 genes (Matsufuji et al., 1995; Mitchell and Judd, 1998; Gandre et al., 2003). Something analogous is seen with the orthologs of mammalian antizyme 3. Phylogenetic evidence suggests the presence of an alternative upstream translation initiation site which is a CUG triplet instead of the standard AUG codon (Ivanov and Atkins, 2007). This prediction is also supported by some experimental data (Fitzgerald et al., 2006). Initiation at the CUG would lead to 48 amino acids N-terminal extension of the human antizyme 3 protein. The role of the alternative initiation in antizyme 3 genes is unknown at the moment. There is strong phylogenetic evidence that alternative initiation on a standard AUG codon is also utilized by the antizyme homolog in flies (Ivanov and Atkins, 2007). It is currently not known if the choice of initiation codon is regulated in any of the examples above.

There are strong hints that regulation of antizyme expression, at least in vertebrates, also occurs through signals in the 3′ UTR. Human antizyme 1 mRNA has two alternative polyadenylation sites (Matsufuji et al., 1990; Ivanov et al., 1998a). Polyadenylation at the downstream site extends the length of the mRNA by about 150 nucleotides. The ratio between the two products varies between different tissues suggesting possible regulation (Ivanov et al., 1998a). These alternative polyadenylation sites are conserved between human and mouse, again suggesting physiological function.

Careful examination of the 3′ UTRs of different antizyme mRNAs identifies several highly conserved segments (Ivanov and Atkins, 2007). In antizyme 1 the segment is about 90 nucleotides long and is almost completely conserved between human and chicken. The proximity of the conserved segment to the alternative polyadenylation site suggests possible role in that process. Antizyme 2 mRNA has at least two conserved segments in its 3′ UTR which are also considerably longer than that of either antizyme 1 or antizyme 3. Once again, their role in regulation of the gene expression is currently unknown.

13.2 The Antizyme Gene Family

Antizyme homologs are widespread in the eukaryotic domain present in at least three of its four kingdoms – protists, fungi, and animals. According to the current understanding of eukaryotic radiation, plants diverged from other eukaryotes after the last common ancestor of protists, fungi, animals, and plants. Therefore, it appears that the absence of antizyme genes in plants is due to a gene loss in one of their common ancestors. This conclusion is strengthened by the identification of an antizyme gene in photosynthetic organisms like *Euglena gracilis*, thought to have radiated from the plant lineage after the last common ancestor of plants, protists, and fungi. Within animals, fungi, and protists, antizyme is present almost universally. In the few exceptions where an antizyme gene cannot be identified it is not clear if that is due to incomplete available sequence or because of genuine gene loss. However, it is known that in *S. pombe* and *S. cerevisiae* the gene encoding antizyme is not essential, indicating that at least in some fungi loss of the gene is possible (Ivanov et al., 2000b; Palanimurugan et al., 2004).

Studies have identified a gene analogous to that for eukaryotic antizyme in the bacteria *Selenomonas ruminantium* (Yamaguchi et al., 2006) but because its synthesis does not require frameshifting and because it is clearly not a homolog of the eukaryotic counterpart it will not be discussed here.

There are over a dozen known examples of antizymes from protists. Because of taxonomic uncertainty in this kingdom, it is difficult to place them in specific protist phyla. However, it is noteworthy that all examples so far come from protozoans.

Fungal antizymes are known from four separate phyla – Ascomycota, Basidiomycota, Glomeromycota, and Zygomycota. As will be discussed below, antizyme genes from the Pezizomycotina, subphylum of Ascomycota, have a number of distinct characteristics that put them in a group of their own. In total, sequences from over 100 fungal antizyme genes are currently known.

Antizymes have been identified in at least a dozen animal phyla: Annelida, Arthropoda, Chordata, Cnidaria, Echinodermata, Mollusca, Myxozoa, Nematoda, Platyhelminthes, Priapulida, Onychophora, and Rotifera (Ivanov and Atkins, 2007; and IPI unpublished results). Depending on certain protein and ribosomal frameshifting characteristics the antizymes in animals can be grouped into several distinct categories. The biggest division is between vertebrates and invertebrates. In total over 150 antizymes are known from vertebrates and another 150 plus from invertebrates.

With one exception fungi and protists appear to have only one antizyme paralog. One paralog is also the rule in invertebrate animals. The exceptions are two antizyme paralogs each in one nematode, one earthworm and one arthropod species. In these exceptional cases the paralogous pair appears to result from comparatively recent gene duplication events. Vertebrates, on the other hand, all have multiple paralogs of antizyme. Some paralogs, like antizyme 1 and antizyme 2, are present in all. Some like AZL and AZS (Saito et al., 2000), themselves both orthologs of antizyme 1, are present only in a superclass of verebrates, in that case bony fish (Osteichthyes). Some, like antizyme 3, are confined to a subgroup – i.e., terrestrial tetrapods (Ivanov and Atkins, 2007). Yet others, like AZR, are present only in a family or a subfamily, in that case the subfamily Leuciscinae of bony fish (Ivanov et al., 2007). In total, mammals have three antizyme paralogs; reptiles probably also three; birds, amphibians, cartilaginous fishes and lampreys have two. Most fishes have three although zebrafish and related fishes have four. The record, six, is held by fish in the Salmonidae family where a whole genome duplication has occurred in the last 100 million years.

13.3 Shift Sites

The original investigation on antizyme cloning, expression, and regulation was performed on rat antizyme 1 (Matsufuji et al., 1990, 1995; Miyazaki et al., 1992). Sequencing of the transframe peptide identified the frameshift site unambiguously at the last two codons of ORF1: UCC-UGA(-U). After decoding of the serine UCC codon the next amino acid added is aspartate encoded by GAU in the +1 frame (Matsufuji et al., 1995). The only other case where the endogenous antizyme frameshift site was verified experimentally is the one in *S. cereviciae* where mass spectroscopy of the transframe product identified the site as GCG-UGA(-C) (Ivanov et al., 2006). In all other antizymes the frameshift site is inferred. The remarkable conservation of these sites makes the task relatively straightforward. In one example, 12 nucleotides, including the frameshift, are completely conserved between antizyme 1 in humans and the antizyme gene in *S. pombe* (Ivanov et al., 2000b).

By coincidence the primordial antizyme frameshift site appears to have been the originally identified shift site – i.e., UCC-UGA. In many evolutionary branches, in addition to the aforementioned *S. cerevisiae*, the shift site has altered identity. The shift site can be broken up in two parts – the "P-site" and "A-site" codons (referring to the place they occupy in the ribosome during the frameshift event) – and those would be considered separately.

In addition to the primordial "P-site" codon UCC, naturally occuring variations include the codons UUU, GUU, AUU, CCC, CCG, GCG, and GCC (Ivanov and Atkins, 2007, and unpublished results). The codons UUU and CCC followed by UGA supplied with a mammalian antizyme 5′ element (see next section) have been shown experimentally to support +1 frameshifting (Petros et al., 2005). The same is also true for the codon CUU which is the P-site codon required for +1 frameshifting of *Escherichia coli* RF2 though it has so far not been observed in a naturally occurring antizyme frameshift site (Petros et al., 2005; Bekaert et al., 2006). It is interesting to note that the naturally occurring antizyme frameshift site UUU-UGA can support +1 frameshifting in *E. coli* when substituting for the RF2 site CUU-UGA (Weiss et al., 1987). Phylogenetic consideration suggests that the individual non-UCC shift codons have emerged more than once. For example, the UUU codon has evolved at least five independent times, GUU three, and CCC at least two (Ivanov and Atkins, 2007; and IPI unpublished results).

Why different branches have evolved P-site codons different from the standard/primordial UCC is not entirely clear but two possible explanations have been put forward. One, and the most obvious, is that some of these sequences are almost universally shifty in eukaryotes. Recurring shift sites like UCC (-UGA), UUU (-UGA), GUU (-UGA), and CCC (-UGA) almost certainly fall in that category. Other sites have likely evolved to take advantage of peculiarities of the translational machinery in the organism or groups of organisms that they occur in. One such example is the antizyme frameshift site in *S. cerevisiae* – GCG-UGA. *S. cerevisiae* lacks the $tRNA^{Ala}$ with an anticodon CGC (Sundararajan et al., 1999). As a result the codon GCG is decoded by the near-cognate $tRNA^{Ala}$ with an anticodon $^{3'}CGI^{5'}$ that forms suboptimal pairing making the complex unstable and prone to frameshift. In fact a P-site GCG is part of a +1 shift site of the TY3 retrotransposon in *S. cerevisiae* and is especially shifty if followed by a slow-to-decode codon in the A-site (Vimaladithan and Farabaugh, 1994). In support of the idea that suboptimal pairing on GCG in the ribosomal P-site is key, expression of the $tRNA^{Ala}$ with anticodon CGC suppresses TY3 +1 frameshifting (Farabaugh et al., 1993; Sundararajan et al., 1999). Why the *S. cerevisiae* antizyme has evolved in this direction is puzzling considering that the standard antizyme UCC-UGA frameshift site is perfectly capable of directing frameshifting to the +1 frame in that organism (Matsufuji et al., 1996).

In vivo transfection experiments, with the antizyme 1 frameshift cassette from mammals, that involve mutating the UCC P-site codon to different codons, including codons like AAA that are poor shifters, do not alter significantly the inducing effect of polyamines on the frameshifting. This result strongly suggests that the P-site codon is not part of the polyamine sensor.

The A-site codon in antizyme frameshift sites, invariably a stop codon, shows surprising levels of conservation. Early mutagenesis experiments, in vitro and later in vivo, indicated that all stop codons support efficient frameshifting. UGA is the optimal but UAG and UAA are almost as efficient, ~80% and ~70% of UGA efficiency, respectively, in in vitro experiments and ~90% and 60%, respectively, in in vivo transfection experiments; in both cases the experiments were performed under high polyamine conditions (Matsufuji et al., 1995; Petros et al., 2005). These findings correspond to the known termination efficiency of each of the three termination codons in eukaryotes – UGA being the least efficient, UAG showing intermediate efficiency, and UAA the most efficient termination codon (Cridge et al., 2006). Over 97% of all antizyme frameshift sites have UGA in their A-site. How this modest increase in efficiency, obtained by having a UGA in that position, can provide such strong selective advantage is unclear at the moment.

That the efficiency of termination is inversely related to the efficiency of frameshifting is confirmed by in vitro experiments manipulating the levels of eRF1 and eRF3 proteins, both proteins essential for eukaryotic termination (Karamysheva et al., 2003). Addition of eRF1 or eRF1 in combination with eRF3 decreases the levels of antizyme frameshifting with the two proteins together having synergistic effect.

Experiments to examine the role the different stop codons play in mediating the polyamine induction of frameshifting show that UGA is most responsive, UAG is intermediate, and UAA is least responsive. For spermidine the folds induction are 5.3, 4, and 3.5, respectively. This by itself accounts for the differences seen for total levels of translation with the three stop codons. In other words the length of pausing during decoding the A-site codon is a key determinant in mediating the polyamine induction of frameshifting. Such a conclusion is strengthened by examining the results of substituting the A-site stop codon with any sense codon where polyamine induction is completely lost (Petros et al., 2005). These codons would be decoded at least an order of magnitude faster than even the UAA stop codon. This result strongly suggests that the A-site codon, unlike the P-site codon, is part of the polyamine sensor.

There is at least one definitive case where an antizyme gene has lost the requirement of frameshifting for full-length protein expression – this is the antizyme gene from *Tetrahymena thermophila*. In another protist, *Capsaspora owczarzaki*, frameshifting is also clearly not involved, although other recoding possibilities have not been ruled out. The available antizyme sequences from Zygomycota, a fungal phylum, strongly suggest that the antizyme genes there also do not require frameshifting. All other eukaryotic antizyme genes appear to require +1 frameshifting for functional protein expression.

Antizyme frameshifting efficiency is often enhanced by the presence of *cis*-acting elements, both 5′ and 3′ of the frameshifting site. An overview of the current knowledge about them is presented below.

13.4 5′ Stimulators of Frameshifting

The *cis*-acting elements present upstream of the frameshift sites of antizyme mRNA that enhance the efficiency of antizyme frameshifting are very enigmatic. In mammalian antizyme 1 mRNAs a region of approximate 50 nucleotides immediately upstream of the frameshift site is required for optimal levels of frameshifting.

Extensive phylogenetic analysis reveals that the 5′ region known to be required for optimal frameshifting in mammalian antizyme 1 mRNA (Matsufuji et al., 1996; Ivanov et al., 1998b) can be divided into three sections (or modules) with discrete pattern of conservation (Ivanov and Atkins, 2007) (Fig. 13.2). The modules appear to have emerged in a defined evolutionary sequence with the one proximal to the frameshift site evolving first, the middle second, and the distal last. The proximal module, also known as module A, which is 6 nucleotides long, must have emerged very early in antizyme evolution and is present in antizyme genes from all three eukaryotic kingdoms in which the gene has been identified. The middle module of antizyme 1 genes, also known as module B, is 8 nucleotides long and appears to have been added in early metazoan evolution as it is present in most antizymes from

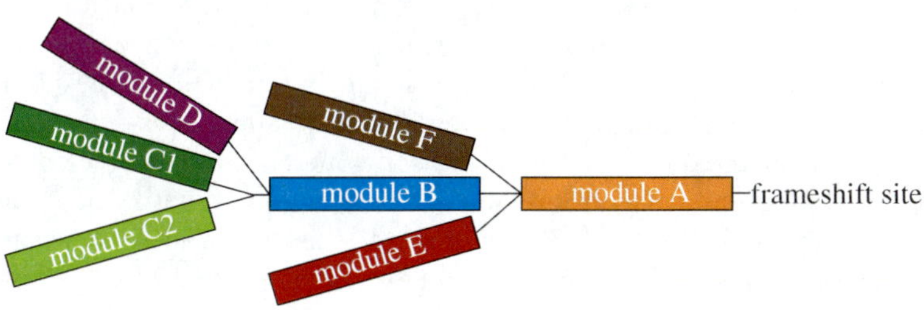

Fig. 13.2 Schematic representation of the modular nature of 5′ stimulatory elements of antizyme genes

animals. The distal module in antizyme 1 mRNA, also known as module C1, is approximately 20 nucleotides long and is specific to the orthologs of this gene, indicating that it emerged in early vertebrate evolution. A homologous distal module, module C2, is present and conserved in orthologs of antizyme 2 genes.

Instead of the middle and distal modules described above, based on strong nucleotide conservation in the corresponding regions, alternative 5′ modules can be predicted in other branches of the antizyme gene family (Ivanov and Atkins, 2007). For example, instead of the distal module C many invertebrate antizyme mRNAs have a highly conserved region in the same position known as module D. Most nematode antizyme mRNAs have dispensed with module B and instead have a different and much larger region of conservation in its place known as module E. The antizymes from the Pezizomycotina subphylum of Ascomycotal fungi have a highly conserved region, known as module F, overlapping the positions of vertebrate modules B and C (Ivanov and Atkins, 2007). Apart from modules A, B, and C1, for which direct experimental evidence exists, it is not clear how many of the other modular regions of high nucleotide conservation can in fact enhance the efficiency of antizyme frameshifting.

A special case of conserved sequence just upstream of the shift site and with a possible frameshifting role, is observed in antizyme mRNAs from Basidiomycota and specifically from mushrooms and close relatives. Alignment of these sequences demonstrates high level of conservation in the region corresponding to the 42 nucleotides just upstream of the shift site. Careful analysis, however, shows that the conservation is in fact at the amino acid level. The peptide encoded by this region has the sequence YYYSTTFSGGP(G/E)WR (Ivanov and Atkins, 2007). Although currently there is absence of experimental data, it is tempting to speculate that as part of the nascent peptide it might enhance the frameshifting efficiency at the adjacent shift site in a manner similar to that of the nascent peptide involved in the translational bypassing during decoding of bacteriophage T4 gene 60 (Weiss et al., 1990; Herr et al., 2001). In that case ribosomes perform the equivalent of +50 (nucleotides) frameshifting.

The mouse antizyme 3 mRNA, which like all orthologs of antizyme 3 lacks all three 5′ modules, has a triplication of the frameshift site resulting in two pseudo-frameshift sites, with the sequence GC-UCC-UGC, preceding a conventional frameshift site GC-UCC-UGA. Because at least one pseudo-frameshift site is conserved in most orthologs of antizyme 3 it appears likely that it has a physiological role. Although this has not been tested it seems conceivable that the pseudo-frameshift site can in fact support residual ribosomal +1 frameshifting that is additive to the frameshifting at the conventional shift site.

The most detailed experiments with the antizyme 5′ stimulatory element have been done with the sequences from mammalian antizyme 1, antizyme 2, and also the gene from the fission yeast *S. pombe*. In vitro translation experiments show that a 5′ deletion of antizyme 1 mRNA up to and excluding the last 17 sense codons of ORF1 (i.e., 5′ modules A, B, and C1) has no discernable effect on frameshifting efficiency. Deletion that removes almost all of ORF1, including modules C1 and B, leaving only the last three sense codons of ORF1 (i.e., module A), leads to 62%

reduction in frameshifting levels compared to wild-type control (Matsufuji et al., 1995). In vivo transfection experiments with antizyme 1 and 2 show that inactivating all three modules leads to ~90% reduction in frameshift efficiency (Howard et al., 2001). In another set of experiments the mammalian antizyme 1 frameshift cassette was expressed in *S. cerevisiae* and *S. pombe* yeast sells and the effects of serial 5′ deletions measured. When the experiment is performed in *S. cerevisiae,* deleting module C1 and B has no effect on frameshifting efficiency relative to a full-length control. Deleting all three modules, however, leads to nearly 95% reduction in frameshift efficiency (Matsufuji et al., 1996). In *S. pombe* deleting the equivalent to module C1 only leads to ~60% reduction in frameshifting efficiency relative to a control that has all three modules intact. Deleting modules C1 and B leads to ~70% reduction of frameshift efficiency. Deleting all three modules, C1, B, and A, leads to more than 90% reduction in frameshift efficiency (Ivanov et al., 1998b). The experiments described above are consistent with a modular nature for the 5′ stimulatory element in mammalian antizymes 1, at least when tested in a heterologous fission yeast system, with the three components having additive effect. In addition they demonstrate that module A can work autonomously of modules B and C1, not an unexpected result considering that many endogenous fungal antizymes have module A by itself. The heterologous experiments also show that modules A and B together can function autonomously (i.e., in the absence) of C1. The observation that modules B and C1 though absent from the endogenous *S. pombe* antizyme can nevertheless enhance +1 frameshifting in that organism suggests that these two modules target a conserved feature of eukaryotic translation.

Additional support for the importance of the 5′ elements for antizyme frameshifting efficiency comes from another heterologous experiment in which the antizyme frameshift site of the mollusc *Crassostrea gigas* is tested in transfection experiments with a mammalian cell line. In that case a deletion that removes modules D and B but leaves module A essentially intact leads to 87% reduction in +1 frameshifting efficiency (Ivanov et al., 2004). Finally, data are also available for the 5′ stimulatory element of the endogenous antizyme gene from *S. pombe*. Serial deletions with that sequence demonstrate that 5′ deletion up to, but excluding, the last three sense codons (i.e., module A) has little or no effect on frameshift efficiency. Deleting the next two codons, which results in inactivation of module A, leads to 75–80% reduction in frameshifting efficiency (Ivanov et al., 2000b).

The pattern of nucleotide conservation of modules A, B, C1, C2, D, E, and F is consistent with the 5′ stimulatory element working at the level of nucleotide sequence and not the peptide encoded by it. This conclusion is also supported by the results of a heterologous experiment in which mammalian antizyme 1 sequence (i.e., modules A, B, and C1) is expressed in *S. pombe* (Ivanov et al., 1998b).

Transfection experiments have probed the role of mammalian antizyme 5′ element in mediating the polyamine frameshift stimulations (Howard et al., 2001; Petros et al., 2005). Deleting the entire 3′ element modules A, B, and C1/C2 leads to significant loss of (spermidine) polyamine stimulation – from ~8-fold to only ~2-fold. This result is consistent with the hypothesis that the 5′ element is an integral part of the polyamine sensor. What has not been tested is whether this is a property associated with all three mammalian modules or if only a subset is required for mediating the polyamine effect (the subset, for example, may be only the near universally conserved module A).

13.5 3′ Stimulators of Frameshifting

The 3′ stimulators of antizyme frameshifting were first identified by mutagenesis experiments. Deleting the region 3′ of the frameshift site in the originally cloned rat antizyme 1 leads to ∼60% reduction of frameshifting efficiency in in vitro reticulocyle lysate translation experiments (Matsufuji et al., 1995). Similar experiments have identified 3′ stimulatory elements in the antizyme mRNA in *S. pombe* in yeast and *C. gigas* in invertebrates (Ivanov et al., 2000b; Ivanov et al., 2004). The vast number of antizyme genes allows for detailed phylogenetic analysis of the 3′ region that has suggested or confirmed numerous stimulators of frameshifting. The 3′ stimulator of frameshifting in *C. gigas* was originally suggested in just such a search.

The first 3′ stimulatory element identified was a RNA pseudoknot structure – now known as class I antizyme pseudoknot (Fig. 13.3A). Site-directed mutagenesis indicated that disrupting either stem of the class I pseudoknot results in frameshifting similar to complete 3′ deletion (Matsufuji et al., 1995). At the same time compensatory mutations in either stem result in frameshifting close to wild-type levels. Phylogenetic analysis indicates that this RNA pseudoknot is present in all vertebrate orthologs of antizyme 1. A slightly different homologous RNA pseudoknot is also present and stimulates frameshifting in the orthologs of vertebrate antizyme 2 (Ivanov et al., 1998a). Since tunicate, cephalochordate, and invertebrate antizyme mRNAs show no trace of this RNA pseudoknot it appears that it emerged in early vertebrate evolution. The class I pseudoknots in both antizyme 1 and 2 are contained within a region that extends approximately 60 nucleotides downstream of the ORF1 stop codon. Stem 1 of the pseudoknot, which starts 2–3 nucleotides downstream

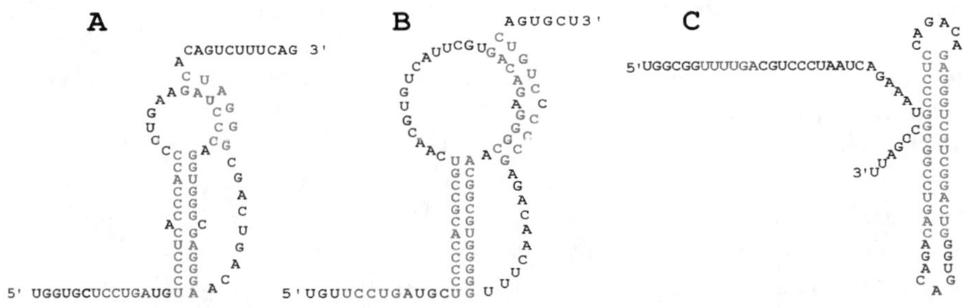

Fig. 13.3 Secondary structure of some known and putative RNA structures required for optimal antizyme frameshifting. Base pairing nucleotides are shown in *green*. The frameshift site is shown in *magenta*. (**A**) Class I RNA pseudoknot from vertebrates is represented by the sequence from mouse antizyme 1. (**B**) Class I RNA pseudoknot from invertebrates is represented by the sequence from oyster antizyme. (**C**) The highly conserved RNA structure of antizymes from Basidiomycota. The putative structured loop 2 is shown in *blue*

of the stop codon, is 10–12 base pairs long and in all but one case is disrupted in the middle by a mismatch, usually A-C. Curiously, despite the conservation of the mismatch, altering it to a G-C base pair, if anything, increases the efficiency of frameshifting (Matsufuji et al., 1995, 1996). Stem 2 of the pseudoknot is always 6 nucleotides long and is separated by a 1 nucleotide hinge in the antizyme 1 pseudoknots or is apparently directly stacked on top of stem 1 in antizyme 2 pseudoknots. The role of the hinge is unclear because deleting it has little effect on frameshift efficiency (Matsufuji et al., 1995). The loops of class I pseudoknots are relatively short – 6–9 nucleotides – and sequences are variable indicating that they have a limited functional role.

Deleting the pseudoknot region in antizyme 1 reduces frameshifting efficiency 61% in rabbit reticulocyte lysate and ~40% in in vivo transfection experiments (Matsufuji et al., 1995; Howard et al., 2001). Just like the 5′ element from mammalian antizyme 1 mRNA, the 3′ pseudoknot has also been tested in *S. cerevisiae* and *S. pombe*, two organisms whose endogenous antizyme mRNAs lack an RNA pseudoknot stimulatory structure. Surprisingly, the antizyme 1 pseudoknot is not only functional in the two yeasts, but in fact frameshifting with the exogenous shift cassette is more dependent on it than it is in rabbit reticulocyte lysate or in mammalian cells in vivo. When the RNA is deleted frameshifting is reduced 84% in *S. pombe* and 97% in *S. cerevisiae* (Ivanov et al., 1998b; Matsufuji et al., 1996).

Vertebrate antizyme 1 pseudoknot is able to stimulate frameshifting to the +1 frame, through both +1 and −2 shift, when an antizyme 1 frameshift cassette is tested in the yeasts *S. cerevisiae* and *S. pombe* even though the endogenous antizymes in these organisms lack an RNA pseudoknot, suggesting that the pseudoknot affects a fundamental aspect of ribosomal translation (Matsufuji et al., 1996; Ivanov et al., 1998b). In *S. cerevisiae* the frameshifting is predominately −2 in the wild-type context. The spacer length between the stop codon and the beginning of the pseudoknot is important for the direction of the frameshift. In the wild-type context the ratio of −2 to +1 frameshifting is approximately 10:1. If the length of the spacer is increased by 3 nucleotides over-all frameshifting to the +1 frame is reduced 30%. However, the proportion of the +1 shift increases, and the ratio of −2 to +1 frameshifting becomes 1:1.

The importance of class I pseudoknots for polyamine stimulation of frameshifting has been studied both in vitro and in vivo (Matsufuji et al., 1995; Howard et al., 2001; Petros et al., 2005). In both cases deleting the pseudoknot by itself has little effect on the ability of polyamines to stimulate the frameshifting, indicating that it is not part of the polyamine sensor. However, these experiments also show that combining a pseudoknot deletion with a 5′ stimulator deletion leads to complete loss of (spermidine) polyamine stimulation compared to 2-fold stimulation with the 5′ deletion alone (Petros et al., 2005). Therefore, even though class I antizyme RNA pseudoknot does not appear to be essential for the polyamine stimulation, it may facilitate the stimulation mediated through the 5′ element.

An entirely different, and apparently older, RNA pseudoknot stimulates frameshifting in antizyme genes from many invertebrates. This pseudoknot, known as class II antizyme pseudoknot (Fig. 13.38), is present in no less than six animal phyla – Annelida, Arthropoda, Mollusca, Nematoda, Platyhelminthes, and Onychophora, attesting to its great antiquity (Ivanov and Atkins 2007). Class II antizyme pseudoknot exists in two forms called class IIa and class IIb. Class IIa is the ancestral form and is the one better studied. This pseudoknot encompasses a larger region than class I antizyme RNA pseudoknot. It extends as much as 90 nucleotides downstream of the stop codon. The expansion is mostly due to the enlarged size of loop 1 which is also highly variable in length – ranging from 10 to over 35 nucleotides. Stem 1 of class II pseudoknots is 10–12 base pairs long, similar to the size of stem 1 in class I pseudoknots, or approximately one helical turn of double-stranded RNA. Unlike class I pseudoknot, its base pairing is uninterrupted by mismatches or bulges. However, in a case of unusual symmetry, stem 2 of class IIa, but not class I, pseudoknot often is interrupted by a mismatch. Just like the bulge of class I pseudoknots changing the mismatch to a base pair leads to even greater stimulation of +1 frameshifting (Ivanov et al., 2004). Class IIa pseudoknots share with class I pseudoknot from antizyme 1 orthologs the presence of a hinge nucleotide in the junction between stem 1 and 2. This hinge is even larger in class IIb pseudoknots where it is 4 nucleotides. Unlike the hinge in class I pseudoknots, however, mutating the hinge of class IIa pseudoknots leads to severe reduction of frameshifting efficiency (Ivanov et al., 2004).

Although class I and II antizyme RNA pseudoknots share many superficial characteristics they may stimulate frameshifting by distinct mechanisms. The biggest known functional distinction between the two is in the way they respond to nucleotide sequence alteration of stem 1. While changing the sequence of stem 1 in class I pseudoknot seems unimportant, so long as base pairing is maintained (Matsufuji et al., 1995, 1996), changes in the stem 1 nucleotide sequence of class II pseudoknot from *C. gigas* result in partial loss of frameshift stimulation even when compensatory restoring base pairing are made (Ivanov et al., 2004).

It appears that some antizyme mRNAs with the ancestral class II pseudoknot have dispensed with the requirement that the 3′ structure is a pseudoknot and instead have only conserved a simple stem-loop structure corresponding to stem 1. This is especially puzzling since in experiments with *C. gigas* pseudoknot disruption of stem 2 has the same effect as deleting the entire structure (Ivanov et al., 2004).

Although a class II pseudoknot was present in the common ancestors of both nematodes and arthropods, many extant antizymes belonging to these phyla have lost the stimulator. Curiously, as discussed below, some of the arthropod antizymes that have lost class II pseudoknot have potentially evolved alternative downstream stimulators.

Experiments with the mRNA of *S. pombe* antizyme identified a region extending up to 150 nucleotides downstream of the stop codon that enhances +1 frameshift efficiency up to 10-fold. The same region is highly conserved in the distantly related *S. octosporus* and *S. japonicus*. Unlike the 3′ stimulators so far discussed, the one in *S. pombe* does not appear to be an identifiable conventional RNA structure. As such the nature of that element is currently a mystery.

In addition to the downstream elements described above supported by direct experimental evidence, there are several others for which strong circumstantial evidence exists from comparative phylogenetic analysis. The foremost among these is an RNA secondary structure downstream of most Basidiomycotal antizyme frameshift sites (Fig. 13.3c). Analysis of the 3′ regions from these antizyme genes identifies two highly conserved stem loops that occur without a spacer in between, suggesting that they coaxially stack on each other. The base pairing of both stems is well supported by compensatory co-variations (Ivanov and Atkins, 2007; and IPI unpublished data). In addition there is an indication that the loop of the second stem is highly structured and possibly forms a triloop. Similar triloops have been observed in a number of RNA–RNA ternary interactions (Mitrasinovic, 2006). Surprisingly, the first stem loop starts 17–20 nucleotides downstream of the ORF1 stop codon. By contrast the two RNA pseudoknot structures described above start 2–5 nucleotides downstream of the stop codon. This means that the putative structure would be just outside the 80S ribosomes when the stop codon of ORF1 is being decoded.

Another putative downstream stimulator is present in antizyme mRNAs from the Pezizomycotina subphylum of Ascomycotal fungi. This includes the antizyme genes from *Neurospora crassa*, *Aspergillus nidulans*, *Gibberella zeae*, and others (Ivanov and Atkins, 2007). One feature of this putative element is a downstream stem-loop structure which is well supported by compensatory co-variations and starts 31–41 nucleotides downstream from the stop codon of ORF1. The stem is usually 10 base pairs, or approximately one helical turn, but could be as short as 4 and as long as 14. The loop region could be surprisingly long and varies in length from 18 to 130 nucleotides. A second and much more unusual feature is a highly conserved nucleotide sequence, GGAAGARUGUGAGAGRUCUUUYUGYGA, starting exactly 16 nucleotides downstream from the end of the stem loop already described. Despite its high conservation this RNA sequence does not appear to fold into a secondary structure or belong to a known RNA functional motif. Since both the stem loop and the conserved primary sequence would be well outside a ribosome decoding the stop codon of ORF1 it is not entirely clear how this region might affect the frameshifting though long distance downstream stimulatory elements are known.

The identity of the nucleotide just 3′ of a stop codon is known to influence the efficiency of translation termination in both bacteria and eukaryotes (Tate et al., 1996). In most antizyme genes the nucleotide least likely to promote efficient termination is present at that position (Ivanov and Atkins, 2007). For example, in *S. cerevisiae* the nucleotide following a stop codon that is least likely to support efficient termination is "C." This is also the nucleotide present immediately following the stop codon of antizyme ORF1 from that organism. "C" is also the least efficient 3′ context in the nematode *C. elegans* and once again it is the nucleotide present following the stop codon of it antizyme ORF1.

Another 3′ element that is well conserved and likely to play a role in stimulating antizyme frameshifting is a pyrimidine-rich sequence with the consensus UCCCU starting 3 nucleotides downstream of the stop codon of ORF1. This sequence is present in antizyme mRNA from both fungi and animals. Curiously, it appears to have emerged, and to have been subsequently conserved, more than once in the evolution of antizyme genes. Work in yeast has shown that the 6 nucleotides following a stop codon can reduce the efficiency of translation termination (Skuzeski et al., 1991; Namy et al., 2001). It seems likely the pyrimidine-rich sequence following the stop codon of ORF1 of antizyme mRNA has a similar effect; however, that hypothesis has not been tested.

13.6 Mechanism of Frameshifting

Experiments on frameshift sites of antizyme genes combined with analysis of extensive phylogenetic data allow us to build a tentative model for the mechanism of the frameshifting. In all cases where it is required for antizyme mRNA expression, frameshifting occurs on the last two codons of ORF1. As mentioned above, during the frameshift the penultimate codon of ORF1 is in the P-site of the ribosome while the A-site is occupied by the stop codon of ORF1. This broadly speaking puts antizyme frameshifting in the category of "P-site" recoding events. These events have two defining features. One is that the P-site is occupied by a tRNA whose interaction with the mRNA counterpart codon is either suboptimal or in some way causes perturbation in the ribosome interfering with standard translation. Obligatory slow decoding of the codon in the A-site is the second characteristic feature of this type of recoding event.

There is direct evidence that at least in budding yeast, suboptimal base pairing in the P-site is important for the antizyme frameshifting. As discussed above the gene encoding the cognate $tRNA^{Ala}$ that decodes P-site codon in the *S. cerevisiae* antizyme frameshift site, GCG, is missing from the genome and the codon is instead decoded by a near-cognate tRNA. Expression of cognate tRNA for the CGC codon inhibits +1 frameshifting in TY3 which also has CGC P-site codon (Sundararajan et al., 1999).

Over the years two alternative hypotheses have been put forward to explain the events taking place in the P-site of such recoding events. One is that the interaction between the P-site codon and the P-site tRNA is less than perfect with the wobble position either weak or nonexistent. This suboptimal base pairing leads to dissociation between tRNA and mRNA at significantly higher than normal frequency. Initially it was also believed that for subsequent re-pairing to mRNA at a triplet in another frame, the new tRNA–mRNA interaction should have higher base pairing potential. Although this indeed seems to be the case in most examples of tRNA slippage in the P-site, there are cases where re-paring with a new codon is recoded with a single Watson–Crick base pair or no base pairing at all (Hansen et al., 2003; Herr et al., 2004). The alternative hypothesis postulates that a specific feature of the wobble juxtaposition in the P-site results in occlusion of the immediately downstream nucleotide from the A-site so that the next available triplet in the ribosomal A-site consists of the 3 nucleotides just downstream of it (Farabaugh et al., 1993).

This hypothesis is incompatible with P-site slippage. The primordial antizyme mRNA zero frame P-site codon, UCC, is usually decoded by the $tRNA^{Ser}$ with anticodon $^{3'}$AGI$^{5'}$. A +1 tRNA slip would result in A-C mismatch in the first position, G-C Watson–Crick base pair in the second, and I-U wobble base pair in the third. Direct evidence that slippage can occur on the UCC codon in the P-site comes from heterologous experiments where the mammalian antizyme 1 frameshift cassette is tested in yeast – *S. cerevisiae* and also *S. pombe* (Matsufuji et al., 1996; Ivanov et al., 1998b). Sequencing the transframe peptides unequivocally shows that in the two yeasts both +1 and −2 frameshifting can occur. The −2 shift can be explained only by P-site slippage resulting in A-G mismatch in the first position, G-C Watson–Crick base pair in the second, and I-U wobble base pair in the third – or nearly identical to the base pairing potential of a +1 slippage. Also, it seems unlikely that the same shift site can support P-site events through different mechanisms. Therefore, it appears that at least in that context, the +1 frameshift occurs through P-site slippage.

Like all other P-site recoding events it is essential that the A-site of antizyme mRNA frameshift sites is a slow-to-decode codon. In almost all naturally occurring antizyme shift sites this is the UGA stop codon – the slowest to decode codon in eukaryotes. In the few exceptions it is one of the other two stop codons – UAG or UAA. Alterations substituting the stop codon with any sense severely reduce antizyme frameshifting. The inefficiency of decoding of the A-site stop codon is further enhanced by its 3′ context – both the immediately downstream nucleotide and perhaps the pyrimidine-rich sequence UCCCU nearby.

How other *cis*-acting elements stimulate the frameshifting is less well understood. At least the proximal A-module of the 5′ stimulatory elements is expected to be in the ribosome during frameshifting. This strongly suggests that whatever its function, module A almost certainly interacts physically with the ribosome in one way or another – i.e., either through mRNA–rRNA or mRNA–ribosomal protein contacts. It is currently not known if module A mediates the activities of the other 5′ stimulatory modules or if the latter can function autonomously.

RNA stem loops and pseudoknots are often present as stimulators of −1 frameshifting and translational readthrough events. In those cases they are believed to cause a translational pause while the ribosome is decoding the frameshift site. Such a pause is often dependent on the thermodynamic stability of the RNA structure or its resistance to the activity of ribosomal helicases. In an alternative mode some stimulatory stem loops and pseudoknots are proposed to work as springs that push the ribosome sitting over the frameshift site in a −1 direction. The mechanism of action of class I and II antizyme RNA pseudoknots is currently unknown. There are two key findings that might provide revealing insight into their function. In both class I and II pseudoknots the start of stem 1 relative to the shift site appears fixed. In class I pseudoknots it is 2–3 nucleotides downstream of the stop codon of ORF1. In class II pseudoknots it is 3–4 nucleotides downstream of the stop codon. Increasing the distance between the stop codon and the beginning of the pseudoknot decreases its effectiveness. The second finding comes from heterologous expression of mammalian antizyme frameshift cassette in *S. cerevisiae*. As discussed above this frameshift cassette supports both +1 and −2 frameshifting with a ratio of approximately 1:10. Inserting an additional 3 nucleotides between the stop codon and the RNA pseudoknot shifts the ratio of +1 and −2 frameshifting to 1:1. This latter observation is consistent with the pseudoknot working in a manner similar to a pulling or a pushing spring.

A role of polyamines in translation has been known for more than 30 years (Atkins et al., 1975). Not much, however, is known about the mechanism of polyamines stimulation of antizyme frameshifting. The current data suggest that the 3′ RNA pseudoknot is not part of the polyamine sensor. The same is also true for the P-site codon. Another set of experiments suggest that the stop codon of ORF1 is perhaps a key component of the sensor with the 5′ element or at least one of its modules, playing a facilitating role. It is tempting to speculate that module A, the part most likely to be entirely within the ribosome, and also the only 5′ module conserved from yeast to mammals, plays a crucial role in this process.

Acknowledgments We thank John F. Atkins for writing the introductory section. Salary support for I.P.I. was derived from funds from Science Foundation Ireland.

References

Atkins JF, Lewis JB, Anderson CW, Gesteland RF (1975) J Biol Chem 250:5688–5695
Auvinen M, Paasinen A, Andersson LC, Hölttä E (1992) Nature 360:355–358
Balasegaram M, Balasegaram S, Malvy D, Millet P (2008) PLoS Negl Trop Dis 2:e234
Bekaert M, Atkins JF, Baranov PV (2006) Bioinformatics 22:2463–2465
Blume-Peytavi U, Hahn S (2008) Dermatol Ther 21:329–339
Burns MR, Graminski GF, Weeks RS, Chen, O'Brien TG (2009) J Med Chem 52:1983–1993
Coffino P (2001) Nat Rev Mol Cell Biol 2:188–194
Cohen S (1998) A Guide to the Polyamines. Oxford University Press, New York, 559pp
Cridge AG, Major LL, Mahagaonkar AA, Poole ES, Isaksson LA, Tate WP (2006) Nucleic Acids Res 34:1959–1973
Dakin HD (1906) J Biol Chem 1:171–176
Fairlamb AH, Blackburn P, Ulrich P, Chait BT, Cerami A (1985) Science 227:1485–1487
Farabaugh PJ, Zhao H, Vimaladithan A (1993) Cell 74:93–103
Fitzgerald C, Sikora C, Lawson V, Dong K, Cheng M, Oko R, van der Hoorn FA J (2006) J Biol Chem 281:38172–80
Fong WF, Heller JS, Canellakis ES (1976) Biochim Biophys Acta 428:456–465
Fujita K, Murakami Y, Hayashi S (1982) Biochem J 204:647–652
Gandre S, Bercovich Z, Kahana C (2003) Mitochondrion 2:245–256
Gerner EW, Meyskens FL (2009) Clin Cancer Res 15:758–761
Gruendler C, Lin Y, Farley J, Wang T (2001) J Biol Chem 276:46533–46543
Hansen TM, Baranov PV, Ivanov IP, Gesteland RF, Atkins JF (2003) EMBO Rep 4:499–504
Heller JS, Fong WF, Canellakis ES (1976) Proc Natl Acad Sci USA 73:1858–1862
Herr AJ, Nelson CC, Wills NM, Gesteland RF, Atkins JF (2001) J Mol Biol 309:1029–1048
Herr AJ, Wills NM, Nelson CC, Gesteland RF, Atkins JF (2004) J Biol Chem 279:11081–11087
Hogarty MD, Norris MD, Davis K, Liu X, Evageliou NF, Hayes CS, Pawel B, Guo R, Zhao H, Sekyere E, Keating J, Thomas W, Cheng NC, Murray J, Smith J, Sutton R, Venn N, London WB, Buxton A, Gilmour SK, Marshall GM, Haber M (2008) Cancer Res 68:9735–9745
Howard MT, Shirts BH, Zhou J, Carlson CL, Matsufuji S, Gesteland RF, Weeks RS, Atkins JF (2001) Genes Cells 6:931–941

Huang Y, Higginson DS, Hester L, Park MH, Snyder SH (2007) Proc Natl Acad Sci USA 104:4194–4199
Ike A, Ohta H, Onishi M, Iguchi N, Nishimune Y, Nozaki M (2004) FEBS Lett 559:159–164
Ivanov IP, Gesteland RF, Atkins JF (1998a) Genomics 52:119–129
Ivanov IP, Gesteland RF, Matsufuji S, Atkins JF (1998b) RNA 4:1230–1238
Ivanov IP, Rohrwasser A, Terreros DA, Gesteland RF, Atkins JF (2000a) Proc Natl Acad Sci USA 97:4808–4813
Ivanov IP, Matsufuji S, Murakami Y, Gesteland RF, Atkins JF (2000b) EMBO J 19:1907–1917
Ivanov IP, Anderson CB, Gesteland RF, Atkins JF (2004) J Mol Biol 339:495–504
Ivanov IP, Gesteland RF, Atkins JF (2006) RNA 12:332–337
Ivanov IP, Atkins JF (2007) Nucleic Acids Res 35:(1842)–1858
Ivanov IP, Pittman AJ, Chien CB, Gesteland RF, Atkins JF (2007) Gene 387:87–92
Ivanov IP, Loughran G, Atkins JF (2008) Proc Natl Acad Sci USA 105:10079–10084
Kajiwara K, Nagawawa H, Shimizu-Nishikawa S, Ookuri T, Kimura M, Sugaya E (1996) Biochem Biophys Res Commun 219:795–799
Kanerva K, Lappalainen J, Mäkitie LT, Virolainen S, Kovanen PT, Andersson LC (2009) PLoS One. 4.e6858
Karamysheva ZN, Karamyshev AL, Ito K, Yokogawa T, Nishikawa K, Nakamura Y, Matsufuji S (2003) Nucleic Acids Res 31:5949–5956
Lim SK, Gopalan G (2007) Oncogene 26:6593–603
López-Contreras AJ, López-Garcia C, Jiménez-Cervantes C, Cremades A, Peñafiel R (2006) J Biol Chem 281:30896–30906
López-Contreras AJ, Sánchez-Laorden BL, Ramos-Molina B, de la Morena ME, Cremades A, Peñafiel R (2009) J Cell Biochem 107:732–740
Mäkitie LT, Kanerva K, Sankila A, Andersson LC (2009) Histochem Cell Biol [Epub ahead of print]
Mangold U, Hayakawa H, Coughlin M, Münger K, Zetter BR (2008) Oncogene 27:604–613
Matsufuji S, Miyazaki Y, Kanamoto R, Kameji T, Murakami Y, Baby TG, Fujita K, Ohno T, Hayashi S (1990) J Biochem 108:365–371
Matsufuji S, Matsufuji T, Miyazaki Y, Murakami Y, Atkins JF, Gesteland RF, Hayashi S (1995) Cell 80:51–60
Matsufuji S, Matsufuji T, Wills NM, Gesteland RF, Atkins JF (1996) EMBO J 1996 15:1360–1370
Mitchell JL, Judd GG, Bareyal-Leyser A, Ling SY (1994) Biochem J 299 (Pt 1):19–22
Mitchell JL, Judd GG (1998) Biochem Soc Trans 26:591–595
Mitrasinovic PM (2006) J Struct Biol 153:207–222
Miyazaki Y, Matsufuji S, Hayashi S (1992) Gene 113:191–197
Murai N, Shimizu A, Murakami, Y, Matsufuji S (2009) J Cell Biochem [Epub ahead of print]
Murakami Y, Matsufuji S, Kameji T, Hayashi S, Igarashi K, Tamura T, Tanaka K, Ichihara A (1992) Nature 360:597–599
Murakami Y, Matsufuji S, Miyazaki Y, Hayashi S (1994) Biochem J 304(Pt 1):183–187
Murakami Y, Ichiba T, Matsufuji S, Hayashi S (1996) J Biol Chem 271:3340–3342
Murakami Y, Suzuki J, Samejima K, Kikuchi K, Hascilowicz T, Murai N, Matsufuji S, Oka T (2009) Exp Cell Res 315:2301–2311
Namy O, Hatin I, Rousset JP (2001) EMBO Rep 2:787–793
Namy O, Galopier A, Martini C, Matsufuji S, Fabret C, Rousset JP (2008) Nat Cell Biol 10: 1069–1075
Newman RM, Mobascher A, Mangold U, Koike C, Diah S, Schmidt M, Finley D, Zetter BR (2004) J Biol Chem 279, 41504–511
Palanimurugan R, Scheel H, Hofmann K, Dohmen RJ (2004) EMBO J 23:4857–4867
Park MH, Lee YB, Joe YA (1997) Biol Signals 6:115–123. Review
Petros LM, Howard MT, Gesteland RF, Atkins JF (1995) Biochem Biophys Res Commun 338:1478–1489

Porat Z, Landau G, Bercovich Z, Krutauz D, Glickman M, Kahana C (2008) J Biol Chem 283:4528–4534
Poulin R, Coward JK, Lakanen JR, Pegg AE (1993) J Biol Chem 268:4690–4698
Rosenheim O (1924) Biochem J 18:1253–1262
Saini P, Elyer DE, Green R, Dever TE (2009) Nature 459:118–121
Saito T, Hascilowicz T, Ohkido I, Kikuchi Y, Okamoto H, Hayashi S, Murakami Y, Matsufuji S (2000) Biochem J 345:99–106
Sakata K, Kashiwagi K, Igarashi K (2000) Biochem J 347(Pt 1):297–303
Sanderson L, Dogruel M, Rodgers J, Bradley B, Thomas SA (2008) J Neurochem 107:1136–1146
Schnier J, Schwelberger HG, Smit-McBride Z, Kang HA, Hershey JW (1991) Mol Cell Biol 11:3105–3114
Seiler N, Raul F (2005) J Cell Mol Med 9:623–642
Simoneau AR, Gerner EW, Nagle R, Ziogas A, Fujikawa-Brooks S, Yerushalmi H, Ahlering TE, Lieberman R, McLaren CE, Anton-Culver H, Meyskens FL (2008) Cancer Epidemiol Biomarkers Prev 17:292–299
Skuzeski JM, Nichols LM, Gesteland RF, Atkins JF (1991) J Mol Biol 218:365–373
Snapir Z, Keren-Paz A, Bercovich Z, Kahana C (2008) Biochem J 410:613–619
Sundararajan A, Michaud WA, Qian Q, Stahl G, Farabaugh PJ (1999) Mol Cell 4:1005–1015
Suzuki T, He Y, Kashiwagi K, Murakami Y, Hayashi S, Igarashi K (1994) Proc Natl Acad Sci USA 91:8930–8934
Takeuchi J, Chen H, Hoyt MA, Coffino P (2008) Biochem J 410:401–407
Tang H, Ariki K, Ohkido M, Murakami Y, Matsufuji S, Li Z, Yamamura K (2009) Genes Cells 14:79–87
Tate WP, Poole ES, Dalphin ME, Major LL, Crawford DJ, Mannering SA (1996) Biochimie 78:945–952
Tosaka Y, Tanaka H, Yano Y, Masai K, Nozaki M, Yomogida K, Otani S, Nojima H, Nishimune Y (2000) Genes Cells 5:265–276
Vimaladithan A, Farabaugh PJ (1994) Mol Cell Biol 14:8107–8116
Wang X, Levic S, Gratton MA, Doyle KJ, Yamoah EN, Pegg AE (2009) J Biol Chem 284:930–937
Weiss RB, Dunn DM, Atkins JF, Gesteland RF (1987) Cold Spring Harb Symp Quant Biol 52: 687–693
Weiss RB, Huang WM, Dunn DM (1990) Cell 62:117–126
Yamaguchi Y, Takatsuka Y, Matsufuji S, Murakami Y, Kamio Y (2006) J Biol Chem 281: 3995–4001
Zhang J, Wang Y, Zhou Y, Cao Z, Huang P, Lu B (2005) FEBS Lett 579:559–66
Zhu C, Lang DW, Coffino P (1999) J Biol Chem 274:26425–26430

Chapter 14
Sequences Promoting Recoding Are Singular Genomic Elements

Pavel V. Baranov and Olga Gurvich

Abstract The distribution of sequences which induce non-standard decoding, especially of shift-prone sequences, is very unusual. On one hand, since they can disrupt standard genetic readout, they are avoided within the coding regions of most genes. On the other hand, they play important regulatory roles for the expression of those genes where they do occur. As a result, they are preserved among homologs and exhibit deep phylogenetic conservation. The combination of these two constraints results in a characteristic distribution of recoding sequences across genomes: they are highly conserved at specific locations while they are very rare in other locations. We term such sequences singular genomic elements to signify their rare occurrence and biological importance.

Contents

In this chapter we describe the distribution of several known shift-prone patterns and stimulatory signals and how they relate to the concept of singular genomic elements. We also discuss how their characteristic distribution can be utilized for identification of novel recoded genes and describe studies where such strategies have been employed.

P.V. Baranov (✉)
Biochemistry Department, University College Cork, Ireland
e-mail: brave.oval.pan@gmail.com

J.F. Atkins, R.F. Gesteland (eds.), *Recoding: Expansion of Decoding Rules Enriches Gene Expression*, Nucleic Acids and Molecular Biology 24,
DOI 10.1007/978-0-387-89382-2_14,

14.1 Singular Genomic Elements

A characteristic property of all biological systems is diversity and specialization of their component parts that play distinctive functional roles. The tendency for specialization and uniqueness is profound on the genomic level, as the existence of identical multiple copies of the same gene (unless they relate to mobile elements) is rare. Functional specialization of gene products demands similarly specific regulation of their biosynthesis and processing. Such specificity can be achieved through a combinatorial effect of several regulatory mechanisms acting on different levels of gene expression – from initiation of transcription to posttranslational modifications, where similar regulatory sequences occur in groups of functionally related genes. Specificity is gained through differential combination of these sequences which could be idiosyncratic for a particular gene. However, it is attractive to imagine a simpler scheme where a unique regulatory element would be responsible for the regulation of a specific gene. Such a sequence could respond to changes in particular cellular conditions associated with expression of the regulated gene and so provide feedback control. Indeed such regulatory elements are known and they are characteristically distributed across genomes. Their occurrence at random locations in a single genome is avoided while their occurrence at specific genomic locations across several species is preserved. Such a distribution is easy to explain. Suppose we have a genomic feature F that specifically regulates expression of a gene G. The feature F then should be avoided in all locations where it may have an effect on expression of genes other than G. On the other hand, since association of the feature F with the gene G is beneficial for the organism, such association would likely be preserved during speciation and therefore it will occur in orthologs of the gene G. In other words these regulatory sequence elements are avoided and hence underrepresented across a single genome or across particular types of sequences (e.g., those coding for proteins) in a single genome. However, they are present in orthologous genes from multiple related organisms. Here we introduce the term *singular genomic element* to denote such elements. There are a number of different biologically active nucleotide sequences that exhibit properties of singular genomic elements. Examples are unique sites of restriction, sites encoding unique protease cleavage sites, cases of transcriptional slippage discussed in Chapter 19, or even miRNA targets (Farh et al., 2005). Nonetheless, perhaps, the most striking type of known singular genomic elements is sequences promoting recoding events. Such elements interfere with standard genetic decoding and increase the chances of erroneous translation. Thus, their occurrence in the protein coding sequence of most genes is detrimental. At the same time they do play important roles in those genes that utilize non-standard decoding in their expression and consequently undergo purifying selection during evolution in their corresponding locations. In this chapter we discuss different examples of sequences implicated in recoding events and their distribution across different regions of single genomes and among orthologous genes. We also discuss how searches for singular genomic elements could be used as a strategy for identification of new cases of recoding events and novel genes that are expressed via recoding mechanisms.

14.2 Sequences Promoting Ribosomal Frameshifting as Singular Genomic Elements

14.2.1 +1 Frameshifting Cassette in Bacterial Release Factor 2 mRNA

The *Escherichia coli* gene *prfB* encodes release factor 2 (RF2) and was among the first discovered chromosomal genes requiring programmed ribosomal frameshifting for their expression (Craigen et al., 1985; Craigen and Caskey, 1986). In bacteria, two class-I release factors are responsible for recognition of codons specifying termination of translation, RF1 and RF2 (reviewed Kisselev and Buckingham, 2000). These factors are semi-specific, they both recognize UAA stop codons. UAG is recognized exclusively by RF1, while UGA recognition is specific to RF2 (Scolnick et al., 1968; Capecchi and Klein, 1970). In *E. coli* and most (∼87%) other bacteria, RF2 is encoded in two overlapping ORFs (Kisselev and Buckingham, 2000; Bekaert et al., 2006). While the main portion of RF2 protein is encoded in the second long ORF (see Fig. 14.1), this ORF does not have its own translation initiation site. Initiation of translation takes place at the start of first short ORF whereas the second ORF can be translated only if elongating ribosomes shift reading frame in the +1 direction at the end of the first ORF. The nucleotide sequence and its conservation across RF2 genes from multiple bacterial species are illustrated in Fig. 14.2. The shift cassette consists of a number of modular elements that are responsible for the stimulatory effects on frameshifting. The ribosomal frameshifting itself takes place at a CUU codon followed by U, i.e., CUU_U (where the underlined space separates codons). When the CUU codon is located at the P-site, $tRNA^{Leu}$ repositions relative to mRNA by shifting 1 nucleotide toward the 3′ -end (+1 frameshift) so that its anticodon forms base pairs with the overlapping UUU codon. Consequently the new frame now corresponds to C_UUU (Curran, 1993). As can be seen from Fig. 14.2, this sequence is nearly universally conserved with the exception of the first nucleotide C, where in some bacteria it is U. In those bacteria, the repositioning tRNA is likely to be $tRNA^{Phe}$ which shifts from one Phe codon to another so that transition of the reading frame occurs from UUU_U to U_UUU.

The second modular element that exhibits similarly astonishing conservation is the stop codon that overlaps the frameshift site (Fig. 14.2). The stop codon is nearly

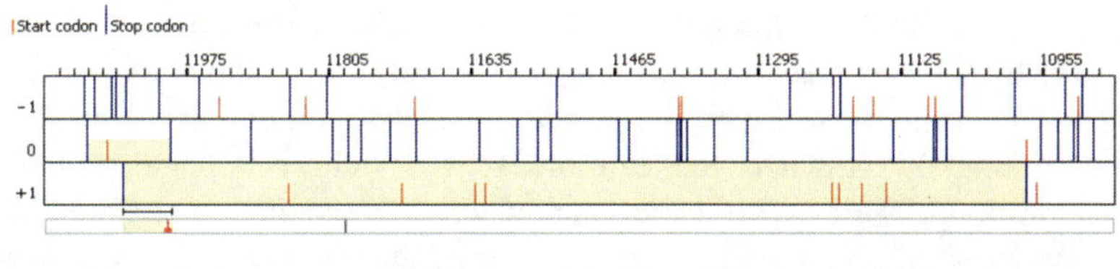

Fig. 14.1 FSfinder2 (Moon et al., 2007) plot of ORF organization in the *E. coli* gene encoding release factor 2 (RF2). The coding sequence is highlighted in *pale yellow* and consists of a first short ORF in the "0" frame and a long second ORF in the "+1" frame

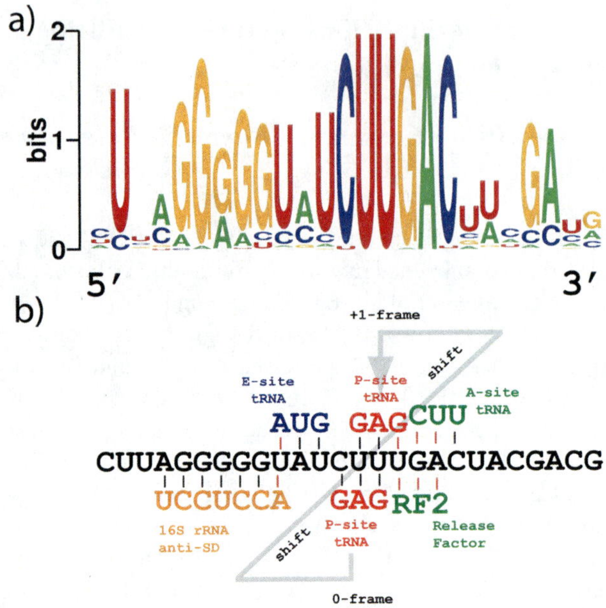

Fig. 14.2 (**a**) Diagram of RF2 frameshift site conservation, the height of symbols indicates conservation of nucleotides, while their weight shows the relative frequency of nucleotides at corresponding positions. The diagram was built using WebLogo3 (Crooks et al., 2004), sequences of RF2 frameshift sites were obtained using ARFA (Bekaert et al., 2006). (**b**) Sequence of *E. coli* K12 RF2 frameshift site aligned to the diagram above. Interactions with different ligands which play roles in the ribosomal frameshifting are indicated with vertical strokes, red strokes correspond to competing interactions

always UGA (with few exceptions where it is UAA). This stop codon is the key element responsible for sensitivity of frameshifting efficiency to the cellular concentration of RF2. When ribosomes approach the end of the first ORF and the stop codon occupies the ribosomal A-site, either of two major events occur: termination of translation or +1 slippage of P-site tRNA which directs translation to the longer ORF. These two events are in competition, so that increasing termination efficiency results in decreasing frameshifting efficiency and vice versa. As termination efficiency is directly influenced by the concentration of release factors, frameshifting efficiency also depends on the concentration of release factors. Since UGA is not recognized by RF1, frameshifting efficiency is solely dependent on the concentration of RF2. This mechanism creates an elegant regulatory feedback loop, as illustrated in Fig. 14.3a, where the level of RF2 biosynthesis depends on the cellular concentration of RF2. With those cases where the stop codon at the frameshift site is UAA, it is likely that frameshifting senses the cumulative concentration of both factors, as illustrated in Fig. 14.3b. Such indiscriminate sensing of the concentration of both release factors could be beneficial as well, since in this case

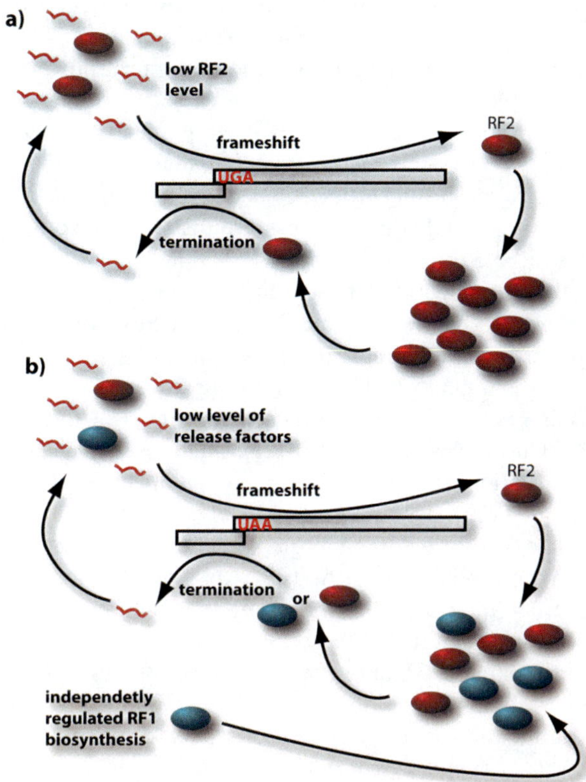

Fig. 14.3 Regulatory feedback provided for RF2 biosynthesis by the frameshifting mechanism. (**a**) The first ORF has a UGA stop codon. The regulation is autonomous and the level of RF2 biosynthesis depends on its own concentration. (**b**) The first ORF has a UAA stop codon. The level of RF2 biosynthesis depends on the concentration of both release factors, RF1 and RF2. RF1 and RF2 likely compensate for each other

low cellular concentration of RF1 may be compensated for by increased synthesis of RF2. Perhaps this is particularly beneficial for those bacteria where UGA and UAA codons are more frequently used than UAG codons. However, whether there is indeed a correlation between occurrence of UAA in the RF2 frameshifting cassette and differential utilization of stop codons in the corresponding bacterial genomes has not been investigated.

All other stimulatory elements in the RF2 frameshifting cassette are not responsible for the sensitivity of frameshifting to release factor concentration. However, they are responsible for elevation of the absolute level of frameshifting efficiency, which in their absence would be insignificant even at low concentrations of release factors. The element whose role in the frameshifting mechanism is relatively easy to understand is the identity of the nucleotide 3′ adjacent to the stop codon. Unlike all sense codons that are recognized by RNA molecules via complementary interactions, stop codons are recognized by protein molecules. The recent analysis of crystal structure of the ribosome complex with RF2 reveals details of RF2 interactions with the UGA stop codon in mRNA (Weixlbaumer et al., 2008). Unfortunately, the crystal structure does not provide information on interactions of RF2 with the mRNA region downstream of the stop codon which seems to interact with release factors as evident from earlier cross-linking studies (Poole et al., 1998). While these interactions do not play a role in stop codon discrimination, they do affect termination efficiency. Since frameshifting efficiency negatively correlates with termination efficiency, it is not surprising that the weakest termination context has been selected in the RF2 frameshift site during its evolution (Major et al., 1996). It can be seen in Fig. 14.2 that the 3′ nucleotide adjacent to the stop codon is nearly always C, which has been shown to be the most inefficient context codon for termination in eubacterial organisms (Mottagui-Tabar and Isaksson, 1998; Pavlov et al., 1998).

Another important stimulatory element in the RF2 frameshifting cassette is the internal Shine–Dalgarno (SD) sequence located upstream of the shift site (Weiss et al., 1987, 1988; Curran and Yarus, 1988). Normally SD sequences are used for the initiation of translation in bacteria and are located upstream of initiator codons (Shine and Dalgarno, 1975). The increase in local concentration of initiating ribosomes around initiator sites is achieved through interactions between the SD and the corresponding complementary region of 16S rRNA, termed anti-Shine–Dalgarno (anti-SD). The internal SD 5′ of the frameshifting site could serve the same purpose, and initiation of translation at the UUG codon (which is a part of the frameshifting site) has been demonstrated (Baranov et al., 2002), although no potential functional role for this internal initiation event has been implicated. It could be that this is simply an unintentional side effect caused by sequence constraints of the RF2 frameshifting cassette. Irrespective of internal initiation, the main role of the internal SD is clearly to target elongating ribosomes. One particular important aspect of the SD stimulatory effect on frameshifting efficiency is the location of the SD relative to the frameshift site (Weiss et al., 1987). The length of the spacer between the SD sequence and the P-site tRNA during the frameshift is shorter than the distance between the SD and initiator codons (Ma et al., 2002). It is reasonable to assume that the distance between an SD and an initiator codon is optimal for the relaxed conformation of the ribosomal RNA during the initiation. If so, the shorter distance between the internal SD and the shift site should create tension in the ribosomal RNA between the anti-SD and the decoding center of the ribosome. Such tension likely acts in a manner of a compressed spring, whose relaxation is achieved by a progressive movement of tRNA with the decoding center of the ribosome toward the 3′-end of mRNA. This movement would explain the stimulatory effect of an SD on +1 frameshifting. Accordingly it is known that an internal SD stimulates frameshifting in the opposite direction when the spacer is longer than the optimal for initiation, in which case RNA likely acts as a stretched spring that alleviates tRNA movement toward the 5′-end of mRNA (Atkins et al., 2001). The conservation of the SD sequence and its location is illustrated in Fig. 14.2. Since base pairing between the SD and rRNA does not have to be perfect to cause the effect, there is a certain degree of flexibility in the RF2 frameshift stimulatory SD sequences; hence, its conservation is less profound than that of the shift site and the stop codon.

While the size of the spacer separating the shift site from the internal SD sequence is crucially important for its stimulatory effect, the identity of the spacer is not inconsequential either (Baranov et al., 2002). During frameshifting the spacer corresponds to the codon located in the ribosomal E-site. It has been suggested that there is a competition between the anti-SD and E-site tRNA for interactions with the corresponding part of mRNA. This interference of the SD with normal occupation of the E-site codon by the E-site tRNA affects fidelity of the ribosome (Baranov et al., 2002; Marquez et al., 2004; Sanders and Curran, 2007). Consequently, as the affinity of different tRNAs for the E-site fluctuates (Lill and Wintermeyer, 1987), it is not surprising that the identity of the spacer affects frameshifting efficiency.

Analysis of the distribution of sequences similar to the RF2 frameshifting cassette in bacterial genomes in terms of its "singularity" is meaningless, due to its size and complexity. If we represent the RF2 frameshift cassette as some kind of a roughly estimated consensus sequence such as GRGGNNNYTT-Stop-C, the probability of its appearance in random sequences of the same length is 1/16,384. Since we are interested only in those stop codons that are really used for the termination of translation, then the probability of such a sequence in a genome similar to *E. coli* (~4,000 genes) will be about 0.2 and the probability of two such sequences in such a genome will be only ~0.05. Even if a deviation of a single nucleotide in the above consensus sequence is allowed, the probability of two random occurrences of such sequences in a genome of a size similar to that of *E. coli* would be less than 1/2. In other words, the fact that the above consensus sequence does not occur at the end of any other *E. coli* gene does not indicate evolutionary selection against such sequences. As for the individual modular stimulatory signals constituting the RF2 frameshifting cassette, they are insufficient to trigger ribosomal frameshifting with comparable efficiency and hence they are relatively frequent in the genomes. Nonetheless, some tendency for their avoidance can be illustrated using the following simple and perhaps somewhat naïve measures. For example, while C nucleotides constitute a 0.25 fraction of the *E. coli* K12 genome (NC_000913), the fraction of C nucleotides adjacent to the 3′-end of *E. coli* stop codons is 0.17, and 0.14 for those adjacent to UGA, whereas the portion of Cs after any UGA trinucleotide in the *E. coli* genome (N_{TGAC}/N_{TGAN} ratio) is 0.22. This seeming underrepresentation of Cs after stop codons and UGA in particular is, of course, due to its weakening effect on termination of translation. A similar tendency could be sensed for the usage of a codon upstream of stop codons. For example, the proportion of UUU codons among all Phe codons in the *E. coli* K12 genome is 0.66. But the proportion of UUU codons among Phe codons that are located upstream of stop codons is 0.47 and only 0.24 upstream of UGA codons. For CUU similar calculations give the less profound corresponding values of 0.16, 0.17, and 0.13. There is no avoidance of SD-like sequences at the end of *E. coli* genes compared to other locations within mRNA coding sequences. On the contrary, analysis of a larger number of bacterial genomes suggests that SD sequences are even overrepresented at the end of coding sequences, perhaps due to translational coupling where such SD sequences are used for the initiation of downstream genes (PVB, unpublished).

Summarizing, the entire RF2 frameshifting cassette constitutes a relatively large and complex constrained sequence pattern whose random occurrence in small genomes, such as the one in *E. coli*, has a low probability. Smaller and simpler components of the frameshift cassette are relatively ineffective in triggering efficient non-standard translation events; nonetheless they probably can increase the chance of errors and thus some level of selection against such sequences can be detected. In the following section we deal with the analysis of relatively short sequences, so their random occurrence is considerably more likely. Despite their shortness, however, they are sufficient to trigger efficient non-standard translational events.

14.2.2 −1 Frameshifting Cassette in Coronavirus Polyprotein-Encoding Gene

The coronaviral gene encoding the ORF1AB polyprotein consists of two overlapping ORFs and the synthesis of the full length protein product requires programmed ribosomal −1 frameshifting (Brierley et al., 1989). The frameshift cassette consists of the slippery heptamer sequence U_UU.U_AA.C (underlined spaces indicate separation of codons in the initial phase and dots separate codons in the frame after the shift). The frameshifting is stimulated by RNA structures downstream of the slippery sequence. There is a degree of variation among the stimulatory structures. In some viruses the structure is formed by two distant stem loops forming complementary interactions between their apical loops (kissing stem-loop structures) (Herold and Siddell, 1993). In others, it is a classical H-type pseudoknot with variable features, for example, in SARS-CoV there is an important RNA stem-loop structure located within the second loop of the pseudoknot (Baranov et al., 2005; Plant et al., 2005; Su et al., 2005).

Although the presence of a structure is evident in all known coronaviruses and is likely essential to support functional frameshifting efficiency, even in its absence frameshifting is detectable at levels greatly exceeding the average background frequency of frameshift errors (Brierley et al., 1991). The distribution of U_UU.A_AA.C sequences within a 27-way alignment of selected coronaviruses is shown in Fig. 14.4. Apparently there is no strong selection against such sequences in coronaviral genomes. Based on combinatorial codon usage analysis of these representative coronaviral genomes U UU.A AA.C patterns are expected to occur about two times per ORF1AB gene. Indeed the real number of patterns corresponds to this expectation value and varies from 1 to 6 per gene (Fig. 14.4). Nonetheless, the overall distribution clearly illustrates the behavior typical of singular genetic elements where U UU.A AA.C is present in all genomes in a particular location, while other occurrences are distributed in a more random manner. For comparison Fig. 14.4 also shows the distribution of the same nucleotide patterns, but in different reading phases. It is clear that their distribution is less ordered. The existence of U_UU.A_AA.C patterns in locations other than the frameshift site can be explained either by neighboring nucleotide context disfavoring ribosomal frameshifting or by the possibility that such low frameshifting levels (in the absence of a stimulator) at a few locations can be tolerated by viruses.

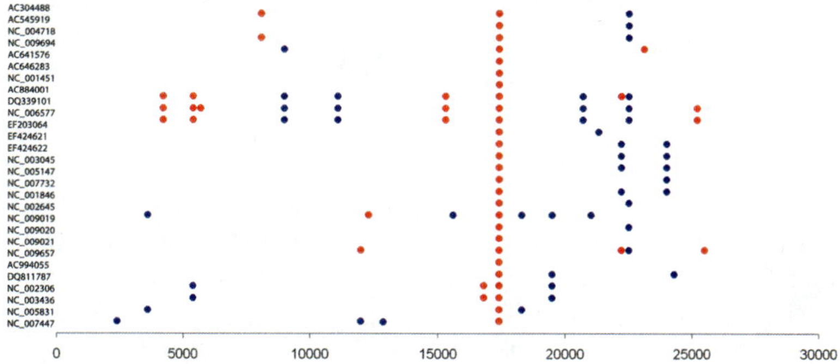

Fig. 14.4 Distribution of UUUAAAC patterns across multiple alignments of coronavirus *orfAB*. Red spots correspond to the patterns in the shift-prone phase U_UUA_AAC, blue spots correspond to the same pattern in other reading phases. The sequences for the alignment were extracted from the CoVDB (Huang et al., 2008). Genbank accession numbers are given within the figure. Initially, the nucleotide sequences were translated and aligned with ClustalW and then the alignment obtained was back-translated and processed with custom-designed Perl scripts

14.3 Cars and Ribosomes, Fast and Furious: Role of mRNA in the Accuracy of Translation

One striking difference between erroneous frameshifting and programmed frameshifting lies in their efficiencies. The translational apparatus is able to decode mRNA with remarkable accuracy; misincorporation of an amino acid due to recognition of incorrect tRNAs occurs with frequencies in the range of 10^{-3}–10^{-5} depending on the exact type of error. These estimates come from a number of studies in *E. coli*, reviewed in Parker (1989). This high accuracy for amino acid incorporation is observed despite the fact that not all such errors are necessarily harmful, since substitution of a single amino acid in a protein does not necessarily lead to its inactivation. The extent of tolerance to misincorporation errors is best illustrated by *Candida albicans* where CUG codons are decoded as both Leu and Ser due to ambiguous aminoacylation of the corresponding tRNA (Moura et al., 2007). In contrast, errors in processivity, such as frameshift errors, pose a greater danger during translation since they result in alterations not of just a single amino acid but of the entire sequence following such an error. It is reasonable to expect that the decoding apparatus should be able to prevent such errors with even greater accuracy. Indeed, it has been estimated that background levels of frameshifting errors fluctuate in the range of 10^{-5}–10^{-7} (Kurland, 1979; Parker, 1989). At the 2007 ribosomal meeting in Cape Cod, Mons Ehrenberg summarized his talk with the following statement: "Ribosomes are very fast and very accurate and this is the summary of my talk." It would be hard and perhaps juvenile to argue with such a statement as it would be hard to argue with commercials advertising modern cars saying that they are fast and safe. Cars are, but the traffic is not, at least not always. The safety and speed of traffic depends not only on cars but also on road conditions. By analogy we can describe mRNAs as the roads for the ribosomal traffic. We will argue that the observed accuracy of translation relies not only on the properties of the ribosome but also on mRNA sequence. Under certain circumstances mRNA can force translating ribosomes to alter their behavior so that translation can no longer be considered accurate.

Frameshifting occurs with strikingly high efficiencies at certain recoding sites exceeding background levels by 10^6 and under certain conditions could be even more efficient than standard triplet translation. Of course, such efficiency is frequently achieved by an ensemble of complex stimulatory signals that have evolved to increase frameshifting efficiency at a local site. This was described above for RF2 mRNA frameshifting and is also evident from many other examples throughout this book. However, even relatively simple sequences such as the heptameric C.UU_A.GG_C in yeast transposon Ty1 cause frameshifting with efficiency comparable to that of standard translation at the same site without additional stimulators (Belcourt and Farabaugh, 1990). Other simple short sequences are also shift-prone and can lead to frameshifting events of lower efficiency, but still much greater than the background levels. Evidently the accuracy of translation in terms of reading frame maintenance is highly dependent on mRNA nucleotide context. Why is there such dependence and why do ribosomes not translate all sequences with a similar accuracy?

A plausible explanation may lie in the fact that the ribosomes as we know them have evolved to achieve the global optimum compromise between speed and fidelity of translation (Kurland et al., 1996). It is possible to increase fidelity of the ribosome by introducing certain mutations leading to hyper-accurate ribosomes. However such improved accuracy has a cost, the speed of translation is reduced, and this diminishes the potential benefit from the higher accuracy of translation. Hyper-accurate ribosomes are usually streptomycin dependent, as addition of streptomycin presumably increases the speed of translation by decreasing its accuracy. Can the ribosome be modified further to increase the speed of translation without losing accuracy? The potential for further improvement lies in mRNA sequences and the set of tRNAs used to decode them. The solution is alteration of the codon bias and the set of unequally distributed tRNAs. To understand how this could help improve both accuracy and speed consider a simple model. The probability of incorporation of a particular tRNA_k at a particular codon_k competing with a set of N tRNAs can be represented as

$$\frac{a_k T_k}{\sum_{i=1}^{N} a_i T_i}$$

where a is the tRNA affinity toward codon_k in the ribosomal A-site and T is its local concentration. An increase of tRNA_k concentration will increase the probability of its incorporation at codon_k as will a decrease in concentration of other tRNAs, even though their affinities (that are partially determined by the ability of the ribosome to discriminate between them) remain the same. If all codons were distributed equally in mRNA sequences, there would be no benefit from such a manipulation. However, a global positive effect can be achieved if there is codon usage bias, with some codons being abundant and others being rare. In this case, corresponding manipulation of the set of tRNAs will lead to improved accuracy and speed of global translation. But this would not come without a cost: decreased accuracy and speed of translation of rare codons. Of course the above scenario is a simplification compared to the real situation since the affinity of the tRNAs to their codons is also context dependent. Further, for frameshifting errors, the probability of its occurrence depends also on the specific combination of codons in the ribosome and the probability of rearrangement of tRNAs in the ribosome relative to mRNA (Baranov et al., 2004; Liao et al., 2008). Consequently a biased occurrence of combinations of codons is also evident (Fedorov et al., 2002; Moura et al., 2007). These simple considerations illustrate the concept of how biases in codon usage and their combination can be used for the benefit of global translation accuracy and speed. This, of course, does not mean that such biases exist purely to increase the efficiency of global translation. There are other, perhaps even more important contributing factors, such as GC content, biases in the usage of amino acids, mutational bias (Bernardi and Bernardi, 1986; Wan et al., 2004). In fact it has been possible to predict codon usage biases for a hundred microbial organisms purely based on a combination of GC content and nucleotide mutational bias, obtained from the analysis of intergenic regions (Chen et al., 2004). However, irrespective of the evolutionary reasons underlying the existence of codon bias, there is a relationship between codon bias and relative tRNA abundance (Ikemura, 1981). It is clear that the translational apparatus, for at least the set of tRNAs used for mRNA decoding, has adapted to these biases, since higher codon biases associate with conserved and highly expressed genes (Stoletzki and Eyre-Walker, 2007). Such adaptation results in decreased accuracy of translation of mRNAs that do not show bias, as is evident from highly erroneous expression of heterologous translation (Kurland and Gallant, 1996), e.g., during synthesis of human proteins in bacterial species whose translational apparatus has not been modified specifically for such purposes (Gustafsson et al., 2004). Moreover, accuracy of translation depends not only on simple codon bias but also on a bias among co-occurring codons in mRNA. This fact has been recently utilized to design a synthetic poliovirus whose genome was modified to encode native capsid protein with CDS consisting of underrepresented codon pairs (Coleman et al., 2008). Such virus triggers host immune response, but reduced translation rates alter virus viability, suggesting an elegant method for immunization.

This explains why certain relatively simple sequences can be particularly prone to frameshift errors and why they are rare in most coding regions. However, the situation is not always so simple as we will see in the following sections.

14.4 Strategies for Searching Recoding Cases as Singular Elements

A number of studies have attempted to search for new cases of programmed frameshifting based on the assumption that the sequences that promote ribosomal frameshifting should behave like singular genomic elements and as such be avoided in the coding regions unless the triggered frameshifting is positively selected for. The simplest idea is to search for further occurrences of sequences, of the type known to be utilized for programmed ribosomal frameshifting, throughout the coding regions of completed genomes. Although this approach limits the search to motifs already known to trigger frameshifting and will not increase our knowledge of frameshift-prone sequences, it could reveal novel cases of utilization of these sequences for gene expression purposes. To analyze the frequency of occurrence of sequences capable of stimulating -1 frameshifting in *Saccharomyces cerevisiae*, Jacobs et al. (2007) searched for viral consensus slippery sites X_XX.Y_YY.Z, where XXX represents any three identical nucleotides, YYY represents AAA or UUU, Z $\neq$ G. With this approach they identified 10,340 slippery sites in the 6,353 annotated coding sequences of the yeast genome, 6,016 of which are followed by at least one pseudoknot motif. According to statistical analyses employed by the authors these signals are underrepresented in the *S. cerevisiae* genome. Of the 6,353 yeast ORFs, 1,275 contain at least one strong and statistically significant -1 frameshift signal [in a recent study Theis et al. (2008) have argued that in some cases there are alternative structures that are more stable than the predicted pseudoknots]. Eight out of nine sequences, selected for experimental verification using artificial genetic constructs, supported efficient levels of frameshifting in vivo. The authors hypothesized that many other frameshift candidates found in their study could lead to significant levels of frameshifting. If frameshifting indeed takes place at those locations, in the vast majority of cases it would result in production of truncated and most likely dysfunctional products. The authors hypothesized that the role of frameshifting could be regulatory (see the following section). It is unclear how beneficial such a regulation might be for the cells and no data on phylogenetic conservation of these sequences have been provided.

In a different work (Gurvich et al., 2003), the *E. coli* K12 genome was searched for occurrences of the very well-known prokaryotic slippery sequence A_AA.A_AA.G. Frameshifting at A_AA.A_AA.G is utilized for expression of the γ subunit of DNA polymerase III, while the τ subunit is expressed by standard translation from the same gene (*dnaX*) (Blinkowa and Walker, 1990; Flower and McHenry, 1990; Tsuchihashi and Kornberg, 1990). Frameshifting at this sequence is also utilized by a number of insertion sequence elements in *E. coli* (Hu et al., 1996; Baranov et al., 2006). Seventy instances of this sequence have been found in 68 *E. coli* genes. Twelve genes have been chosen for experimental analysis and all of them have been shown to support -1 frameshifting at levels above background. The authors used comparative phylogenetic analysis to address potential utilization of any of those sequences for gene expression purposes. Apart from the *dnaX* gene, six IS2-like elements and the *ydaY* gene of unknown function, utilize A_AA.A_AA.G for gene expression. Although the number of occurrences is quite high, according to the statistical analysis this sequence is underrepresented in coding regions, and thus does behave as a singular element. The distribution of three other known shift-prone sequences in *E. coli* K12, CCC_UGA (Gurvich et al., 2003), AGG_AGG, and AGA_AGA (Gurvich et al., 2005), was also examined. All three sequences trigger +1 frameshifting in *E. coli*. Frameshifting at C.CC_U.GA occurs through near-cognate recognition of the CCC codon by $tRNA^{Pro}{}_{5'U*GG3'}$(where U* designates the cmo^5U34 modification) (O'Connor, 2002). Because of suboptimal base pairing with the CCC codon, this tRNA is prone to shift into the +1 frame to re-pair to mRNA at the cognate CCU codon. As with RF2 mRNA frameshifting, that on C.CC_U.GA is in direct competition with termination mediated by RF2 and its efficiency is increased due to slow decoding of the termination codon. Although not known to be utilized for gene expression in *E. coli*, frameshifting at C.CC_U.GA is employed for expression of antizyme genes in some eukaryotes (Ivanov and Atkins, 2007) and for expression of the *tsh* gene of *Listeria monocytogenes* phage PSA (Zimmer et al., 2003). Nineteen genes in *E. coli* K12 end with C.CC_TGA and in half of them frameshifting occurs at above 1% (Gurvich et al., 2003).

Frameshifting on A.GG_A.GG and A.GA_A.GA is due to limited abundance of the cognate arginine $tRNA^{Arg}{}_{3'UCC5'}$ and $tRNA^{Arg}{}_{3'UCU*5'}$ (where U* is 5-methylaminomethyl-2-thiouridine), respectively. Due to sequestration of the sparse tRNA by the first of the tandem codons, its availability for the second codon is drastically reduced. When the second codon occupies the A-site of the translating ribosome the longer-than-usual time for arrival of the cognate tRNA increases the chance for dissociation of the peptidyl-tRNA which may re-pair to mRNA in the overlapping +1 frame (or potentially −1 frame as has been shown for an A.GA_A.GA tandem by Lainé et al. (2008)). Frameshifting to the new frame is greatly favored by availability of the tRNA cognate to the new codon in the +1 frame. The A.GG_A.GG and A.GA_A.GA tandems were originally reported to trigger up to 50% frameshifting (Spanjaard and van Duin, 1988; Spanjaard et al., 1990). Although such high levels of frameshifting are likely due to overexpression of the mRNAs containing these sequences (Gurvich et al., 2005) and due to the use of streptomycin-resistant strains, in which ribosomes translate the mRNA more slowly making them prone to +1 frameshifting at the rare codons (Sipley and Goldman, 1993). Nevertheless, even at the lowest possible expression level of the transgene, frameshifting at A.GA_A.GA (and likely A.GG_A.GG) occurs at about 1% level (Gurvich et al., 2005). All three frameshift-prone sequences C.CC_U.GA, A.GG_A.GG, and A.GA_A.GA are not underrepresented in *E. coli* and in fact C.CC_U.GA is significantly overrepresented. However, none of these sequences including A_AA.A_AA.G, occur in the subset of highly expressed genes in *E. coli* (Karlin et al., 2001). This means that although not significantly underrepresented in coding regions, overall these sequences are selected against in highly expressed ORFs and in the way they behave as singular elements in highly expressed genes. In contrast to the Jacobs et al. study, Gurvich et al. suggested that the occurrence of these frameshift candidates in protein coding regions does not have a functional role, since they do not exhibit phylogenetic conservation. Gurvich et al. argued that frameshifting above background level in lowly expressed genes could easily be tolerated by cells, since only a few aberrant protein molecules would be produced as a result of frameshifting. Therefore, the presence of shift-prone sequences in certain locations can be explained not by their beneficial effects but by the lack of strong selection against such sequences. Future studies are expected to resolve the contrasting interpretations.

The most general ab initio study related to singular elements supporting frameshifting was performed by Shah et al. (2002) where the distribution of all heptamers occurring in coding regions of the yeast *S. cerevisiae* genome was analyzed. A fraction of the least abundant and the most underrepresented heptamers have been tested for their ability to trigger ribosomal frameshifting. All sequences tested stimulated ribosomal frameshifting at above background levels with some of them promoting highly efficient frameshifting. Notably, the heptamer sequences C.UU_A.GU_U and C.UU_A.GG_C used to trigger programmed ribosomal frameshifting for expression of EST3 (Morris and Lundblad, 1997; Taliaferro and Farabaugh, 2007) and ABP140 (Asakura et al., 1998), respectively, are ranked among the least represented in coding regions of *S. cerevisiae*. While this approach appeared to have good predictability for sequences supporting +1 frameshifting in yeast, it failed in predicting sequences that would stimulate −1 frameshifting. The authors suggested this could be because the sequences utilized for −1 programmed frameshifting in yeast do not stimulate frameshifting at sufficiently high efficiency without additional *cis*-acting elements.

Frameshift-prone sequences do not necessarily exhibit properties of singular elements. In certain organisms frameshifting could be highly abundant. This seems to be the case in the ciliate *Euplotes*. To date there are eight different types of genes identified in *Euplotes* that utilize +1 frameshifting for their expression (Klobutcher, 2005). Only about 90 genes have been sequenced in *Euplotes* and the current estimate is that about 10% of the *Euplotes* genes require frameshifting for expression. Interestingly, three of these genes require multiple frameshift events for expression. Of these eight genes, five encode enzymes, one encodes a protein associated with the RNA component of telomerase, and two have unknown function. None of these genes is expected to be highly expressed, even though the subset of sequenced genes in *Euplotes* is biased toward highly expressed genes. All genes share the same sequence A.AA_U.AA_A (A.AA_U.AG_A for one gene) within the overlap of the upstream and downstream ORFs. Thus, it is likely that frameshifting takes place at these sequences and its mechanism is likely the same for all genes. The frameshift propensity of A.AA_U.AA_A heptanucleotide in *Euplotes* entails inefficient translation termination at the UAA stop codon and slippage of the $tRNA^{Lys}$ from the AAA codon to AAU. Ineffective translation termination at the UAA codon in *Euplotes* is proposed to be linked to UGA stop codon reassignment to cysteine (Klobutcher and Farabaugh, 2002; vallabhaneni et al., 2009). Such reassignment is complemented by changes in eukaryotic release factor one (eRF1), so that it no longer recognizes UGA codons. However, such changes may have rendered eRF1 less potent in recognition of UAA and UAG stop codons as well. If so, translation termination in *Euplotes* might be generally slow and inefficient, consequently favoring a competing process of +1 frameshifting. As a result it might be that *Euplotes* is tolerant of arising frameshift mutations that would be compensated by +1 frameshifting on A.AA_U.AR_A. Conservation of the 3′ A adjacent to the stop codon is believed to weaken the stop codon as a signal for termination and the conservation of AAA 5′ of the stop codon is explained by an unknown special feature of the corresponding $tRNA^{Lys}$ that makes it shift prone compared to other

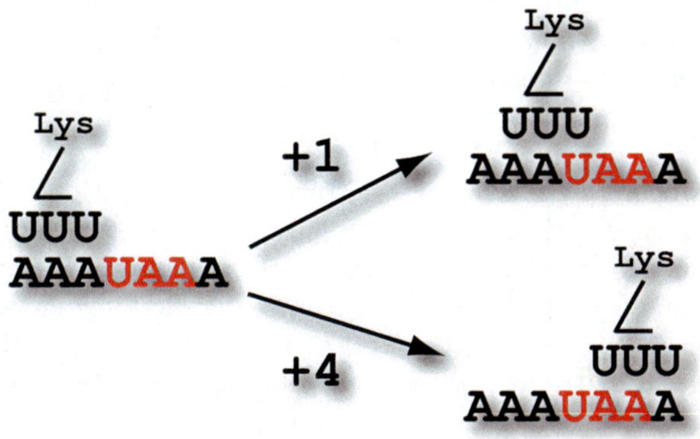

Fig. 14.5 Tentetive alternative mechanisms of frameshifting in the ciliate *Euplotes*. The stop codon in the frameshift site is shown in red. $tRNA^{Lys}$ could be repositioned in two alternative ways, by a +1 shift (*above*) or a +4 shift (*below*)

tRNAs (Klobutcher and Farabaugh, 2002). Otherwise it is unclear why frameshift sequences (such as U.UU_U.AA_A) with other slippage-prone codons 5′ of the stop codon have not been found. An alternative explanation for why frameshifting does not occur at other X.XX_U.AA_A (where X≠A) would be that ribosomes shift +4 (or bypass 1 nucleotide), see Fig. 14.5. Slow decoding of the UAA could facilitate repositioning of the P-site $tRNA^{Lys}$ 4 nucleotides downstream to re-pair to mRNA at the AAA codon in the +1 frame. For the A.AA_U.AG_A sequence, repositioning of the $tRNA^{Lys}$, which likely has the anticodon xm^5s^2UUU (Björk et al., 2007), would re-pair to mRNA at the AGA codon, which has only slightly lower thermodynamic stability than cognate pairing (see Fig. 14.5). Though such a shift mechanism would make frameshifting at A.AG_U.AA_A equally plausible, no such sequences have been identified as potential frameshift sites. Direct sequencing or mass spectrometry is essential to decipher the exact mechanism, since +1 frameshifting would yield two lysines corresponding to this site, while +4 shift would result in incorporation of a single lysine. Mass spectroscopic analysis has been carried out only in the analysis of *p45*-encoded telomerase component of *Euplotes*; however, no peptides matching the ORF junction has been detected (Aigner et al., 2000).

14.5 Possible Functions of Products Generated by Low-Level Aberrant Translation

As has been shown by several studies described above, shift-prone sequences, although somewhat underrepresented throughout the genome and absent in highly expressed genes, are frequent in coding sequences. In a few distinct cases specific functional consequences of frameshifting can be envisioned. However, such cases are rare and in general the frameshifting on frameshift-prone sequences will result in premature termination and production of a nonfunctional peptide that gets degraded. Most likely such frameshift events occur without any specific functional role and constitute minor faults of the translation process. Nevertheless, some general impact of such erroneous frameshifting on regulation of different cellular processes has been proposed. Some authors suggest that erroneous frameshifting can posttranscriptionally regulate mRNA stability, since encountering a premature termination codon by translating ribosome would trigger mRNA degradation through nonsense-mediated decay (NMD) pathway (Jacobs et al., 2007). However, the growing evidence suggests that in higher eukaryotes NMD can be triggered only during the first, so-called pioneer round of translation [review in Chang et al. (2007)]. If frameshifting occurs at a level of about 1%, then an mRNA containing such a frameshift site would be degraded through the NMD pathway only in 1% of the cases. On the other hand, in *S. cerevisiae* where NMD is inefficient and can be triggered after a number of translations of the PTC-containing mRNA, some downregulation of the mRNAs containing frameshift sites is feasible.

A consequence of erroneous frameshifting is production of an aberrant peptide. In some cases, when frameshifting occurs near the end of the coding region, the peptide synthesized might retain its function and could be utilized along with the products of standard translation (Mejlhede et al., 1999). In all other cases it is generally assumed that nonfunctional peptides get degraded. However, the exact fate is indeed unknown. Peptides produced by erroneous frameshifting can be potentially utilized as cryptic epitopes in the immune system. Two such cases have been described in the literature to date. One was identified in a patient with Reuter's syndrome. There, a transframe peptide produced via frameshifting from the IL-10 gene served as cryptic epitope to activate cytotoxic T cells (Saulquin et al., 2002). Intriguingly, the authors speculated that the frameshifting in the IL-10 could be of pathophysiological relevance since the preliminary data suggested recognition of the same epitope in another rheumatoid arthritis patient. Another example was identified in the herpes simplex virus (HSV) *tk* gene which encodes thymidine kinase (TK). Thymidine kinase is crucial for reactivation of the virus from a latent phase and is a target for antiviral therapy with the drug acyclovir. An acyclovir-resistant mutant has the insertion of a single *G* nucleotide in a run of 7 *G*'s in *tk* gene, resulting in a run of 8 *G*'s (Horsburgh et al., 1996). This frameshift mutation results in synthesis of nonfunctional TK and the mutant is resistant to acyclovir, which has to be phosphorylated by TK and subsequently by host kinases to an active form that interferes with viral replication (Elion, 1982). However, low levels of functional TK that are crucial for viral propagation are synthesized via ribosomal frameshifting on the run of 8 *G* (Griffiths et al., 2006; Besecker et al., 2007). In the wild-type *tk* gene the run of 7 *G* also causes about 1% frameshifting and the truncated peptide serves as a cryptic epitope and can trigger an immune response (Zook et al., 2006).

14.6 Conclusions

As we demonstrated in this chapter, sequences responsible for highly efficient alterations of standard genetic readout are sometimes underrepresented in protein coding regions of genomes. When such sequences play crucial roles for gene expression, e.g., required for the biosynthesis of functional gene products, they exhibit deep phylogenetic conservation. Such sequences can be classified as singular genetic elements. Yet, there are a substantial number of sequences prone to low-level aberrant translational events and their underrepresentation in coding sequences is less pronounced. Even though the negative impact of such sequences in gene expression is less critical and their genomic locations are not strictly conserved, the subsequent non-canonical translational events have important functional implications, such as fine-tuning of expression levels during posttranscriptional regulation or production of epitopes for an immune response.

Acknowledgments P.V.B. thanks Science Foundation Ireland for Support.

References

Aigner S, Lingner J, Goodrich KJ, Grosshans TA, Shevchenko A, Mann, M, Cech TR (2000). *Euplotes* telomerase contains a La motif protein produced by apparent translational frameshifting. EMBO J 19:6230–6239

Asakura T, Sasaki T, Nagano F, Satoh A, Obaishi H, Nishioka H, Imamura H, Hotta K, Tanaka K, Nakanishi, H, Takai, Y (1998) Isolation and characterization of a novel actin filament-binding protein from *Saccharomyces cerevisiae*. Oncogene 16:121–130

Atkins JF, Baranov PV, Fayet O, Herr AJ, Howard MT, Ivanov IP, Matsufuji S, Miller WA, Moore B, Prere MF, Wills NM, Zhou J, Gesteland RF (2001) Overriding standard decoding: implications of recoding for ribosome function and enrichment of gene expression. Cold Spr Harb Symp Quant Biol 66:217–232

Baranov PV, Fayet O, Hendrix RW, Atkins JF (2006) Recoding in bacteriophages and bacterial IS elements. Trends Genet 22:174–181

Baranov PV, Gesteland RF, Atkins JF (2002) Release factor 2 frameshifting sites in different bacteria. EMBO Rep 3:373–377

Baranov PV, Gesteland RF, Atkins JF (2004) P-site tRNA is a crucial initiator of ribosomal frameshifting. RNA 10:221–230

Baranov PV, Henderson CM, Anderson CB, Gesteland RF, Atkins JF, Howard MT (2005) Programmed ribosomal frameshifting in decoding the SARS-CoV genome. Virol 332:498–510

Bekaert M, Atkins JF, Baranov PV (2006) ARFA: A program for annotating bacterial release factor genes, including prediction of programmed ribosomal frameshifting. Bioinformatics 22: 2463–2465

Belcourt MF, Farabaugh PJ (1990) Ribosomal frameshifting in the yeast retrotransposon Ty: tRNAs induce slippage on a 7 nucleotide minimal site. Cell 62:339–352

Bernardi G, Bernardi G (1986) Compositional constraints and genome evolution. J Mol Evol 24:1–11

Besecker MI, Furness CL, Coen DM, Griffiths A (2007) Expression of extremely low levels of thymidine kinase from an acyclovir-resistant herpes simplex virus mutant supports reactivation from latently infected mouse trigeminal ganglia. J Virol 81:8356–60

Björk GR, Huang B, Persson OP, Bystrom AS (2007) A conserved modified wobble nucleoside (mcm5s2U) in lysyl-tRNA is required for viability in yeast. RNA 13:1245–1255

Blinkowa AL, Walker JR (1990) Programmed ribosomal frameshifting generates the *Escherichia coli* DNA polymerase III gamma subunit from within the tau subunit reading frame. Nucl Acids Res 18:1725–1729

Brierley I, Digard, P, Inglis SC (1989) Characterization of an efficient coronavirus ribosomal frameshifting signal: requirement for an RNA pseudoknot. Cell 57:537–547

Brierley I, Rolley NJ, Jenner AJ, Inglis SC (1991) Mutational analysis of the RNA pseudoknot component of a coronavirus ribosomal frameshifting signal. J Mol Biol 220:889–902

Capecchi MR, Klein HA (1970) Release factors mediating termination of complete proteins. Nature 226:1029–1033

Chang YF, Imam JS, Wilkinson MF (2007) The nonsense-mediated decay RNA surveillance pathway. Ann Rev Biochem 76:51–74

Chen SL, Lee W, Hottes AK, Shapiro L, McAdams HH (2004) Codon usage between genomes is constrained by genome-wide mutational processes. Proc Natl Acad Sci USA 101: 3480–3485

Coleman JR, Papamichail D, Skiena S, Futcher B, Wimmer, E, Mueller S (2008) Virus attenuation by genome-scale changes in codon pair bias. Science 320:1784–1787

Craigen WJ, Caskey CT (1986) Expression of peptide chain release factor 2 requires high-efficiency frameshift. Nature 322:273–275

Craigen WJ, Cook RG, Tate WP, Caskey CT (1985) Bacterial peptide chain release factors: conserved primary structure and possible frameshift regulation of release factor 2. Proc Natl Acad Sci USA 82:3616–3620

Crooks GE, Hon G, Chandonia JM, Brenner SE (2004) WebLogo: a sequence logo generator. Genome Res 14:1188–1190

Curran JF (1993) Analysis of effects of tRNA:message stability on frameshift frequency at the *Escherichia coli* RF2 programmed frameshift site. Nucleic Acids Res 21:1837–1843

Curran JF, Yarus M (1988) Use of tRNA suppressors to probe regulation of *Escherichia coli* release factor 2. J Mol Biol 203:75–83

Elion GB (1982) Mechanism of action and selectivity of acyclovir. Am J Med 73:7–13

Farh KK, Grimson A, Jan C, Lewis BP, Johnston WK, Lim LP, Burge CB, Bartel DP (2005) The widespread impact of mammalian Micro RNAs on mRNA repression and evolution. Science 310:1817–1821

Fedorov A, Saxonov S, Gilbert W (2002) Regularities of context-dependent codon bias in eukaryotic genes. Nucleic Acids Res 30:1192–1197

Flower AM, McHenry CS (1990) The gamma subunit of DNA polymerase III holoenzyme of *Escherichia coli* is produced by ribosomal frameshifting. Proc Natl Acad Sci USA 87: 3713–3717

Griffiths A, Link MA, Furness CL, Coen DM (2006) Low-level expression and reversion both contribute to reactivation of herpes simplex virus drug-resistant mutants with mutations on homopolymeric sequences in thymidine kinase. J Virol 80:6568–6574

Gurvich OL, Baranov PV, Gesteland RF, Atkins JF (2005) Expression levels influence ribosomal frameshifting at the tandem rare arginine codons AGG_AGG and AGA_AGA in *Escherichia coli.* J Bacteriol 187:4023–4032

Gurvich OL, Baranov PV, Zhou J, Hammer AW, Gesteland RF, Atkins JF (2003) Sequences that direct significant levels of frameshifting are frequent in coding regions of *Escherichia coli.* EMBO J 22:5941–5950

Gustafsson C, Govindarajan S, Minshull J (2004) Codon bias and heterologous protein expression. Trends Biotechnol 22:346–353

Herold J, Siddell SG (1993) An 'elaborated' pseudoknot is required for high frequency frameshifting during translation of HCV 229E polymerase mRNA. Nucl Acids Res 21:5838–5842

Horsburgh BC, Kollmus H, Hauser, H, Coen DM (1996) Translational recoding induced by G-rich mRNA sequences that form unusual structures. Cell 86:949–959

Hu ST, Lee LC, Lei GS (1996) Detection of an IS2-encoded 46-kilodalton protein capable of binding terminal repeats of IS2. J Bacteriol 178:5652–5659

Huang Y, Lau SK, Woo PC, Yuen KY (2008) CoVDB: a comprehensive database for comparative analysis of coronavirus genes and genomes. Nucl Acids Res 36:D504–511

Ikemura T (1981) Correlation between the abundance of *Escherichia coli* transfer RNAs and the occurrence of the respective codons in its protein genes. J Mol Biol 146:1–21

Ivanov IP, Atkins JF (2007) Ribosomal frameshifting in decoding antizyme mRNAs from yeast and protists to humans: Close to 300 cases reveal remarkable diversity despite underlying conservation. Nucleic Acids Res 35:1842–1858

Jacobs JL, Belew AT, Rakauskaite R, Dinman JD (2007) Identification of functional, endogenous programmed −1 ribosomal frameshift signals in the genome of *Saccharomyces cerevisiae*. Nucl Acids Res 35:165–174

Karlin S, Mrazek J, Campbell, A, Kaiser, D (2001) Characterizations of highly expressed genes of four fast-growing bacteria. J Bacteriol 183:5025–5040

Kisselev LL, Buckingham RH (2000) Translational termination comes of age. Trends Biochem Sci 25:561–566

Klobutcher LA (2005) Sequencing of random *Euplotes crassus* macronuclear genes supports a high frequency of +1 translational frameshifting. Eukaryotic Cell 4:2098–2105

Klobutcher LA, Farabaugh PJ (2002) Shifty ciliates: frequent programmed translational frameshifting in euplotids. Cell 111:763–766

Kurland, C (1979) Reading frame errors on ribosomes. In: Celis J, Smith JD (eds) Nonsense mutations and tRNA suppressors, Academic Press, London, pp 97–108

Kurland C, Gallant J (1996) Errors of heterologous protein expression. Curr Opinion Biotechnol 7:489–493

Kurland CG, Hughes D, Ehrenberg M (1996) Limitations of translational accuracy. In *Escherichia coli* and *Salmonella typhimurium*: Cellular and molecular biology, ASM Press, Washington, DC, pp 979–1004

Lainé S, Thouard A, Komar AA, Rossignol JM (2008) Ribosome can resume the translation in both +1 or −1 frames after encountering an AGA cluster in Escherichia coli. Gene 412:95–101

Liao PY, Gupta P, Petrov AN, Dinman JD, Lee KH (2008) A new kinetic model reveals the synergistic effect of E-, P- and A-sites on +1 ribosomal frameshifting. Nucl Acids Res 36:2619–2629

Lill R, Wintermeyer W (1987) Destabilization of codon-anticodon interaction in the ribosomal exit site. J Mol Biol 196:137–148

Ma J, Campbell A, Karlin S (2002) Correlations between Shine-Dalgarno sequences and gene features such as predicted expression levels and operon structures. J Bacteriol 184: 5733–5745

Major LL, Poole ES, Dalphin ME, Mannering SA, Tate WP (1996) Is the in-frame termination signal of the *Escherichia coli* release factor–2 frameshift site weakened by a particularly poor context? Nucl Acids Res 24:2673–2678

Marquez V, Wilson DN, Tate WP, Triana-Alonso, F, Nierhaus KH (2004) Maintaining the ribosomal reading frame: the influence of the E site during translational regulation of release factor 2. Cell 118:45–55

Mejlhede N, Atkins JF, Neuhard, J (1999) Ribosomal −1 frameshifting during decoding of *Bacillus subtilis cdd* occurs at the sequence CGA AAG. J Bacteriol 181:2930–2937

Moon S, Byun Y, Han K (2007) FSDB: a frameshift signal database. Computat Biol Chem 31: 298–302

Morris DK, Lundblad, V (1997) Programmed translational frameshifting in a gene required for yeast telomere replication. Curr Biol 7:969–976

Mottagui-Tabar, S, Isaksson LA (1998) The influence of the 5′ codon context on translation termination in *Bacillus subtilis* and *Escherichia coli* is similar but different from *Salmonella typhimurium.* Gene 212:189–196

Moura G, Pinheiro M, Arrais J, Gomes AC, Carreto L, Freitas A, Oliveira JL, Santos MA (2007) Large scale comparative codon-pair context analysis unveils general rules that fine-tune evolution of mRNA primary structure. PLoS ONE 2:e847

O'Connor M (2002) Imbalance of tRNA(Pro) isoacceptors induces +1 frameshifting at near-cognate codons. Nucl Acids Res 30:759–765

Parker J (1989) Errors and alternatives in reading the universal genetic code. Microbiol Rev 53:273–298

Pavlov MY, Freistroffer DV, Dincbas V, MacDougall J, Buckingham RH, Ehrenberg, M (1998) A direct estimation of the context effect on the efficiency of termination. J Mol Biol 284: 579–590

Plant EP, Perez-Alvarado GC, Jacobs JL, Mukhopadhyay B, Hennig, M, Dinman JD (2005) A three-stemmed mRNA pseudoknot in the SARS coronavirus frameshift signal. PLoS Biol 3:e172

Poole ES, Major LL, Mannering SA, Tate WP (1998) Translational termination in *Escherichia coli*: three bases following the stop codon crosslink to release factor 2 and affect the decoding efficiency of UGA-containing signals. Nucl Acids Res 26:954–960

Sanders CL, Curran JF (2007) Genetic analysis of the E site during RF2 programmed frameshifting. RNA 13:1483–1491

Saulquin X, Scotet E, Trautmann L, Peyrat MA, Halary F, Bonneville, M, Houssaint, E (2002) +1 Frameshifting as a novel mechanism to generate a cryptic cytotoxic T lymphocyte epitope derived from human interleukin 10. J Exper Med 195:353–358

Scolnick E, Tompkins R, Caskey, T, Nirenberg, M (1968) Release factors differing in specificity for terminator codons. Proc Natl Acad Sci USA 61:768–774

Shah AA, Giddings MC, Parvaz JB, Gesteland RF, Atkins JF, Ivanov IP (2002) Computational identification of putative programmed translational frameshift sites. Bioinformatics 18: 1046–1053

Shine J, Dalgarno L (1975) Determinant of cistron specificity in bacterial ribosomes. Nature 254:34–38

Sipley J, Goldman E (1993) Increased ribosomal accuracy increases a programmed translational frameshift in *Escherichia coli*. Proc Natl Acad Sci USA 90:2315–2319

Spanjaard RA, Chen K, Walker JR, van Duin J (1990) Frameshift suppression at tandem AGA and AGG codons by cloned tRNA genes: assigning a codon to argU tRNA and T4 tRNA(Arg). Nucl Acids Res 18:5031–5036

Spanjaard RA, van Duin J (1988) Translation of the sequence AGG-AGG yields 50% ribosomal frameshift. Proc. Natl Acad Sci USA 85:7967–7971

Stoletzki N, Eyre-Walker A (2007) Synonymous codon usage in *Escherichia coli*: selection for translational accuracy. Mol Biol Evol 24:374–381

Su MC, Chang CT, Chu CH, Tsai CH, Chang KY (2005) An atypical RNA pseudoknot stimulator and an upstream attenuation signal for −1 ribosomal frameshifting of SARS coronavirus. Nucl Acids Res 33:4265–4275

Taliaferro D, Farabaugh PJ (2007) An mRNA sequence derived from the yeast EST3 gene stimulates programmed +1 translational frameshifting. RNA 13:606–613

Theis C, Reeder J, Giegerich R (2008) KnotInFrame: prediction of −1 ribosomal frameshift events. Nucl Acids Res 36:6013–6020

Tsuchihashi Z, Kornberg A (1990) Translational frameshifting generates the gamma subunit of DNA polymerase III holoenzyme. Proc Natl Acad Sci USA 87:2516–2520

Vallabhaneni H, Fan-Minogue H, Bedwell DM, Farabaugh PJ (2009) Connection between stop codon reassignment and frequent use of shifty stop frameshifting. RNA 15:889–897

Wan XF, Xu D, Kleinhofs, A, Zhou, J (2004) Quantitative relationship between synonymous codon usage bias and GC composition across unicellular genomes. BMC Evol Biol 4:19

Weiss RB, Dunn DM, Atkins JF, Gesteland RF (1987) Slippery runs, shifty stops, backward steps, and forward hops: −2, −1, +1, +2, +5, and +6 ribosomal frameshifting. Cold Spr Harb Symp Quant Biol 52:687–693

Weiss RB, Dunn DM, Dahlberg AE, Atkins JF, Gesteland RF (1988) Reading frame switch caused by base-pair formation between the 3′ end of 16S rRNA and the mRNA during elongation of protein synthesis in *Escherichia coli*. EMBO J 7:1503–1507

Weixlbaumer A, Jin H, Neubauer C, Voorhees RM, Petry S, Kelley AC, Ramakrishnan V (2008) Insights into translational termination from the structure of RF2 bound to the ribosome. Science 322:953–956

Zimmer M, Sattelberger E, Inman RB, Calendar, R, Loessner MJ (2003) Genome and proteome of *Listeria monocytogenes* phage PSA: an unusual case for programmed + 1 translational frameshifting in structural protein synthesis. Mol Microbiol 50:303–317

Zook MB, Howard MT, Sinnathamby G, Atkins JF, Eisenlohr LC (2006) Epitopes derived by incidental translational frameshifting give rise to a protective CTL response. J Immunol 176:6928–6934

Chapter 15
Mutants That Affect Recoding

Jonathan D. Dinman and Michael O'Connor

Abstract Fast, accurate translation is a hallmark of protein synthesis. However, recoding systems subvert the normal decoding mechanisms to redefine codons or shift translation into alternate reading frames. Genetic analyses in yeast and bacteria have enhanced our understanding of non-standard decoding and of normal decoding processes. Recent crystallographic efforts have provided unprecedented insights into the mechanisms underlying translation and the selection of cognate tRNAs. Ribosomal components and factors involved in decoding have also been identified genetically, through selections for mutants that alter the fidelity of protein synthesis, or impact upon various recoding phenomena. The availability of high-resolution structures of ribosomal complexes together with advances in biochemistry now allow the effects of many of these mutant components to be explained in molecular terms.

Contents

J.D. Dinman (✉)
Department of Cell Biology and Molecular Genetics, 2135 Microbiology Building, University of Maryland, College Park MD 20742, USA
e-mail: dinman@umd.edu

J.F. Atkins, R.F. Gesteland (eds.), *Recoding: Expansion of Decoding Rules Enriches Gene Expression*, Nucleic Acids and Molecular Biology 24,
DOI 10.1007/978-0-387-89382-2_15,

15.1 Introduction

Translation systems in living cells are fast, accurate and complex. Among the most complex feats performed during protein synthesis is the selection of a correct tRNA and its faithful translocation through the ribosome. Our current understanding of this process derives from the combined approaches of genetics, biochemistry, and structural biology. Genetics is well suited to the dissection of complex processes. Thus, the isolation and analysis of mutants that alter the fidelity of translation, as well as the analysis of recoding phenomena that circumvent the normal decoding process, have provided unique insights into the rules that govern accurate translation in wild-type cells. Recent crystallographic investigations have advanced hugely our understanding of the mechanisms of tRNA selection, translocation, and termination. In addition, these structural investigations are beginning to shed light on how mutant translational components influence decoding. Extensive genetic analyses of decoding-related processes have been carried out in yeast and *Escherichia coli*. As befits a highly conserved process, there is remarkable overlap in the types of yeast and bacterial mutants that affect decoding. Nonetheless, each organism has produced mutants that reflect the unique features of the differing protein synthesis systems. This chapter reviews the range of mutants that have been recovered in yeast and bacteria and suggests how they may influence decoding in the light of our current knowledge of translation.

15.2 Bacterial Mutants Affecting the Fidelity of Decoding

E. coli has served as a model organism for genetic and biochemical analyses of many fundamental processes, including translation. The high degree of conservation of translation components has meant that structural and functional data on protein synthesis systems from other organisms can usually be applied successfully to the *E. coli* system. The recent availability of crystal structures of *E. coli* ribosomes (Schuwirth et al., 2005) offers the advantages of carrying out genetic, structural, and biochemical analyses on translational components from a single organism. Historically, much of the early genetic and biochemical experiments focused on the protein components of the ribosome. Moreover, since the ribosomal RNA (rRNA) genes are present in seven copies per chromosome, conventional, genetic approaches could not readily be used to generate *E. coli* rRNA mutants. With the switch in focus from the protein to RNA components of the ribosome, various strategies were devised to facilitate genetic analyses of *E. coli* rRNAs. This has culminated in the generation of strains of *E. coli* that lack chromosomal *rrn* operons and express all rRNA from a plasmid-borne *rrn* operon. These strains, as well as the development of "specialized ribosomes" have greatly facilitated genetic and biochemical analysis of rRNA function in *E. coli* (Asai et al., 1999; Lee et al., 1996). While most of the ribosomal mutants affecting the accuracy of decoding have been isolated in *E. coli*, many interesting mutants have also been identified in its close relative, *Salmonella enterica* serovar Typhimurium (Björkman et al.,

1999). In addition, the availability of high-resolution crystal structures of ribosomes from *Thermus thermophilus* has spurred the development of genetic systems in that organism (Gregory et al., 2005). While much of this review draws from the *E. coli* literature, relevant data from these other bacterial systems are also included.

15.2.1 The Ribosome as a "Recognition Screen" for tRNAs

15.2.1.1 The First Ribosomal Accuracy Mutants

The idea that the ribosome plays an active part in the discrimination between correct and incorrect tRNAs was formulated by Gorini in the 1960s and 1970s. The starting point for these studies was the analysis of streptomycin-resistant mutants that decreased the inherent leakiness of an amber mutation in the *argF* gene and which Gorini attributed to ambiguous reading by wild-type tRNAs (Gorini, 1971). The ribosomal component responsible for streptomycin resistance and hyperaccuracy was identified as ribosomal protein S12 and additional mutants were isolated that restricted miscoding to varying levels (Ozaki et al., 1969). Subsequently, streptomycin was found to stimulate the incorporation of near-cognate amino acids in vitro (Davies et al., 1964). Additional components that reversed S12's effect on decoding were subsequently isolated and identified as ribosomal proteins S4 and S5 (Zimmermann et al., 1971; Piepersberg et al., 1975). When only the altered S4 or S5 was present in a strain, the mutant ribosomes showed increased misreading in vivo and in vitro and were designated *ram* for ribosomal ambiguity (Gorini, 1971).

The pioneering experiments described above set the stage for much subsequent analyses of ribosomal accuracy. Variations of the selections used by Gorini have yielded many more mutants affecting the accuracy of decoding, both in *E. coli, S. enterica, T. thermophilus* and in other model organisms.

15.2.2 The Decoding Mechanism, Streptomycin and Ribosomal Proteins S4, S5, and S12; Insights from Crystallography

High-resolution crystallographic analyses of 30S ribosome complexes by Ramakrishnan and coworkers have allowed a detailed view of the decoding mechanism. Correct pairings at the first and second codon–anticodon positions are monitored by interactions with the 16S rRNA residues A1492, A1493, and G530 (Ogle et al., 2003). These residues interact with the minor groove of the codon–anticodon helix and sense the shape and geometry of Watson–Crick pairs. Interactions other than Watson–Crick pairs are tolerated at the third codon–anticodon pair. Residues of ribosomal protein S12 also participate in these interactions; Ser50 interacts with A1492 at the second codon–anticodon pair while Pro48 interacts with the mRNA base at the third codon position. Finally, A1054 of 16S rRNA stacks on the third anticodon base.

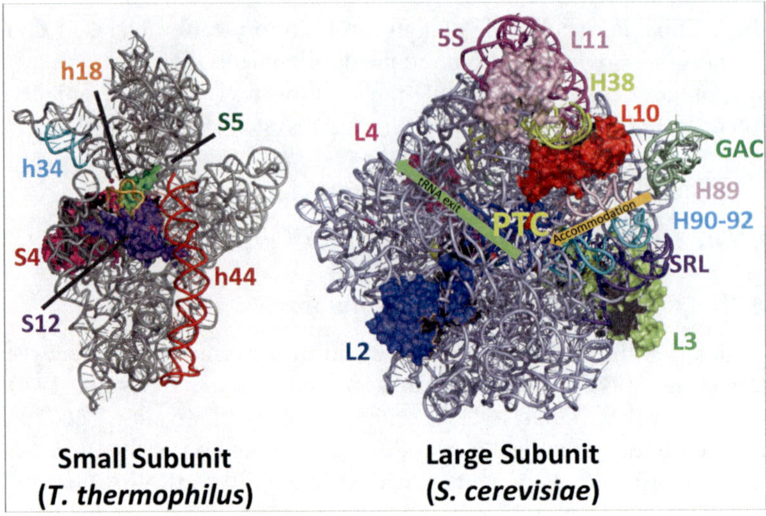

Fig. 15.1 The small (*left*) and large (*right*) ribosomal subunits. The small subunit is the *Thermus thermophilus* 30S structure of Selmer et al. (2006; PDB file 2J00). The large subunit is the yeast 60S structure of Spahn et al. (2004; accession code 1S1I). In both cases, helices are numbered according to the *E. coli* numbering system. In the small subunit, the indicated ribosomal proteins refer to the *Thermus/E. coli* proteins. In the large ribosomal subunit, the proteins are labeled using the yeast numbering system. Yeast L11 and L10 are homologs of bacterial L5 and L16, respectively. The other indicated large subunit proteins have the same nomenclature in yeast and bacteria

In addition to local conformational changes at the decoding center, binding of cognate tRNAs also triggers larger-scale movements of 30S domains. These global movements involve rotation of the head and shoulder of the 30S subunit toward the intersubunit space and platform regions, to form the so-called closed conformation. Near-cognate tRNAs are unable to achieve the closed conformation of the 30S subunit and are preferentially discarded from the ribosome. These studies also provide insights into the mechanism of error stimulation by streptomycin and paromomycin. Streptomycin binds near helices 18 and 27 of 16S rRNA and by establishing interactions between the shoulder region (including S12 and h18 in 16S rRNA; Fig. 15.1) and the h27/h44 regions of the 30S subunit, it can facilitate transition to the closed conformation, even in the absence of a cognate tRNA. Paromomycin binds at the decoding center and by inducing some of the same structural changes normally induced by cognate tRNAs, the antibiotic can compensate in part for lack of complete codon–anticodon complementarity. While several of the mutations in S12 and 16S rRNA that confer resistance to streptomycin are in residues that contact streptomycin others are remote from the streptomycin binding site and may confer resistance by affecting the open–closed equilibrium. As observed with S12 mutations, many of the rRNA mutations conferring streptomycin resistance also have an accuracy phenotype. Thus, base changes in the U13, and C912 regions, which are at/near sites of contact with streptomycin, increase the accuracy of decoding (Pinard et al., 1994; Lodmell et al., 1995).

The structural changes that occur in the 30S subunit upon transition from an open to a closed conformation also help explain the effects of some of the mutations in ribosomal proteins on tRNA selection. Transition to the closed conformation requires (among other rearrangements) disruption of interactions between proteins S4 and S5 and the formation of new sets of interactions between S12 and helices h27 and h44 of 16S rRNA (Fig. 15.1). At least some of the S4 and S5 ram mutations lie on the interface between the two proteins and consequently, it has been proposed that these disruptions facilitate transition to the closed conformation in the mutant ribosomes. Conversely, at least some of the hyperaccurate mutations in S12 disrupt S12–rRNA interactions that are formed upon domain closure. However, the phenotypes of some S4 alleles challenge the generality of this proposal. Björkman et al. (1999) have described novel S4 alleles that also map to the S4/S5 interface yet are restrictive and confer resistance to streptomycin. The isolation of a restrictive C912U mutation in 16S rRNA as a suppressor of a streptomycin-dependent S12 mutation also shows that not all ribosomal mutations that suppress restrictive S12 mutants are *ram* like (Vila-Sanjurjo et al., 2007).

15.2.2.1 Decoding Center rRNA Mutants

Nucleotides G530, A1492, A1493, and C1054 (*E. coli* numbering) are intimately involved in the decoding process and unsurprisingly, mutations at or near these residues impact on the efficiency and accuracy of translation. Base substitutions at G530, A1492, and A1493 are all lethal. However, base substitutions at C1054, as well at C1200 across the helix (h34), have been isolated in genetic screens for nonsense suppressors (Murgola et al., 1988; Gregory and Dahlberg, 1995). Base substitutions in the 530 loop (h18) or loss of the m^7G modification at G527 have been associated with resistance to streptomycin in bacteria and organelles. In *E. coli*, the A523C streptomycin-resistant ribosomes are refractory to antibiotic-induced misreading (Melancon et al., 1988). In addition, base substitutions at positions G517 and G529 in *E. coli* have also been shown to promote stop codon readthrough and frameshifting (O'Connor et al., 1995). The G529U mutation also supports misreading of sense codons during elongation and of non-AUG codons at initiation, suggesting that both A and P sites are affected by this alteration. Defects in tRNA discrimination at both A and P sites are also observed in ribosomes carrying mutations at several different positions in the upper part of h44, near the decoding center (Fig. 15.1; nt C1395,C1400, C1407, or G1505), or lacking the KsgA-associated $m_2{}^6A$ modifications at A1518 and A1519 (O'Connor et al., 1997). While no structural analyses have been carried out on these h44 mutant ribosomes, potentially, the single base changes in 16S rRNA alter the structures of the adjacent A and P sites.

15.2.2.2 The Intersubunit Bridges

In the 70S ribosome, the large and small subunits are held together via a series of non-covalent connections, or bridges, involving RNA–RNA, RNA–protein, and protein–protein interactions (Schuwirth et al., 2005). In addition to holding the subunits together in an appropriate configuration for protein synthesis, the bridges have been proposed to serve as conduits for the transmission of signals between the large and small subunits. A subset of bridge connections are disrupted during translocation, while several other bridges are adjacent to tRNA binding sites. The intersubunit bridges thus have the potential to influence translocation and tRNA selection in multiple ways during translation. The results of mutagenesis support a role for intersubunit bridges in maintaining the accuracy of translation. In a systematic mutagenesis of the 23S rRNA components of seven intersubunit bridges, Liiv and O'Connor (2006) showed that mutations in six bridges (B1a, B2b, B2c, B3, B5, and B7a) affected stop codon readthrough. No bridge B4 mutants were recovered that stimulated readthrough. In a similar study targeting 30S bridge components, disruptions of bridges B5, B6, and B8 also affected stop codon readthrough (Q. Sun and M. O'Connor, unpublished). Other analyses indicate a role for h38 of 23S rRNA and protein S13 (components of bridges B1a and B1b) in regulating translocation and tRNA–ribosome interactions, respectively (Cukras and Green, 2005; Sergiev et al., 2005a; Komoda et al., 2006).

Bridge B2a has long been linked with subunit association, factor-binding, and tRNA selection functions. This bridge is formed by interaction of the h69 stem loop in 23S rRNA (nt 1905–1925) and the upper part (nt 1408–1490) of h44 in 16S rRNA, adjacent to the decoding center. In addition, elements of this stem loop contact the D-arms of both A and P site tRNAs. Several single and multiple base mutations in this loop increase readthrough of all three stop codons and also increase both +1 and −1 frameshifting (O'Connor and Dahlberg, 1995; Hirabayashi et al., 2006). These experiments suggest a role for h69 in tRNA selection via interactions with tRNA and/or by affecting 30S–50S signaling. This conclusion is challenged by the findings of Ali et al. (2006) who showed that mutant ribosomes carrying a complete deletion of h69 (nt 1906–1930) could translate mRNA faithfully in vitro, without detectable misincorporation. However, the mutant ribosomes were defective in RF1-mediated termination, consistent with the interaction of h69 with RF1 (Laurberg et al., 2008). Further links between h69 and RF function come from analyses of *rluD* mutants. RluD is the pseudouridine synthetase responsible for the Ψ1911, Ψ1915, and Ψ1917 modifications. Inactivation of RluD increases stop codon readthrough and reduces cell growth substantially (Ejby et al., 2007). The UGA readthrough and growth phenotypes can be suppressed by *prfB* mutations encoding RF2 that increase the efficiency of termination (Ejby et al., 2007). These results prompt a reevaluation of the genetic data linking h69 with tRNA selection. The effects of h69 mutants on readthrough and at least some instances of frameshifting can be explained by an altered RF–ribosome interaction. Nonetheless, some h69 mutants stimulate frameshifting in the absence of an adjacent stop codon (O'Connor and Dahlberg, 1995) and at least one h69 mutant (ΔA1916) suppresses a missense mutation (O'Connor, 2007), suggesting that h69 plays a role in the decoding of sense codons and frame maintenance. Potentially, some h69 mutations may have differential effects on decoding and termination; the complete h69 deletion used by Ali et al. (2006) could affect RF–ribosome interactions without altering the geometry of tRNA–mRNA–ribosome interactions, while other h69 alterations, such as ΔA1916, may affect both decoding and termination.

Several bridge mutations also show genetic interactions with restrictive S12 alleles. In an extensive search for compensatory mutations that ameliorated the growth defect of a K42N substitution in S12, Maisnier-Patin et al. (2002) recovered several alterations in protein L19, in addition to the expected S4, S5, and S12 alterations. L19 contacts the lower part of h44, as well as the 345 stem loop of 16S rRNA, to form bridges B6 and B8, respectively. In addition, mutations in the rRNA components of bridges B5, B6, and B8 also compensate for the growth defects of S12 mutants (Q. Sun and M. O'Connor, unpublished). These data suggest a link between bridge function and the open/closed transition on the 30S subunit. Although the open–closed transition has only been analyzed on 30S subunits, these rearrangements occur on 70S ribosomes in vivo and must also require considerable rearrangement of the 30S–50S interface. Potentially, loosening of bridge contacts may facilitate domain closure, explaining the accuracy phenotype of some bridge mutants.

15.2.2.3 The Peptidyltransferase Center and Beyond

The active center of the 50S subunit is responsible for both peptide bond formation and peptide release. Despite its remoteness from the decoding center in the 30S subunit, mutations in several of the 23S rRNA residues affecting tRNA positioning and/or peptidyltransferase activity have also been shown to affect stop codon readthrough and frameshifting, linking these two activities of the 70S ribosome.

Biochemical, genetic, and structural studies have shown that peptide bond formation depends critically on the precise alignment of the $CCA^{3'}$ ends of the A and P site tRNA substrates in the active center of the 50S subunit (Brunelle et al., 2006). This positioning involves base pairing interactions between one or more residues of the tRNAs' $CCA^{3'}$ ends and complementary bases in the A- and P-loops, respectively, of 23S rRNA. Disruption of these base pairs, through mutations either in the tRNA, or in the P loop residues G2252 and G2253, or in the A-loop residue U2555, increase frameshifting and readthrough of stop codons (O'Connor and Dahlberg, 1993; O' Connor et al., 1993; Gregory et al., 1994). In contrast, loss of the A-loop Um2552 modification has severe effects on ribosomal function and decreases the levels of programmed frameshifting and stop codon readthrough (Widerak et al., 2005). In addition to the A and P loop mutations, several alterations to 23S rRNA residues close to the sites of interaction with the tRNA's $CCA^{3'}$ end also affect stop codon readthrough and frameshifting. These include mutations in h89 (Fig. 15.1), base substitutions at Ψ2504, D2449, G2447, and A2451 at the active site of the 50S subunit (O'Connor et al., 2001; O'Connor and Dahlberg, 1995; Thompson et al., 2001; M. O'Connor, unpublished). In contrast, mutations at position G2585 decrease miscoding in vitro (Saarma and Remme, 1992). Several of these rRNA mutations have been shown to affect peptide bond formation, and the A2451 mutations in addition show modest defects in peptide release (Youngman et al., 2004). With the exception of the G2585 variants, these mutants have not been tested in a true miscoding assay and it is unclear if the reported increases in frameshifting and readthrough reflect perturbation of the tRNA selection process per se (at the accommodation step, for instance), or an effect on termination, or both.

15.2.3 Factor Interaction Sites

15.2.3.1 The GTPase-Associated Center (GAC), The Sarcin–Ricin Loop (SRL), and L7/L12 Stalk

Binding of the translation factor GTPases (IF2, EF-Tu, EF-G, and RF3) to the 50S ribosomal subunit and subsequent GTP hydrolysis involves their interaction with several distinct elements, including the conserved sarcin–ricin loop in 23S rRNA and the multi-component L7/L12 stalk (Fig. 15.1). The stalk is composed of multiple copies of protein L7 and its N-acetylated derivative, L12, proteins L10 and L11, and the 23S rRNA element to which L10 and L11 bind. The sarcin–ricin loop (SRL) is adjacent to the stalk base and is bound by protein L6. The SRL interacts with the G domains of the translation factor GTPases and alteration of the ribosomal elements that control GTP hydrolysis have the potential to influence the filling of the ribosomal A site by aa-tRNA-EF-Tu.GTP. In a variety of selections, mutations in the SRL, the region of 23S rRNA that binds L10 and L11, as well as in proteins L6, L11, and L7/L12 have been isolated and shown to affect various aspects of translational accuracy.

Cryo-EM analysis of kirromycin-stalled ribosomes carrying ternary complexes shows that the SRL contacts EF-Tu directly (Valle et al., 2002, 2003). Single base mutations at G2661C decrease UGA readthrough and +1 frameshifting while mutations at A2654 and C2666 have the opposite effects (Melancon et al., 1992; O'Connor and Dahlberg, 1996). In vitro analyses by Bilgin and Ehrenberg (1994) showed that the G2661C mutation reduced the binding rate of cognate ternary complexes to the A site, reduced the rate of EF-Tu-dependent GTP hydrolysis, and increased the stringency of initial selection. In addition, the rate of binding of EF-G to the mutant ribosomes was also decreased. Potentially, by delaying ribosome-bound ternary complexes from achieving the GDP state, the G2661 mutations may facilitate dissociation of near-cognate ternary complexes, while the A2654 and C2666 mutants may have the opposite effects.

Selection for gentamicin resistance has allowed the isolation of L6 mutants (Ahmad et al., 1980). In vitro, the L6 mutant ribosomes are refractory to aminoglycoside-induced misreading and in vivo, the mutants decrease the leakiness of the *argF40* amber mutation (Kühberger et al., 1979). While the precise step in translation that is affected by mutant L6 has not been identified, its connection with SLR and factor-binding functions suggests that L6 mutants may affect the accuracy of translation by altering EF-Tu–ribosome interactions.

L7/L12 mutants that affect the fidelity of decoding have been described by Kirsebom and Isaksson (1985). In vivo, the mutants display increased readthrough of all three stop codons and at least one of the mutants shows increased (2.5 fold) levels of misreading in vitro (Kirsebom et al., 1986). Subsequent kinetic analyses of the mutant ribosomes showed that the altered L7/L12 affected the interactions with both EF-Tu and EF-G (Bilgin et al., 1988). In addition to effects on EF-Tu and EF-G, it is also likely that the L7/L12 mutant ribosomes are altered in their interactions with IF2 and RF3, since all four factors have been shown to bind directly to isolated L12 (Helgstrand et al., 2007).

The complex of L11 and its 58 nt 23S rRNA binding site have been termed the GTPase-associated center (GAC; Fig. 15.1), in light of its association with the function of GTPase factors on the ribosome. Biochemical experiments using L11 mutants or L11-depleted ribosomes have also implicated L11 in the termination activities of RF1 and/or RF2 (Tate et al., 1983; Van Dyke et al., 2002; Bouakaz et al., 2006; Sato et al., 2006). While there is general agreement from these biochemical studies that RF1 function is compromised by the loss or alteration of L11, the effect on RF2 is contentious. The structures of 70S ribosome complexes carrying either RF1 or RF2 have been solved at low resolution (Petry et al., 2005). From these structures, it can be seen that binding of either RF causes a reorientation of the N-terminal domain of L11. However, there is no obvious difference in the structures of L11 in the RF1- vs. the RF2-containing complexes. The discrepancies in these various results likely derive from the different assay systems employed, as well as from the possible effects of some L11 mutants on expression of the downstream *rplA* gene encoding protein L1.

L11 mutants have also been isolated on the basis of their effects on ppGpp accumulation in both *E. coli* and *Streptomyces* spp. (Friesen et al., 1974; Parker et al., 1976; Kelly et al., 1991). Similar to other relaxed mutants, these L11 (*relC*) mutants display increased levels of mistranslation during amino acid starvation (Parker et al., 1976).

Mutants in the region of 23S rRNA that interacts with L11 have also been isolated in several genetic selections. The G1093A base substitution has been recovered as a UGA suppressor (Gregory and Dahlberg, 1995; Murgola et al., 1995) and in vitro, the mutant ribosomes show specific defects in RF2–ribosome interactions (Arkov et al., 1998). Mutations in adjacent residues (1094–1098) and at A1067 also show UGA suppressor phenotypes and ribosomes carrying the A1067U mutation display defects in EF-Tu and EF-G functions in vitro (Saarma et al., 1997). X-ray structures of termination complexes show differential interactions of RF1 and RF2 with this rRNA region, perhaps explaining the RF2-specific defects (Petry et al., 2005).

15.2.3.2 The Exit Site

Despite extended efforts, considerable controversy still surrounds the role of the E site in elongation, as well as the nature of codon–anticodon interactions at this site. According to one model advanced by Nierhaus and colleagues, the A and E sites are allosterically coupled (Blaha and Nierhaus, 2001). Occupancy of the E site by a cognate tRNA is proposed to enhance selection of cognate ternary complexes at the A site. The function of the E site has now begun to be addressed using mutants in E site rRNA residues (Sergiev et al., 2005b). In the 30S subunit, C-terminal residues of protein S7 contact position 41 of the E site tRNA (Korostelev et al., 2006). Deletions in this region of S7 promote −1 frameshifting, stop codon readthrough, and missense suppression (Robert and Brakier-Gingras, 2003). On the 50S subunit, the CCA$^{3'}$ end of the E site tRNA interacts with C2493 of 23S rRNA and mutations at this rRNA base stimulate readthrough and frameshifting (Sergiev et al., 2005b). These results are consistent with the allosteric three-site model. However, the C2493G mutant ribosomes are also defective in translocation, which may be related to the increased levels of frameshifting seen in this mutant.

15.2.3.3 Other Ribosomal Protein Mutants

Mutations in the genes encoding ribosomal proteins S15, S17, and S20 have been isolated that affect decoding. These proteins are remote from tRNA and factor-binding sites, but are important for subunit assembly. In many cases, the mutant ribosomes are defective in multiple steps of translation. Incorrectly assembled subunits likely have multiple structural alterations, making it difficult to identify the structural defects that contribute to altered decoding (Bollen et al., 1975; Topisirovic et al., 1977; Yano and Yura, 1989; Ryden-Aulin et al., 1993).

Several of the characterized instances of frameshifting promoted by ribosomal mutants can be attributed to slippage of near-cognate tRNAs into alternate reading frames in the P site (O'Connor et al., 1997). A role for protein L9 in restraining mRNA slippage derives from studies on bypassing (Herr et al., 2001 and Chapter 17 by Wills in this text). Ribosomes carrying altered L9 or lacking L9 altogether promote dissociation and re-pairing at a range of different codons (Leipuviene and Bjork, 2007).

Erythromycin resistance in *E. coli* can arise through mutations or modifications in the 23S rRNA or via alterations to ribosomal proteins L4 or L22. In vivo, the L4 mutant ribosomes supported increases in readthrough of stop codons, +1 and −1 frameshifting, missense suppression, and increasing initiation from non-AUG (O'Connor et al., 2004). The L4 mutant also displays a defect in peptide bond formation and an altered response to several antibiotics. Clearly, multiple steps in translation are altered in the L4 mutant. The K63E substitution in the protein is distant from both the decoding and peptidyltransferase centers and any effects of mutant L4 on these functions must be indirect. Cryo-EM analyses have shown that there are many local and long-range structural alterations (in both subunits) in the L4 mutant ribosomes (Gabashvili et al., 2001). In addition to a narrowing of the tunnel, the h69 and L7/L12 stalk regions were altered in the mutant. Both of these regions have the potential to influence tRNA selection (at A and P sites) via their influence on tRNA and factor interactions with the ribosome.

Since the pioneering work of Gorini and colleagues, genetic selections have identified a large number of sites on the ribosomal components affecting the fidelity of protein synthesis. Biochemical characterization as well as mapping the mutant sites on high-resolution structures of the ribosome can suggest various mechanisms to explain these effects. However, despite the development of sophisticated biochemical techniques to dissect the mechanism of translation, few mutants have been characterized using such assays. Similarly, there is a dearth of information on the structural changes induced by rRNA or protein mutations. From the few structural analyses that have been carried out on mutant ribosomes, both long-range alterations and changes in the vicinity of the mutation are observed (Gabashvili et al., 2001; Gregory and Dahlberg, 1999). Consequently, the possibility that the altered ribosomal structural element that is impinging on the decoding process is distant from the site of mutation cannot be discounted. Structural and biochemical characterization of mutant ribosomes are likely to be highly informative and will undoubtedly provide further insights into the mechanism of ribosome function.

15.3 Mutants Affecting Translational Recoding in Eukaryotes

The fundamental aspects of protein synthesis are shared by all three kingdoms of life: a two-subunit ribosome employs a two-step tRNA selection process to ensure high-fidelity translation of genetic information contained in mRNAs. Translational recoding also occurs in eukaryotic systems and has also exploited to investigate translational fidelity by the eukaryotic protein synthetic apparatus. However, a few important features differentiate translational recoding in eukaryotes from bacteria. These include the larger size of the eukaryotic ribosome and the fact that eukaryotes do not employ a mechanism analogous to the Shine–Dalgarno sequence. Both of these differences affect how the decoding center can be re-positioned relative to *cis*-acting mRNA signals responsible for recoding. Another salient difference lies in the mechanisms employed for translational recoding. In bacteria, it appears that each recoding signal is unique. In eukaryotes, enough examples of *P*rogrammed *R*ibosomal *F*rameshifting (PRF) employed by RNA viruses and retrotransposable elements have been defined to enable the proposal of some general models of PRF. One example of this is the "Integrated Model" of PRF (Harger et al., 2002) in which PRF is placed within the context of the translation elongation cycle. For example, PRF in the 3′ or +1 direction (+1 PRF) is driven by the presence of an unoccupied "hungry codon" in the ribosomal A site, which enables the P site tRNA to slip forward by one base. Therefore, the Integrated Model predicts that +1 PRF must occur after translocation and prior to delivery of aa-tRNA to the A site. In contrast, PRF in the 5′ or −1 direction (–1 PRF) requires the simultaneous slippage of tRNAs occupying the ribosomal A and P sites by one base. Thus, −1 PRF must occur after delivery of aa-tRNA to the A site and not after translocation. Therefore, the Integrated Model of PRF posits that −1 and +1 PRF occur at different stages of the translation elongation cycle. Corollary to this is that +1 and −1 PRF can be used to specifically probe the contributions of both *cis*-acting elements and *trans*-acting factors to translational fidelity during different phases of the translation elongation cycle.

Recoding reflects a "deviation from the norm" in ribosome function: thus it can be exploited as a means to understand how the translational apparatus normally translates genetic information with high fidelity. Genetic approaches have proven very powerful in identifying components of the translational apparatus that affect this function and the genetic and biochemical malleable yeast has provided the model eukaryotic genetic system of choice. Screens for changes in ribosome function and recoding have employed direct quantitative assays of recoding efficiencies, as well as indirect assays including changes in response to antibiotics and changes in the ability of cells to maintain PRF-dependent viruses and virus-like elements. For example, in yeast the Ty retrotransposable elements are dependent on correct levels of +1 PRF, while the dsRNA "killer" virus requires just the right frequency of −1 PRF. In parallel, biochemical and structural methods are used to characterize how changes in ribosome structure affect its various functions and how these changes in turn affect recoding. Together, these synergistic strategies have been used to further our understanding of how ribosomes normally decode genetic information and how they can be reprogrammed by *cis*-acting mRNA elements and *trans*-acting factors. This section discusses known eukaryotic mutants that affect translational recoding within this conceptual framework.

15.3.1 Cis-Acting Elements Affecting Translational Recoding

15.3.1.1 The Ribosome

The bulk of the ribosome consists of ribosomal RNA (rRNA), into which are embedded highly basic ribosomal proteins (RPs). Eukaryotic ribosomes contain four rRNA species (25S, 5.8S, and 5S in the large subunit, and 18S in the small subunit) and approximately 80 RPs depending on the species. It is now clear that the major functions are primarily performed by the rRNAs and that RPs play a more ancillary role. As noted above, genetic methods have been instrumental in identifying components of the translational apparatus that affect recoding. Given that rRNA is the major component of the ribosome, it was reasonable to assume that these molecules should be important for recoding and that rRNA mutants would provide significant insight into the issue. Genetic approaches to rRNA have been complicated by the fact that eukaryotic cells contain >100 copies of the rDNA genes encoding rRNAs, most likely due to the high demand for ribosomes by rapidly growing cells (Warner, 1999). A genetic observation originally made by E. Morgan made it possible to study rDNA mutations in yeast, however. Specifically, treatment of yeast cells with the translational inhibitor hygromycin selects strongly for cells expressing a specific rRNA mutant and selects against cells expressing any wild-type rRNAs. Based on this, hygromycin was used to generate yeast strains lacking chromosomal rDNA genes and in which rRNAs are expressed from episomal plasmids. Although conceptually simple, the practical implementation of this system required almost 20 years due to the large sizes of rDNA-bearing plasmids, high levels of rRNA expression, and high recombination rates (see Rakauskaite and Dinman, 2006).

15.3.1.2 18S rRNA Mutants

The first two yeast rRNA mutants affecting translational fidelity were isolated in yeast 18S rRNA based on their altered sensitivities to streptomycin and paromomycin, two inhibitors of protein synthesis (Chernoff et al., 1994). These were located in the 18S rRNA: *rdn2* (G517A, helix 18) and *rdn4* (C912U, helix 27) (*E. coli* 16S rRNA numbering; Fig. 15.1). Subsequently, this group used oligonucleotide site-directed mutagenesis to generate the *rdn1* series of mutants of C1054 (*E. coli* numbering) in the decoding center (helix 44) of the small subunit (where *rdn1A* = C1054A, *rdn1G* = C1054G, *rdn1T* = C1054U) (Chernoff et al., 1996). In addition to characterizing the effects of these mutants on resistance/hypersensitivity to translational inhibitors, this study also examined both codon-specific and -nonspecific termination suppression, providing the first demonstration of the role of specific rRNA residues in translation termination. Additional alleles in helix 18 (*rdn12A* = C526A), helix 27 (*rdn6* = G888A, *rdn8* = G886A), and helix 44 (*rdn15* = A1491G) were subsequently obtained based on either their ability to suppress nonsense mutations, to act as antisuppressors of the [*PSI*$^+$] prion, or to suppress other mutants of eukaryotic release factor 3 (eRF3), encoded by the *SUP35* gene (Velichutina et al., 2000, 2001). Importantly, the *rdn4*, *rdn6*, and *rdn8* mutants also decreased ribosome affinities for aminoacyl-tRNA (aa-tRNA), linking changes in ribosome biochemistry to function. A later study also showed that *rdn1T* and *rdn2* ribosomes were hyperaccurate and that *rdn1A* promoted decreased peptidyltransferase activity, thus functionally linking the small and large subunits (Konstantinidis et al., 2006). A collaborative study with the Farabaugh laboratory demonstrated that many of these mutants affected +1 PRF directed by either the *Ty*1 or *Ty*3 retrotransposable elements of yeast (Burck et al., 1999). Specifically, *rdn*1*T* inhibited +1 PRF by both recoding signals; *rdn*1*A* only inhibited *Ty*3-directed +1 PRF; and *rdn2* and *rdn4* specifically inhibited *Ty1*-mediated +1 PRF, consistent with the integrated model (Harger et al., 2002). More recently, the Bedwell laboratory identified two additional mutants in the decoding region of helix 44 (Fan-Minogue and Bedwell, 2008). The viable mutants G1645A and G1645C (A1408 in *E. coli*) and A1754G (G1491 in *E. coli*) differentially suppressed the three stop codons and missense mutations in the presence of paromomycin.

15.3.1.3 25S rRNA Mutants

The Liebman group also used random mutagenesis to identify a mutant in the large subunit of yeast. The *rdn5* mutant (C2025U, yeast 25S rRNA numbering) was found by its ability to suppress the *ade1-14*, *trp1-289* (UGA), and *lys2-L63* (UGA), (UAG) alleles (Liu and Liebman, 1996). Interestingly, this mutant was not able to suppress *his7-1* (UAA). Importantly, *rdn5*-1 also suppressed the +1 frameshift *his4-713* allele, demonstrating the involvement of rRNA in translational reading frame maintenance. Follow-up studies demonstrated that *rdn5* inhibited Ty*1*-promoted +1 PRF (Burck et al., 1999) and decreased eEF2-dependent translocation of Ac-Phe-tRNA from the A to P site (Panopoulos et al., 2004). In later studies, two viable yeast 25S rRNA mutants in the vicinity of the peptidyltransferase center (PTC; Fig. 15.1), C2820U and Ψ2922C, did not alter rates of PRF, but did affect recognition of nonsense codons, but only in the presence of the [PSI$^+$] prion (Rakauskaite and Dinman, 2008). Similarly, mutants located in helix 38 (the A site finger) altered nonsense codon recognition but not PRF (Rakauskaite and Dinman, 2006).

15.3.1.4 5S rRNA Mutants

A forward genetic screen in yeast identified the *mof9-1* mutant, which was found to encode 5S rRNA; this was the initial observation that mutant alleles of 5S rRNA could promote semi-dominant effects on both −1 and +1 PRF (Dinman and Wickner, 1995). A follow-up study mutagenized 5S rRNA to near-saturation, identifying multiple semi-dominant alleles affecting nonsense suppression, −1 PRF, and virus maintenance (Smith et al., 2001). These tended to map in three general regions of the molecule: along helices II and III where it interacts with ribosomal proteins L5 and L11; in helix IV where it interacts with 25S rRNA helix 38; and in loop D where it interacts with ribosomal protein L10. A later study identified seven alleles of 5S rRNA viable as the sole forms expressed in yeast (Kiparisov et al., 2005). Six of these promoted increased rates of −1 PRF and killer virus maintenance. That study also more closely analyzed additional semi-dominant 5S rRNA alleles, showing that many affected both −1 and +1 PRF. Interestingly, the naturally occurring *RDN5-1* allele of yeast inhibited −1 PRF. Similarly, the naturally occurring form of 5S rRNA found in *Xenopus* oocytes inhibited −1 PRF in a semi-dominant fashion, but the form expressed in somatic cells did not. None affected +1 PRF. This suggests that cells may regulate −1 PRF (and hence gene expression) via differential expression of 5S rRNA variants.

15.3.1.5 rRNA Base Modification Mutants

Co-transcriptional modifications of rRNAs include 2′ O methylation (Nm) and pseudouridylation (Ψ) (reviewed in Decatur and Fournier, 2002), and Ψ defects have been found to cause human disease (Ruggero et al., 2003). A survey of Ψ- and Nm-deficient yeast strains revealed allele-specific defects in translational fidelity (Baxter-Roshek et al., 2007). Cells unable to Nm modify both Um2920 and Gm2921 (*spb1Da/snr52*Δ) in the A-loop promoted increased rates of −1 PRF and loss of the yeast killer virus, but did not affect +1 PRF. These also promoted hyperaccurate decoding of UAA and UAG, but not of UGA codons. Cells unable to produce Ψ2922 (*snr10*Δ) in the A-loop promoted hyperaccurate decoding of all three termination codons, while those deficient in Ψ2974 (*snr42*Δ) at the base of helix 93 promoted hyperaccurate decoding of UAA and UGA, but not of UAG codons.

15.3.1.6 Ribosomal Proteins and Translational Inhibitors

Studies on the effects of ribosomal proteins (RPs) on translational recoding in eukaryotes have mainly taken advantage of yeast molecular genetics systems and the availability of ribosome-interacting translational inhibitors and have focused on RPs of the large subunit. Ribosomal protein L3 (encoded by *RPL3*) is the most thoroughly investigated of these. Based on the inability of the *mak8-1* allele of *RPL3* to maintain the killer virus (Wickner et al., 1982), an initial study determined that this was due to increased rates of −1 PRF (Meskauskas et al., 2003b). This work went on to show using strains deficient in ribosomal protein L41 that stimulation of

−1 PRF correlated with decreased rates of peptidyltransfer, consistent with previous data demonstrating that peptidyltransferase inhibitors specifically affected −1 but not +1 PRF (reviewed in Dinman et al., 1998; Harger et al., 2002). Additional studies of ∼100 alleles of L3 demonstrated that changes in −1 PRF specifically correlated with decreased peptidyltransferase activity and not to changes in affinity for aa-tRNA or with eEF2 (Meskauskas et al., 2005; Meskauskas and Dinman, 2007, 2008). Interestingly, despite the existence of many alleles of ribosomal protein L10 (physically located across the "aa-tRNA accommodation corridor" from L3; Fig. 15.1), none have been found to significantly affect −1 PRF; this correlates with no observed changes in peptidyltransferase activity, despite the observance of significant changes in ribosomal affinities for aa-tRNA and eEF2 (Petrov et al., 2008). Decreased peptidyltransferase activity promoted by the V48D, L125Q, and H215Y mutants of ribosomal protein L2 (located on the other side of the peptidyltransferase center from L3 and L10; Fig. 17.1) also resulted in elevated levels of −1 PRF, while none significantly affected Ty*1*-mediated +1 PRF (Meskauskas et al., 2008). In contrast, decreased affinity for peptidyl-tRNA enhanced rates of both −1 and +1 PRF in cells expressing the T28A (a.k.a. HA-2) mutant of ribosomal protein L5, demonstrating that slippage of tRNA in the P site is critical for both processes. Consistent with this model, sparsomycin, which increases the affinity of peptidyl-tRNA with the ribosome, inhibited +1 PRF (Dinman et al., 1997). L5 is located on the solvent side of the large subunit "crown." It interacts with 5S rRNA, which in turn interacts with ribosomal protein L11, which forms the intersubunit face of the crown. Preliminary results from the Dinman laboratory indicate that the F96N mutant of L11 promotes increased rates of Ty*1*-mediated +1 PRF. Interestingly, unlike the L5 mutants, this does not correlate with changes in peptidyl-tRNA binding, but rather correlates with decreased affinity for eEF2. This is consistent with the Integrated Model of PRF (Harger et al., 2002), which predicts that decreased affinity for eEF2 would promote slower rates of translocation, which in turn would promote increased residence times of ribosomes paused at the +1 PRF , thus enhancing frameshift rates.

15.3.2 Trans-Acting Elements Affecting Translational Recoding

The most likely places to look for *trans*-acting factors that could influence recoding would be those that are known to interact with the ribosome during elongation. This narrows the immediate search to tRNAs, the elongation factors eEF1 and eEF2, and the two release factors involved in termination (a special case of elongation) eRF1 and eRF3. Translational frameshifting was first identified in yeast in the form of dominant-negative frameshift-suppressing tRNAs. This class of genes was named *SUF* for *SU*ppression of *F*rameshift alleles (Roth, 1981). Although tremendously useful as tools for investigation of translational fidelity, no cases of "intentional" or "programmed" recoding by this class of tRNAs have been characterized to date. eEF1A is the eukaryotic equivalent of prokaryotic EF-Tu: it delivers aa-tRNA to the ribosome. In the event of a cognate tRNA–mRNA interaction, eEF1A hydrolyzes GTP, initiating aa-tRNA accommodation into the peptidyltransferase center (reviewed in Rodnina et al., 2005). Several alleles of eEF1A have been identified that affect either −1 or +1 PRF (Dinman and Kinzy, 1997). Consistent with the Integrated Model of PRF (Harger et al., 2002), no single allele affected both. Also consistent with the Integrated Model, alleles of eEF2 that affected +1 PRF but not −1 PRF have also been characterized (Harger et al., 2001). Similarly, translocation inhibitors such as pokeweed antiviral protein and sordarin stimulated +1 PRF, but did not affect −1 PRF (Harger et al., 2001; Tumer et al., 1998; T. Dever, pers. comm.). However, eEF2 mutants deficient in their ability to be ADP-ribosylated by diphtheria toxin and strains deficient for diphthamide modification enzymes showed apparent increases in −1 PRF (Ortiz et al., 2006). While this is consistent with a "co-translocational" model of −1 PRF (Takyar et al., 2005; Namy et al., 2006), it should be noted that these data were obtained using mono-cistronic reporter vectors, which, as described in greater detail below, have high rates of false positive indications of increased −1 PRF.

Genetic screens have been employed in yeast to identify other *trans*-acting factors affecting PRF. The first such screens, designed to detect increased rates of −1 PRF, identified the *MOF* (*M*aintenance *O*f *F*rame) and the *IFS* (*I*ncreased *F*rame *S*hifting) complementation groups (Dinman and Wickner, 1994, 1995; Lee et al., 1995). Upon cloning of the *MOF4/IFS1/UPF1* (Cui et al., 1996; Lee et al., 1995) and *MOF2/SUI1* (Cui et al., 1998a, b) genes, it became apparent that the design of the screens could not distinguish between increased rates of −1 PRF and stabilization of the mono-cistronic reporter mRNAs employed for the screens (Harger and Dinman, 2004). This problem led to the creation of bicistronic reporters for PRF, which internally control for differences in mRNA stability (Bidou et al., 2000; Harger and Dinman, 2003). Rescreening of the original nine *MOF* genes with bicistronic reporters revealed that only three, *mof1-1*, *mof6-1*, and *mof9-1*, actually enhanced −1 PRF, while the remaining *mof* and *ifs* mutants only affected mRNA stability.

The *MOF1* gene has not yet been cloned, and *MOF9* was found to encode an allele of 5S rRNA (see above). *MOF6* is an allele of *RPD3*, best known as a histone deacetylase (Meskauskas et al., 2003a). Increased −1 PRF also correlated with decreased peptidyltransferase activity in ribosomes harvested from cells expressing the *mof6-1* allele. Interestingly, deletions of genes encoding proteins that target the Rpd3p to the nucleolus but not to the nucleus also promoted delays in rRNA biogenesis and resulted in increased −1 PRF. We have suggested that targeting of the deacetylase activity of Rpd3p to the compartment where ribosome biogenesis occurs may allow it to remove acetyl groups from the N-termini of ribosomal proteins, enhancing their affinity for rRNA during the process of ribosome assembly, and that loss of this activity results in ribosome biogenesis defects that ultimately affect peptidyltransferase activity and hence rates of −1 PRF (Dinman, 2009).

The ribosome-associated chaperone complex (RAC) has also been implicated in −1 PRF (Muldoon-Jacobs and Dinman, 2006). Deletion of Ssb1p/Ssb2p or of Ssz1p/Zuo1p resulted in specific inhibition of −1 PRF but had no effects on +1 PRF. Quantitative measurements of growth profiles showed that translational inhibitors exacerbated underlying growth defects in these mutants. It was suggested that impaired chaperone activity may causes nascent peptides to back up into the exit tunnel of the ribosome, mispositioning the peptidyl-tRNA 3′ end. This in turn would inhibit the full accommodation of the aa-tRNA in the A site, inhibiting peptidyltransferase activity, and thus promoting increased −1 PRF.

Polyamines have also been implicated in regulating +1 but not −1 PRF. Autoregulation of ornithine decarboxylase antizyme expression by polyamines via +1 PRF is described in greater detail elsewhere in this book. Polyamines have also been shown to affect Ty*1*-directed +1 PRF (Balasundaram et al., 1994a, b). High levels of +1 PRF resulted from the combined effects of both spermidine deprivation and increased levels of intracellular putrescine consequent to derepression of the gene for ornithine decarboxylase (*SPE1*) in spermidine-deficient strains. However, examination of the polyamine biosynthetic pathway suggests that lack of spermidine would lead to depletion of arginine, the primary precursor of polyamines. This would decrease the abundance of Arg-$tRNA^{CCU}$, the cognate 0-frame A site tRNA at the Ty*1* slippery site, which would in turn promote increased frequencies of the +1 frameshift. Thus, the effect of polyamine depletion on Ty*1* frameshifting may be an indirect consequence of arginine metabolism. Finally, RNase L, an endoribonuclease that requires 2′ −5′ oligoadenylates to cleave single-stranded RNA, has been shown to affect human ODC antizyme-mediated +1 PRF via its interaction with eRF3 (Le Roy et al., 2005). Specifically, interaction of eRF3 with RNase L promoted increased translation readthrough efficiency at premature termination codons and increased ODC antizyme-mediated +1 PRF. These findings suggested that RNase L may be involved in regulating gene expression by modulating translation termination.

15.4 Concluding Comments

Genetic investigations over the last four decades have generated countless mutations in ribosomal components and translation factors. These studies have helped define the mechanisms of decoding, antibiotic resistance, subunit association, and ribosome assembly. The availability of X-ray crystal structures of ribosomal complexes at various stages of translation has now allowed the effects of some of these mutations to be described in structural terms. Genetic and structural analyses have been complemented by biochemical studies that have elucidated the distinct steps of translation in further detail. While novel genetic selections will undoubtedly uncover additional interesting mutants, ribosome structures and biochemical data are themselves now being used as the starting point for genetic studies. The function of distinct structural features of the ribosome, such as the intersubunit bridges and the large subunit ribosomal stalk, can be tested through mutagenesis and characterization of the mutant ribosomes. Moreover, in vitro characterization of ribosomal mutants is an essential step in the validation of mechanistic models derived from biochemical experiments. The development of site-directed mutagenesis and recombineering techniques as well as genetic systems for analyses of rRNA in yeast and bacteria allows for the construction of essentially any desired ribosomal mutation.

While in vivo genetic selections are limited by the viability of the respective mutant, strategies that permit purification of poorly functional ribosomes (from mixed populations of mutant and wild-type ribosomes) as well as techniques that permit in vitro selection of altered ribosomes have extended the range of mutants that can be analyzed. Along with biophysical, structural, and biochemical investigations, genetics continues to provide new and unanticipated insights into the translation process.

Acknowledgments Work in the authors' laboratories was supported by grants GM058859 and AI064307 from the National Institutes of Health (to J.D.D.) and MCB0745025 from the National Science Foundation (to M.OC.).

References

Ahmad MH, Rechenmacher A, Böck A (1980) Interaction between aminoglycoside uptake and ribosomal resistance mutations. Antimicrob Agents Chemother 18:798–806

Ali IK, Lancaster L, Feinberg J, Joseph S, Noller HF (2006) Deletion of a conserved, central ribosomal intersubunit. RNA bridge Mol Cell 23:865–874

Arkov AL, Freistroffer DV, Ehrenberg M, Murgola EJ (1998) Mutations in RNAs of both ribosomal subunits cause defects in translation termination. EMBO J 17:1507–1514

Asai T, Zaporojets D, Squires C, Squires CL (1999) An *Escherichia coli* strain with all chromosomal rRNA operons inactivated: complete exchange of rRNA genes between bacteria. Proc Natl Acad Sci USA 96:1971–1976

Balasundaram D, Dinman JD, Tabor CW, Tabor H (1994a) Two essential genes in the biosynthesis of polyamines that modulate +1 ribosomal frameshifting in *Saccharomyces cerevisiae*. J Bacteriol 176:7126–7128

Balasundaram D, Dinman JD, Wickner RB, Tabor CW, Tabor H (1994b) Spermidine deficiency increases +1 ribosomal frameshifting efficiency and inhibits Ty*1* retrotransposition in *Saccharomyces cerevisiae*. Proc Natl Acad Sci USA 91:172–176

Baxter-Roshek JL, Petrov AN, Dinman JD (2007) Optimization of ribosome structure and function by rRNA base modification. PLoS ONE:e174

Bidou L, Stahl G, Hatin I, Namy O, Rousset JP, Farabaugh PJ (2000) Nonsense-mediated decay mutants do not affect programmed −1 frameshifting. RNA 6:952–961

Bilgin N, Ehrenberg M (1994). Mutations in 23 S ribosomal RNA perturb transfer RNA selection and can lead to streptomycin dependence. J Mol Biol 235:813–824

Bilgin N, Kirsebom LA, Ehrenberg M, Kurland CG (1988) Mutations in ribosomal proteins L7/L12 perturb EF-G and EF-Tu functions. Biochimie 70:611–618

Björkman J, Samuelsson P, Andersson DI, Hughes D (1999) Novel ribosomal mutations affecting translational accuracy, antibiotic resistance and virulence of *Salmonella typhimurium*. Mol Microbiol 31:53–58

Blaha G, Nierhaus KH (2001) Features and functions of the ribosomal E site. Cold Spring Harb Symp Quant Biol 66:135–146

Bollen A, Cabezón T, de Wilde M, Villarroel R, Herzog A (1975) Alteration of ribosomal protein S17 by mutation linked to neamine resistance in *Escherichia coli*. I. General properties of neaA mutants. J Mol Biol 99:795–806

Bouakaz L, Bouakaz E, Murgola EJ, Ehrenberg M, Sanyal S (2006) The role of ribosomal protein L11 in class I release factor-mediated translation termination and translational accuracy. J Biol Chem 281:4548–4556

Brunelle JL, Youngman EM, Sharma D, Green R (2006) The interaction between C75 of tRNA and the A loop of the ribosome stimulates peptidyl transferase activity. RNA 12: 33–39

Burck CL, Chernoff YO, Liu R, Farabaugh PJ, Liebman SW (1999) Translational suppressors and antisuppressors alter the efficiency of the *Ty1* programmed translational frameshift. RNA 5:1451–1457

Chernoff YO, Newnam GP, Liebman SW (1996) The translational function of nucleotide C1054 in the small subunit rRNA is conserved throughout evolution: genetic evidence in yeast. Proc Natl Acad Sci USA 93:2517–2522

Chernoff YO, Vincent A, Liebman SW (1994). Mutations in eukaryotic 18S ribosomal RNA affect translational fidelity and resistance to aminoglycoside antibiotics. EMBO J 13:906–913

Cui Y, Dinman JD, Kinzy TG, Peltz SW (1998a). The Mof2/Sui1 protein is a general monitor of translational accuracy. Mol Cell Biol 18:1506–1516

Cui Y, Dinman JD, Peltz SW (1996) *mof4-1* is an allele of the *UPF1/IFS2* gene which affects both mRNA turnover and −1 ribosomal frameshifting efficiency. EMBO J 15:5726–5736

Cui Y, Kinzy TG, Dinman JD, Peltz SW (1998b) Mutations in the *MOF2/SUI1* gene affect both translation and nonsense-mediated mRNA decay. RNA 5:794–804

Cukras AR, Green R. (2005) Multiple effects of S13 in modulating the strength of intersubunit interactions in the ribosome during translation. J Mol Biol 349:47–59.

Davies J, Gilbert W, Gorini L (1964) Streptomycin, suppression, the code. Proc Natl Acad Sci USA 51:883–890

Decatur WA, Fournier MJ (2002) rRNA modifications and ribosome function. Trends Biochem Sci 27:344–351

Dinman JD (2009) The eukaryotic ribosome: current status and challenges. J biol Cgen 284:11761–11765

Dinman JD, Kinzy TG (1997) Translational misreading: Mutations in translation elongation factor 1a differentially affect programmed ribosomal frameshifting and drug sensitivity. RNA 3: 870–881

Dinman JD, Ruiz-Echevarria MJ, Czaplinski K, Peltz SW (1997) Peptidyl transferase inhibitors have antiviral properties by altering programmed −1 ribosomal frameshifting efficiencies: development of model systems. Proc Natl Acad Sci USA 94:6606–6611

Dinman JD, Ruiz-Echevarria MJ, Peltz SW (1998) Translating old drugs into new treatments: Identifying compounds that modulate programmed −1 ribosomal frameshifting and function as potential antiviral agents. Trends Biotechnol 16:190–196

Dinman JD, Wickner RB (1994) Translational maintenance of frame: mutants of *Saccharomyces cerevisiae* with altered −1 ribosomal frameshifting efficiencies. Genetics 136:75–86

Dinman JD, Wickner RB (1995) 5S rRNA is involved in fidelity of translational reading frame. Genetics 141:95–105

Ejby M, Sørensen MA, Pedersen S (2007) Pseudouridylation of helix 69 of 23S rRNA is necessary for an effective translation termination Proc Natl Acad Sci USA 104:19410–19415

Fan-Minogue H, Bedwell DM (2008) Eukaryotic ribosomal RNA determinants of aminoglycoside resistance and their role in translational fidelity. RNA 14:148–157

Friesen, JD, Fiil NP, Parker JM, Haseltine WA (1974) A new relaxed mutant of *Escherichia coli* with an altered 50S ribosomal subunit. Proc Natl Acad Sci USA 71:3465–3469

Gabashvili IS, Gregory ST, Valle M, Grassucci R, Worbs M, Wahl MC, Dahlberg AE, Frank J (2001) The polypeptide tunnel system in the ribosome and its gating in erythromycin resistance mutants of L4 and L22. Mol Cell 8:181–188

Gorini L (1971) Ribosomal discrimination of tRNAs. Nat New Biol 234:261–264

Gregory ST, Carr JF, Rodriguez-Correa D, Dahlberg AE (2005) Mutational analysis of 16S and 23S rRNA genes of Thermus thermophilus. J Bacteriol 187:4804–4812

Gregory ST, Dahlberg AE (1995) Nonsense suppressor and antisuppressor mutations at the 1409–1491 base pair in the decoding region of *Escherichia coli* 16S rRNA. Nucl Acids Res 23: 4234–4238

Gregory ST, Dahlberg AE (1999) Erythromycin resistance mutations in ribosomal proteins L22 and L4 perturb the higher order structure of 23 S ribosomal RNA. J Mol Biol 289: 827–834

Gregory ST, Lieberman KR, Dahlberg AE (1994) Mutations in the peptidyl transferase region of E. coli 23S rRNA affecting translational accuracy. Nucl Acids Res 22:279–284

Gregory ST, Carr JF, Rodriguez-Correa D, Dahlberg AE (2005) Mutational analysis of 16S and 23S rRNA genes of Thermus thermophilus. J Bacteriol 187:4804–4812

Harger JW, Dinman JD (2003) An in vivo dual-luciferase assay system for studying translational recoding in the yeast *Saccharomyces cerevisiae*. RNA 9:1019–1024

Harger JW, Dinman JD (2004) Evidence against a direct role for the Upf proteins in frameshfiting or nonsense codon readthrough. RNA 10:1721–1729

Harger JW, Meskauskas A, Dinman JD (2002) An 'integrated model' of programmed ribosomal frameshifting and post-transcriptional surveillance. TIBS 27:448–454

Harger JW, Meskauskas A, Nielsen N, Justice MC, Dinman JD (2001) Ty1 retrotransposition and programmed +1 ribosomal frameshifting require the integrity of the protein synthetic translocation step. Virology 286:216–224

Helgstrand M, Mandava CS, Mulder FA, Liljas A, Sanyal S, Akke M (2007) The ribosomal stalk binds to translation factors IF2, EF-Tu, EF-G and RF3 via a conserved region of the L12 C-terminal domain. J Mol Biol 365:468–479

Herr AJ, Nelson CC, Wills NM, Gesteland RF, Atkins JF (2001) Analysis of the roles of tRNA structure, ribosomal protein L9, the bacteriophage T4 gene 60 bypassing signals during ribosome slippage on mRNA. J Mol Biol 309:1029–1048

Hirabayashi N, Sato NS, Suzuki T (2006) Conserved loop sequence of helix 69 in *Escherichia coli* 23 S rRNA is involved in A-site tRNA binding and translational fidelity. J Biol Chem 281:17203–17211

Kelly KS, Ochi K, Jones GH (1991) Pleiotropic effects of a relC mutation in Streptomyces antibioticus. J Bacteriol 173:2297–3003

Kiparisov S, Petrov A, Meskauskas A, Sergiev PV, Dontsova OA, Dinman JD (2005) Structural and functional analysis of 5S rRNA. Mol Genet Genomics 27:235–247

Kirsebom LA, Amons R, Isaksson LA (1986) Primary structures of mutationally altered ribosomal protein L7/L12 and their effects on cellular growth and translational accuracy. Eur J Biochem 156:669–675

Kirsebom LA, Isaksson LA (1985) Involvement of ribosomal protein L7/L12 in control of translational accuracy. Proc Natl Acad Sci USA 82:717–721

Komoda T, Sato NS, Phelps SS, Namba N, Joseph S, Suzuki T (2006) The A-site finger in 23 S rRNA acts as a functional attenuator for translocation. J Biol Chem 281:32303–32309

Konstantinidis TC, Patsoukis N, Georgiou CD, Synetos D (2006) Translational Fidelity Mutations in 18S rRNA Affect the Catalytic Activity of Ribosomes and the Oxidative Balance of Yeast Cells. Biochemistry 45:3225–3533

Korostelev A, Trakhanov S, Laurberg M, Noller HF (2006) Crystal structure of a 70S ribosome-tRNA complex reveals functional interactions and rearrangements. Cell 126:1065–1077

Kühberger R, Piepersberg W, Petzet A, Buckel P, Böck A (1979) Alteration of ribosomal protein L6 in gentamicin-resistant strains of *Escherichia coli*. Effects on fidelity of protein synthesis. Biochemistry 18:187–193

Laurberg M, Asahara H, Korostelev A, Zhu J, Trakhanov S, Noller HF (2008) Structural basis for translation termination on the 70S ribosome. Nature 454:852–857.

Lee SI, Umen JG, Varmus HE (1995) A genetic screen identifies cellular factors involved in retroviral −1 frameshifting. Proc Natl Acad Sci USA 92:6587–6591

Le Roy F, Salehzada T, Bisbal C, Dougherty JP, Peltz SW (2005) A newly discovered function for RNase L in regulating translation termination. Nat Struct Mol Biol 12:505–512

Lee K, Holland-Staley CA, Cunningham P R (1996) Genetic analysis of the Shine-Dalgarno interaction: selection of alternative functional mRNA-rRNA combinations. RNA 2: 1270–1285

Leipuviene R, Björk GR (2007) Alterations in the two globular domains or in the connecting alpha-helix of bacterial ribosomal protein L9 induces +1 frameshifts. J Bacteriol 189: 7024–7031

Liiv A, O'Connor M (2006) Mutations in the intersubunit bridge regions of 23 S rRNA. J Biol Chem 281:29850–29862

Liu R, Liebman SW (1996) A translational fidelity mutation in the universally conserved sarcin/ricin domain of 25S yeast ribosomal RNA. RNA 2:254–263

Lodmell JS, Gutell RR, Dahlberg AE (1995) Genetic and comparative analyses reveal an alternative secondary structure in the region of nt 912 of *Escherichia coli* 16S rRNA. Proc Natl Acad Sci USA 92:10555–10559

Maisnier-Patin S, Berg OG, Liljas L, Andersson DI (2002) Compensatory adaptation to the deleterious effect of antibiotic resistance in Salmonella typhimurium. Mol Microbiol 46: 355–366

Melançon P, Lemieux C, Brakier-Gingras L (1988) A mutation in the 530 loop of *Escherichia coli* 16S ribosomal RNA causes resistance to streptomycin. Nucleic Acids Res 16:9631–9639

Melançon P, Tapprich WE, Brakier-Gingras L (1992) Single-base mutations at position 2661 of *Escherichia coli* 23S rRNA increase efficiency of translational proofreading. J Bacteriol 174:7896–7901

Meskauskas A, Baxter JL, Carr EA, Yasenchak J, Gallagher JEG, Baserga SJ, Dinman JD (2003a) Delayed rRNA processing results in significant ribosome biogenesis and functional defects. Mol Cell Biol 23:1602–1613

Meskauskas A, Dinman JD (2007) Ribosomal protein L3:Gatekeeper to the A-site. Mol Cell 25:877–888

Meskauskas A, Dinman JD (2008) Ribosomal protein L3 functions as a 'rocker switch' to aid in coordinating of large subunit-associated functions in eukaryotes and Archaea. Nucl Acids Res 36:6175–6186

Meskauskas A, Harger JW, Jacobs KLM, Dinman JD (2003b) Decreased peptidyltransferase activity correlates with increased programmed −1 ribosomal frameshifting and viral maintenance defects in the yeast *Saccharomyces cerevisiae*. RNA 9:982–992

Meskauskas A, Petrov AN, Dinman JD (2005) Identification of functionally important amino acids of ribosomal protein L3 by saturation mutagenesis. Mol Cell Biol 25:10863–10874

Meskauskas A, Russ JR, Dinman JD (2008) Structure/function analysis of yeast ribosomal protein L2. Nucleic Acids Res 36:1826–1835

Muldoon-Jacobs KL, Dinman JD (2006) Specific effects of ribosome-tethered molecular chaperones on programmed −1 ribosomal frameshifting. Eukaryot Cell 5:762–770

Murgola EJ, Hijazi KA, Göringer HU, Dahlberg AE (1988) Mutant 16S ribosomal RNA: a codon-specific translational suppressor. Proc Natl Acad Sci USA 85:4162–4165

Murgola EJ, Pagel FT, Hijazi KA, Arkov AL, Xu W, Zhao SQ (1995) Variety of nonsense suppressor phenotypes associated with mutational changes at conserved sites in *Escherichia coli* ribosomal RNA. Biochem Cell Biol 73:925–931

Namy O, Moran SJ, Stuart DI, Gilbert RJ, Brierley I (2006) A mechanical explanation of RNA pseudoknot function in programmed ribosomal frameshifting. Nature 441:244–247

O'Connor M (2007) Interaction between the ribosomal subunits:16S rRNA suppressors of the lethal DeltaA1916 mutation in the 23S rRNA of *Escherichia coli*. Mol Genet Genomics 278:307–315

O'Connor M, Brunelli CA, Firpo MA, Gregory ST, Lieberman KR, Lodmell JS, Moine H, Van Ryk DI, Dahlberg AE (1995) Genetic probes of ribosomal RNA function Biochem Cell Biol 73:859–868

O'Connor M, Dahlberg AE (1993)Mutations at U2555, a tRNA-protected base in 23S rRNA, affect translational fidelity. Proc Natl Acad Sci USA 90:9214–9218

O'Connor M, Dahlberg AE (1995) The involvement of two distinct regions of 23 S ribosomal RNA in tRNA selection. J Mol Biol 254:838–847

O'Connor M, Dahlberg AE (1996) The influence of base identity and base pairing on the function of the alpha-sarcin loop of 23S rRNA. Nucleic Acids Res 24:2701–2705

O'Connor M, Gregory ST, Dahlberg AE (2004) Multiple defects in translation associated with altered ribosomal protein L4. Nucleic Acids Res 32:5750–5756

O'Connor M, Lee WM, Mankad A, Squires CL, Dahlberg AE (2001) Mutagenesis of the peptidyltransferase center of 23S rRNA: the invariant U2449 is dispensable Nucleic Acids Res 29:710–715

O'Connor M, Thomas CL, Zimmermann RA, Dahlberg AE (1997) Decoding fidelity at the ribosomal A and P sites: influence of mutations in three different regions of the decoding domain in 16S rRNA. Nucleic Acids Res 25:1185–1193

O'Connor M, Willis NM, Bossi L, Gesteland RF, Atkins JF (1993) Functional tRNAs with altered 3' ends. EMBO J 12:2559–2566

Ogle JM, Carter AP, Ramakrishnan V (2003) Insights into the decoding mechanism from recent ribosome structures. Trends Biochem Sci 28:259–266

Ortiz PA, Ulloque R, Kihara GK, Zheng H, Kinzy TG (2006) Translation elongation factor 2 anticodon mimicry domain mutants affect fidelity and diphtheria toxin resistance. J Biol Chem 281:32639–32648

Ozaki M, Mizushima S, Nomura M (1969) Identification and functional characterization of the protein controlled by the streptomycin-resistant locus in E. coli. Nature 222:333–339

Panopoulos P, Dresios J, Synetos D (2004) Biochemical evidence of translational infidelity and decreased peptidyltransferase activity by a sarcin/ricin domain mutation of yeast 25S rRNA. Nucleic Acids Res 32:5398–5408

Parker, J, Watson, R.J, Friesen J D, Fiil N,P (1976) A relaxed mutant with an altered ribosomal protein L11. Mol Gen Genet 144:111–114

Petrov AN, Meskauskas A, Roshwalb SC, Dinman JD (2008) Yeast ribosomal protein L10 helps coordinate tRNA movement through the large subunit. Nucleic Acids Res 36:6187–6198

Petry S, Brodersen DE, Murphy FV. 4th, Dunham CM, Selmer M, Tarry MJ, Kelley AC, Ramakrishnan V (2005) Crystal structures of the ribosome in complex with release factors RF1 and RF2 bound to a cognate stop codon. Cell 123:1255–1266

Piepersberg W, Böck A, Yaguchi M, Wittmann HG (1975) Genetic position and amino acid replacements of several mutations in ribosomal protein S5 from *Escherichia coli*. Mol Gen Genet 143:43–52

Pinard R, Côté M, Payant C, Brakier-Gingras L (1994) Positions 13 and 914 in *Escherichia coli* 16S ribosomal RNA are involved in the control of translational accuracy. Nucleic Acids Res 22:619–624

Rakauskaite R, Dinman JD (2006) An arc of unpaired "hinge bases" facilitates information exchange among functional centers of the ribosome. Mol Cell Biol 26:8992–9002

Rakauskaite R, Dinman JD (2008) rRNA mutants in the yeast peptidyltransferase center reveal allosteric information networks and mechanisms of drug resistance. Nucl Acids Res 36: 1497–1507

Robert F, Brakier-Gingras L (2003) A functional interaction between ribosomal proteins S7 and S11 within the bacterial ribosome. J Biol Chem 278:44913–44920

Rodnina MV, Gromadski KB, Kothe U, Wieden HJ (2005) Recognition and selection of tRNA in translation. FEBS Lett 579:579–942

Roth JR (1981) Frameshift suppression. Cell 24:601–602

Ruggero D, Grisendi S, Piazza F, Rego E, Mari F, Rao PH, Cordon-Cardo C, Pandolfi PP (2003) Dyskeratosis congenita and cancer in mice deficient in ribosomal RNA modification. Science 299:259–262

Rydén-Aulin M, Shaoping Z, Kylsten P, Isaksson LA (1993) Ribosome activity and modification of 16S RNA are influenced by deletion of ribosomal protein S20. Mol Microbiol 7: 983–892

Saarma U, Remme J (1992) Novel mutants of 23S RNA: Characterization of functional properties. Nucleic Acids Res 20:3147–3152

Saarma U, Remme J, Ehrenberg M, Bilgin N (1997) An A to U transversion at position 1067 of 23 S rRNA from *Escherichia coli* impairs EF-Tu and EF-G function. J Mol Biol 272: 327–335

Sato H, Ito K, Nakamura Y (2006) Ribosomal protein L11 mutations in two functional domains equally affect release factors 1 and 2 activity. Mol Microbiol 60:108–120

Schuwirth BS, Borovinskaya MA, Hau CW, Zhang W, Vila-Sanjurjo A, Holton JM, Cate JH (2005) Structures of the bacterial ribosome at 3.5 A resolution. Science 310:827–834

Selmer M, Dunham CM, Murphy FV, Weixlbaumer A, Petry S, Kelley AC, Weir JR, Ramakrishnan V (2006) Structure of the 70S ribosome complexed with mRNA and tRNA. Science 313: 1935–1942

Sergiev PV, Kiparisov SV, Burakovsky DE, Lesnyak DV, Leonov AA, Bogdanov AA, Dontsova OA (2005a) The conserved A-site finger of the 23S rRNA: just one of the intersubunit bridges or a part of the allosteric communication pathway? J Mol Biol 353:116–123

Sergiev PV, Lesnyak DV, Kiparisov SV, Burakovsky DE, Leonov AA, Bogdanov AA, Brimacombe R, Dontsova OA (2005b) Function of the ribosomal E-site: A mutagenesis study. Nucleic Acids Res 33:6048–6056

Smith MW, Meskauskas A, Wang P, Sergiev PV, Dinman JD (2001) Saturation mutagenesis of 5S rRNA in *Saccharomyces cerevisiae*. Mol Cell Biol 21:8264–8275

Spahn CM, Gomez-Lorenzo MG, Grassucci RA, Jorgensen R, Andersen GR, Beckmann R, Penczek PA, Ballesta JP, Frank J (2004) Domain movements of elongation factor eEF2 and the eukaryotic 80S ribosome facilitate tRNA translocation. EMBO J 23:1008–1019

Takyar S, Hickerson RP, Noller HF (2005) mRNA helicase activity of the ribosome. Cell 120: 19–58

Tate WP, Schulze H, Nierhaus KH (1983) The *Escherichia coli* ribosomal protein L11 suppresses release factor 2 but promotes the release factor 1 activities in peptide chain termination. J Biol Chem 258:12816–12820

Thompson J, Kim DF, O'Connor M, Lieberman KR, Bayfield MA, Gregory ST, Green R, Noller HF, Dahlberg AE (2001) Analysis of mutations at residues A2451 and G2447 of 23S rRNA in the peptidyltransferase active site of the 50S ribosomal subunit. Proc Natl Acad Sci USA. 98:9002–9007

Topisirovic L, Villarroel R, De Wilde M, Herzog A, Cabez6n T, Bollen A (1977) Translational fidelity in *Escherichia coli*: contrasting role of neaA and ramA gene products in the ribosome functioning. Mol Gen Genet 151:89–94

Tumer NE, Parikh B, Li P, Dinman JD (1998) Pokeweed antiviral protein specifically inhibits Ty1 directed +1 ribosomal frameshifting and Ty1 retrotransposition in *Saccharomyces cerevisiae*. J Virol 72:1036–1042

Valle M, Sengupta J, Swami NK, Grassucci RA, Burkhardt N, Nierhaus KH, Agrawal RK, Frank J (2002) Cryo-EM reveals an active role for aminoacyl-tRNA in the accommodation process. EMBO J 21:3557–3567

Valle M, Zavialov A, Li W, Stagg SM, Sengupta J, Nielsen RC, Nissen P, Harvey SC, Ehrenberg M, Frank J (2003) Incorporation of aminoacyl-tRNA into the ribosome as seen by cryo-electron microscopy. Nat Struct Biol 10:899–906

Van Dyke N, Xu W, Murgola EJ (2002) Limitation of ribosomal protein L11 availability in vivo affects translation termination. J Mol Biol 319:329–339

Velichutina IV, Dresios J, Hong JY, Li C, Mankin A, Synetos D, Liebman SW (2000) Mutations in helix 27 of the yeast *Saccharomyces cerevisiae* 18S rRNA affect the function of the decoding center of the ribosome. RNA 6:1174–1184

Velichutina IV, Hong JY, Mesecar AD, Chernoff YO, Liebman SW (2001) Genetic interaction between yeast *Saccharomyces cerevisiae* release factors and the decoding region of 18 S rRNA. J Mol Biol 305:715–727

Vila-Sanjurjo A, Lu Y, Aragonez JL, Starkweather RE, Sasikumar M, O'Connor M (2007) Modulation of 16S rRNA function by ribosomal protein S12. Biochim Biophys Acta (1769):462–471

Warner JR (1999) The economics of ribosome biosynthesis in yeast. Trends Biochem Sci 24: 437–440

Wickner RB, Porter-Ridley S, Fried HM, Ball SG (1982) Ribosomal protein L3 is involved in replication or maintenance of the killer double-stranded RNA genome of *Saccharomyces cerevisiae*. Proc Natl Acad Sci USA 79:4706–4708

Widerak M, Kern R, Malki A, Richarme G (2005) U2552 methylation at the ribosomal A-site is a negative modulator of translational accuracy. Gene 347:109–114

Yano R, Yura T (1989) Suppression of the *Escherichia coli* rpoH opal mutation by ribosomes lacking S15 protein. J Bacteriol 171:1712–1717

Youngman EM, Brunelle JL, Kochaniak AB, and Green R (2004) The active site of the ribosome is composed of two layers of conserved nucleotides with distinct roles in peptide bond formation and peptide release. Cell 117:589–599

Zimmermann RA, Garvin RT, Gorini L (1971) Alteration of a 30S ribosomal protein accompanying the ram mutation in *Escherichia coli*. Proc Natl Acad Sci USA 68:2263–2267

Chapter 16
The E Site and Its Importance for Improving Accuracy and Preventing Frameshifts

Markus Pech, Oliver Vesper, Hiroshi Yamamoto, Daniel N. Wilson, and Knud H. Nierhaus

Abstract The ribosome contains three tRNA binding sites, the A, P, and E sites. Although the E site is separated from the A via the intervening P site, there is striking communication between these sites. This cross-talk plays an important role for the accuracy of the decoding process. Codon–anticodon interaction at the E site seems to be the signal to switch into the post-translocational (POST) state characterized by a low affinity of the A site. This low-affinity state forces the ternary complexes aminoacyl-tRNA•EF-Tu•GTP to enter the A site via the decoding center preventing the selection of non-cognate aminoacyl-tRNAs and incorporation of the incorrect amino acid. This has the important consequence that only 1 in 400 misincorporations affects protein function. Another aspect of the allostery between A and E sites is that during elongation there are always at least two tRNAs present on the ribosome at the same time. Since the tRNAs are firmly bound by the ribosome whereas the mRNA is held predominantly via codon–anticodon interaction, it is the movement of the tRNAs during translocation that pulls the mRNA through the ribosome. In fact, the six base pairs of two adjacent codon–anticodon interactions are instrumental for maintaining the reading frame, and there is evidence that without the codon–anticodon interaction of the E-tRNA the reading frame would be lost at least after the incorporation of about 50 amino acids into the nascent chain.

Contents

M. Pech (✉)
Max-Planck-Institut für Molekulare Genetik, Ihnestr. 73, D-14195 Berlin, Germany
e-mail: nierhaus@molgen.mpg.de

J.F. Atkins, R.F. Gesteland (eds.), *Recoding: Expansion of Decoding Rules Enriches Gene Expression*, Nucleic Acids and Molecular Biology 24,
DOI 10.1007/978-0-387-89382-2_16,

16.1 Introduction: All Ribosomes Have Three tRNA Binding Sites

All ribosomes have three tRNA binding sites, the A, P, and E sites. The elongation phase is the central functional phase of the ribosome, and in the course of this phase a tRNA moves through all three tRNA binding sites on the ribosome in the order A → P → E. Our current understanding of the elongation cycle is illustrated in Fig. 16.1 (for review see Wilson and Nierhaus 2006). We start with a ribosome already carrying two tRNAs: a peptidyl-tRNA at the P site (P for peptidyl-tRNA) with the synthesized nascent polypeptide chain and a deacylated tRNA at the E site (E for exit). This ribosome state is called post-translocational state or POST state (Fig. 16.1a). An aminoacyl-tRNA (aa-tRNA) enters the ribosome at the A site (A for aminoacyl-tRNA) in the form of the ternary complex aa-tRNA•EF-Tu•GTP. EF-Tu is one of two universal elongation factors, both of which are G proteins. Selection

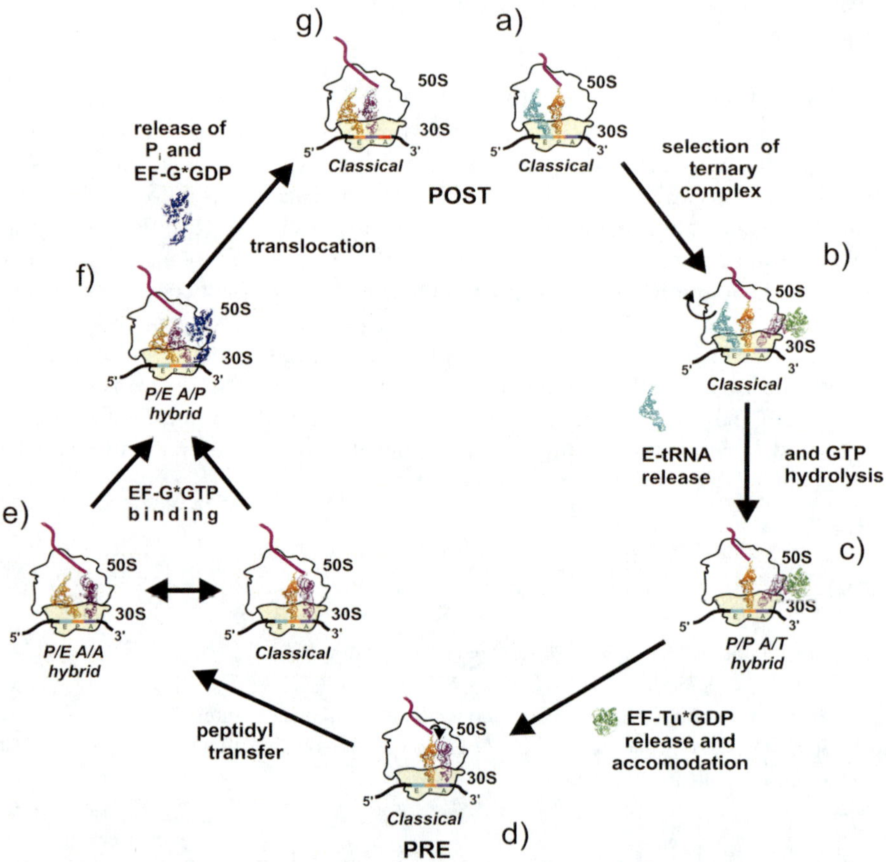

Fig. 16.1 The elongation cycle of protein synthesis. For explanations, see text

occurs at the decoding center of the A site on the basis of the codon of the mRNA exposed at this site (Fig. 16.1b). Decoding of the aa-tRNA occurs while it is still bound with the ternary complex, thus allowing codon–anticodon interaction at the A site but preventing most of the contacts of the aa-tRNA with the A site outside the decoding center. Following decoding, three tightly coupled steps ensue (Fig. 16.1c and d): (i) The deacylated tRNA is released from the E site of the ribosome; (ii) the ribosome triggers the GTPase activity of EF-Tu, which leads to a conformational change within EF-Tu, and the low-affinity EF-Tu•GDP dissociates from the ribosome; and (iii) the aa-tRNA moves fully into the A site (accommodation). Now the ribosome is in the pre-translocational state (PRE state) characterized by tRNAs at the A and P sites. The peptidyl residue is transferred from the P-tRNA (the tRNA in the P site) to the adjacent aa-tRNA yielding the peptidyl-tRNA at the A site prolonged by one amino acid and leaving a deacylated tRNA at the P site (Fig. 16.1e). This PRE state seems to be in equilibrium with a number of hybrid states where the tRNAs can move on the 50S subunit but maintain codon–anticodon interaction on the 30S subunit (Munro et al. 2007), namely A/P (tRNA is located at A site on 30S and P site on 50S) and P/E hybrid sites. Next the second elongation factor EF-G•GTP binds and pushes the equilibrium toward the hybrid states (Valle et al. 2003). The energy for this tRNA movement might be paid by the binding energy of this large factor (Fig. 16.1f). The tRNA movement is accompanied (or caused) by a ratchet movement (forward ratcheting) involving a rotation of the 30S subunit by about 4° relative to the 50S subunit and movement of the head by 10–11° in a counterclockwise direction (Frank and Agrawal 2000; Spahn et al. 2001). After the ribosome has triggered GTP hydrolysis of EF-G, release of the P_i leads to both the release of EF-G•GDP from the ribosome and the full translocation, or movement of the tRNAs on the 30S, to bring the two tRNAs into the classical P and E sites (back-ratcheting; POST state; Fig. 16.1 g). The ribosome is now ready to enter the next round of the elongation cycle.

16.2 Features of the E Site

As the description of the elongation cycle in the preceding section implies, the three classical ribosomal tRNA binding sites A, P, and E are characterized by a strikingly different capacity to bind different kinds of tRNAs: (i) The A site binds aminoacyl-tRNA and peptidyl-tRNA in each elongation cycle and deacylated tRNA only under defined stress conditions, the stringent response (see Wendrich et al. 2002). (ii) The P site accepts both deacylated tRNAs and peptidyl-tRNAs during an elongation cycle as well as aminoacyl-tRNAs during initiation. (iii) In contrast, the E site is the most specific since it binds exclusively deacylated tRNA. The structural explanation for this is a hydrogen-bond network around the ultimate A76 of an E-tRNA, with nucleotide C2394 playing a central role (Fig. 16.2). This nucleotide is located at the 3′ base of helix H88, and whenever this helix landmark is present in the 23S-type rRNA, the corresponding C residue is observed, including in mitochondrial ribosomes.

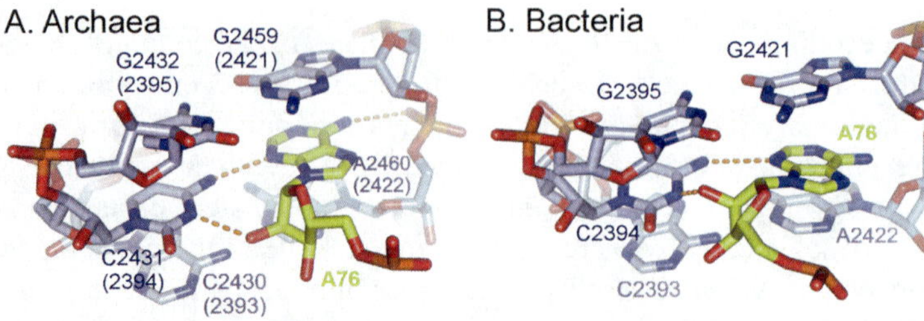

Fig. 16.2 The ultimate residue A76 of the E-tRNA in a network of hydrogen bonds, which exclusively allows the presence of a deacylated tRNA at the E site. (**A**) A76 of the E-tRNA stacks in between 23S rRNA nucleotides G2421 and A2422 (*E. coli* nomenclature) and hydrogen bonds with universally conserved C2394 ((Schmeing et al. 2003), modified). (**B**) A corresponding situation is seen in bacterial ribosomes (large subunit). Again, the universally conserved C2394 plays a decisive role (from Selmer et al. 2006)

At the other tip of the L-shaped tRNA, i.e., the anticodon site, the situation is clearer: tRNAs in all three sites as well as in hybrid states undergo codon–anticodon interactions. Codon–anticodon interaction at the A site must occur, since this is the decoding site where the stereochemistry of the codon–anticodon interaction of the mRNA–tRNA duplex is monitored. In the 1970s codon–anticodon interaction at the P site was still a controversial issue and was not settled until 1979 (Lührmann et al. 1979; Wurmbach and Nierhaus 1979). Shortly afterward the E site was detected (Rheinberger and Nierhaus 1980; Rheinberger et al. 1981), but it took another 10 years for acceptance by the scientific community, with demonstration of an E site in archaeal and eukaryotic ribosomes (Saruyama and Nierhaus 1986; El'skaya et al. 1997; Triana-Alonso et al. 1995, respectively). Despite this, the existence of codon–anticodon interaction continued to be discussed, and the issue was only settled recently, when previous biochemical evidence (Rheinberger et al. 1986; Triana-Alonso et al. 1995) could be confirmed by genetic and structural data (Jenner et al. 2007; Liao et al. 2008; Sanders and Curran 2007). (For the mode of codon–anticodon recognition at A and E sites see Section 16.3.)

Although the E site is separated from the A *via* the intervening P site, there is a striking communication between these sites. This cross-talk plays an important role for two accuracy aspects:

(i) Codon–anticodon interaction at the E site seems to be the signal to switch into the post-translocational (POST) state characterized by a low affinity of the A site (Geigenmüller and Nierhaus 1990). As we will point out in Section 16.3, the low-affinity A site is instrumental for both accurate aminoacyl-tRNA selection and for preventing misincorporation of an amino acid from a non-cognate aminoacyl-tRNA, the chemical nature of which is distinctly different from the cognate one and thus more likely to affect folding, stability, and function of the mature protein. However, if the E site is so important for accuracy of decoding, there should be an accuracy problem at the codon just following the initiation codon AUG, where the initiator (f)Met-tRNA is present at the P site and the E site is free. How the bacterial ribosome solves this problem is discussed in Section 16.4.

(ii) A and E sites interact in a sense of negative cooperativity (Rheinberger and Nierhaus 1983, 1986b; Triana-Alonso et al. 1995). This means not only that an E-tRNA triggers a low-affinity state of the A site, but that occupation of the A site also induces a low-affinity state of the E site, leading to a loss of the E-tRNA from the ribosome (Dinos et al. 2005). A consequence of this interaction is that on average two tRNAs are on the ribosome, either at A and P site in the pre-translocational (PRE) state or at P and E sites in the POST state (Remme et al. 1989; Warner and Rich 1964). Since the tRNAs are firmly bound by the ribosome, whereas the mRNA is fixed only *via* codon–anticodon interaction during the elongation phase (Alexeeva et al. 1996), the movement of the tRNA during translocation pulls the mRNA through the ribosome. In fact, the six base pairs of two adjacent codon–anticodon interactions are instrumental for maintaining the reading frame; without the codon–anticodon interaction of the E-tRNA the reading frame would be lost after synthesis of a short polypeptide. If one considers the frequencies of codons allowing frameshifts in either direction without losing codon–anticodon contacts and taking into account the measured frequency of frameshifts in the absence of an E-tRNA (Marquez et al. 2004; Section 16.5), one can estimate that the reading frame is lost at least after the incorporation of about 50 amino acids into the nascent peptide chain.

16.3 A Cognate E-tRNA Prevents Misincorporation of Non-cognate Amino Acids

The binding of aa-tRNAs to the ribosome is dictated by the complementarity between the anticodon of the tRNA and the codon of the mRNA. To ensure a high fidelity of translation, the correct stereochemistry of the mRNA–tRNA codon–anticodon interaction is monitored by components of the small ribosomal subunit (reviewed by Ogle and Ramakrishnan 2005). During this decoding, the first and second nucleotide positions (in terms of the codon) of the mRNA–tRNA duplex are closely monitored, whereas interaction at the third or wobble position is less strictly recognized. The misincorporation of wrong amino acids into polypeptide chains usually occurs through the binding of near-cognate aa-tRNAs, i.e., those tRNAs carrying an anticodon similar to that of the cognate aa-tRNA, rather than non-cognate aa-tRNAs, which carry dissimilar anticodons.

Proteins are surprisingly tolerant to misincorporation, with only 1 in ~400 misincorporations being deleterious for the protein's activity (reviewed by Kurland et al. 1990). The reason for this is an intimate co-evolution of the ribosomal decoding center and the logic of the genetic code. Consider Fig. 16.3, where the codon lexicon is presented in the form of the codon sun and the chemical natures of the amino acids are indicated with colored spots. Acidic, basic, and polar uncharged amino

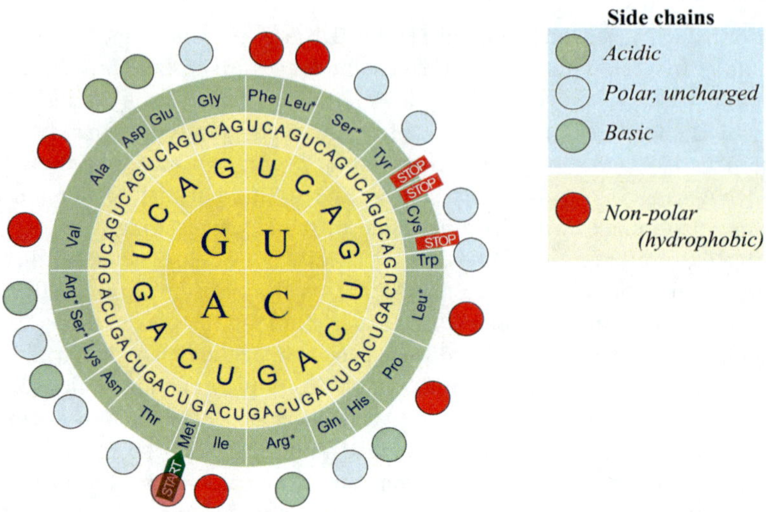

Fig. 16.3 The codon sun. The central circle represents the first nucleotide of a codon followed by two rings representing the second and third nucleotide. The outer ring indicates the amino acids coded for. In addition, the chemical nature of the amino acids is shown with a color code. Two classes are distinguished, class I comprises acidic, basic, and uncharged polar residues; class II consists of the hydrophobic amino acids

acids are usually found at the surface of the proteins, a change within this class of amino acids usually does not have consequences for folding, structure, or function of a protein. The second class of amino acids is the non-polar amino acids residing in the interior of a protein. A change from one class to the other is, however, usually detrimental for the fate of a protein and thus has to be prevented. A further inspection of Fig. 16.3 reveals that misreading of the last nucleotide of a codon results in an amino acid change within a class in all cases (e.g., GAU to GAA, the acidic Asp to the acidic Glu) and thus has no serious consequence. Misreading of the first position changes quite often the class (e.g., CCU to UCU, the hydrophobic Pro to the polar Ser) as does a change of the middle position in most cases (UUC to UCC, the hydrophobic Phe to the polar Ser). For example, the universally conserved residue A1493 flips into the shallow groove of the first base pair of codon–anticodon and forms hydrogen bonds with the 2′OH groups of the participating nucleotides. Non-Watson–Crick pairs would form H-bonds of lower energy if at all thus defining the recognition mode. The middle base pair is checked even more strictly by involving A1492, G530, and the Ser50 residue of the ribosomal protein S12 (Fig. 16.4; Ogle and Ramakrishnan 2005). This provides a structural basis for the known fact that the middle position is practically never misread and the first position very rarely if at all, even under error-inducing conditions such as high magnesium or the presence of aminoglycosides, and are thus considered as being non-cognate (for review and references see Szaflarski et al. 2008). In contrast, the third or wobble position has more freedom to accommodate incorrect base pairings, and thus is often misread. It

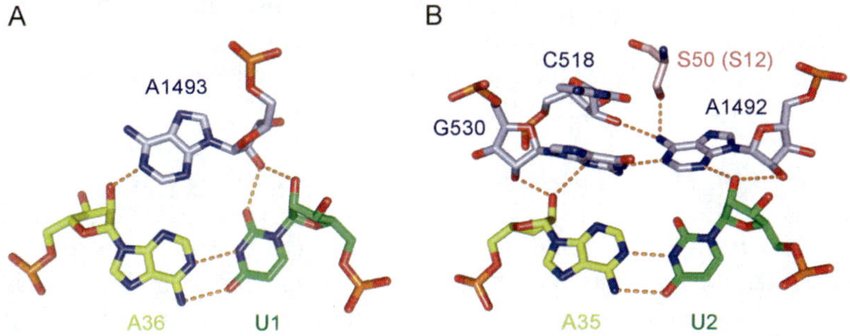

Fig. 16.4 Codon–anticodon interaction at the decoding center of the A site in the first two positions of the codon. (**A**) In the first position, A1493 binds in the minor groove of the A36-U1 base pair. (**B**) In the second position, G530 and A1492 (both *blue*) act in concert to monitor the A35-U2 base pair. According to Ogle et al. (2001), modified

follows that the type of base pair, whether A:U or G:C, is of no importance for the accuracy of decoding – an observation that was a surprise 25 years ago (Andersson et al. 1984).

The terms cognate, near-cognate, and non-cognate are also defined functionally, viz., the misincorporation of near-cognate amino acids in vivo and in vitro require higher GTP consumption than for cognate, whereas non-cognate amino acids are never incorporated and no GTP is consumed (Geigenmüller and Nierhaus 1990; Nierhaus 1990), or if incorporation is observed, the rate is greatly reduced (Cochella and Green 2005; Daviter et al. 2006).

Aa-tRNAs can, however, occupy the A site without being subjected to the decoding process. One example is seen with Ala-tmRNA, the tRNA moiety of which does not even have an anticodon, but still binds efficiently to the A site in complex with EF-Tu•GTP and the SmpB protein (Moore and Sauer 2007). Bypassing the decoding process can also happen when the ribosome has an empty E site, i.e., a peptidyl-tRNA occupies the P site while the A and E sites are free. In such a situation, even a non-cognate aa-tRNA can enter the A site leading to an incorporation of the non-cognate amino acid into the nascent peptide chain (Di Giacco et al. 2008; Geigenmüller and Nierhaus 1990). After translocation a peptidyl-tRNA resides at the P site and the E site is tightly occupied by a deacylated tRNA (Marquez et al. 2004; Rheinberger and Nierhaus 1983). The E-tRNA is released through an active mechanism, whereby interaction of a ternary complex aminoacyl-tRNA•EF-Tu•GTP at the A site is coupled to the release of the E-tRNA (Rheinberger and Nierhaus 1986b; Triana-Alonso et al. 1995). E-tRNA release follows the decoding step, but occurs before accommodation of the aa-tRNA into the A site (Dinos et al. 2005).

The presence of an E-tRNA has been shown to be important for – as mentioned above – preventing the selection of non-cognate aa-tRNAs (Geigenmüller and Nierhaus 1990). In the latter experiment, Geigenmüller et al. demonstrated that when the E site was unoccupied, the non-cognate acidic Asp (codon GAC/U) could be misincorporated in place of the cognate aromatic hydrophobic Phe (codon

UUU/C); however, no misincorporation of Asp was observed when the E site was occupied. Since tRNA near-cognate to the E site could not prevent the incorporation of a non-cognate amino acid, the conclusion was that codon–anticodon interaction at the E site is required to prevent misincorporation at the A site (Geigenmüller and Nierhaus 1990). This is consistent with the genetic (Leger et al. 2007; Sanders and Curran 2007), biochemical (Gnirke et al. 1989; Rheinberger and Nierhaus 1986a; Triana-Alonso et al. 1995), and structural evidence (Jenner et al. 2007) demonstrating the likelihood of codon–anticodon interaction at the E site.

We note that the recognition of the codon–anticodon duplex in the E site is very different from that at the decoding site in the A site, where the kind of Watson–Crick base pair A:U versus G:C does not play a role. In contrast, the type of base pair at the E site appears to be important, since it has been demonstrated that the stability of codon–anticodon interaction influences the affinity of the E-tRNA, which in turn is inversely proportional to the accuracy at the A site (Liao et al. 2008; Sanders and Curran 2007).

A more recent demonstration of E site occupancy preventing the misincorporation of non-cognate amino acids is depicted in Fig. 16.5 (Di Giacco et al. 2008).

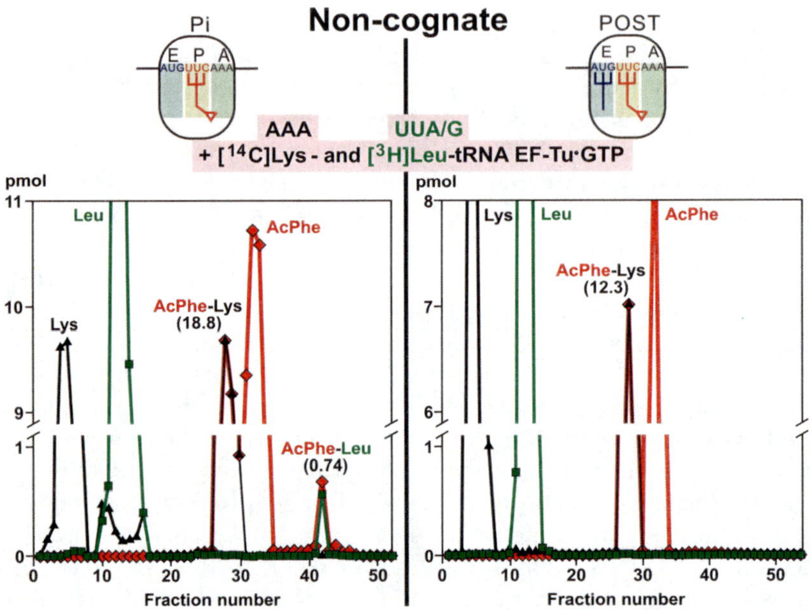

Fig. 16.5 Non-cognate misincorporation levels. The influence of the E-tRNA: HPLC analysis of dipeptides formed by the addition of a stoichiometric mixture of ternary complexes containing cognate [^{14}C]Lys-tRNA and non-cognate [^{3}H]Leu-tRNA to either (i) Pi-state ribosomes (*left*) containing AcPhe-tRNA at the P site or (ii) POST-state ribosomes (*right*) carrying AcPhe-tRNA at the P site and deacylated [^{32}P]tRNAfMet at the E site, generated via EF-G-dependent translocation. The codons are given above the amino acids. According to Di Giacco et al. (2008)

Here, *Escherichia coli* 70S ribosomes carry an AcPhe-tRNA at the P site and display an AAA codon at the A site. A stoichiometric mixture of the cognate basic hydrophilic Lys-tRNA and the non-cognate hydrophobic Leu-tRNA (codon UUA/G) was added and the dipeptides formed were identified via HPLC chromatography. In the absence of an E-tRNA a significant amount of the deleterious dipeptide AcPhe-Leu is observed, whereas in the presence of an E-tRNA only the cognate product AcPhe-Lys can be detected.

So how does the presence of an E-tRNA influence decoding at the A site? An occupied E site dramatically increases (almost three-fold) the activation energy barrier for A site occupation, namely from ~40 to ~115 kJ/mol (in a physiological buffer with 3–6 mM Mg^{2+} and polyamines; Schilling-Bartetzko et al. 1992). These findings were incorporated into the allosteric three-site model (Nierhaus 1990; Rheinberger and Nierhaus 1983; Triana-Alonso et al. 1995) stating that the A and E sites are reciprocally linked, such that occupation of the E site induces a low-affinity A site, and vice versa. This model explains why in native polysomes from both eukaryotes and bacteria precisely two tRNAs per ribosome are observed (Remme et al. 1989; Warner and Rich 1964).

The next question is how the low-affinity A site excludes the selection of non-cognate aa-tRNAs? One possibility is that the low-affinity A site restricts the binding of the ternary complex (aa-tRNA•EF-Tu•GTP) with the ribosome to only the interaction between the A site codon and the anticodon of the tRNA, until successful decoding is completed. In this model, contacts outside of the codon–anticodon interaction would not contribute to the selection precision because they are common to all ternary complexes, regardless of cognate or non-cognate, and therefore would allow even non-cognate aa-tRNAs to interfere with the selection process, as well as leading to the occasional misincorporation before the decoding potential of the codon–anticodon interaction has been exploited (Nierhaus 1993). Indeed, cryo-electron microscopic (cryo-EM) studies reveal that during A site decoding the incoming ternary complex aa-tRNA•EF-Tu•GTP binds in an initial A/T state, where codon–anticodon interaction is checked in the decoding center of the A site before the aa-tRNA fully moves into the classic A site (Stark et al. 2002; Valle et al. 2002, 2003). Interestingly, the anticodon loop is kinked relative to the anticodon stem by ~40° to allow decoding, while simultaneously preventing interaction of the tRNA outside the anticodon loop with the A site (Fig. 16.6A–D). However, EF-Tu interaction with the ribosome visualized in these complexes probably reflects a state after the decoding process has been completed, and therefore it is unclear whether EF-Tu interacts with the ribosome prior to or during the selection process. We assume that EF-Tu contacts the ribosome only after the decoding process (Nierhaus 1993), but we note that this point remains controversial (Cochella and Green 2005; Daviter et al. 2006).

Since non-cognate aa-tRNAs in the cell are in five- to ten-fold excess over a cognate aa-tRNA, it is clear that without the beneficial effects of a cognate E-tRNA and in particular codon–anticodon interaction at the E site, the synthesis of a protein of a length of about 400 amino acids with an undisturbed structure and function would be improbable.

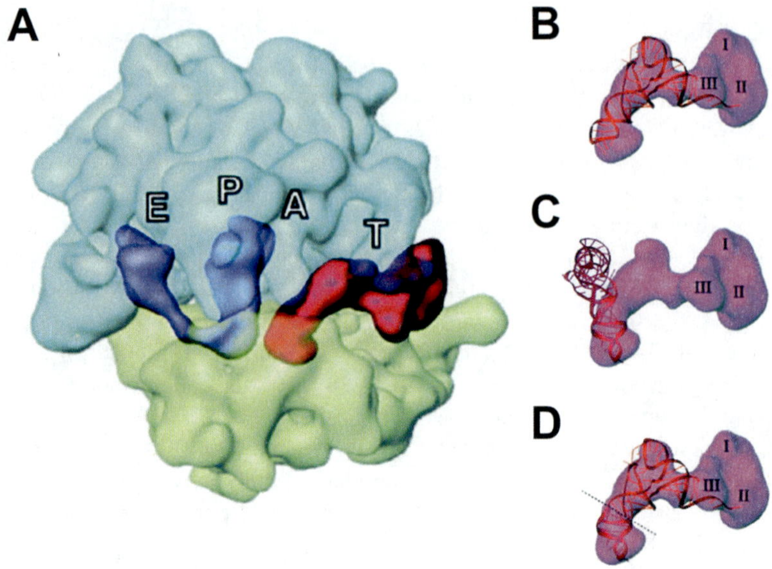

Fig. 16.6 The ternary complex interacting with the decoding center of the A site as seen by using cryo-EM. (**A**) The ribosome position of the ternary complex during the decoding process (A/T site). (**B–D**) Fitting the aminoacyl-tRNA within the ribosomal-bound ternary complex. To satisfactorily fit the crystal structure of a tRNA into the corresponding cryo-EM density requires the introduction of a kink codon stem of the aminoacyl-tRNA (according to Valle et al. (2002), modified)

16.4 Shine–Dalgarno Sequence Can Take Over the Function of the E-tRNA

If an occupied E site is important for translational fidelity by reducing near-cognate or preventing non-cognate misincorporations at the A site, as explained by the allosteric three-site model, this raises the question as to how accuracy is maintained when the first aa-tRNA binds to the A site directly following the initiation phase. This is a unique situation, in which ribosomes contain only one tRNA, namely an initiator tRNA bound at the P site, referred to as a Pi state. Therefore, directly following initiation the binding of ternary complex to the ribosome and decoding at the A site occur with an empty E site and according to the allosteric three-site model, should be error prone. However, there is a strong codon bias at the second position for GCN codons in highly expressed genes (Tats et al. 2006), i.e., the codon directly following the start codon, and this position has been shown to have a strong influence on the efficiency of translation initiation (Stenstrom et al. 2001). Indeed, stable cognate codon–anticodon interaction at this position has been proposed to be important for preventing premature peptidyl-tRNA drop-off (Tats et al. 2006). Furthermore, the first few N-terminal amino acids modulate the stability of proteins as well as providing determinants for the cleavage of the N-terminal formylmethionine residue from nascent peptide (Solbiati et al. 1999; Varshavsky 1996).

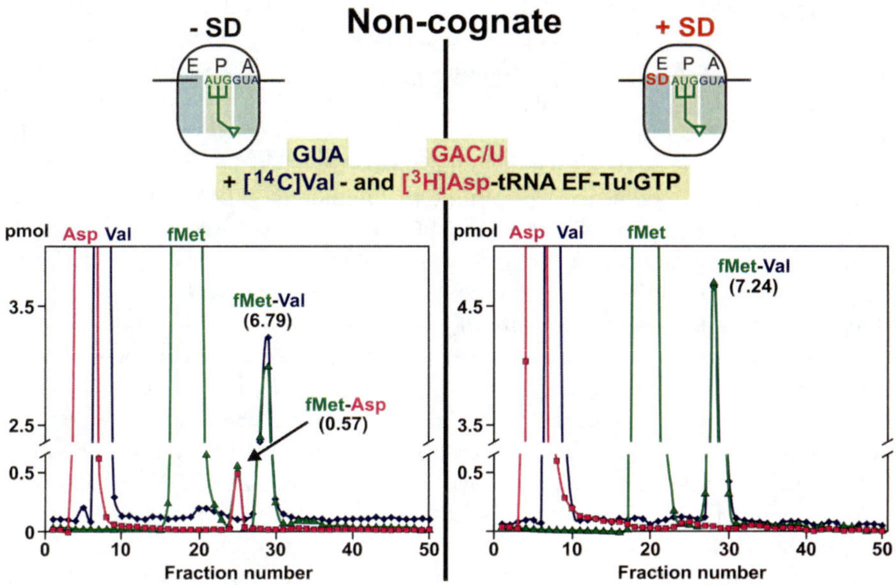

Fig. 16.7 Non-cognate misincorporation levels. The influence of SD on the selection of non-cognate aminoacyl-tRNA in the presence of MVF-mRNA: HPLC analysis of dipeptides. After filling the P site with fMet-tRNA, a mixture of ternary complexes was added containing cognate [^{14}C]Val-tRNA (codon GUA) and non-cognate [^{3}H]Asp-tRNA (GAC/U). In the absence of the SD sequence, an error of 7.7% was observed (*left*), whereas in its presence, the formation of non-cognate fMet-Asp is not observed (*right*)

Collectively, this suggests that accurate decoding at the second position is important for gene expression and therefore bacteria must have developed a mechanism to ensure accurate decoding at the A site in the absence of an E-tRNA.

Recently, we have demonstrated that the presence of a Shine–Dalgarno (SD) sequence located in the 5′ untranslated region of an mRNA can functionally compensate for the lack of a cognate tRNA at the E site, a situation that occurs directly following the initiation phase of translation. In these experiments, ribosomes were programmed with an fMet-tRNA at the P site, which exposed a GUA codon cognate for Val-tRNA at the A site (Fig. 16.7). Similar to the experiment in Fig. 16.5, a stoichiometric mixture of cognate hydrophobic Val-tRNA and non-cognate acidic hydrophilic Asp-tRNA was added. The HPLC analysis revealed that the presence of the Shine–Dalgarno interaction suppresses the formation of non-cognate fMet-Asp in a similar way to the presence of an E-tRNA (Di Giacco et al. 2008). This demonstrates that the SD sequence confers similar beneficial effects as an E-tRNA, in terms of accuracy during the selection of ternary complexes aa-tRNA•EF-Tu•GTP at the decoding center: The selection of the near-cognate aa-tRNA is moderately improved by a factor of two (Di Giacco et al. 2008), but – most significant – the misincorporation of non-cognate amino acids – in many cases detrimental – is abolished.

Recent X-ray crystallography studies have visualized the interaction of the SD sequence with the anti-SD sequence located in the 3′-end of the 16S rRNA on the ribosome (Jenner et al. 2007; Kaminishi et al. 2007; Korostelev et al. 2007; Yusupova et al. 2006). These studies reveal that the SD helix sits in a pocket located between the head and the platform of the 30S subunit, adjacent to but not directly in the E site. The SD–anti-SD interaction probably reduces the time necessary for mRNA–ribosome programming, since it helps to guide the mRNA from an initial stand-by site into a position whereby the AUG start codon is correctly positioned in the presence of initiator tRNA (de Smit and van Duin 2003; Gualerzi et al. 2001; Kaminishi et al. 2007). Interestingly, the conformation of the mRNA in the E site appears to be influenced by the state of the ribosome. In the initiation state, the mRNA is considered to be in a structurally constrained conformation, such that codon–anticodon interaction would not be possible in the E site (Jenner et al. 2007; Yusupova et al. 2006). However, following initiation the whole SD helix rotates on the ribosome toward the E site, which leads to conformational relaxation in the mRNA, such that the A-form helix adopted by the E codon of the mRNA now allows codon–anticodon interaction at the E site (Jenner et al. 2007). The recent observation that the SD helix appears to fix the orientation of the head of the 30S subunit (Korostelev et al. 2007) might provide the first structural hint as to how the SD helix (or E-tRNA) influences A site accuracy; however, further work utilizing both in vitro and in vivo experimental systems will be required to fully elucidate this mechanism.

A recent analysis of 162 completely sequenced prokaryotic genomes (141 of bacterial origin) revealed that an astonishingly large percentage (46%) of mRNAs do not contain a SD sequence, with the corresponding value for *E. coli* mRNA being 39% (Chang et al. 2006). However, it is well documented that a SD sequence is preferentially found in highly expressed genes (Ma et al. 2002), suggesting that accuracy during the first decoding step may be more important for high expression, although it is unclear exactly why. In addition to the SD sequence, an optimal spacer length of ~6 nt between SD and the initiator AUG codon, an A/U rich enhancer upstream the SD sequence, and the absence of strong secondary structures around the SD sequence are important determinants for high expression (for review and references, see Vimberg et al. 2007).

We do not know if or how the eukaryotic 80S ribosomes overcome the accuracy problem during the first decoding step, since their mRNAs do not contain SD sequences. Eukaryotic translation systems require about 12 initiation factors, some of which are composed of several different subunits (for reviews, see Gebauer and Hentze (2004); Pestova et al. (2001)), whereas bacterial systems use only three monomeric initiation factors. Thus, we can only speculate that the more complicated system required for the formation of both the 40S and subsequent 80S initiation complexes solves the accuracy problem of the first aa-tRNA selection.

Finally, archaeal mRNAs often contain an identifiable SD, but their set of initiation factors is similar – although somewhat simpler – to that in eukaryotes (Londei 2005). Curiously, archaea contain a number of leaderless mRNAs, i.e., mRNAs that have no 5′ untranslated region (and therefore no SD sequence) and start directly with an AUG start codon. Based on the findings presented here, we would predict that leaderless mRNAs are error prone at the step of forming the initial dipeptide. Whether the corresponding proteins can tolerate an increased error at the N-terminus or whether another mechanism operates remains unknown. At least in bacteria, only a fraction (below 0.1%) of mRNAs are leaderless and do not comprise mRNAs of essential genes (Moll et al. 2002), therefore the accuracy problem might not pose a significant problem toward cell viability in these cases.

In summary, the SD sequence, in addition to its canonical function related to mRNA positioning, has a second important function. This is seen in the fact that the SD–anti-SD interaction can functionally replace the E-tRNA to confer accurate decoding of the codon following the AUG. Specifically, the SD sequence reduces near-cognate misincorporation and precludes the selection of non-cognate aa-tRNAs, thereby protecting the cell from amino acid substitutions detrimental to protein folding, stability, and function.

16.5 Maintaining the Reading Frame

The ribosome must ensure that the binding of the tRNAs remains faithful to the codon of the mRNA displayed at the A site and that the correct reading frame of the mRNA is maintained during translation (reviewed by Wilson and Nierhaus 2003). During translation the error frequency associated with frameshift events is extremely low and has been estimated to be lower than 1 per 30,000 incorporations of amino acids (Jorgensen and Kurland 1990). However, in specific mRNAs there are loci designated recoding sites, where the efficiency of these frameshift events is significantly higher (reviewed by Atkins et al. 2000). A classic example of such a site is located within the mRNA of the *E. coli* prfB gene, which encodes the peptide chain release factor 2 (RF2). Translation of the full-length and active RF2 protein requires a +1 frameshift at the 26th position of the mRNA, in order to bypass an in-frame UGA stop codon. In fact, this programmed frameshifting site acts as an auto-regulatory mechanism, since RF2 terminates translation at UGA stop codons. Therefore, when the intracellular levels of RF2 are high, termination at the 26th position in the *prfB* mRNA predominates producing an inactive truncated RF2 protein that is rapidly degraded. However, when RF2 levels are low, the stop codon is bypassed via the +1 frameshifting event, leading to the production of full-length protein. What is extraordinary is that the frameshifting efficiency was determined to be ~30% (Curran and Yarus 1988; Weiss et al. 1987, 1988) and could be modulated to occur with up to 100% efficiency (Donly et al. 1990), i.e., frameshifting on the *prfB* mRNA occurs with a frequency that is more than four orders of magnitude higher than normal.

Several features have been identified that contribute to this efficiency; frameshifting is facilitated because (i) translation termination occurs slowly at a weak UGAC stop signal (Major et al. 1996; Poole et al. 1995), particularly when the intracellular level of RF2 is low; (ii) of the weak G:U wobble base pair of the oligopeptidyl-$tRNA^{Leu}$ at the P site, which promotes slippage into the new +1 frame (Curran 1993); (iii) a perfect realignment of the peptidyl-tRNA at the P site with the new aminoacyl-tRNA Asp-tRNA in the new frame is acquired after the frameshifting (Curran 1993); and (iv) a Shine–Dalgarno (SD)-like sequence precedes the UGA stop codon, which has complementary to the anti-SD sequence found at the 3′ end of 16S rRNA (Weiss et al. 1988).

We have noted that the complementarity between the SD-like sequence of the mRNA and the anti-SD sequence of the 16S rRNA extends into the ribosomal E site. This prompted us to establish an in vitro translation system that allows both the efficiency of frameshifting to be measured and the extent of deacylated tRNA release from the ribosomal E site. We found that the SD–anti-SD interaction enhances frameshifting by causing the release of the deacylated tRNA from the ribosomal E site. Indeed, we could show by monitoring dipeptide formation within a model of the *prfB* +1 frameshift window that the presence of a tRNA at the E site, and probably codon–anticodon interaction at this site, prohibits slippage of the tRNAs in the +1 frame and also stable binding of the A site tRNA out-of-frame (Marquez et al. 2004). This suggests that the occupation of the E site by a tRNA is instrumental for maintaining the reading frame, and that modulation of this dependence is exploited for the highly efficient feedback regulation of the translation of the RF2 mRNA.

It is likely that codon–anticodon interaction at the E site plays a decisive role for the observed effects. There is a growing body of evidence for the presence of such interaction. (i) Chasing experiments of labeled E-tRNAs from the ribosome are only effective, if the chase tRNA carries an anticodon complementary to the E site codon (reviewed in Blaha and Nierhaus 2001). (ii) The distances between anticodons of adjacent tRNAs on the ribosome are comparable between tRNAs at A and P sites and tRNAs at P and E sites (20±3 and 16±3 Å, cryo-electron-microscopic study, Agrawal et al. 2000). Since simultaneous codon–anticodon interaction of tRNAs at A and P sites is a generally accepted feature, the same feature should therefore hold for tRNAs at P and E sites. (iii) The X-ray structure of 70S ribosomes during the initiation and elongation phase demonstrated that after initiation the E site codon adopted a classical A-helical conformation ready to form codon–anticodon interaction (Jenner et al. 2007). (iv) Recently two groups demonstrated in vivo that the strength of codon–anticodon interaction at the E site is inversely proportional to the accuracy/frameshift efficiency in a system containing the RF2 frameshift window (Liao et al. 2008; Sanders and Curran 2007).

It follows that codon–anticodon interaction at the E site is a standard feature during protein synthesis, essential for maintaining the reading frame. One can estimate that without an E-tRNA translation would run into a frameshift after incorporation of 20–50 amino acids making it prohibitively difficult to synthesize proteins of a length of 300–500 amino acids, the average length of proteins.

References

Agrawal RK, Spahn CMT, Penczek P, Grassucci RA, Nierhaus KH, Frank J (2000) Visualization of tRNA movements on the *Escherichia coli* 70S ribosome during the elongation cycle. J Cell Biol 150:447–459

Alexeeva EV, Shpanchenko OV, Dontsova OA, Bogdanov AA, Nierhaus KH (1996) Interaction of mRNA with the *Escherichia coli* ribosome: Accessibility of phosphorothioate-containing mRNA bound to ribosomes for iodine cleavage. Nucl Acids Res 24: 2228–2235

Andersson SGE, Buckingham RH, Kurland CG (1984) Does codon composition influence ribosome function? The EMBO J 3: 91–94

Atkins JA, Herr AJ, Massire C, O'Connor M, Ivanov I, Gesteland RF (2000) Poking a hole in the sanctity of the triplet code: Inferences for framing. In: Garret RA, Douthwaite SR, Liljas A, Matheson AT, Moore PB, Noller HF (eds) The ribosome: Structure, function, antibiotics, and cellular interactions. ASM Press, American Society for Microbiology, Washington, DC, pp 369–384

Blaha G, Nierhaus KH (2001) Features and functions of the ribosomal E site. Cold Spring Harbor Symposia on Quantitative Biology 65:135–145

Chang B, Halgamuge S, Tang SL (2006) Analysis of SD sequences in completed microbial genomes: Non-SD-led genes are as common as SD-led genes. Gene 373:90–99

Cochella L, Green R (2005) Fidelity in protein synthesis. Curr Biol 15:R536–R540

Curran JF (1993) Analysis of effects of tRNA:message stability on frameshift frequency at the *Escherichia coli* RF2 programmed frameshift site. Nucl Acids Res 21:1837–1843

Curran JF, Yarus M (1988) Use of tRNA Suppressors To Probe Regulation Of *Escherichia coli* release factor 2. J Mol Biol 203:75–83

Daviter T, Gromadski KB, Rodnina MV (2006) The ribosome's response to codon-anticodon mismatches. Biochimie 88:1001–1011

de Smit MH, van Duin J (2003) Translational standby sites: how ribosomes may deal with the rapid folding kinetics of mRNA. J Mol Biol 331:737–743

Di Giacco V, Marquez V, Qin Y, Pech M, Triana-Alonso FJ, Wilson DN, Nierhaus KH (2008) Shine-Dalgarno interaction prevents incorporation of noncognate amino acids at the codon following the AUG. Proc Natl Acad Sci USA 105:10715–10720

Dinos G, Kalpaxis DL, Wilson DN, Nierhaus KH (2005) Deacylated tRNA is released from the E site upon A site occupation but before GTP is hydrolyzed by EF-Tu. Nucl Acids Res 33: 5291–5296

Donly BC, Edgar CD, Adamski FM, Tate WP (1990) Frameshift autoregulation in the gene for *Escherichia coli* release factor-2 – Partly functional mutants result in frameshift enhancement. Nucl Acids Res 18:6517–6522

El'skaya AV, Ovcharenko GV, Palchevskii SS, Petrushenko ZM, Triana-Alonso FJ, Nierhaus KH (1997) Three tRNA binding sites in rabbit liver ribosomes and role of the intrinsic ATPase in 80S ribosomes from higher eukaryotes. Biochemistry 36:10492–10497

Frank J, Agrawal RK (2000) A ratchet-like inter-subunit reorganization of the ribosome during translocation. Nature 406: 318–322

Gebauer F, Hentze MW (2004) Molecular mechanisms of translational control. Nat Rev Mol Cell Biol 5:827–835

Geigenmüller U, Nierhaus KH (1990) Significance of the third tRNA binding site, the E site, on *E. coli* ribosomes for the accuracy of translation: an occupied E site prevents the binding of non-cognate aminoacyl-transfer RNA to the A site. EMBO J 9: 4527–4533

Gnirke A, Geigenmüller U, Rheinberger H-J, Nierhaus KH (1989) The allosteric three-site model for the ribosomal elongation cycle. J Biol Chem 264:7291–7301

Gualerzi CO, Brandi L, Caserta E, Garofalo C, Lammi M, La Teana A, Petrelli D, Spurio R, Tomsic J, Pon CL (2001) Initiation factors in the early events of mRNA translation in bacteria. Cold Spring Harb Symp Quant Biol 66:363–76

Jenner L, Rees B, Yusupov M, Yusupova G (2007) Messenger RNA conformations in the ribosomal E site revealed by X-ray crystallography. EMBO Rep 8:846–850

Jorgensen F, Kurland CG (1990) Processivity errors of gene expression in *Escherichia coli.* J Mol Biol 215:511–521

Kaminishi T, Wilson DN, Takemoto C, Harms JM, Kawazoe M, Schluenzen F, Hanawa-Suetsugu K, Shirouzu M, Fucini P, Yokoyama S (2007) A snapshot of the 30S ribosomal subunit capturing mRNA via the Shine-Dalgarno interaction. Structure 15:289–297

Korostelev A, Trakhanov S, Asahara H, Laurberg M, Lancaster L, Noller HF (2007) Interactions and dynamics of the Shine Dalgarno helix in the 70S ribosome. Proc Natl Acad Sci USA 104:16840–16843

Kurland CG, Jørgensen F, Richter A, Ehrenberg M, Bilgin N, Rojas A.-M (1990) Through the accuracy window. In: Dahlberg A, Hill WE, Garrett RA, Moore PB, Schlessinger D, Warner JR (eds) The ribosome-structure, function, and evolution. American Society for Microbiology, Washington, DC, pp 513–526

Leger M, Dulude D, Steinberg SV, Brakier-Gingras L (2007) The three transfer RNAs occupying the A, P and E sites on the ribosome are involved in viral programmed –1 ribosomal frameshift. Nucl Acids Res 35:5581–5592

Liao PY, Gupta P, Petrov AN, Dinman JD, Lee KH (2008) A new kinetic model reveals the synergistic effect of E-, P- and A-sites on +1 ribosomal frameshifting. Nucl Acids Res 36:2619–2629

Londei P (2005) Evolution of translational initiation: new insights from the archaea. FEMS Microbiol Rev 29: 185–200

Lührmann R, Eckhardt H, Stöffler G (1979) Codon-anticodon interaction at the ribosomal peptidyl-site. Nature 280:423–425

Ma J, Campbell A, Karlin S (2002) Correlations between Shine-Dalgarno sequences and gene features such as predicted expression levels and operon structures. J Bacteriol 184:5733–5745

Major LL, Poole ES, Dalphin ME, Mannering SA, Tate WP (1996) Is the in-frame termination signal of the *Escherichia coli* release factor-2 frameshift site weakened by a particularly poor context? Nucl Acids Res 24: 2673–2678

Marquez V, Wilson DN, Tate WP, Triana-Alonso F, Nierhaus KH (2004) Maintaining the ribosomal reading frame: The influence of the E site during translational regulation of release factor 2. Cell 118:45–55

Moll I, Grill S, Gualerzi CO, Blasi U (2002) Leaderless mRNAs in bacteria: surprises in ribosomal recruitment and translational control. Mol Microbiol 43:239–246

Moore SD, Sauer RT (2007) The tmRNA system for translational surveillance and ribosome rescue. Annu Rev Biochem 76:101–124

Munro JB, Altman RB, O'Connor N, Blanchard SC (2007) Identification of two distinct hybrid state intermediates on the ribosome. Mol Cell 25:505–517

Nierhaus KH (1990) The allosteric three-site model for the ribosomal elongation cycle: features and future. Biochemistry 29:4997–5008

Nierhaus KH (1993) Solution of the ribosomal riddle: How the ribosome selects the correct aminoacyl-tRNA out of 41 similar contestants. Mol Microbiol 9:661–669

Ogle JM, Brodersen DE, Clemons Jr, WM, Tarry MJ, Carter AP, Ramakrishnan V (2001) Recognition of cognate transfer RNA by the 30S ribosomal subunit. Science 292:897–902

Ogle JM, Ramakrishnan V. 2005 Structural insights into translational fidelity. Annu Rev Biochem 74:129–177

Pestova TV, Kolupaeva VG, Lomakin IB, Pilipenko EV, Shatsky IN, Agol VI, Hellen CU T (2001) Molecular mechanisms of translation initiation in eukaryotes. Proc Natl Acad Sci USA 98:7029–7036

Poole ES, Brown CM, Tate WP (1995) The identity of the base following the stop codon determines the efficiency of in vivo translational termination in *Escherichia coli*. EMBO J 14:151–158

Remme J, Margus T, Villems R, Nierhaus KH (1989) The third ribosomal tRNA-binding site, the E site, is occupied in native polysomes. Eur J Biochem 183:281–284

Rheinberger H.-J, Nierhaus KH (1980) Simultaneous binding of the 3 tRNA molecules by the ribosome of *E coli*. Biochem Internatl 1:297–303

Rheinberger H.-J, Nierhaus KH (1983) Testing an alternative model for the ribosomal peptide elongation cycle. Proc Natl Acad Sci USA 80:4213–4217

Rheinberger H.-J, Nierhaus KH 1986a. Adjacent codon-anticodon interactions of both tRNAs present at the ribosomal A and P or P and E sites. FEBS Lett 204:97–99

Rheinberger H.-J, Nierhaus KH 1986b Allosteric interactions between the ribosomal transfer RNA-binding sites A and E. J Biol Chem 261:9133–9139

Rheinberger H.-J, Sternbach H, Nierhaus KH (1981) Three tRNA binding sites on *Escherichia coli* ribosomes. Proc Natl Acad Sci USA 78:5310–5314

Rheinberger H.-J, Sternbach H, Nierhaus KH (1986) Codon-anticodon interaction at the ribosomal E site. J Biol Chem 261:9140–9143

Sanders CL, Curran JF (2007) Genetic analysis of the E site during RF2 programmed frameshifting. RNA 13: 1483–1491

Saruyama H, Nierhaus KH (1986) Evidence that the three-site model for ribosomal elongation cycle is also valid in the archaebacterium *Halobacterium halobium*. Mol Gen Genet 204: 221–228

Schilling-Bartetzko S, Bartetzko A, Nierhaus KH (1992) Kinetic and thermodynamic parameters for transfer RNA binding to the ribosome and for the translocation reaction. J Biol Chem 267:4703–4712

Schmeing TM, Moore PB, Steitz TA (2003) Structures of deacylated tRNA mimics bound to the E site of the large ribosomal subunit. RNA 9: 1345–1352

Selmer M, Dunham C, Murphy FV 4th, Weixlbaumer A, Petry S, Kelley A, Weir J, Ramakrishnan V (2006) Structure of the 70S ribosome complexed with mRNA and tRNA. Science 313: 1935–1942

Solbiati J, Chapman-Smith A, Miller JL, Miller CG, Cronan JE, Jr (1999) Processing of the N termini of nascent polypeptide chains requires deformylation prior to methionine removal. J Mol Biol 290:607–614

Spahn CM, Blaha G, Agrawal RK, Penczek P, Grassucci RA, Trieber CA, Connell SR, Taylor DE, Nierhaus KH, Frank J (2001) Localization of the ribosomal protection protein Tet(O) on the ribosome and the mechanism of tetracycline resistance. Mol Cell 7: 1037–1045

Stark H, Rodnina MV, Wieden HJ, Zemlin F, Wintermeyer W, Van Heel M (2002) Ribosome interactions of aminoacyl-tRNA and elongation factor Tu in the codon-recognition complex. Nat Struct Biol 15:15–20

Stenstrom CM, Jin H, Major LL, Tate WP, Isaksson LA (2001) Codon bias at the 3′-side of the initiation codon is correlated with translation initiation efficiency in *Escherichia coli*. Gene 263: 273–284

Szaflarski W, Vesper O, Teraoka Y, Plitta B, Wilson DN, Nierhaus KH (2008) New features of the ribosome and ribosomal inhibitors: Non-enzymatic recycling, misreading and back-translocation. J Mol Biol 380:193–205

Tats A, Remm M, Tenson T (2006) Highly expressed proteins have an increased frequency of alanine in the second amino acid position. BMC Genomics 7:28

Triana-Alonso FJ, Chakraburtty K, Nierhaus KH (1995) The elongation factor 3 unique in higher fungi and essential for protein biosynthesis is an E site factor. J Biol Chem 270: 20473–20478

Valle M, Sengupta J, Swami NK, Grassucci RA, Burkhardt N, Nierhaus KH, Agrawal RK, Frank J (2002) Cryo-EM reveals an active role for aminoacyl-tRNA in the accommodation process. EMBO J 21: 3557–3567

Valle M, Zavialov A, Sengupta J, Rawat U, Ehrenberg M, Frank J (2003) Locking and unlocking of ribosomal motions. Cell 114: 123–134

Varshavsky A (1996) The N-end rule: Functions, mysteries, uses – Inaugural paper. Proc Natl Acad Sci USA 93:12142–12149

Vimberg V, Tats A, Remm M, Tenson T (2007) Translation initiation region sequence preferences in *Escherichia coli*. BMC Mol Biol 8: 100

Warner JR, Rich A (1964) The number of soluble RNA molecules on reticulocyte polyribosomes. Proc Natl Acad Sci USA 51: 1134–1141

Weiss RB, Dunn DM, Atkins JF, Gesteland RF (1987) Slippery runs, shifty stops, backward steps, and forward hops: 2, 1, +1, +2, +5, and +6 ribosomal frameshifting. Cold Spring Harbor Symp Quant Biol 52:687–693

Weiss RB, Dunn DM, Dalhberg AE, Atkins JF, Gesteland RF (1988) Reading frame switch caused by base-pair formation between the 3′ end of 16S rRNA and the mRNA during elongation of protein synthesis in *Escherichia coli*. EMBO J 7:1503–1507

Wendrich TM, Blaha G, Wilson DN, Marahiel MA, Nierhaus KH (2002) Dissection of the mechanism for the stringent factor RelA. Mol Cell 10:779–788

Wilson DN, Nierhaus KH (2003) The ribosome through the looking glass. Angew Chem Int Ed Engl 42:3464–3486

Wilson DN, Nierhaus KH (2006) The E-site story: the importance of maintaining two tRNAs on the ribosome during protein synthesis. Cell Mol Life Sci 63:2725–2737

Wurmbach P, Nierhaus KH (1979) Codon-anticodon interaction at the ribosomal P(peptidyl)-tRNA site. Proc Natl Acad Sci USA 76:2143–2147

Yusupova G, Jenner L, Rees B, Moras D, Yusupov M (2006) Structural basis for messenger RNA movement on the ribosome. Nature 444:391–394

Part III
Discontiguity

Chapter 17
Translational Bypassing – Peptidyl-tRNA Re-pairing at Non-overlapping Sites

Norma M. Wills

Abstract Ribosomal bypassing can lead to the translational fusion of non-contiguous ORFs. It involves dissociation of codon:anticodon pairing in the ribosomal P-site followed by mRNA slippage and re-pairing of the retained tRNA anticodon to mRNA at a non-overlapping codon. It is frame independent. The most studied case involves the bypassing of 50 non-coding nucleotides between codons 46 and 47 of phage T4 gene *60* where half the translating ribosomes successfully accomplish the feat. A nascent peptide signal encoded 5′ of the start of the coding gap facilitates the initial codon:anticodon dissociation. An mRNA structure forms in the ribosomal A-site. Only when sequence participating in this structure has passed the ribosomal P-site does the potential for anticodon re-pairing to mRNA at a matched codon arise. After such re-pairing, normal decoding of the A-site codon mediates resumption of standard translation.

Contents

N.M. Wills (✉)
Department of Human Genetics, University of Utah, Salt Lake City, UT 84112-5330, USA
e-mail: nwills@genetics.utah.edu

J.F. Atkins, R.F. Gesteland (eds.), *Recoding: Expansion of Decoding Rules Enriches Gene Expression*, Nucleic Acids and Molecular Biology 24,
DOI 10.1007/978-0-387-89382-2_17,

17.1 Introduction

Most, if not all, programmed frameshifting involves codon:anticodon dissociation and anticodon re-pairing to mRNA in a new frame. The newly engaged codon overlaps the original zero frame codon – most programmed frameshifting involves a single nucleotide shift, although –2 shifts are also known. In bypassing, however, peptidyl-tRNA anticodon re-pairing is not to an overlapping codon, but is to a separate codon so a shift in reading frame may or may not be involved. The key feature is synthesis of a single protein from two separated ORFs. Extensive mRNA movement can be involved and whether there is a substantive difference in the ribosome–mRNA relationship from single nucleotide frameshifting is just one of the many points of interest.

17.2 Non-programmed Bypassing

In the absence of stimulatory recoding signals, low-level error bypassing can be detected at certain sequences, especially when there is a slow-to-decode A-site codon. The first demonstration of such non-programmed bypassing, called tRNA hopping at the time, was discovered in *Escherichia coli* cells using test sequences fused to the *lacZ* gene (Weiss et al., 1987). In one of the sequences, the slow-to-decode codon, a stop codon, was flanked by leucine codons. This sequence, CUU UAG CUA, produced 1% of the level of β-galactosidase as the stop-free control. Protein sequencing showed that the 9 nt sequence was decoded as a single leucine residue leading to the conclusion that peptidyl-$tRNA^{Leu}$ dissociated from CUU, the "take-off" site, "hopped" over the UAG stop codon, and paired to the cognate CUA, the "landing" site, whereupon translation continued (Fig. 17.1). In experiments of this type two other examples of low-level dissociation and re-pairing were found. In one example, $tRNA^{Leu}$ hops forward by 5 nt on the sequence AAC UCA AUC (zero frame codons separated by spaces and re-pairing site underlined). Mutants of $tRNA_1^{Val}$ containing an extra nucleotide in their anticodon loop showed

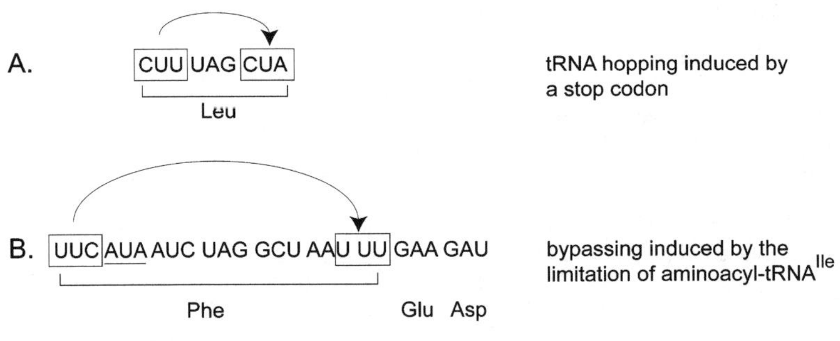

Fig. 17.1 tRNA hopping or bypassing. (**A**) A stop hop induced by a slow-to-decode stop codon in the A-site. One amino acid, Leu, inserted by the 9 nt sequence. (**B**) Bypassing induced by a slow-to-decode isoleucine codon, underlined, due to limitation of its cognate aminoacyl-tRNA. Peptidyl-tRNA re-pairing to UUU 15–17 nt 3′ of the UUC take-off codon

increased hopping from GUG to GUU in the sequence GUG UAA GUU (O'Connor et al., 1989). A particular substitution mutant of $tRNA_2^{Gly}$, C40→G, also increased hopping over stop codons (Herr et al., 2001a).

A-site codons that are slow to decode due to limitation of aminoacyl-tRNA can also promote bypassing (Fig. 17.1). This was first encountered in overexpression of a mammalian gene in the *E. coli* heterologous system (Kane et al., 1992). Such "hungry" codon bypassing has been studied extensively by Gallant and Lindsley (1998) [following insightful in-depth studies of amino acid starvation effects on single nucleotide frameshifting (Lindsley and Gallant, 1993 and Chapter 14)]. Peptidyl-tRNA re-pairing occurred at a matched codon and the distance bypassed could be as large as 40 nt although the efficiency of bypassing decreased as the distance increased. The presumption that the anticodon of peptidyl-tRNA scans the transiting mRNA for potential complementarity was confirmed with constructs having competing landing sites between the take-off site and the original landing site (Gallant et al., 2003). Linear scanning of mRNA by the peptidyl-tRNA anticodon in 70S ribosomes was further substantiated in a study which also involved re-pairing to mRNA, though at lower efficiency at poorly matched codons (Herr et al., 2004). When an impediment to the scanning ribosomes, a stem-loop structure, was introduced between the take-off and landing sites, bypassing was decreased (Gallant et al., 2003).

A ribosomal pause is a common feature of non-programmed bypassing. This allows sufficient time for peptidyl-tRNA dissociation from the take-off codon and the initiation of forward ribosome movement on the mRNA. It follows that the stability of the anticodon:codon interaction should be a major determinant for take-off and for landing. An extensive study performed by Gallant et al. (2004) showed that matched codons with A or U in the first two positions were the most efficient for bypassing, suggesting that the anticodon:codon dissociation at the initiation of bypassing is a limiting factor (Fig. 17.2).

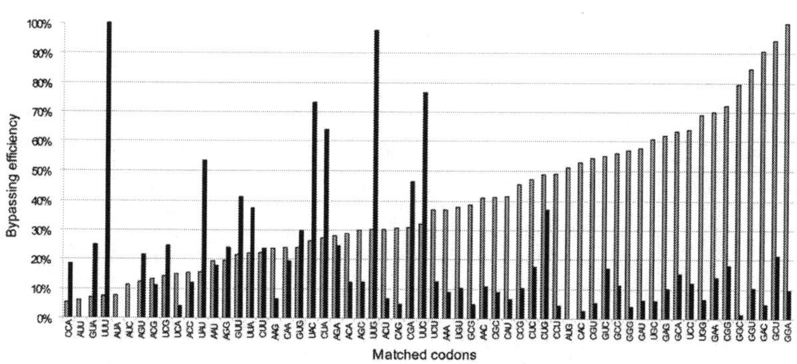

Fig. 17.2 Bypassing efficiencies of matched sets of take-off and landing codons. Black bars, bypassing induced by limitation of aminoacyl-tRNA (non-programmed bypassing); gray bars, bypassing in the T4 gene *60* context (programmed bypassing) (from Bucklin et al., 2005). Efficiencies are reported as a percentage of the most efficient codon for each context

17.3 Programmed Bypassing

Huang et al. (1988) remarkably discovered a coding gap between two separated ORFs expressing a fusion product. Elegant follow-up studies by Weiss et al. (1990) revealed the recoding signals in this first case of programmed bypassing. The two ORFs are in bacteriophage T4 gene *60* that encodes a topoisomerase subunit. Sequencing of phage DNA and RNA isolated from T4-infected cells determined that 50 nt separates codons 46 and 47 which are in two discrete open reading frames that jointly would encode the expected 18 kDa protein. N-terminal sequencing of the full-length protein and its peptides confirmed that the 50 nt region is not translated and that amino acids for codons 46, glycine, and 47, leucine, are contiguous in the protein.

The 50 nt coding gap contains stop codons in all three frames, one (UAG) immediately after codon 46, GGA. At the end of the coding gap is another GGA triplet followed by codon 47. The UAG immediately following the take-off GGA codon suggests that this could be viewed as an exaggerated "stop hop" with landing considerably downstream. This comparison is thin, however, for a number of reasons.

GGA codons are not particularly prone to bypassing, since when stimulated by a 3′ hungry codon, bypassing is only 10% as efficient as UUU and UUC, the most favorable codons (Gallant et al., 2004). Also, part of the coding gap can potentially form a stem-loop structure (see below) that should impede ribosomal movement and, hence, scanning by the peptidyl-tRNA, and a competitive potential landing site, GGG, is present between the take-off and the landing site. [GGG and GGA are decoded by $tRNA_2^{Gly}$ (Murgola and Pagel, 1980), the bypassing peptidyl-tRNA. If this GGG were to act as a landing site, bypassing would be unproductive due to the presence of a stop codon in the same frame 8 codons downstream.] A long distance must be bypassed (50 nt) which should make the process extremely inefficient, but successful bypassing occurs 50% of the time (Maldonado and Herr, 1998; Herr et al., 2000).

What features of the gene *60* mRNA sequence (programming signals) and translational components are important for this remarkably efficient bypassing event? Figure 17.3 shows various elements of the gene *60* sequence that have been identified as contributors to efficient bypassing: (1) the UAG immediately 3′ to the take-off site, (2) matched take-off and landing codons, (3) a region of the nascent peptide, (4) a Shine–Dalgarno-like sequence, GAG, located 6 nt 5′ to the landing site, (5) a stem-loop structure comprising sequences both 5′ and 3′ of the take-off site. Structural features of the rest of the coding gap are also likely crucial. Each of these elements is considered separately below.

17.4 The UAG Codon Following Take-Off Site

The GGA take-off site in gene *60* is followed by a UAG stop codon and its importance was demonstrated by greatly diminished bypassing when UAG was replaced by either GGU or UAC sense codons (Weiss et al., 1990). Whether a delay in arrival

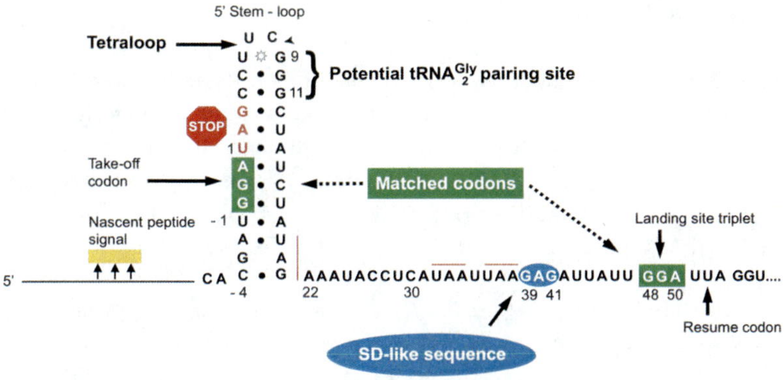

Fig. 17.3 Features important for translational bypassing in decoding T4 gene *60*. The nascent peptide signal is indicated by the *yellow box*; the matched take-off and landing codons, GGA, are shown in *white letters* in *dark green boxes*; the UAG stop codon immediately 3′ of the take-off site is in *red letters* next to the stop sign and stop codons within the coding gap are overlined in *red*. A Shine–Dalgarno-like sequence is shown in the *blue oval*. The translational resume codon is indicated. A potential tRNA$^{Gly}_2$ pairing site is shown by a *bracket* (from Wills et al., 2008)

of its cognate release factor 1 (RF1) or, alternatively, an abortive termination was the relevant feature associated with the stop codon was tested by altering the levels of RF1 (Herr et al., 2000). Lowering the levels of active RF1 with non-permissive temperature growth of an RF1 ts mutant had no effect on bypassing efficiency (but greatly elevated simple stop hopping with another construct), whereas elevated levels of RF1 did reduce bypassing. Consequently, it appears that a stable termination complex does not form because the initiation of bypassing occurs much faster. As discussed below, a stem-loop structure (that includes the UAG in the stem) in the 5′ portion of the coding gap is an important feature in the prevention of efficient RF1 recognition.

17.5 Matched Take-Off and Landing Codons

In gene *60*, codon 46, the take-off codon is GGA and the three nucleotides preceding the resume codon, the position of the landing codon, are GGA. With the matched take-off/landing pairs of GGA (WT) or GCA, bypassing is highly efficient, whereas with unmatched take-off/landing pairs, GGA/GCA or GCA/GGA, bypassing efficiencies were greatly reduced (Weiss et al., 1990). The relative efficiencies of 61 matched codons in place of the GGAs were tested by Bucklin et al. (2005) (Fig. 17.2). The relative efficiency of bypassing with the different matched codon sets reflected re-pairing potential at the landing site in contrast to the result with non-programmed bypassing. (This observation supports the contention that take-off is 100% efficient.). Matched take-off and landing codons with G or C in the first two positions were the most efficient. Of all codons tested, GGA, the wild-type gene *60* take-off and landing sites, has the highest bypassing efficiency.

17.6 Peptidyl-tRNA$_2^{Gly}$

A genetic approach was undertaken in *E. coli* to identify mutants in translational components that decrease bypassing efficiency (Herr Atkins and Gesteland, 1999). Multiple independent mutations [called *byp* (*bypassing*) mutants] were recovered and all were in *glyT*, the sole gene encoding tRNA$_2^{Gly}$, the peptidyl-tRNA involved in gene *60* bypassing. Mutations occurred at three positions that are universally conserved in tRNAs, G18, G19, and C56. One mutation was found in the anticodon stem at C40 (Fig. 17.4).

Mutants of *glyT* (*sufS*) in *Salmonella enterica* (*typhimurium*) that promote –1 frameshifting at G GGA (–1 codon underlined) had previously been isolated (Riyasaty and Atkins, 1968; O'Mahony et al., 1989; Pagel et al., 1992). Of several *sufS* alleles tested, 60+U (and insertion of an extra U between U59 and U60), C61→U, and C62→A reduced bypassing to the same extent as the *byp* mutants (Fig. 17.4).

The *byp* mutant C40→G most likely destabilizes the anticodon stem and/or alters stacking interactions in the anticodon loop. The other tRNA$_2^{Gly}$ mutants contain substitutions or an addition in the dihydrouracil (D) loop or the ribothymidine (T) stem or loop. Interactions between bases in these regions are crucial for stabilizing the

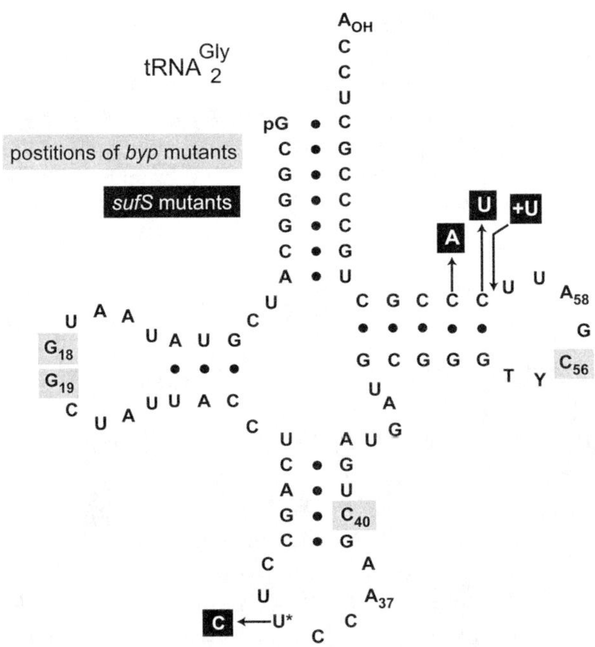

Fig. 17.4 tRNA$_2^{Gly}$ mutants that affect bypassing. Black boxes indicate the positions of *sufS* mutants in *Salmonella enterica* (*typhimurium*) and gray boxes indicate the positions of *byp* mutants isolated in *E. coli*

elbow structure. Increased flexibility of the elbow region is thought to decrease codon–anticodon stability and this effect is manifest at landing where peptidyl-tRNA anticodon–codon re-pairing occurs. Increased flexibility of the elbow region may also interfere with important conformational changes induced by ternary complex binding in the A-site (Schuette et al., 2009) and/or disrupt interaction with the L1 stalk during translocation (see below).

17.7 Nascent Peptide Effect

Part of the nascent peptide sequence encoded upstream of the take-off site is important for bypassing (Weiss et al., 1990), exerting a nine-fold stimulatory effect (Herr et al., 2001a). The nascent peptide signal which stimulates bypassing affects destabilization of peptidyl-tRNA anticodon:codon pairing at take-off and imposes stringency on peptidyl-tRNA re-pairing to mRNA at landing (Herr et al., 2000, 2001a, b, 2004).

The important region of the nascent peptide was localized to the specific amino acids K_{-31}YKLQNNVRRSIKSSS$_{-16}$ (distance from the take-off site indicated) of the WT sequence (Weiss et al., 1990). Substitutions of each amino acid in the region were constructed where the remainder of the peptide was unchanged (Larsen et al., 1995; M. O'Connor, G. Loughran and J. Atkins, unpublished). The largest effect occurred with substituting Tyr$_{-30}$ (30 residues before the amino acid esterified to peptidyl-tRNA at the time of take-off) where bypassing was decreased to 30% of WT. Interestingly, substitution of Tyr$_{-30}$ with another aromatic amino acid, Phe, had no effect on bypassing.

The peptide exit tunnel, the birth canal, of the ribosome can accommodate a fully extended peptide of approximately 28 amino acids and earlier studies showed that *E. coli* ribosomes can protect more than 40 amino acids (Tsalkova et al., 1998). This indicates that the gene *60* nascent peptide exerts its effect while still within the ribosome since it is positioned 16–31 amino acids distant from the take-off site. One study has shown that the N-terminus of the gene *60* nascent peptide can be cross-linked to small subunit proteins S1–S4; however, this may not be related to bypassing since a similar sized *ompA* peptide showed an almost identical cross-linking pattern (Choi et al., 1998).

17.8 Shine–Dalgarno Sequence Within the Coding Gap

From studies of programmed frameshifting involved in expression of *E. coli* release factor 2 and *dnaX,* it is clear that the anti-Shine–Dalgarno (SD) sequence in translating ribosomes scans mRNA and can pair with internal SD sequences. After pairing occurs, the ribosome continues translation until the hybrid ruptures when there are approximately 14 nt between the 3′ end of the SD sequence and the P-site codon. A minimal distance between the SD and the frameshift site stimulates +1 frameshifting and a maximal distance stimulates −1 frameshifting (Weiss et al., 1987; Larsen et al., 1994).

A minimal SD sequence, $GAG_{39–41}$, is located 6 nt 5′ to the GGA landing site in gene *60* (Fig. 17.3) (Wills et al., 2008). In contrast to where SD sequences serve to stimulate single nucleotide frameshifting, the 6 nt spacing corresponds to the optimal distance between an SD and an initiation codon (Chen et al., 1994; Jin et al., 2006). Ribosome swapping experiments demonstrated the direct interaction between the anti-SD sequence of 16S rRNA and the SD-like sequence, $GAG_{39–41}$, of gene *60*. However, while the disruption of the SD:anti-SD interaction causes a notable decrease in landing efficiency, ~40% of WT, there is a very small effect on landing site selection (Wills et al., 2008).

17.9 RNA Structure of the Coding Gap and Landing Fidelity

A stem-loop structure in the 5′ portion of the coding gap was found to be important for bypassing (Weiss et al., 1990; Herr et al., 2000). It comprised the upper 5 base pairs and loop of the structure shown in Fig. 17.3. Mutations which disrupted base pairing reduced bypassing while compensatory double mutations restored bypassing to a high level. Altering the tetraloop sequence at the top of the stem or increasing loop size also reduced bypassing as did extending the length of the top of the stem. Mutations that disrupted potential base pairing of the part of the stem that includes the take-off site decreased bypassing to a limited extent suggesting that this pairing is not crucial. Results of experiments that monitored two competing recoding events, bypassing and readthrough of the UAG codon, reinforced the importance of the 5-base stem-loop structure. Base pairing of the UAG codon is postulated to preclude RF1 acceptance in the A-site implying that stem formation occurs after or coincident with peptidyl-tRNA dissociation from mRNA (Herbst et al., 1994; Herr et al., 2000).

Recent work showed that the stem loop is more extensive – comprising a 12 base pair stem (with one mismatch) and a tetraloop shown in Fig. 17.3 (Wills et al., 2008). In two cases where pairing of the first 4 or 7 bases is disrupted, bypassing efficiency is drastically reduced, while stronger pairing potential results in elevated bypassing efficiency over WT (manuscript in preparation). The fidelity of landing in several of the stem-loop mutants has also been examined. In cases where base pairing in the stem is disrupted or the tetraloop is altered, landing is observed at the WT position, but varying amounts occur at the GGG triplet 9–11 nt 3′of the take-off site (Wills et al., 2008). When the strengthened stem-loop construct is tested, landing occurs only at the WT position indicating that the structure is involved in masking the potential GGG landing site from bypassing peptidyl-tRNA.

When and where does RNA structure influence bypassing? In simplistic terms, the stem loop must exert its effect either before or after decoding of the take-off site GGA since it must be single-stranded to be read by the incoming tRNA. All

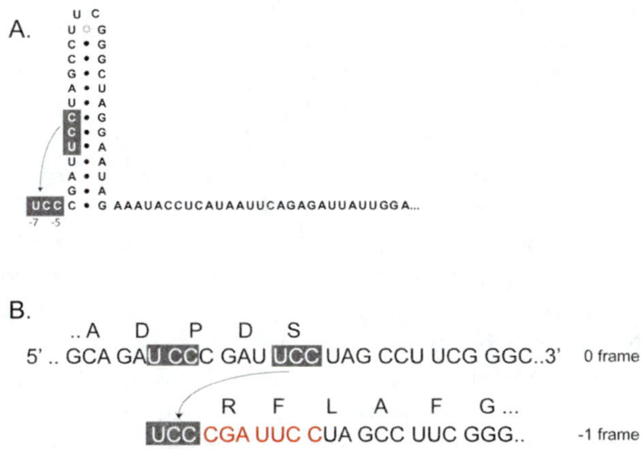

Fig. 17.5 Backward bypassing or "loop-de-loop." (**A**) Sequence and potential secondary structure of a construct designed to detect backward bypassing. The take-off codon, UCC, and a matched codon 7–5 nt 5′ are shown in *white letters* in *dark gray boxes*. (**B**) A region of the sequence shown in two alternate reading frames. Peptidyl-tRNA dissociates from the UCC codon in the zero frame and re-pairs to UCC in the –1 reading frame. The seven nucleotides shown in red are translated twice

evidence points toward it acting after take-off. If the structure acted before take-off, it would presumably be sensed by the leading edge or unwinding site of the ribosome and a "memory" of the stimulation would have to be carried until GGA codon 46 is decoded. Additionally, this scenario would require certain coding gap mutants to alter the memory such that landing could occur within the gap. However, formation of RNA structure after decoding of the take-off GGA is consistent with masking both the UAG from RF1 recognition and GGG_{9-11} from peptidyl-tRNA recognition.

Further support for structure formation after take-off comes from the demonstration of "backward bypassing" in modified gene *60* cassettes (Wills et al., 2008). The underlying premise is that formation of the stem-loop structure requires RNA to be drawn in from both the 5′ and 3′ directions (see Fig. 17.3). If the structure forms in the A-site, then the 3 nt preceding the base of the stem loop would occupy the P-site. When a codon matched to the take-off site is introduced immediately 5′ to the base of the stem loop (Fig. 17.5), backward bypassing is detected to that site (as well as to other sites in the forward direction). The discovery of backward bypassing fulfills a long-standing dream for ribosome gymnast fanciers: that one day ribosomes would be found to "loop-de-loop" with a single ribosome going back to re-translate part of the mRNA that it has just translated in a different frame!

The WT landing site, GGA_{48-50}, is preceded by two AUU triplets. When the WT take-off and landing sites are changed to AUU such that three, tandem AUU triplets are present at the end of the coding gap (and an additional one at nt 34–36 in the WT sequence), landing occurs only at the AUU triplet positioned at the WT landing site (nt 48–50) (Wills et al., 2008). In a similar experiment, the WT take-off and landing sites were changed to UCC and the landing site was followed by five additional UCC triplets. Landing again occurred only at the UCC triplet positioned at the WT landing site. From these experiments, it is apparent that landing does not occur at matched codons either immediately 5′ or 3′ of the normal landing position when in competition with the same codon at the WT landing position.

In apparent contrast to these results, when two additional GAG sequences are introduced into the coding gap, the predominant landing triplet utilized is the second GAG, i.e., the first available landing site 3′ of an SD sequence. This indicated that an SD can have a significant effect on landing site selection with a non-WT coding gap, most likely by facilitating the initiation of scanning by the anticodon of peptidyl-tRNA. A probable explanation for the contrast is that the entire coding gap is involved in RNA structure such that in WT, the SD plays a lesser role in landing site selection than in the case where multiple sequence substitutions disrupt structure and the SD effect on landing site selection is amplified. Consistent with this, mutant sequences 3′ of the stem loop can also promote landing at the normally "hidden" GGG codon (O'Connor, Wills and Atkins, manuscript in preparation).

17.10 Ribosomal Protein L9

Debilitation of one of the recoding signals was used to genetically identify a relevant ribosomal component. The stem loop at the 5′ end of the coding gap was extended by 15 base pairs such that bypassing decreased to 0.3% of the WT level (Fig. 17.6). This mutant, as part of a gene *60-lacZ* fusion, was used in a selection for chromosomal mutants that restored bypassing. One mutation, which caused a Ser to Phe change at position 92 in ribosomal protein L9, showed a 10-fold restoration of bypassing (Herbst et al., 1994). Large subunit protein L9 has two globular domains separated by a long α-helix whose length is phylogenetically conserved (Hoffman et al., 1996). The L9 N-terminal domain anchors L9 to domain V of 23S rRNA (Adamski et al., 1996; Lieberman et al., 2000; Berk and Cate, 2007) close to the 2250 loop near the E-site and the base of the L1 stalk (Valle et al., 2003). Remarkably, crystallographic and cryo-EM studies show the rest of L9 projecting outward (Yusupov et al., 2001; Schuwirth et al., 2005 and Fig. 17.6); however, this may be in response to EF-G binding in the GTP state (Spahn et al., 2001). Interestingly (especially for bypassing), L9 can be cross-linked to peptidyl-tRNA when aminoacyl-tRNA occupies the A-site (Graifer et al., 1989). The position of the C-terminal domain of L9, however, raises the intriguing possibility that it could interact with the trailing ribosome (inter-ribosomally) as well as, or in addition to, within its own ribosome. The proximity of L9 to the L1 stalk raises the possibility that mutants in L9 may influence the E-site "gating function" of the L1 stalk. A direct interaction between the L1 stalk and deacylated P-site tRNA is thought to be involved in translocation from the P- to the E-site by movement and rotation of the L1 stalk by 30–40 Å and 15–20°, respectively (reviewed in Korostelev et al., 2008). L9 mutants may also indirectly influence base pair formation between G2252 of 23S

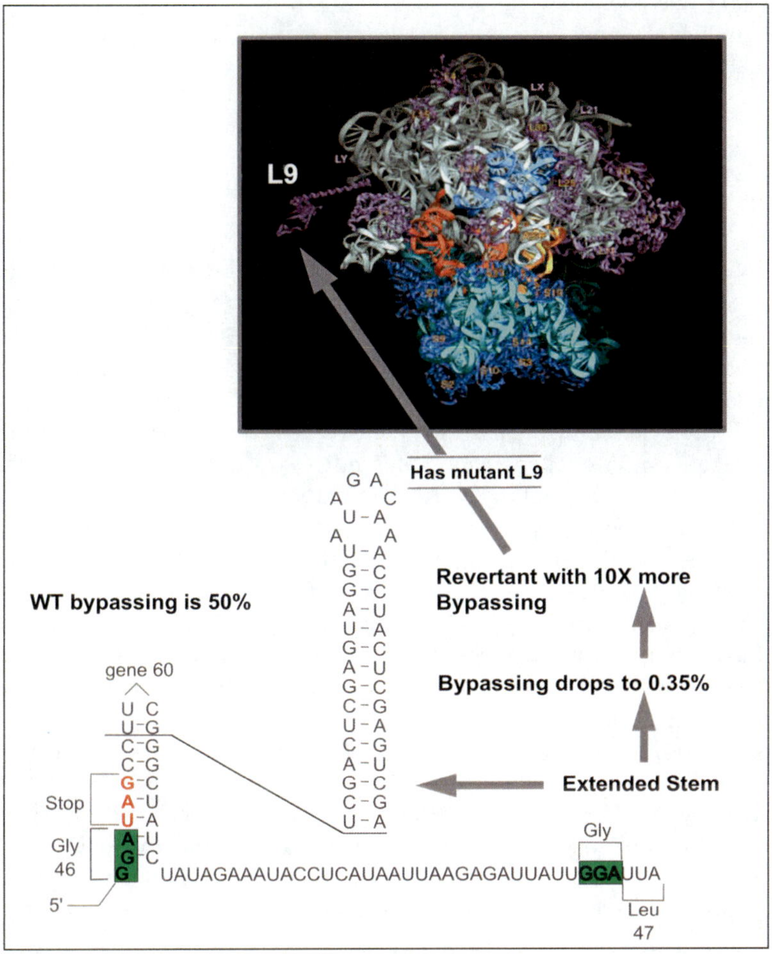

Fig. 17.6 Isolation of a mutant in ribosomal protein L9 that partially rescues bypassing in a debilitated gene *60* cassette. The wild-type stem loop including the 5′ portion of the coding gap is shown at the left and results in 50% bypassing. A 15 base-pair extension at the top of the stem reduces bypassing to 0.3%. A mutant of ribosomal protein L9 that has phenylalanine at position 92 instead of serine partially restores bypassing to 3%. The position of L9 in the ribosome is shown at the right. Its N-terminal domain is bound near the base of the L1 stalk near the mRNA exit site. A-, P-, and E-site tRNAs are shown in yellow, orange and red, respectively. Large submit proteins are shown in purple. (Picture of ribosome from Berk and Cate, 2007)

rRNA (in the 2250 loop) and C74 of peptidyl-tRNA which is critical for efficient peptidyl transfer (Samaha et al., 1995).

The interplay between L9 and the gene *60* elements important for bypassing has been rigorously studied by Herr et al. (2000; 2001a). From these experiments, it was established that one of the functions of L9 is to preclude forward mRNA movement.

In gene *60* bypassing, the stem-loop structure in the 5′ part of the coding gap interferes with this L9 activity thereby allowing forward mRNA movement through the ribosome.

17.11 Model for Gene *60* Bypassing

A central part of a current model for bypassing is the conclusion that the stem loop exerts its effect on bypassing after decoding of GGA. After codon 46, GGA, is decoded and standard ribosome unlocking occurs, the GGA enters the ribosomal P-site and the 3′ adjacent UAG moves into the A-site (Fig. 17.7, panel A). Several likely interconnected events then proceed. Since the UAG stop codon in the A-site is slow to decode, a pause facilitates 3′ nucleotides entering the A-site, perhaps aided by its distortion caused by the nascent peptide signal. The presence of complementary sequences and a tetraloop causes this mRNA to immediately form a hairpin structure which requires "pulling" mRNA initially from the 3′ direction into the A-site. While this is occurring, or closely following it, the P-site codon, GGA, and the anticodon of peptidyl-tRNA dissociate, facilitated by the effect(s) of the nascent "special" peptide sequence 16–31 amino acids from the peptidyl transfer site and by continuing stem formation as it now "pulls" mRNA from the 5′ and 3′ directions. Base pairing of the UAG within the stem-loop structure precludes release factor 1 (RF1) access (Fig. 17.7, panel B). The rest of the coding gap mRNA then enters the A-site. The space normally occupied by release factor or tRNA is filled by coding gap mRNA structure (Fig. 17.7, panel B) (although some of the structure may be in the inter-subunit space). Occupancy of the A-site tRNA position, in this case by mRNA structure, is indirectly sensed (via ratcheting?) by ribosomal protein L9 (Fig. 17.7, panel B). L9 then influences the L1 stalk/protuberance movement (Schuwirth et al., 2005). This movement directly or more likely indirectly (see below) helps release mRNA for forward mRNA slippage.

Fig. 17.7 Model for programmed bypassing. (**A**) The A-, P-, and E-sites of the ribosome are filled with RNA or shown by a dotted outline. The indirect influence of the segment of the nascent peptide (*pale yellow*) on peptidyl-tRNA anticodon:GGA "take-off" codon (*green flag*) dissociation is indicated by a dotted line. The UAG (*red flag*) in the A-site causes a pause which permits extra mRNA (*dark blue*) to enter the A-site, where it forms a structure diagrammed in B. The SD-like GAG sequence in the coding gap (*dark blue dashes in the mRNA*) and the landing site codon GGA (white letters on green flag) are indicated. (**B**) Intra-mRNA pairing drags mRNA initially from both the 5′ and 3′ directions to allow formation of the 5′ stem loop. Occupancy of the A-site by structure precludes entry by release factor 1 (*pale green*) and permits E-site tRNA exit mediated by L9 (*purple*). 3′ RNA movement "resolves" the structure in the A-site without peptidyl-tRNA scanning. (**C**) Return to linear mRNA and pairing of GAG (*gray flag*) 6 nt 5′ of the end of the coding gap to the 3′ end of 16S rRNA (*light blue*) contributes to the initiation of peptidyl-tRNA scanning and pairing to the landing site, GGA (*green flag*). Standard decoding resumes at the adjacent 3′ codon, UUA (*gray flag*) (from Wills et al., 2008)

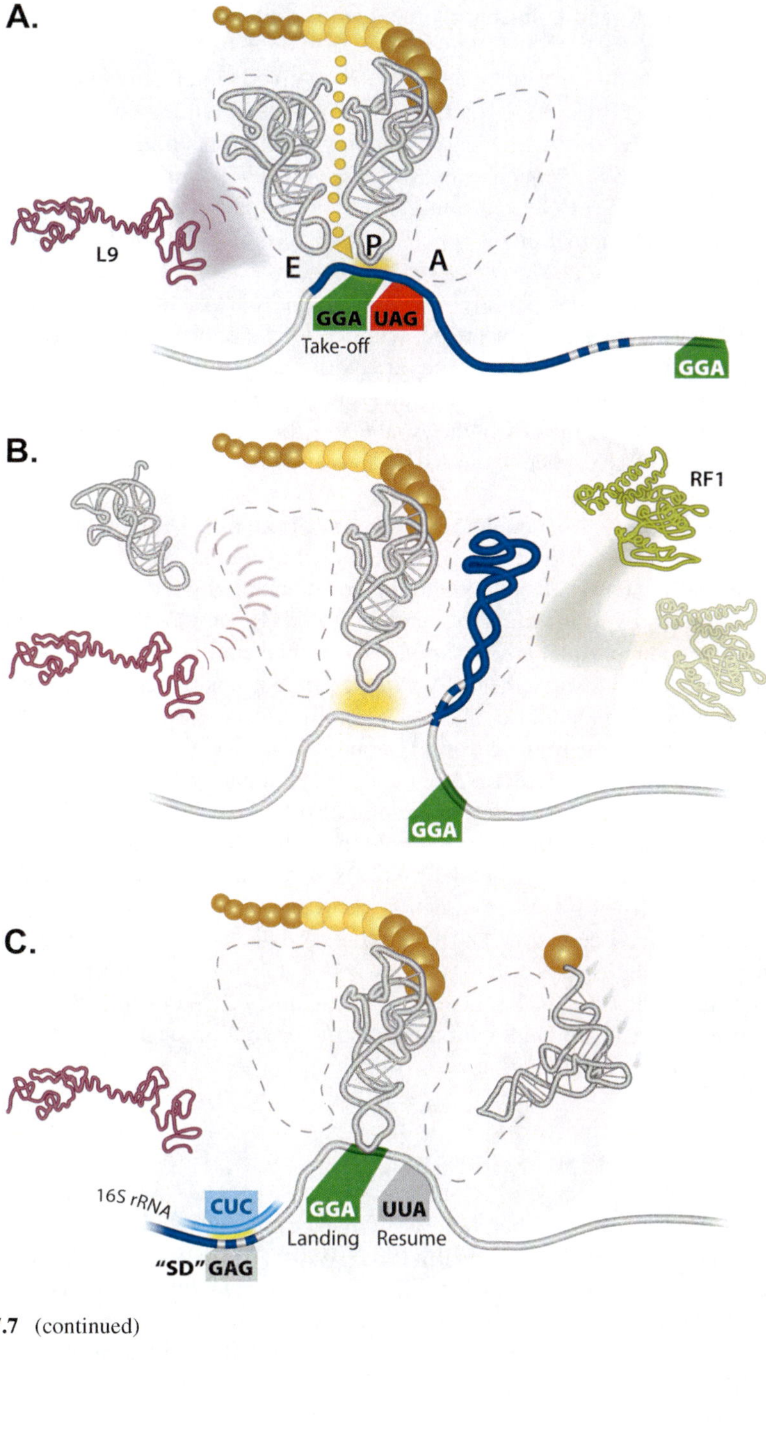

Fig. 17.7 (continued)

As the coding gap mRNA exits the A-site and passes through the P-site, it is not scanned by the peptidyl-tRNA anticodon, perhaps because the normal mRNA kink between the A- and P-sites is altered due to A-site mRNA structure. After this mRNA progresses through the E-site, it is scanned by the anti-SD sequence near the 3′ end of 16S rRNA. Formation of a weak rRNA anti-SD interaction with a complementary sequence, GAG, in mRNA (Fig. 17.7, panel C) indirectly contributes to the initiation of peptidyl-tRNA anticodon scanning of the transiting mRNA. The 6 nt distance between the SD sequence and GGA_{48-50} would position the landing codon in the P-site with the mRNA returning to its normal path. [Other work has shown that an SD:anti-SD interaction influences the path of intra-ribosomal mRNA (Jenner et al., 2007; Yusupova et al., 2006).] However, dissipation of the intra-ribosomal mRNA structure facilitates peptidyl-tRNA scanning to a greater extent than the SD sequence. Continuing effects of the nascent peptide help ensure stringency of anticodon re-pairing to mRNA, and so, the fidelity of landing. When the peptidyl-tRNA anticodon pairs with the landing site triplet, GGA, the resume codon, UUA (codon 47), is present as linear mRNA in the A-site (Fig. 17.7, panel C) and cognate tRNA enters the vacant tRNA space in the A-site allowing resumption of standard peptidyl transfer and translation.

An indirect effect of L9 could be in allowing E-site tRNA departure. Nierhaus has proposed an allosteric model wherein the ribosome senses A-site occupancy before permitting E-site tRNA exit to ensure that the anticodons of two tRNAs are paired to mRNA to maintain the reading frame (Review, Wilson and Nierhaus, 2006). While some of the many in vitro tests of aspects of the allosteric model have yielded controversial results, the mRNA in the E-site is positioned with the potential for anticodon pairing (Jenner et al., 2007), and in vitro (Márquez et al., 2004) and in vivo experiments on frameshifting (Baranov et al., 2002; Sanders and Curran, 2007) are supportive of E-site tRNA anticodon:codon pairing. In addition, perturbation of the small subunit E-site by mutation of protein S7 increases both +1 and –1 frameshifting, further implicating the E-site in reading frame maintenance (Devaraj et al., 2009). As the gene *60* model involves A-site mRNA structure rather than tRNA, it potentially provides a distinction between ribosome movements involving A-site tRNA space occupancy and those due to delivery of the aminoacyl-tRNA EF-Tu ternary complex.

The issue of what drives the forward movement of mRNA during bypassing and the associated internal unwinding of mRNA structure required has not been resolved and requires further investigation.

17.12 Significance of Bypassing

There have been several other reports of bypassing and while some have been shown to be incorrect (Tuohy et al., 1994; Wills et al., 1997), other candidates require further analysis (Manch-Citron et al., 1999). Doubtless, further examples will emerge since it has been reported that the C-terminal sequence of rabbit β-globin shows low-level hopping over its gene terminator with landing at several different sites (Chittum et al., 1998) and that spontaneous bypassing can occur without aminoacyl-tRNA starvation (Lindsley et al., 2005).

The emphasis so far in studying bypassing has not been on utility in gene expression but rather on how it relates to ribosome function. Isolation of mutants of L9 and establishing the role for L9 in restraining forward mRNA slippage is one of the highlights, but unusual features of the nascent peptide signal and the site of formation of the relevant mRNA structure are among other notable features.

Acknowledgments This work was supported by NIH grant ROI GM079523.

References

Adamski FM, Atkins JF, Gesteland RF (1996) Ribosomal protein L9 interactions with 23S rRNA: The use of translational bypass assay to study the effect of amino acid substitutions. J Mol Biol 261:357–371

Baranov PV, Gesteland RF, Atkins JF (2002) Release factor 2 frameshifting sites in different bacteria. EMBO Reports 3:373–377

Berk V, Cate JH (2007) Insights into protein biosynthesis from structures of bacterial ribosomes. Curr Opin Struct Biol 17:302–309

Bucklin DJ, Wills NM, Gesteland RF, Atkins JF (2005) P-site pairing subtleties revealed by the effects of different tRNAs on programmed translational bypassing where anticodon re-pairing to mRNA is separated from dissociation. J Mol Biol 345:39–49

Chen H, Bjerknes M, Kumar R, Jay E (1994) Determination of the optimal aligned spacing between the Shine-Dalgarno sequence and the translation initiation codon of *Escherichia coli* mRNAs. Nucl Acids Res 22:4953–4957

Chittum HS, Lane WS, Carlson BA, Roller PP, Lung, F-DT, Lee BJ, Hatfield DL (1998) Rabbit β-globin is extended beyond its UGA codon by multiple suppressions and translational reading gaps. Biochemistry 37:10866–10870

Choi KM, Atkins JF, Gesteland RF, Brimacombe R (1998) Flexibility of the nascent polypeptide chain within the ribosome – Contacts from the peptide N-terminus to a specific region of the 30S subunit. Eur J Biochem 255:409–413

Devaraj A, Shoji S, Holbrook E.D, Fredrick K (2009) A role for the 30S subunit E site in maintenance of the translational reading frame. RNA 15:255–265

Gallant J, Bonthuis P, Lindsley D (2003) Evidence that the bypassing ribosome travels through the coding gap. Proc Natl Acad Sci USA 100:13430–13435

Gallant J, Bonthuis P, Lindsley D, Cabellon J, Gill G, Heaton K, Kelley-Clarke B, MacDonald L, Mercer S, Vu H, Worsley A (2004) On the role of the starved codon and the takeoff site in ribosome bypassing in *Escherichia coli*. J Mol Biol 342:713–724

Gallant JA, Lindsley D (1998) Ribosomes can slide over and beyond "hungry" codons, resuming protein chain elongation many nucleotides downstream. Proc Natl Acad Sci USA 95: 13771–13776

Graifer DM, Babkina GT, Matasova NB, Vladimirov SN, Karpova GG, Vlassov VV (1989) Structural arrangement of tRNA binding sites on *Escherichia coli* ribosomes, as revealed from data on affinity labelling with photoreactive tRNA derivatives. Biochim Biophys Acta 1008:146–156

Herbst KL, Nichols LM, Gesteland RF, Weiss RB (1994) A mutation in ribosomal protein L9 affects ribosomal hopping during translation of gene *60* from bacteriophage T4. Proc Nat. Acad Sci USA 91:12525–12529

Herr AJ, Atkins JF, Gesteland RF (1999) Mutations which alter the elbow region of $tRNA_2^{Gly}$ reduce T4 gene 60 translational bypassing efficiency. EMBO J 18:2886–2896

Herr AJ, Gesteland RF, Atkins JF (2000) One protein from two open reading frames: Mechanism of a 50nt translational bypass. EMBO J 19:2671–2680

Herr AJ, Nelson CC, Wills NM, Gesteland RF, Atkins JF (2001a) Analysis of the roles of tRNA structure, ribosomal protein L9, and the bacteriophage T4 gene 60 bypassing signals during ribosome slippage on mRNA. J Mol Biol 309:1029–1048

Herr AJ, Wills NM, Nelson CC, Gesteland RF, Atkins JF (2001b) Drop-off during ribosome hopping. J Mol Biol 311:445–452

Herr AJ, Wills NW, Nelson CC, Gesteland RF, Atkins JF (2004) Factors that influence selection of coding resumption sites in translational bypassing. J Biol Chem 279:11081–11087

Hoffman DW, Davies C, Gerchman SE, Kycia JH, Porter SJ, White SW, Ramakrishnan V (1994) Crystal structure of prokaryotic ribosomal protein L9: A bi-lobed RNA-binding protein. EMBO J 13:205–212

Hoffman DW, Cameron CS, Davies C, White SW, Ramakrishnan V (1996) Ribosomal protein L9: A structure determination by the combined use of X-ray crystallography and NMR spectroscopy. J Mol Biol 264:1058–1071

Huang WM, Ao S, Casjens S, Orlandi R, Zeikus R, Weiss R, Winge D, Fang M (1988) A persistent untranslated sequence within bacteriophage T4 DNA topoisomerase gene *60*. Science 239:1005–1012

Jenner L, Rees B, Yusupov, M, Yusupova G (2007) Messenger RNA conformations in the ribosomal E site revealed by X-ray crystallography. EMBO Reports 8:846–850

Jin H, Zhao Q, de Valdivia EIG, Ardell DH, Stenström M, Isaksson LA (2006) Influences on gene expression in vivo by a Shine-Dalgarno sequence. Mol Microbiol 60:480–492

Kane JF, Violand BN, Curran DF, Staten NR, Duffin KL, Bogosian, G. (1992) Novel in-frame two codon translational hop during synthesis of bovine placental lactogen in a recombinant strain of *Escherichia coli*. Nucl Acids Res 24:6707–6712

Korostelev A., Ermolenko DN, Noller HF (2008) Structural dynamics of the ribosome. Curr Opin Chem Biol 12:674–683

Larsen B, Peden J, Matsufuji S, Matsufuji T, Brady K, Maldonado R, Wills NM, Fayet O, Atkins JF, Gesteland RF (1995) Upstream regulators for recoding Biochem. Cell Biol 73:1123–1129

Larsen, B, Wills NM, Gesteland RF, Atkins JF (1994) rRNA-mRNA base pairing stimulates a programmed –1 ribosomal frameshift. J Bacteriol 176:6842–6851

Lieberman KR, Firpo MA, Herr AJ, Nguyenle T, Atkins JF, Gesteland RF, Noller HF (2000) The 23 S rRNA environment of ribosomal protein L9 in the 50 S ribosomal subunit. J Mol Biol 297:1129–1143

Lindsley D, Gallant J (1993) On the directional specificity of ribosome frameshifting at a "hungry" codon. Proc Natl Acad Sci USA 90:5469–5473

Lindsley D, Gallant J, Doneanu C, Bonthuis P, Caldwell S, Fontelera A (2005) Spontaneous ribosome bypassing in growing cells. J Mol Biol 349:261–272

Maldonado R, Herr AJ (1998) Efficiency of T4 gene 60 translational bypassing. J Bacteriol 180:1822–1830

Manch-Citron JN, Dey A, Schneider R, Nguyebn NY (1999) The translational hop junction and the 5′ transcriptional start site for the *Prevotella loescheii* adhesion encoded by *plaA*. Curr Microbiol 38:22–26

Márquez V, Wilson DN, Tate WP, Triana-Alonso F, Nierhaus KH (2004) Maintaining the ribosomal reading frame: the influence of the E site during translational regulation of release factor 2. Cell 118:45–55

Murgola EJ, Pagel FT (1980) Codon recognition by glycine transfer RNAs of *Escherichia coli* in vivo. J Mol Biol 138:833–844

O'Connor M, Gesteland RF, Atkins JF (1989) tRNA hopping: enhancement by an expanded anticodon. EMBO J 8:4315–4323

O'Mahony DJ, Mims BH, Thompson S, Murgola EJ, Atkins JF (1989) Glycine tRNA mutants with normal anticodon loop size cause –1 frameshifting. Proc Natl Acad Sci USA 86: 7979–7983

Pagel FT, Tuohy TMF, Atkins JF, Murgola EJ (1992) Doublet translocation at GGA is mediated directly by mutant $_2$. J Bacteriol 174:4179–4182

Riyasaty S, Atkins JF (1968) External suppression of a frameshift mutant in *Salmonella.* J Mol Biol 34:541–557

Rodnina MV, Fricke R, Wintermeyer W (1994) Transient conformational states of aminoacyl-tRNA during ribosome binding catalyzed by elongation factor Tu. Biochemistry 33: 12267–12275

Samaha RR, Green R, Noller HF (1995) A base pair between tRNA and 23 S rRNA in the peptidyl transferase center of the ribosome. Nature 377:309–314

Sanders CL, Curran JF (2007) Genetic analysis of the E site during RF2 programmed frameshifting. RNA 13:1483–1491

Schuette JC, Murphy FV, Kelley AC, Weir JR, Giesebrecht J, Connell SR, Loerke J, Mielke T, Zhang W, Penczek PA, Ramakrishnan V, Spahn CM (2009) GTPase activation of elongation factor EF-Tu by the ribosome during decoding. EMBO J 28:755–765

Schuwirth BS, Borovinskaya MA, Hau CW, Zhang W, Vila-Sanjurjo A, Holton JM, Cate JH (2005) Structure of the bacterial ribosome at 3.5 Å resolution. Science 310:827–834

Spahn CMT, Blaha, G., Agrawal RK, Penczek P, Grassucci RA, Trieber CA, Connell SR, Taylor DE, Nierhaus KH, Frank J (2001) Localization of the ribosomal protection protein Tet(O) on the ribosome and the mechanism of tetracycline resistance. Mol Cell 7:1037–1045

Tsalkova T, Odom OW, Kramer G, Hardesty B (1998) Different conformations of nascent peptides on ribosomes. J Mol Biol 278:713–723

Tuohy TMF, Kidd T, Gesteland RF, Atkins JF (1994) Uninterrupted translation through putative 12-nucleotide coding gap in sequence of *carA*: business as usual. J Bacteriol 176:265–267

Valle M, Zavialov A, Sengupta J, Rawat U, Ehrenberg M, Frank J (2003) Locking and unlocking of ribosomal motions. Cell 114:123–134

Weiss RB, Dunn DM, Dahlberg AE, Atkins JF, Gesteland RF (1988) Reading frame switch caused by base-pair formation between the 3' end of 16S rRNA and the mRNA during elongation of protein synthesis in *Escherichia coli.* EMBO J 7:1503–1507

Weiss RB, Dunn DM, Atkins JF, Gesteland RF (1987) Slippery runs, shifty stops, backward steps, and forward hops: −2,−1,+1, +2, +5 and +6 ribosomal frameshifting. Cold Spring Harbor Symp. Quant Biol 52:687–693

Weiss RB, Huang WM, Dunn DM (1990) A nascent peptide is required for ribosomal bypass of the coding gap in bacteriophage T4 gene *60*. Cell 62:117–126

Wills NM, Ingram JA, Gesteland RF, Atkins JF (1997) Reported translational bypass in a *trpR'-lacZ'* fusion is accounted for by unusual initiation and +1 frameshifting. J Mol Biol 271: 491–498

Wills NM, O'Connor M, Nelson CC, Rettberg CC, Huang WM, Gesteland RF, Atkins JF (2008) Translational bypassing without peptidyl-tRNA anticodon scanning of coding gap mRNA. EMBO J 27:2533–2544

Wilson DN, Nierhaus KH (2006) The E-site story: the importance of maintaining two tRNAs on the ribosome during protein synthesis. Cell Mol Life Sci 63:2725–2737

Yusupov M, Yusupova G, Baucom A, Lieberman K, Earnest TN, Cate JH, Noller HF (2001) Crystal structure of the ribosome at 5.5 Å resolution. Science 292:883–896

Yusupova G, Jenner L, Rees B, Moras D and Yusupov M (2006) Structural basis for messenger RNA movement on the ribosome. Nature 444:391–394

Chapter 18
trans-Translation

Kenneth C. Keiler and Dennis M. Lee

Abstract *trans*-Translation is an extreme version of recoding in which the translating ribosome is diverted onto a specialized RNA, producing a protein encoded in two distinct RNA molecules. The specialized RNA that is used *in trans*, called tmRNA or SsrA, has properties of both a tRNA and an mRNA. tmRNA bound to a small protein, SmpB, can enter the A-site of substrate ribosomes and accept the nascent polypeptide, acting like a tRNA. The mRNA is removed from the ribosome, and an open reading frame within tmRNA is inserted in the decoding center and translated. The product of *trans*-translation is a protein encoded in part from the original mRNA and in part from tmRNA. This reaction is the only known example of translation from two physically distinct messages. One use of *trans*-translation is to release ribosomes that are stalled at the end of damaged mRNAs. However, *trans*-translation can also be induced in response to signals in the mRNA or nascent polypeptide, and by specific cleavage of the mRNA, suggesting that *trans*-translation can be used for regulation as well as quality control.

Contents

K.C. Keiler (✉)
Department of Biochemistry and Molecular Biology, Penn State University, 401 Althouse Laboratory, University Park, PA 16802, USA
e-mail: kkeiler@psu.edu

J.F. Atkins, R.F. Gesteland (eds.), *Recoding: Expansion of Decoding Rules Enriches Gene Expression*, Nucleic Acids and Molecular Biology 24,
DOI 10.1007/978-0-387-89382-2_18,

18.1 Introduction

trans-Translation is an extreme version of recoding in which the translating ribosome is diverted onto a specialized RNA, producing a protein encoded in two distinct RNA molecules. The specialized RNA that is used *in trans*, called tmRNA or SsrA, has properties of both a tRNA and an mRNA (Tyagi and Kinger, 1992, Komine et al., 1994, Ushida et al., 1994, Tu et al., 1995). tmRNA bound to a small protein, SmpB, can enter the A-site of substrate ribosomes and accept the nascent polypeptide, acting like a tRNA. The mRNA is removed from the ribosome, and an open reading frame within tmRNA is inserted in the decoding center and translated. The product of *trans*-translation is a protein encoded in part from the original mRNA and in part from tmRNA (Tu et al., 1995, Keiler et al., 1996). *trans*-Translation is the only known example of translation from two physically distinct messages.

The basic model for the *trans*-translation reaction is shown in Fig. 18.1 (Keiler, 2008). The complex of tmRNA and SmpB forms a structure that mimics alanyl-tRNA, and the 3′ end of tmRNA is charged with alanine. The alanyl-tmRNA–SmpB complex enters the A-site of a substrate ribosome with the mRNA engaged and peptidyl-tRNA in the P-site. The nascent polypeptide is transferred to tmRNA in what is assumed to be a normal transpeptidation reaction. The tmRNA open reading frame then displaces the mRNA in the decoding center, and translation resumes using tmRNA as a message. Termination at a stop codon within tmRNA releases the ribosome and a protein containing the tmRNA-encoded peptide tag at its C terminus. The peptide tag contains recognition determinants for several intracellular proteases, targeting the tagged protein for rapid degradation.

One use of *trans*-translation is to release ribosomes that are stalled at the end of damaged mRNAs. When the ribosome reaches the end of an mRNA and there is no stop codon to terminate translation, *trans*-translation releases the ribosome and promotes degradation of both the damaged mRNA and the nascent polypeptide, which is likely to be incomplete (Keiler et al., 1996). Because all components of the stalled translation complex are removed, these reactions are considered translation quality control. *trans*-Translation can also be induced in response to signals in the mRNA or nascent polypeptide and by specific cleavage of the mRNA (Keiler, 2007). The intentional targeting of some translation reactions for *trans*-translation suggests that this pathway can be used for regulation as well as quality control.

The conservation and abundance of the *trans*-translation machinery in bacteria indicate that it confers an evolutionary advantage. tmRNA, SmpB, and other factors

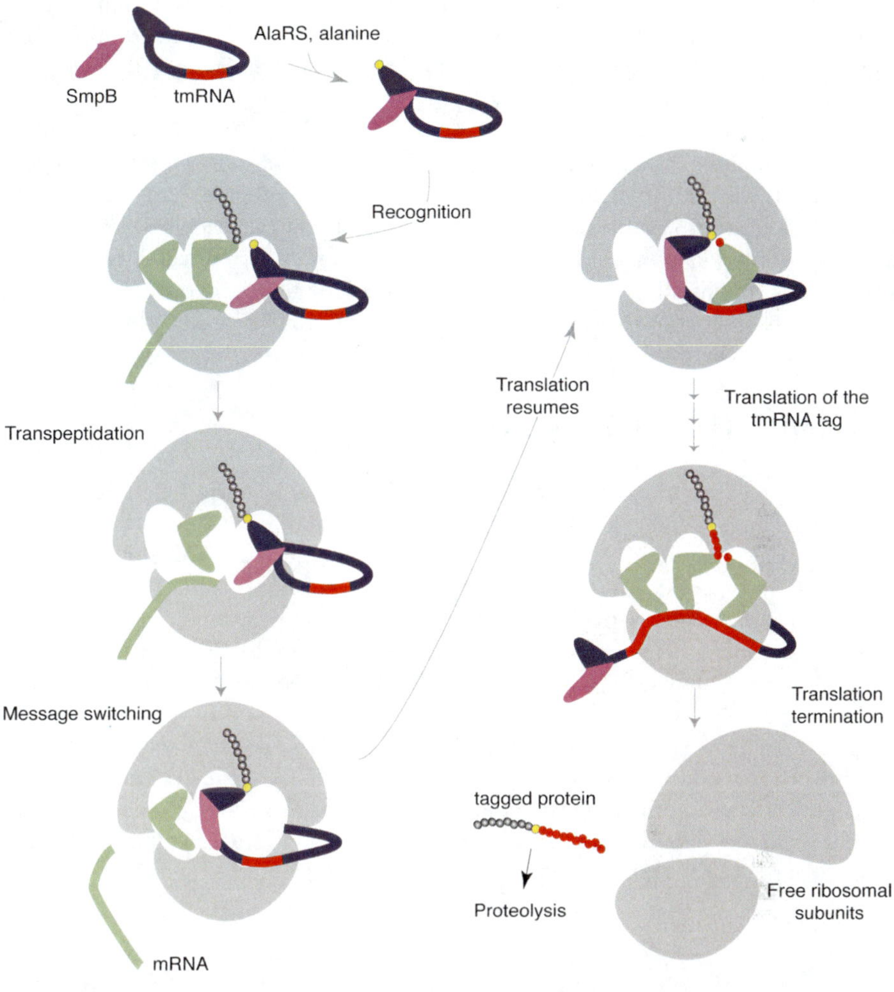

Fig. 18.1 Model of the *trans*-translation mechanism. tmRNA (*blue*) binds SmpB (*purple*) and is aminoacylated by alanyl-tRNA synthetase (AlaRS). The alanyl-tmRNA–SmpB recognizes ribosomes (*gray*) at the 3′ end of an mRNA (*green line*) and enters the A-site. The nascent polypeptide is transpeptidated from the tRNA in the P-site to alanyl-tmRNA. The mRNA is then replaced with the tag reading frame of tmRNA (red line) as the message in the ribosomal mRNA channel, and translation resumes, decoding the tag reading frame. Translation terminates at a stop codon in the tag reading frame, releasing the ribosomal subunits and the tagged protein, which is targeted for proteolysis. EF-Tu and other general translation factors participate in the reaction but are not shown for clarity

required for *trans*-translation have been identified in all bacterial genome sequences, as well as in some plastid genomes of eukaryotes and some bacteriophage genomes (Keiler et al., 2000, Gimple and Schon, 2001, Pedulla et al., 2003, Gueneau de Novoa and Williams, 2004, Jacob et al., 2005). No tmRNA has been found in the nuclear genomes of eukaryotes or in archaea, so these organisms are unlikely to use *trans*-translation. In bacteria, tmRNA is one of the most abundant RNAs in the cell, found at concentrations 5–10% of rRNA (Lee et al., 1978, Keiler et al., 2000, Moore and Sauer, 2005). Moreover, physiological measurements in *Escherichia coli* indicate that approximately 1 in 200 nascent polypeptides is tagged by *trans*-translation before it is released from the ribosome (Moore and Sauer, 2005, Lies and Maurizi, 2008). The ubiquity, abundance, and activity of *trans*-translation systems suggest that this pathway is important for bacterial physiology. In fact, several bacteria are known to require *trans*-translation for normal physiological responses, in particular under conditions where the gene expression program is dramatically changed (Keiler, 2007). This chapter will describe how recoding by *trans*-translation occurs, what signals lead to *trans*-translation during normal translation reactions, and the physiological role of this reaction in bacteria.

18.2 tmRNA–SmpB Structure

The structure of the tmRNA–SmpB complex allows it to perform functions of both a tRNA and an mRNA (Fig. 18.2). The 5′ and 3′ ends of tmRNA fold into a structure that resembles alanyl-tRNA, including an acceptor stem with a CCA overhangs at the 3′ end and a TΨC arm (Tyagi and Kinger, 1992, Komine et al., 1994, Ushida et al., 1994). Like alanyl-tRNA, the third base pair of the acceptor stem is a G:U wobble pair, which is the recognition determinant for alanyl-tRNA synthetase (Hou and Schimmel, 1988, McClain and Foss, 1988). tmRNA also has a D-loop that makes contacts with the TΨC arm similar to those observed in a tRNA (Bessho et al., 2007). However, the base pairing in the D-arm is not retained. The rest of the tmRNA secondary structure does not resemble a tRNA. Instead of an anticodon stem, tmRNA has a much longer sequence containing a specialized reading frame encoding the tag peptide. In most species, the tag reading frame is flanked by one pseudoknot at the 5′ end, and one to three pseudoknots at the 3′ end (Felden et al.,

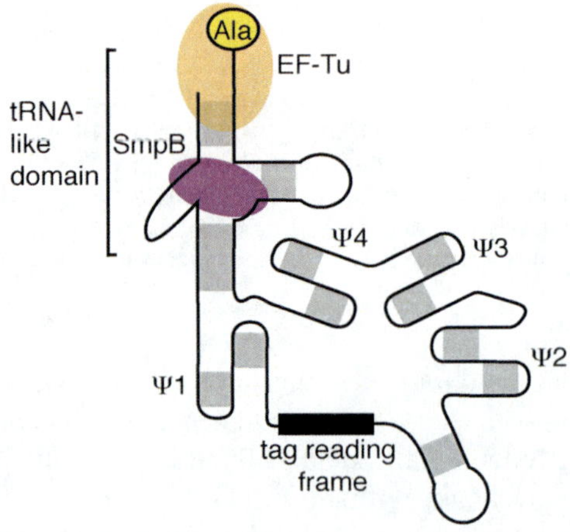

Fig. 18.2 Model of tmRNA secondary structure. The 5′ and 3′ ends of tmRNA fold into a structure similar to the acceptor arm and TΨC arm of a tRNA. The tRNA-like domain is charged with alanine and binds EF-Tu (*orange oval*) and SmpB (*purple oval*). The rest of the tmRNA structure contains several pseudoknots (labeled with Ψ) and the tag reading frame (*black box*)

1997, Williams and Bartel, 1996, Gueneau de Novoa and Williams, 2004). Although these pseudoknots are phylogenetically conserved, they are dispensable for *trans*-translation activity. The 5′ pseudoknot can be replaced with a hairpin, and the 3′ pseudoknots can each be replaced with single-stranded RNA without eliminating tmRNA activity (Tanner et al., 2006, Nameki et al., 2000). The role of these pseudoknots is still speculative, but they are likely to play a role in tmRNA folding and stability.

The tag reading frame does not have a canonical initiation sequence, and de novo translation initiation has not been observed on this sequence, suggesting that it is optimized for use in *trans*-translation. Although many features of the tag reading frame are conserved, few are absolutely required for *trans*-translation. The first codon of the tag sequence that is read, sometimes called the "resume codon," encodes alanine in most species, but other codons can be substituted without disrupting tagging (Williams et al., 1999, Lee et al., 2001, Konno et al., 2007). Phylogenetic analyses show that a stop codon frequently occurs two bases 5′ of the resume codon, but mutations in these bases do not eliminate tagging indicating that they are not required to specify the resume codon (Williams et al., 1999). The lack of sequence requirements for translation initiation in the tag reading frame suggests that structural cues are used to orient the resume codon in the decoding center of the A-site. There also appear to be few constraints on the sequences that can be encoded in the tag reading frame after the resume codon. The tag sequence can be mutated or truncated, and the encoded peptide tags will be added to the nascent polypeptide (Keiler et al., 1996, Withey and Friedman, 1999, Tanner et al., 2006). Nevertheless, phylogenetic conservation of the tag sequences suggests that the tag peptide is important for the physiology of *trans*-translation.

Although tmRNA does not have a tRNA-like anticodon stem, SmpB binds to tmRNA, and mimics the folded structure of the anticodon stem (Fig. 18.3) (Bessho

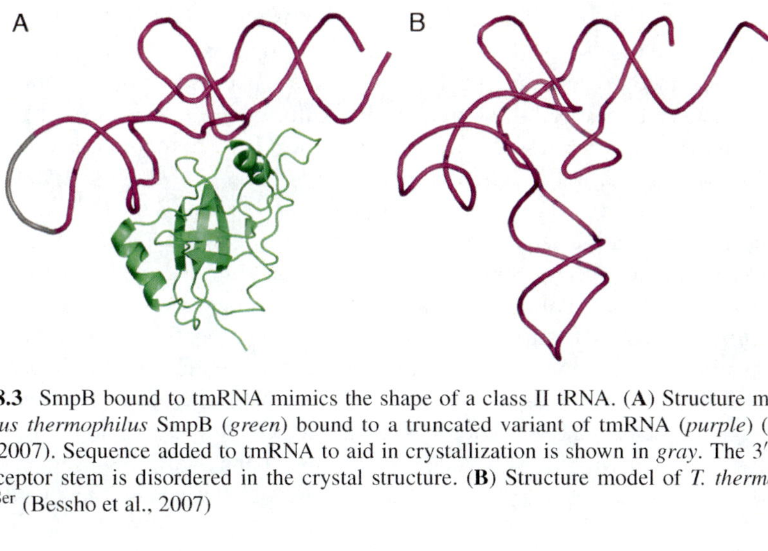

Fig. 18.3 SmpB bound to tmRNA mimics the shape of a class II tRNA. (**A**) Structure model of *Thermus thermophilus* SmpB (*green*) bound to a truncated variant of tmRNA (*purple*) (Bessho et al., 2007). Sequence added to tmRNA to aid in crystallization is shown in *gray*. The 3′ end of the acceptor stem is disordered in the crystal structure. (**B**) Structure model of *T. thermophilus* $tRNA^{Ser}$ (Bessho et al., 2007)

et al., 2007). This interaction is required for *trans*-translation activity and is important for tmRNA structure and stability in some bacteria (Karzai et al., 1999, Wiegert and Schumann, 2001, Keiler and Shapiro, 2003b, Hong et al., 2005, Sundermeier and Karzai, 2007). SmpB binds to the tRNA-like domain of tmRNA in a 1:1 complex with nanomolar affinity under physiologically relevant conditions (Karzai et al., 1999, Barends et al., 2001, Wower et al., 2002). Crystal structures of SmpB bound to tmRNA fragments show that SmpB extends from the tRNA-like domain at an angle similar to the anticodon stem, and SmpB side chains replace some of the base-stacking interactions usually formed by the anticodon stem (Bessho et al., 2007). Biochemical and genetic experiments indicate that SmpB also mimics the interactions of a tRNA anticodon stem within the ribosome. Cryoelectron micrographs of tmRNA–SmpB entering the ribosome show SmpB near the decoding center, where the anticodon stem of a tRNA would be (Valle et al., 2003, Kaur et al., 2006). Chemical probing experiments indicate that SmpB remains associated with tmRNA in a similar structure when the complex moves to the P-site and E-site, consistent with a persistent tRNA-like conformation (Ivanova et al., 2005a, Ivanova et al., 2007). Mutations in the C-terminal tail of SmpB eliminate transpeptidation of the nascent polypeptide onto alanyl-tmRNA and subsequent steps of *trans*-translation, but do not disrupt binding with tmRNA or initial interactions with the ribosome (Jacob et al., 2005, Sundermeier et al., 2005, Dulebohn et al., 2006). These data indicate the C-terminal tail of SmpB specifically interacts with the ribosome to promote message switching. *trans*-Translation in vitro requires only a substrate ribosome, the general translation factors, and tmRNA–SmpB, so the tmRNA–SmpB complex provides all specific functions needed for the *trans*-translation reaction.

18.3 tmRNA Charging

tmRNA must be charged with alanine for *trans*-translation activity, and this alanine becomes the first residue of the tag peptide added to the nascent protein (Himeno et al., 1997). tmRNA contains all the determinants necessary for recognition by alanyl-tRNA synthetase (AlaRS), and AlaRS will charge tmRNA in vitro and in vivo (Komine et al., 1994, Ushida et al., 1994). SmpB is not required for recognition of tmRNA by AlaRS, but it enhances the rate of charging (Barends et al., 2001). Modeling studies using the crystal structures of tmRNA–SmpB and AlaRS suggest that SmpB directly contacts AlaRS during charging (Bessho et al., 2007).

tmRNAs from all species appear to use AlaRS for charging, and this preference is likely due to the specificities of aminoacyl-tRNA synthetases rather than a role for alanine in the *trans*-translation mechanism or in the tag peptide. Most aminoacyl-tRNA synthetases recognize features of the tRNA anticodon stem (Giege et al., 1998), and thus would not be able to recognize tmRNA. The only other synthetase that recognizes determinants predominantly in the acceptor stem is histidine-tRNA synthetase (HisRS). Mutations in tmRNA that replace the AlaRS determinants with sequences similar to histidinyl-tRNA can be charged with histidine in vitro by

HisRS, but charging of this tmRNA variant with histidine in vivo is very inefficient (Nameki et al., 1999).

The charged tRNA-like domain is recognized by EF-Tu in the same fashion as a charged tRNA (Rudinger-Thirion et al., 1999, Barends et al., 2001). EF-Tu binds to the tRNA-like domain of tmRNA and protects the aminoacyl bond from hydrolysis. EF-Tu and SmpB bind different faces of the tRNA-like domain and can bind simultaneously (Barends et al., 2001). The ternary complex of alanyl-tmRNA, SmpB, and EF-Tu enters the A-site of substrate ribosomes to initiate *trans*-translation.

18.4 Interaction with the Ribosome

Models of the alanyl-tmRNA–SmpB–EF-Tu complex entering the A-site of a stalled ribosome have been constructed based on cryo-EM and structural probing data (Valle et al., 2003, Kaur et al., 2006, Ivanova et al., 2007). tmRNA–SmpB is accommodated in a structure similar to that of charged tRNA during translation elongation: the alanylated acceptor arm of tmRNA is near the peptidyl transfer center, SmpB extends to the decoding center, and EF-Tu is near the GTPase center (Valle et al., 2003, Kaur et al., 2006). However, tmRNA–SmpB is much larger than a tRNA, and much of the tmRNA sequence remains outside the ribosome during accommodation (Valle *et al.*, 2003, Kaur et al., 2006). This extra sequence may play a role in substrate selectivity by preventing tmRNA–SmpB from entering nonsubstrate ribosomes, as described below.

After accommodation, subsequent peptidyl transfer and translocation steps of tmRNA–SmpB are probably similar to canonical translation elongation. However, the tmRNA resume codon must be positioned in the decoding center of the A-site before the next tRNA is accommodated (Williams et al., 1999, Lee et al., 2001, Konno et al., 2007). Details of the removal of the mRNA from the ribosome and the placement of the tag reading frame in the mRNA channel are just beginning to come to light. Experiments in vitro indicate that the mRNA is rapidly released after transpeptidation of the nascent polypeptide onto alanyl-tmRNA (Ivanova et al., 2005b). Presumably the resume codon is positioned in the A-site during or immediately after this translocation step. The challenges of positioning the resume codon in the A-site and resuming translation on tmRNA are similar in some respects to those faced by programmed translational bypassing mechanisms, such as the 50 nucleotide bypass in phage T4 gene *60* (discussed in Chapter 17). In both cases, the ribosome must stop elongation without releasing the peptidyl-tRNA in the P-site and restart translation at a precise codon at a distant location. Unlike gene *60*, there is no Shine–Dalgarno-like sequence upstream of the tmRNA resume codon to assist in orienting the message. It is likely that structural cues in tmRNA are responsible for fixing the position of the resume codon, but these structures have not been identified.

Another challenge faced by *trans*-translation is to move the large and highly structured tmRNA through the ribosome. The pseudoknot structure is significantly larger than a single-stranded mRNA and cannot be modeled into the mRNA channel without significant structural change to the ribosome. To allow the resume codon to reach the A-site, pseudoknot 1 must either move through the mRNA channel, accompanied by a change in the ribosome structure, or the pseudoknot must unfold. Even after the resume codon is correctly positioned, there is a unique topological problem in translating tmRNA. Because the 5′ and 3′ ends are paired in the tRNA-like domain, the molecule is circular as it passes through the ribosome. Chemical probing studies indicate that the tRNA-like structure is maintained throughout *trans*-translation (Ivanova et al., 2007), so the 5′ and 3′ ends do not unfold to allow a linearized tmRNA to be translated. Instead, tmRNA must "double back" through the mRNA channel or another part of the ribosome. Structures of *trans*-translation complexes after accommodation will be required to determine how these topological problems are solved during message switching. After message switching, translation continues through the tag reading frame and terminates at a stop codon (Keiler et al., 1996). At this point, the tagged protein, ribosome, and mRNA have been dissociated and recoding is complete.

As described in the following sections, the ultimate outcomes of *trans*-translation are proteolysis of the tagged protein, degradation of the substrate mRNA, and release of the ribosome. Thus, all components of the substrate translational complex are removed or recycled for further use.

18.5 Proteolysis of Tagged Proteins

The peptide tag added to proteins during *trans*-translation targets them for rapid proteolysis (Keiler et al., 1996). In fact, most tagged proteins are degraded with a half-life of less than 2 min in vivo (Keiler et al., 1996, Gottesman et al., 1998). In *E. coli*, the peptide tag contains overlapping recognition determinants for at least four proteases and one proteolytic adaptor protein (Fig. 18.4) (Keiler et al.,

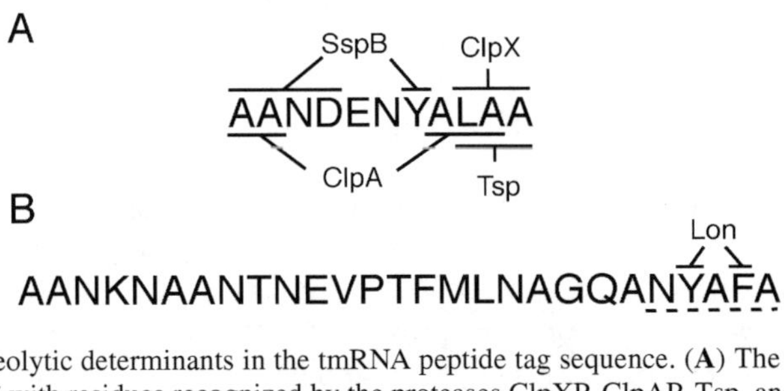

Fig. 18.4 Proteolytic determinants in the tmRNA peptide tag sequence. (**A**) The tmRNA peptide tag from *E. coli* with residues recognized by the proteases ClpXP, ClpAP, Tsp, and the proteolytic adaptor SspB. FtsH also degrades tagged proteins, but the residues required for FtsH recognition are not known. (**B**) The *M. florum* tmRNA tag peptide with residues required for Lon proteolysis. Residues conserved in *Mycoplasma* species that may also be important for Lon recognition are indicated by the *dotted line*

1996, Herman et al., 1998, Flynn et al., 2001), however, the ATP-dependent protease ClpXP degrades ~90% of the cytoplasmic *trans*-translation substrates during exponential growth (Farrell et al., 2005, Lies and Maurizi, 2008). The ClpX subunits, which are responsible for substrate recognition, bind the C-terminal LAA sequence of the peptide tag. Mutations in tmRNA that replace the LAA tag codons with codons for charged or polar residues, such as LDD, result in tagged proteins that are not recognized and degraded by ClpXP (Gottesman et al., 1998). *E. coli* and other species also contain a proteolytic adaptor protein, SspB, which enhances proteolysis of tagged proteins by ClpXP (Levchenko, 2000; Lessner et al., 2007, Chien et al., 2007). SspB binds to residues in the N-terminal part of the tag sequence (Fig. 11.4), and also binds to ClpX, tethering the substrate to the protease and increasing the rate of degradation (Levchenko, 2000). Two other ATP-dependent cytoplasmic proteases, ClpAP and Lon, also degrade tagged proteins under some conditions (Gottesman et al., 1998, Choy et al., 2007).

Because some proteins are directed to the membrane or cytoplasm by N-terminal signal sequences and secretion is initiated before translation is complete, some tagged proteins will not have access to cytoplasmic proteases. In the periplasm, the Tsp protease recognizes the C-terminal LAA sequence and specifically degrades tagged proteins (Keiler et al., 1996). The integral membrane protease FtsH (also called HflB) can also recognize the tmRNA peptide tag and degrades some tagged proteins in vivo (Herman et al., 1998). Thus, *E. coli* contains proteases in all compartments of the cell capable of degrading proteins tagged during *trans*-translation.

The tmRNA tag has evolved to ensure tagged proteins are rapidly degraded. *Mycoplasma* species do not have ClpXP, ClpAP, or Tsp, but tmRNAs in these bacteria have an unusual tag peptide ending with residues that more closely match known substrate recognition sequences for Lon (Fig. 18.4) (Gur and Sauer, 2008). In one *Mycoplasma* species, *Mesoplasma florum*, Lon efficiently degrades proteins with the *M. florum* tag peptide, but not proteins with the *E. coli* tag peptide. The conserved coupling of proteolysis with *trans*-translation suggests that it confers a significant selective advantage (Gur and Sauer, 2008). Rapid proteolysis of tagged proteins could serve to both prevent the accumulation of incomplete or unfolded proteins made from truncated mRNAs and eliminate proteins specifically targeted to *trans*-translation for regulatory reasons.

18.6 Degradation of the Substrate mRNA

The mRNA that is being translated when tmRNA enters the stalled ribosome is rapidly degraded in conjunction with *trans*-translation. This degradation depends on tmRNA (Yamamoto et al., 2003, Mehta et al., 2006, Richards et al., 2006). Model mRNAs lacking a stop codon are degraded more rapidly in cells containing tmRNA than in Δ*ssrA* cells, but the degradation of otherwise identical mRNAs that have a stop codon is not altered by the presence of tmRNA (Yamamoto et al., 2003, Mehta et al., 2006, Richards et al., 2006). Several mechanisms have been proposed for this degradation, and each may operate on different messages. For some messages, tmRNA-dependent degradation is eliminated by deletion of the gene encoding RNase R, a conserved 3′–5′ exonuclease (Richards et al., 2006). However, degradation of other mRNAs does not require RNase R activity, so multiple ribonucleases may be involved (Yamamoto et al., 2003). mRNA degradation may also be accelerated by clearing ribosomes from the 3′ end of the message, thereby exposing it to the general exoribonuclease activities in the cell.

18.7 Signals for Recoding by *trans*-Translation

Ribosomes that have reached the end of an mRNA without terminating translation are efficiently targeted for *trans*-translation. Ribosomes can translate to the 3′ end of an mRNA either because the mRNA has no in-frame stop codons or because the stop codons were not decoded correctly. For many *trans*-translation substrates, including the first observed substrate, recombinant IL-6 in *E. coli*, mRNAs truncated before the stop codon are observed (Tu et al., 1995). These truncated mRNAs can be produced by physical damage, nucleolytic cleavage, or premature termination of transcription. For example, cloning a strong transcriptional terminator before the stop codon produces tagging on a variety of mRNAs (Keiler et al., 1996). A ribosome will also reach the 3′ end of the mRNA if the in-frame stop codon is not decoded correctly due to mechanisms such as ribosomal frameshifting or readthrough. In fact, *trans*-translation activity increases both in the presence of antibiotics that promote readthrough and in strains with suppressor tRNAs (Abo et al., 2002, Ueda et al., 2002). Targeting these terminally stalled translational complexes for *trans*-translation recycles the ribosome and removes the incomplete or incorrect protein and mRNA from the cell.

mRNA sequences that slow translation termination or elongation can also promote *trans*-translation (Collier et al., 2002, Hayes et al., 2002b). In *E. coli*, the *rbsK* and *yjgR* genes end with multiple rare arginine codons (AGG or AGA), and translation of these messages results in tagging of the proteins near the C terminus (Hayes et al., 2002b). Mutations that speed translational elongation or termination decrease the frequency of *trans*-translation for these genes. For example, when one of the rare arginine codons is replaced by the frequently used CGU arginine codon, tagging is dramatically decreased. Likewise, tagging is decreased when the inefficient UGA stop codon is replaced by the efficient UAA stop codon. Tagging is also reduced when translation of the *rbsK* sequence is accelerated by overproduction of $tRNA_5^{Arg}$, which decodes AGG and AGA (Hayes et al., 2002b). Similar results were observed when rare arginine codons were engineered into exogenous proteins: tagging occurred with high frequency at the arginine residues encoded by rare codons, tagging was enhanced by depletion of $tRNA_5^{Arg}$, and tagging was inhibited by overproduction of $tRNA_5^{Arg}$ (Roche and Sauer, 1999, Hayes et al., 2002b).

Amino acid sequences in the nascent polypeptide can also promote *trans*-translation. The *ybeL* gene has no rare codons and an efficient termination sequence but has a C-terminal Glu–Pro sequence and is tagged with high frequency after the proline residue (Hayes et al., 2002a). Mutations in the C-terminal dipeptide showed that Asp, Ile, Val, and Pro in the penultimate position also produced high levels of tagging, and the C-terminal Pro was required for tagging activity. This signal appears to work in other proteins, because insertion of a C-terminal proline residue was sufficient to target an unrelated protein, thioredoxin, for *trans*-translation. Experiments with proline analogs demonstrated that it is the chemical nature of the protein sequence that triggers *trans*-translation. In proline auxotrophs, azetidine is charged to $tRNA^{Pro}$ and inserted at proline codons. Incorporation of azetidine at the C terminus of YbeL dramatically decreased the amount of tagging. The interactions of the C-terminal sequence that promote tagging are not known, but it was suggested that these sequences may slow translation termination. In support of this idea, tagging of YbeL was increased by mutations that changed the stop codon to an inefficient termination sequence and by deletion of the cognate release factor, but overexpression of the cognate release factor decreased tagging (Hayes et al., 2002a).

Studies with the LacI protein identified another peptide, LESG, that promotes *trans*-translation when it is at the C terminus (Sunohara et al., 2002). Although this sequence is not normally found at the C terminus of any endogenous *E. coli* proteins, when it is encoded at the 3′ end of *lacI* or *crp*, the protein is efficiently tagged. Similar to the Xaa–Pro sequence at the end of YbeL, the codons used to encode LESG do not affect *trans*-translation, indicating that it is the nascent polypeptide that is important and not the mRNA. Inefficient translation termination signals increased the amount of LESG-directed tagging, suggesting that the ribosome stalling at the stop codon is an important part of the LESG signal (Sunohara et al., 2002).

Identification of tagged proteins in *Caulobacter crescentus* revealed another potential mechanism for targeting proteins for *trans*-translation. Seventy-three proteins tagged by *trans*-translation under exponential growth conditions were identified, and 46 of these substrates share a 16 nucleotide motif upstream of the tagging site (Hong et al., 2007). Mutations in this motif decrease tagging of the encoded protein, indicating that the motif is involved in substrate selectivity by an unknown mechanism (Hong et al., 2007).

18.8 mRNA Cleavage as a Signal for *trans*-Translation

Ribosomes stalled at the 3′ end of an mRNA are targeted for *trans*-translation, but several lines of investigation suggest that most elongating ribosomes are not substrates for *trans*-translation. In *E. coli*, there is an excess of tmRNA–SmpB complex compared to the amount of tagging that is observed, and this extra tmRNA–SmpB does not interfere with translation (Moore and Sauer, 2005). In addition, overproduction of tmRNA and SmpB does not increase the amount of tagging in the cell (Moore and Sauer, 2005). These data suggest that *trans*-translation is not in competition with translation elongation.

A mechanistic explanation for why *trans*-translation does not compete with elongation comes from biochemical experiments measuring the rates of *trans*-translation in vitro. Tagging was most efficient using ribosomes that contained no more than 6 nucleotides extending 3′ of the P-site codon, and very little tagging was observed with more than 15 nucleotides past the P-site (Ivanova et al., 2004). Crystal structures of ribosomes bound to mRNA indicate that this amount of mRNA could be contained within the ribosome, without extending past the leading edge (Yusupova et al., 2001, Jenner et al., 2005). Based on these data, it was proposed that either tmRNA–SmpB is sterically excluded from the ribosome when the mRNA extends past the leading edge or the ribosome changes conformation to allow tmRNA–SmpB entry when the leading edge passes the 3′ end of the mRNA (Moore and Sauer, 2007). Either of these models would explain why *trans*-translation does not compete with translation elongation but occurs efficiently on ribosomes at the 3′ end of the mRNA.

Many of the protein and mRNA signals for *trans*-translation described above act by promoting mRNA cleavage, thereby preventing elongation and placing the ribosome at the 3′ end of the mRNA. Investigation of the mechanism for targeting YbeL for *trans*-translation led to the surprising discovery that the mRNA is cleaved at the stop codon (Hayes and Sauer, 2003). This cleavage explains the high tagging rate of YbeL, because it produces an mRNA that lacks an in-frame stop codon and thus is a good substrate for *trans*-translation. The LESG sequence acts in a similar manner. Northern blots show that *crp* mRNA is truncated when LESG is encoded immediately before the stop codon (Sunohara et al., 2004a). In these examples, the amount of mRNA cleavage depends on the rate of translation, so conditions that decrease the rate of elongation or termination promote mRNA cleavage followed by *trans*-translation. Rare codons, depletion of tRNAs, and depletion of release factors have all been shown to promote mRNA cleavage followed by *trans*-translation (Ivanova et al., 2004, Sunohara et al., 2004a, Sunohara et al., 2004b, Garza-Sanchez et al., 2006, Li et al., 2006, Li et al., 2007). In some cases, the mRNA is cut in the A-site of the ribosome, and in others the mRNA is degraded by exoribonucleases up to the leading edge of the ribosome to generate a substrate for *trans*-translation (Hayes and Sauer, 2003, Ivanova et al., 2004, Sunohara et al., 2004a, Sunohara et al., 2004b, Garza-Sanchez et al., 2006, Li et al., 2006).

Several exoribonucleases have been implicated in 3′ mRNA trimming, but the source of A-site mRNA cleavage is not known. Small protein toxins, such as RelE, MazF, and ChpBK, promote A-site cleavage of mRNAs in response to amino acid starvation and other stresses (Christensen and Gerdes, 2003, Christensen et al., 2003, Pedersen et al., 2003). Efficient recovery from toxin stress requires tmRNA, suggesting that the stalled translation complexes generated by mRNA cleavage are released by *trans*-translation (Christensen and Gerdes, 2003, Pedersen et al., 2003). However, some A-site cleavage occurs in the absence of all known toxins, raising the possibility that the ribosome itself contains a nuclease activity (Hayes and Sauer, 2003). However, this nuclease activity has not yet been demonstrated.

It is possible that all *trans*-translation substrates are generated by a ribosome stalled at the 3′ end of an mRNA either by translation of the ribosome to the end of the mRNA or by cleavage of the mRNA at the ribosome. However, not all stalled translational complexes result in *trans*-translation. In particular, ribosomes paused as part of regulatory pathways are not targeted for *trans*-translation. Programmed pauses on the *tnaC* mRNA (Gong et al., 2007) and on the *secMA* mRNA do not lead to tagging (Garza-Sanchez et al., 2006). In both cases, there is a tRNA present in the A-site, which may inhibit mRNA cleavage in the paused complex. mRNA cleavage and *trans*-translation will occur on these complexes if the paused ribosome is not properly released (Collier et al., 2002, Hayes and Sauer, 2003), suggesting that *trans*-translation provides a mechanism to resolve defective paused complexes.

18.9 Regulation of *trans*-Translation Activity

In addition to the mechanisms for controlling the generation of *trans*-translation substrates, bacteria can control the amount of tmRNA and SmpB in the cell to modulate tagging activity. In most bacteria, tmRNA and SmpB levels change in response to the amount of *trans*-translation substrates, but in at least one case, *C. crescentus*, the availability of tmRNA and SmpB is likely to control whether *trans*-translation occurs (Keiler and Shapiro, 2003a, Hong et al., 2005).

In most species that have been investigated, such as *E. coli* and *Bacillus subtilis*, tmRNA is stable (Lee et al., 1978, Ushida et al., 1994). In fact, the name SsrA was given to this RNA because it was a small, stable RNA (Lee et al., 1978). Because degradation of tmRNA is slow, its abundance is controlled by production of new molecules. With the exceptions described below, tmRNA is transcribed from the *ssrA* gene as pre-tmRNA, which contains extensions at the 5′ and 3′ ends of the mature sequence (Fig. 18.5) (Chauhan and Apirion, 1989, Subbarao and Apirion, 1989). Processing to the mature RNA is similar to tRNA processing: RNase P makes an endonucleolytic cleavage to generate the mature 5′ end, and a combination of several exonucleases removes the 3′ extension leaving the terminal CCA (Komine et al., 1994, Ushida et al., 1994, Li et al., 1998, Lin-Chao et al., 1999). In species that add CCA to the 3′ end of tRNAs after processing, CCA is also added to tmRNA (Ushida et al., 1994). No evidence for control of tmRNA processing has been reported, and pre-tmRNA does not accumulate in wild-type cells.

In *E. coli* and *B. subtilis* growing vegetatively, tmRNA transcription and levels are fairly constant, with ~700 molecules of tmRNA and SmpB per cell (Moore and Sauer, 2005). However, tmRNA levels vary under conditions when substrates for *trans*-translation are expected to increase or decrease. In *E. coli*, *ssrA* and *smpB* are transcribed from σ^{70}-dependent promoters (Chauhan and Apirion, 1989). In *B. subtilis*, the *ssrA* and *smpB* genes are in an operon with *secG*, *yvaK,* and *rnr* (the gene encoding RNase R) (Shin and Price, 2007). No functional connection between *trans*-translation and SecG or YvaK is known, but these genes are transcribed from a common σ^{A}-dependent promoter. σ^{A} is the housekeeping sigma factor in *B. subtilis*, and transcription from this promoter is responsible for most *ssrA* and *smpB* expression during vegetative growth (Shin and Price, 2007). In addition to the σ^{A} promoter, *ssrA* and *smpB* are transcribed from a stress-responsive σ^{B}-dependent promoter, and the two promoters are induced by ethanol stress (Shin and Price, 2007). Finally, *ssrA* is independently controlled by a heat-shock promoter (Shin and Price, 2007). This

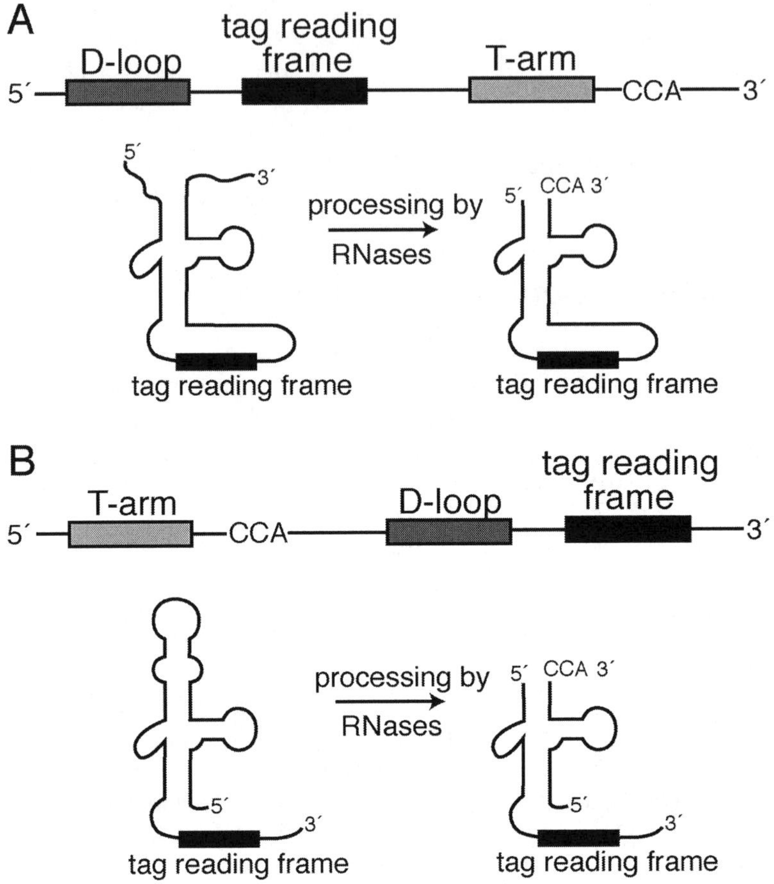

Fig. 18.5 Processing of one-piece and two-piece tmRNAs (**A**) Transcription of the *ssrA* gene followed by processing of pre-tmRNA by RNases at the 5′ and 3′ ends produces mature tmRNA. (**B**) Transcription of a circularly permuted *ssrA* gene results in a pre-tmRNA in which the tRNA-like 5′ and 3′ ends are connected by a loop. Processing of pre-tmRNA removes the loop and results in a two-piece mature tmRNA. The relative positions of the D-loop, tag reading frame, and T-arm in the linear and circularly permuted *ssrA* genes are indicated

array of promoters increases *ssrA* and *smpB* transcription during a wide variety of environmental stresses, as would be expected if increased *trans*-translation activity were required under stress conditions. Other bacteria also upregulate tmRNA in response to stress. *Synechocystis* species and *Thermatoga maritima* increase tmRNA levels under antibiotic stress, and *T. maritima* also increases tmRNA levels during biofilm formation (de la Cruz and Vioque, 2001, Montero et al., 2006). Increasing tmRNA and SmpB levels in response to stress can be rationalized by the quality control function of *trans*-translation. Stresses that are expected to disrupt mRNA metabolism are expected to produce more *trans*-translation substrates, so tmRNA and SmpB levels would be increased to provide more *trans*-translation capacity.

tmRNA and SmpB are regulated in a very different manner in *C. crescentus*, where the levels of both molecules vary as a function of the cell cycle. The *ssrA* gene in *C. crescentus* is circularly permuted, so the tRNA-like 3′ end is upstream of the tmRNA-like 5′ end (Fig. 18.5) (Keiler et al., 2000). The gene is transcribed as a single RNA, which folds into a structure similar to canonical tmRNAs, but with a loop connecting the ends of the mature molecule. The loop in pre-tmRNA is processed to produce a mature tmRNA composed of two RNA chains (Keiler et al., 2000). Transcription of *ssrA* and *smpB* increases during the transition from G1 to S phase, decreases after DNA replication is initiated, and increases again late in S phase prior to cell division (Keiler and Shapiro, 2003a). Likewise, the levels of both tmRNA and SmpB increase fourfold just before DNA replication initiates and are almost completely removed from the cell in early S phase (Keiler and Shapiro, 2003a, Hong et al., 2005).

In addition to transcriptional control of tmRNA and SmpB expression, both tmRNA and SmpB are specifically degraded as a function of the cell cycle. tmRNA is degraded by RNase R with a half-life of ~5 min during early S phase (Keiler and Shapiro, 2003a, Hong et al., 2005). This degradation rate is sufficient to remove all of the tmRNA each round of the cell cycle. tmRNA is stable in G1 phase and late S phase, when tmRNA levels increase (Keiler and Shapiro, 2003a). The timing of degradation is controlled by the abundance of SmpB and not by RNase R, which is constitutively expressed (Hong et al., 2005). SmpB inhibits tmRNA degradation by RNase R in vitro and deletion of *smpB* results in constant degradation of tmRNA in vivo. In wild-type cells, SmpB is proteolyzed in early S phase, exposing tmRNA to RNase R degradation (Hong et al., 2005). The factors responsible for the cell cycle-dependent proteolysis of SmpB have not been identified. It is also not yet known why tmRNA and SmpB levels change during the cell cycle, but clearly no *trans*-translation can occur when tmRNA and SmpB are absent. Temporal regulation of *trans*-translation may play a role in genetic control of the cell cycle.

RNase R degradation starts at the non tRNA-like 3′ end of the two-piece *C. crescentus* tmRNA (Hong et al., 2005). Degradation from this end makes sense, because the tRNA-like 3′ end is charged with alanine and likely to be resistant to exoribonucleases. This degradation may also explain why the circularly permuted version of the *ssrA* gene was retained through evolution: it provides an opportunity to control the turnover of tmRNA. In fact, all α-proteobacteria have the permuted *ssrA* gene, and two other bacterial lineages have permutations that occurred independently during evolution. (Keiler et al., 2000, Gaudin et al., 2002, Sharkady and Williams, 2004) Therefore, the permuted gene appears to have a selective advantage in some species.

18.10 Physiology of *trans*-Translation

Mutations in *ssrA* and *smpB* have been isolated or engineered in a wide variety of bacterial species. Disruption of *ssrA* in *Neisseria gonorrhoeae* (Huang et al., 2000) and *Shigella flexneri* (Keiler, unpublished) is lethal, and no disruptions of *ssrA* or *smpB* were isolated in saturating mutagenesis of *Haemophilus influenzae*, *Mycoplasma pneumonia*, and *Mycoplasma genitalium* (Hutchison et al., 1999, Akerley et al., 2002). In other species, mutations that disrupt *trans*-translation produce a variety of phenotypes, some associated with stress and others highly specific.

18.11 Stress Phenotypes

Several of the observed phenotypes are consistent with a decreased availability of ribosomes, as might be expected if stalled ribosomes are not released efficiently in the absence of *trans*-translation. The slow recovery from toxin stress in *E. coli* described above is one example (Christensen and Gerdes, 2003, Pedersen et al., 2003). In addition, many species lacking *ssrA*or *smpB* are more sensitive to heat shock, cold shock, ethanol stress, amino acid starvation, antibiotic exposure and are slow to recover from stationary phase (Table 18.1). In *E. coli,* disruption of *ssrA* results in elevated levels of the heat-shock proteins DnaK, GroEL, and ClpYQ, suggesting that in the absence of *trans*-translation cells are constitutively stressed (Munavar et al., 2005). This stress could explain why growth of *E. coli* is slowed at high temperature in the absence of tmRNA; cells that are already stressed are less able to adapt to additional challenges. Disruption of tmRNA in *E. coli* also decreases the level of the stress-responsive σ factor RpoS (Ranquet and Gottesman, 2007), potentially hampering the stress response.

18.12 Effects on Regulatory Pathways

Other phenotypes associated with *ssrA* and *smpB* mutations suggest that individual genetic pathways are disproportionately affected by the absence of *trans*-translation. For two of these specific pathways, lactose utilization in *E. coli* and activation of sporulation in *B. subtilis*, the molecular role of *trans*-translation is known. These mechanisms are described below, followed by other phenotypes that may be specifically regulated by *trans*-translation.

E. coli preferentially grow on glucose, but once glucose is depleted from the medium they will induce expression of the *lac* genes and metabolize lactose. In cells deleted for *ssrA,* induction of the *lac* operon is delayed (Abo et al., 2000). Although these mutants can utilize lactose, they are out-competed by wild-type cells that rapidly turn on the *lac* genes. Regulation of the *lac* operon is controlled in part by the *lac* repressor, LacI. *trans*-Translation is required for an autoregulatory mechanism that limits the amount of LacI activity. LacI represses transcription by binding to two sites in the *lac* operator, *O1* and *O3*, looping the DNA between these sites and inhibiting the binding of RNA polymerase (Fig. 18.6). The *lacI* gene is immediately upstream of the *lac* operon, and the *O3* site is within the *lacI*-coding region. Binding

Table18.1 Phenotypes of strains lacking tmRNA activity

Species	Phenotype	References
Neisseria gonorrhoeae	Lethal	Huang et al. (2000)
Shigella flexneri	Lethal	Keiler, unpublished
Haemophilus influenzae	Lethal	Akerley et al. (2002)
Mycoplasma species	Lethal	Hutchison et al. (1999)
Escherichia coli	Slow growth	Oh and Apirion (1991), Karzai et al. (1999)
	Slow recovery from carbon starvation	Oh and Apirion (1991)
	Decreased antibiotic resistance	Abo et al. (2002)
	Temperature sensitive	Komine et al. (1994), Oh and Apirion (1991)
	Decreased motility	Komine et al. (1994)
	Slow *lac* operon induction	Abo et al. (2000)
	Constitutive heat shock	Munavar et al. (2005)
	Slow recovery from toxin stress	Christensen and Gerdes (2003), Christensen et al. (2003), Pedersen et al. (2003)
Bacillus subtilis	Decreased stress survival	Muto et al. (2000), Shin and Price (2007)
Bacillus subtilis ATCC 6051	Stabilizes KinA allowing normal sporulation	Kobayashi et al. (2008)
Caulobacter crescentus	Delayed DNA replication initiation, plasmid maintenance disrupted	Keiler and Shapiro (2003b)
Bradyrhizobium japonicum	Symbiosis decreased	Ebeling et al. (1991)
Phage P22 (*in S. enterica*)	No lytic development	Julio et al. (2000)
Phage λ*imm*P22 (*in E. coli*)	No lytic development	Withey and Friedman (1999), Karzai et al. (1999)
Phage Mu (*in E. coli*)	No lytic development	Ranquet et al. (2001), Karzai et al. (1999)
Phage T4 (*in E. coli*)	No lysis inhibition	Slavcev and Hayes (2003)
Salmonella enterica	Decreased virulence	Julio et al. (2000), Baumler et al. (1994)
Yersinia pseudotuberculosis	Slow recovery from carbon starvation, low motility, decreased virulence	Okan et al. (2006)
Synechocystis species	Decreased antibiotic resistance	de la Cruz and Vioque (2001)

of the LacI protein to the *O1* and *O3* sites prevents RNA polymerase that initiated on *lacI* from completing the transcript, producing a *lacI* mRNA that lacks the last few sense codons and the stop codon (Abo et al., 2000). Protein made from this mRNA is tagged by tmRNA and rapidly degraded. At high concentrations, LacI also represses transcription of the *lacI* gene, so *trans*-translation is likely required only for a few transcripts generated after *O1* and *O3* are bound but before *lacI* transcription is shut off, thereby preventing accumulation of excess LacI protein in the cell. In mutants lacking *trans*-translation activity, truncated LacI protein is not tagged and degraded; instead, it accumulates in the cell. This truncated LacI contains the

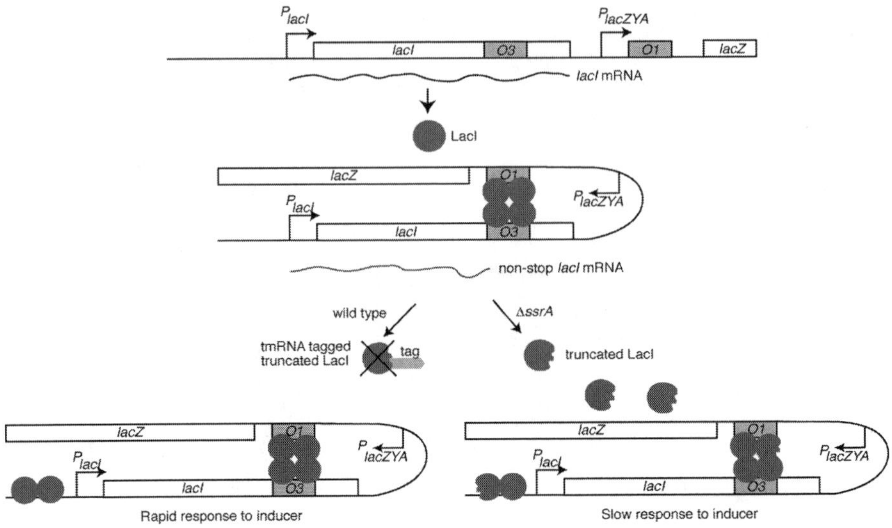

Fig. 18.6 Modulation of LacI activity by *trans*-translation. LacI binds to operator sites upstream of the *lac* operon to repress transcription. One of the operator-binding sites, *O3,* is within the *lacI* gene. Binding of LacI (*circles*) to *O3* results in a nonstop *lacI* mRNA. In wild-type cells, translation of the nonstop *lacI* mRNA leads to *trans*-translation and degradation of the tagged LacI protein. In cells lacking tmRNA, truncated LacI accumulates and delays induction when lactose or IPTG is added

DNA-binding domain and can repress the *lac* operon (Abo et al., 2000). Mutants lacking *trans*-translation respond slowly to both lactose and the inducer IPTG, presumably because the truncated LacI protein must be inactivated before transcription of the *lac* operon can initiate (Fig. 18.6). Mutation of either *O1* or *O3*, or expression of *lacI* from a different locus that does not have *lac* operator sites, eliminated control by *trans*-translation, indicating that operator binding and DNA looping within the *lacI* gene are required for tagging (Abo et al., 2000). Similar regulatory mechanisms have not been confirmed for other repressors. However, many DNA-binding proteins contain cognate binding sites within their own coding sequence, and could, in principle, be regulated in a *trans*-translation-dependent manner similar to LacI (Roy et al., 2002).

A second example of *trans*-translation regulation of gene expression has been demonstrated for the *kinA* gene in environmental isolates of *B. subtilis*. KinA is one of the kinases that initiate sporulation by phosphorylating SpoOF under specific nutrient conditions. In two environmental isolates, there is a sense codon in place of the stop codon in *kinA*, leaving no in-frame stop codon before the transcription terminator (Kobayashi et al., 2008). Transcription of these variant *kinA* genes produces an mRNA without a stop codon and very little KinA protein accumulates, presumably because it is tagged and degraded. Cells with the variant *kinA* gene do not sporulate in response to KinA-specific signals, but sporulation behavior is restored if *ssrA* is deleted (Kobayashi et al., 2008). These results suggest that *trans*-translation prevents sporulation in the wild-type isolates. The variant *kinA* could provide a mechanism to initiate sporulation in response to KinA signals only when *trans*-translation activity is saturated or specifically inactivated.

18.13 *trans*-Translation Effects on Bacterial Development

In *C. crescentus*, mutations that eliminate *trans*-translation cause a delay in cell cycle progression and morphological differentiation (Keiler and Shapiro, 2003b). *C. crescentus* cells divide asymmetrically into stalked cells that can immediately reinitiate DNA replication and swarmer cells that cannot replicate DNA or divide before differentiating into a stalked cell. During the swarmer to stalked cell differentiation, the transcriptional profile of the cell changes significantly, the cells shed their flagellum and grow a stalk, and initiate DNA replication. In swarmer cells (G_1 phase), DNA replication initiation is blocked by the master regulator CtrA, which binds to the origin of replication to inhibit replication and also controls the transcription of many cell cycle-regulated genes. In wild-type cells, proteolysis of CtrA leads to initiation of DNA replication and progression through the developmental program. In cells lacking tmRNA or SmpB, CtrA proteolysis is uncoupled from DNA replication and differentiation (Keiler and Shapiro, 2003b). Even though CtrA is degraded at the same time as in wild-type cells, the rest of the cell cycle, including the initiation of DNA replication, is delayed. After replication initiates, the cell cycle and developmental programs resume with no further disruption, suggesting that *trans*-translation is specifically required for some event prior to initiation of DNA replication (Keiler and Shapiro, 2003b). DNA replication, recombination, and repair factors are highly overrepresented among substrates for *trans*-translation in *C. crescentus*, so the delay in DNA replication may be due to misregulation of these proteins (Hong et al., 2007).

Mutations in *ssrA* were also identified in a screen for developmental defects in *Bradyrhizobium japonicum* (Ebeling et al., 1991). *B. japonicum* is a plant symbiont that differentiates into a nitrogen-fixing bacteroid in root nodules. Cells lacking tmRNA are able to form root nodules but are unable to differentiate into the bacteroid form (Ebeling et al., 1991). It is not known why tmRNA is required for this differentiation.

18.14 *trans*-Translation Effects on Phage Development

In addition to affecting bacterial physiology, the host *trans*-translation machinery is required for some bacteriophage genetic circuits. Lytic development of phage P22 is decreased in *Salmonella enterica* hosts lacking tmRNA. The efficiency of plating of P22 is 10,000-fold lower in strains lacking *ssrA* than in wild-type *S. enterica*, and induction from lysogens is delayed, but there are no defects in phage adsorption, lysogeny, or the ability to produce viable phage in these strains (Julio et al., 2000). These results suggest that *trans*-translation is required for proper control of lytic development of P22. Similarly, the hybrid phage λ*imm*P22, which has the immunity region of P22 in an otherwise λ genome, is unable to form plaques in *E. coli* strains lacking tmRNA or SmpB (Strauch et al., 1986, Retallack et al., 1994, Karzai et al., 1999). Both P22 and λ*imm*P22 phage lacking the C1 transcriptional activator develop normally even in the absence of tmRNA (Withey and Friedman, 1999, Karzai et al., 1999, Julio et al., 2000), suggesting that *trans*-translation affects phage development through this transcription factor. However, it is not known if *trans*-translation directly regulates C1 or if the effects are indirect.

Mu phage also has developmental defect hosts lacking *trans*-translation (Karzai et al., 1999, Ranquet et al., 2001). Mu lysogens containing a temperature-sensitive allele of the *c* repressor (*c-ts*) cannot be induced for lytic growth in *E. coli* lacking *ssrA* or *smpB* (Karzai et al., 1999, Ranquet et al., 2001).Mu *c* repressor maintains the lysogenic state by binding to the operator of the P_e and P_{cm} promoters, and Mu *c-ts* lysogens are induced at high temperature (Ranquet et al., 2001, Karzai et al., 1999). In wild-type hosts, Mu *c-ts* is tagged, and tagging promotes de-repression (O'Handley and Nakai, 2002). In the absence of tmRNA, truncated forms of the Mu *c-ts* repressor accumulate and repress transcription (Ranquet et al., 2001). These data suggest that production of truncated species of Mu *c* repressor in strains deficient for *trans*-translation disrupts Mu development.

18.15 Virulence Defects

Mutations that disrupt *trans*-translation cause defects in virulence in *S. enterica* and *Yersinia pseudotuberculosis*. *S. enterica ssrA* mutants are unable to proliferate in macrophages and are avirulent in mouse models for infection (Julio et al., 2000). Likewise, mice infected with *Y. pseudotuberculosis* deleted for *ssrA* or *smpB* show no signs of infection and clear the bacteria after 21 days, whereas mice infected with wild-type *Y. pseudotuberculosis* perish within 1 week of infection (Okan et al., 2006). The *Y. pseudotuberculosis ssrA* and *smpB* mutants are also unable to proliferate in macrophages (Okan et al., 2006). The proliferation defect in macrophages is due to misregulation of the transcription factor VirF which controls expression of the type III secretion system and is required for secretion of Yop effectors (Okan et al., 2006). It is not known how *trans*-translation affects VirF activity. The *Y. pseudotuberculosis* mutants are also sensitive to antibiotic, oxidative, and nitrosative stresses and have decreased motility, suggesting a broad requirement for *trans*-translation activity.

18.16 Role of Proteolysis and Ribosome Release in Bacterial Physiology

Many phenotypes caused by *trans*-translation defects are complemented by tmRNA variants that add a proteolysis-resistant peptide to substrate proteins. For example, viability in *N. gohorrhoeae*, stress phenotypes in *E. coli* and *B. subtilis*, and virulence defects in *Y. pseudotuberculosis* are complemented by tmRNA variants that add peptides ending in Asp–Asp or 6 His residues (Huang et al., 2000, Muto et al., 2000, Munavar et al., 2005, Okan et al., 2006). In these cases, it is likely that tagging per se and not the degradation of tagged proteins is important. When there are many *trans*-translation substrates, such as during toxin-induced stasis, *trans*-translation may be required to ensure that there are enough free ribosomes to maintain translation. Under some in vitro conditions, ribosomes stalled at the 3′ end of an mRNA are very stable (Karimi et al., 1999), but experiments using more physiologically relevant conditions suggest that ribosomes will rapidly fall off the 3′ end of mRNAs in vivo (Szaflarski et al., 2008). Nevertheless, turnover of stalled ribosomes may be faster in cells with *trans*-translation activity, facilitating ribosome release. Proteolysis of tagged proteins is required for other phenotypes, such as cell cycle control in *C. crescentus* and motility in *Y. pseudotuberculosis* (Keiler and Shapiro, 2003b, Okan et al., 2006). Continued investigation of these phenotypes and the mechanism underlying *trans*-translation will reveal the importance of tmRNA in both protein quality control and in regulation of cellular processes.

Acknowledgments We thank S. Yokoyama and Y. Bessho for providing us with the coordinates for the model of $tRNA^{Ser}$. We apologize to authors whose work we were not able to cite due to space constraints. The authors were supported by National Institutes of Health grant GM068720.

References

Abo T, Inada T, Ogawa K, Aiba H (2000) EMBO J 19:3762–3769
Abo T, Ueda K, Sunohara T, Ogawa K, Aiba H (2002) Genes Cells **7**:629–638
Akerley BJ, Rubin EJ, Novick VL, Amaya K, Judson N, Mekalanos JJ 2002. Proc Natl Acad Sci USA 99:966–971
Barends S, Karzai AW, Sauer RT, Wower J, Kraal B (2001) J Mol Biol 314:9–21
Baumler AJ, Kusters JG, Stojiljkovic I, Heffron F (1994) Infect Immun 62:1623–1630
Bessho Y, Shibata R, Sekine S, Murayama K, Higashijima K, Hori-Takemoto C, Shirouzu M, Kuramitsu S, Yokoyama S (2007) Proc Natl Acad Sci USA 104:8293–8298
Chauhan AK, Apirion D (1989) Mol Microbiol 3:1481–1485
Chien P, Perchuk BS, Laub MT, Sauer RT, Baker TA (2007) Proc Natl Acad Sci USA 104: 6590–6595
Choy JS, Aung LL, Karzai AW (2007) J Bacteriol 189:6564–6571
Christensen SK, Gerdes K (2003) Mol Microbiol 48:1389–1400
Christensen SK, Pedersen K, Hansen FG, Gerdes K (2003) J Mol Biol 332:809–819
Collier J, Binet E, Bouloc P (2002) Mol Microbiol 45:745–754
de la Cruz J, Vioque A (2001) RNA **7**:1708–1716
Dulebohn DP, Cho HJ, Karzai AW (2006) J Biol Chem 281:28536–28545
Ebeling S, Kundig C, Hennecke H (1991) J Bacteriol 173:6373–6382
Farrell CM, Grossman AD, Sauer RT (2005) Mol Microbiol 57:1750–1761
Felden B, Himeno H, Muto A, McCutcheon JP, Atkins JF, Gesteland RF (1997) RNA 3:89–103
Flynn JM, Levchenko I, Seidel M, Wickner SH, Sauer RT, Baker TA (2001) Proc Natl Acad Sci USA 98:10584–10589
Garza-Sanchez F, Janssen BD, Hayes CS (2006) J Biol Chem 281:34258–34268
Gaudin C, Zhou X, Williams KP, Felden B (2002) Nucleic Acids Res 30:2018–2024
Giege R, Sissler M, Florentz C (1998) Nucleic Acids Res 26:5017–5035
Gimple O A. Schon, (2001) Biol Chem 382:1421–1429

Gong M, Cruz-Vera LR, Yanofsky C (2007) J Bacteriol 189:3147–3155
Gottesman S, Roche E, Zhou Y, Sauer RT (1998) Genes Dev 12:1338–1347
Gueneau de Novoa P, Williams KP (2004) Nucleic Acids Res 32:D104–108
Gur E, Sauer RT (2008) Proc Natl Acad Sci USA 105:16113–16118
Hayes CS, Bose B, Sauer RT (2002a) J Biol Chem 277:33825–33832
Hayes CS, Bose B, Sauer RT (2002b) Proc Natl Acad Sci USA 99:3440–3445
Hayes CS, Sauer RT (2003) Mol Cell 12:903–911
Herman C, Thevenet D, Bouloc P, Walker GC, D'Ari R (1998) Genes Dev 12:1348–1355
Himeno H, Sato M, Tadaki T, Fukushima M, Ushida C, Muto A (1997) J Mol Biol 268:803–808
Hong SJ, Lessner FH, Mahen EM, Keiler KC (2007) Proc Natl Acad Sci USA 104:17128–17133
Hong SJ, Tran QA, Keiler KC (2005) Mol Microbiol 57:565–575
Hou YM, Schimmel P (1988) Nature 333:140–145
Huang C, Wolfgang MC, Withey J, Koomey M, Friedman DI (2000) EMBO J 19:1098–1107
Hutchison CA, Peterson SN, Gill SR, Cline RT, White O, Fraser CM, Smith HO, Venter JC (1999) Science 286:2165–2169
Ivanova N, Lindell M, Pavlov M, Holmberg Schiavone L, Wagner EG, Ehrenberg, M (2007) RNA 13:713–722
Ivanova N, Pavlov MY, Bouakaz E, Ehrenberg M, Schiavone LH (2005a) Nucleic Acids Res 33:3529–3539
Ivanova N, Pavlov MY, Ehrenberg M (2005b) J Mol Biol 350:897–905
Ivanova N, Pavlov MY, Felden B, Ehrenberg M (2004) J Mol Biol 338:33–41
Jacob Y, Sharkady SM, Bhardwaj K, Sanda A Williams KP (2005) J Biol Chem 280:5503–5509
Jenner L, Romby P, Rees B, Schulze-Briese C, Springer M, Ehresmann C, Ehresmann, B., Moras D, Yusupova G, Yusupov M (2005) Science 308:120–123
Julio SM, Heithoff DM, Mahan MJ (2000) J. Bacteriol. 182:1558–1563
Karimi R, Pavlov MY, Buckingham RH, Ehrenberg M (1999) Mol Cell 3:601–609
Karzai AW, Susskind MM, Sauer RT (1999) EMBO J 18:3793–3799
Kaur S, Gillet R, Li W, Gursky R, Frank J (2006) Proc Natl Acad Sci USA 103:16484–16489
Keiler KC, (2008) Annu. Rev Microbiol 62:133–151
Keiler KC, (2007) Curr Opin Microbiol 10:169–175
Keiler KC, Shapiro L (2003a) J Bacteriol 185:1825–1830
Keiler KC, Shapiro L (2003b) J Bacteriol 185:573–580
Keiler KC, Shapiro L, Williams KP (2000) Proc Natl Acad Sci USA 97:7778–7783
Keiler KC, Waller PR, Sauer RT (1996) Science 271:990–993
Kobayashi K, Kuwana R., Takamatsu H (2008) Microbiology 154:54–63
Komine Y, Kitabatake M, Yokogawa T, Nishikawa K, Inokuchi H (1994) Proc Natl Acad Sci USA 91:9223–9227
Konno T, Kurita D, Takada K, Muto A, Himeno H (2007) RNA 13:1723–1731
Lee S, Ishii M, Tadaki T, Muto A, Himeno H (2001) RNA **7**:999–1012
Lee SY, Bailey SC, Apirion D (1978) J Bacteriol 133:1015–1023
Lessner FH, Venters BJ, Keiler KC (2007) J Bacteriol 189:272–275
Levchenko I (2000) Science 289:2354–2356
Li X, Hirano R, Tagami H, Aiba H (2006) RNA 12:248–255
Li X, Yokota T, Ito K, Nakamura Y, Aiba H (2007) Mol Microbiol 63:116–126
Li Z, Pandit S, Deutscher MP (1998) Proc Natl Acad Sci USA 95:2856–2861
Lies M, Maurizi MR (2008) J Biol Chem 283:22918–22929
Lin-Chao S, Wei CL, Lin YT (1999) Proc Natl Acad Sci USA 96:12406–12411
McClain WH, Foss K (1988) Science 240:793–796
Mehta P, Richards J, Karzai AW (2006) RNA 12:2187–2198
Montero CI, Lewis DL, Johnson MR, Conners SB, Nance EA, Nichols JD, Kelly RM (2006) J Bacteriol 188:6802–6807
Moore SD, Sauer RT (2005) Mol Microbiol 58:456–466
Moore SD, Sauer RT (2007) Annu Rev Biochem 76:101–124

Munavar H, Zhou Y, Gottesman S (2005) J Bacteriol 187:4739–4751
Muto A, Fujihara A, Ito KI, Matsuno J, Ushida C, Himeno H (2000) Genes Cells 5:627–635
Nameki N, Felden B, Atkins JF, Gesteland RF, Himeno H, Muto A (1999) J Mol Biol 286:733–744
Nameki N, Tadaki T, Himeno H, Muto A (2000) FEBS Lett 470:345–349
O'Handley D, Nakai H (2002) J Mol Biol 322:311–324
Oh BK, Apirion D (1991) Mol Gen Genet 229:52–56
Okan NA, Bliska JB, Karzai AW (2006) PLoS Pathog 2:e6
Pedersen K, Zavialov AV, Pavlov MY, Elf J, Gerdes K, Ehrenberg M (2003) Cell 112:131–140
Pedulla ML, Ford ME, Houtz JM, Karthikeyan T, Wadsworth C, Lewis JA, Jacobs-Sera D, Falbo J, Gross J, Pannunzio NR, Brucker W, Kumar V, Kandasamy J, Keenan, L, Bardarov S, Kriakov J, Lawrence JG, Jacobs WR, Hendrix RW, Hatfull GF (2003) Cell 113:171–182
Ranquet C, Geiselmann J, Toussaint A (2001) Proc Natl Acad Sci USA 98:10220–10225
Ranquet C, Gottesman S (2007) J Bacteriol 189:4872–4879
Retallack DM, Johnson LL, Friedman DI (1994) J Bacteriol 176:2082–2089
Richards J, Mehta P, Karzai AW (2006) Mol Microbiol 62:1700–1712
Roche ED, Sauer RT (1999) EMBO J 18:4579–4589
Roy S, Sahu A, Adhya S (2002) Gene 285:169–173
Rudinger-Thirion J, Giege R, Felden B (1999) RNA 5:989–992
Sharkady SM, Williams KP (2004) Nucleic Acids Res 32:4531–4538
Shin JH, Price CW (2007) J Bacteriol 189:3729–3737
Slavcev RA, Hayes S (2003) Gene 321:163–171
Strauch MA, Baumann M, Friedman DI, Baron LS (1986) J Bacteriol 167:191–200
Subbarao MN, Apirion D (1989) Mol Gen Genet 217:499–504
Sundermeier TR, Dulebohn DP, Cho HJ, Karzai AW (2005) Proc Natl Acad Sci USA 102: 2316–2321
Sundermeier TR, Karzai AW (2007) J Biol Chem 282:34779–34786
Sunohara T, Abo T, Inada T, Aiba H (2002) RNA 8:1416–1427
Sunohara T, Jojima K, Tagami H, Inada T, Aiba H (2004a) J Biol Chem 279:15368–15375
Sunohara T, Jojima K, Yamamoto Y, Inada T, Aiba H (2004b) RNA 10:378–386
Szaflarski W, Vesper O, Teraoka Y, Plitta B, Wilson D, Nierhaus K (2008) J Mol Biol 380:193–205
Tanner DR, Dewey JD, Miller MR, Buskirk AR (2006) J Biol Chem 281:10561–10566
Tu GF, Reid GE, Zhang JG, Moritz RL, Simpson RJ (1995) J Biol Chem 270:9322–9326
Tyagi JS, Kinger AK (1992) Nucleic Acids Res 20:138
Ueda K, Yamamoto Y, Ogawa K, Abo T, Inokuchi H, Aiba H (2002) Genes Cells 7:509–519
Ushida C, Himeno H, Watanabe T, Muto A (1994) Nucleic Acids Res 22:3392–3396
Valle M, Gillet R, Kaur S, Henne A, Ramakrishnan V, Frank J (2003) Science 300:127–130
Wiegert T, Schumann W (2001) J Bacteriol 183:3885–3889
Williams KP, Bartel DP (1996) RNA 2:1306–1310
Williams KP, Martindale KA, Bartel DP (1999) EMBO J 18:5423–5433
Withey J, Friedman D (1999) J Bacteriol 181:2148–2157
Wower J, Zwieb CW, Hoffman DW, Wower IK (2002) Biochemistry 41:8826–8836
Yamamoto Y, Sunohara T, Jojima K, Inada T, Aiba H (2003) RNA 9:408–418
Yusupova GZ, Yusupov MM, Cate JHD, Noller HF (2001) Cell 106:233–241

Part IV
Transcription Slippage

Chapter 19
Transcript Slippage and Recoding

Michael Anikin, Vadim Molodtsov, Dmitry Temiakov, and William T. McAllister

Abstract Accurate transmission of genetic information during transcription requires that RNA polymerases maintain the correct register of the active site during each cycle of nucleotide incorporation. The RNA:DNA hybrid plays an important role in maintaining this lateral stability, and it has been observed that when the polymerase encounters homopolymeric tracts in the DNA template the transcript and/or the transcription complex may slip along the template, allowing the polymerase to incorporate more or fewer nucleotides than are encoded by the template. This phenomenon has been observed during all phases in the transcription cycle, including initiation, elongation, and termination. Here we review the evidence for transcript slippage in vivo and its implications for miscoding events. In addition, we review experiments that bear upon the mechanistic aspects of transcript slippage and the parameters that may affect its frequency. Aside from its implications for miscoding, transcript slippage may also be involved in regulatory roles during initiation and termination and promote expression of alternative information from the same gene.

Contents

W.T. McAllister (✉)
Department of Cell Biology, School of Osteopathic Medicine, University of Medicine and Dentistry of New Jersey, 42 E. Laurel Rd, UDP 2200 Stratford, NJ 08084, USA
e-mail: mcalliwt@umdnj.edu

J.F. Atkins, R.F. Gesteland (eds.), *Recoding: Expansion of Decoding Rules Enriches Gene Expression*, Nucleic Acids and Molecular Biology 24,
DOI 10.1007/978-0-387-89382-2_19,

Most of the chapters in this book are concerned with recoding events that occur during translation; however, errors made during transcription may also affect the transmission of genetic information. While some of these inaccuracies result from substitution or misincorporation errors, another route to faulty information transfer results from transcript slippage (also referred to as stuttering or pseudo-templated transcription (Jacques and Kolakofsky, 1991).

19.1 The Phenomenon of Transcript Slippage

The transcription process may be divided into three phases. During *initiation* the RNA polymerase (RNAP) binds to the promoter in a sequence-specific manner, melts apart the two strands of the DNA around the start site, and begins to incorporate NTP substrates that are complementary to the template (T) strand of the DNA. During this phase the RNAP remains associated with the upstream region of the promoter while extending the 3′ end of the transcript in the downstream direction. This results in a short RNA:DNA hybrid and a locally denatured region of the DNA in which the nontemplate (NT) strand is unpaired (the transcription bubble). The short hybrid in the initiation complex (IC) is not stable, and dissociation of the nascent RNA results in the release of abortive products without dissociation of the RNAP from the template. Eventually, the IC surmounts a thermodynamic barrier required to clear the promoter and undergoes a transition to form a stable elongation complex (EC). During *elongation* each step of nucleotide incorporation is followed by a translocation event in which the enzyme melts the downstream DNA, the active site moves along the template by 1 bp, the nascent RNA is displaced at the trailing edge by a corresponding interval, and the NT strand is reannealed (Fig. 19.1). This coupling of nucleotide incorporation and translocation results in a transcription bubble in the EC in which the RNA:DNA hybrid is maintained at a fixed length of ~8–9 bp. In the final phase of the transcription cycle, a *termination* signal in the template (or in the nascent RNA) causes destabilization of the EC, release of the product, and dissociation of the complex.

The accuracy of transcription involves a number of factors: discrimination of rNTPs vs. dNTPs, identification of the correct complementary base, and, most importantly for the discussion here, maintenance of the correct register of the active site with the template (precise translocation of the EC along the template) after each cycle of nucleotide incorporation. It is thought that the RNA:DNA hybrid is important for maintaining the lateral stability of the complex, and it is apparent how this would be the case in complexes in which the sequence in the hybrid is not homopolymeric, as a shift in the register of the RNA with the template would result in mismatched base pairs and would be thermodynamically disfavored. However, in cases where the RNA:DNA hybrid consists of a homopolymeric sequence, an aberrant mode of transcription has been observed that appears to involve slippage of the transcript and subsequent elongation without translocation of the active site (Fig. 19.1). This results in the incorporation of extra nucleotides in the RNA that were not encoded in the DNA template.

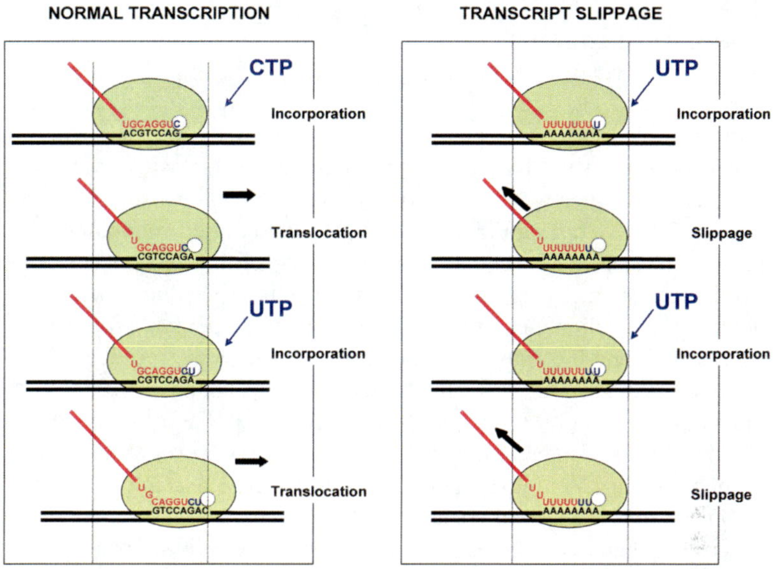

Fig. 19.1 Transcript elongation vs. transcript slippage. During normal transcript elongation each cycle of nucleotide incorporation is followed by movement of the RNAP by 1 nt along the template DNA, maintaining the correct register of the active site with the template. Under these conditions, polymerization is coupled to translocation. During transcript slippage, the transcript slips back along the template without movement of the enzyme and polymerization is uncoupled from translocation (idling)

The phenomenon of transcript slippage was first proposed by Chamberlin and Berg (Chamberlin and Berg, 1962) to account for the observation that during transcription of denatured calf thymus DNA in the presence of ATP as the sole substrate, purified *Escherichia coli* RNAP synthesized poly(A). As this phenomenon was suppressed if any of the other three NTP substrates was present, the authors concluded that it was likely that the transcript was slipping on poly(dT) tracts in the template and that repeated cycles of transcript slippage and incorporation of AMP resulted in the synthesis of poly(A).

Since that time, there have been numerous reports of transcript slippage in vitro, most prominently during initiation of transcription, but also during elongation and termination. The effects of transcript slippage during each of these phases of the transcription cycle are quite different. Slippage during initiation may affect the rate of promoter clearance and hence the rate of transcription of the corresponding gene, and in some cases, this phenomenon provides a basis for regulation; however, slippage during initiation affects only the 5′ end of the transcript and in most cases is not expected to alter the nature of the protein product. In contrast, slippage during elongation is expected to result in an alteration of the reading frame of the product and can be considered to be a recoding event. Lastly, slippage at the end of the transcription unit may be involved in the termination process or in the addition of extra nucleotides (such as poly(A)) to the end of the transcript.

19.2 Evidence for Transcript Slippage During Elongation In Vivo

A number of studies have demonstrated that transcript slippage can be significant in vivo under physiological conditions. In a seminal study by Wagner et al. (Wagner et al., 1990), the authors inserted an A_{11} tract into the beta galactosidase gene (*lacZ*) of *E. coli* (in this chapter we use the convention of indicating the sequence in the RNA product). While this should have disrupted the reading frame, the cells were observed to make lacZ at a level that was nearly 25% of the control. This effect was abolished if the A_{11} tract was interrupted by a single G residue. The authors concluded that transcript slippage in the poly(A) tract was responsible for restoration of the reading frame, and this was confirmed by observation of significant length and sequence heterogeneity in the transcripts. Significantly, slippage was also observed in U_{11} tracts, but not in G_{11} tracts, presumably reflecting the different stabilities of the RNA:DNA hybrids in the transcribing complexes. (However, transcript slippage has been reported during elongation in poly(G) tracts in *Neisseria gonorrhea* (Burch et al., 1997) and during initiation (see below).)

If we assume that the proteins that arise from transcript slippage and restoration of the reading frame are as active as the control, the level of lacZ production observed in the experiments above suggests that, at a minimum, as many as 25% of transcripts made from genes having an A_{11} tract arise from slippage. The actual frequency of slippage is likely to be higher than 25%, as slippage events that do not restore the proper reading frame would not have been detected in this assay.

Ordinarily, one would expect transcript slippage to be a deleterious event, as it would alter the reading frame of the expressed region, and that the occurrence of such homopolymeric tracts would be discriminated against in coding regions. In agreement with this, Baranov et al. (Baranov et al., 2005) found that there is a significant bias against the occurrence of poly(A) or poly(U) tracts greater than 6–7 nt in coding vs. noncoding regions in most bacterial species with A/T-rich genomes. However, in some species there appeared to be little bias, as these bacteria appear to tolerate poly(A) tracts >8 nt. This suggests that the transcription systems in these organisms may be less prone to transcript slippage (ibid) or that the cells have a high tolerance for aberrant products. Consistent with the possibility that different RNAPs may be more or less prone to slippage, Wagner et al. observed that transcription of a *lacZ* gene interrupted with A_{11} tracts did not result in efficient slippage when cloned into yeast and transcribed by pol II (ibid); (however, slippage by pol II has been documented in other eukaryotic systems (see below)). In addition, we have observed that whereas the single subunit T7 RNAP slips efficiently in poly(U) tracts, the related RNAP from yeast mitochondria does not (Molodtsov et al., submitted).

Despite the considerations above, there are circumstances in which transcript slippage could provide an advantage. The first is if a shift in the reading frame would allow the synthesis of two (or more) functional proteins from the same transcription unit. This could be used to optimize the use of limited genetic information or to provide variability in proteins that may be important in responding to changes in growth conditions or host immune surveillance. The second is if slippage restores a reading frame that would otherwise be interrupted by the addition or deletion of a nucleotide(s).

An example of the first situation occurs in the *dnaX* gene of *Thermus thermophilus*, which gives rise to two essential subunits of DNA polymerase III (tau and gamma) that are produced in approximately equimolar abundance (Larsen et al., 2000). These two proteins arise from a transcriptional shift in an A_9 tract within the coding region. Interestingly, a similar shift in reading frame in the *dnaX* gene is observed in *E. coli*; however, in the latter case it is a translational frameshift that results in the differential synthesis of the two proteins (ibid, and refs therein).

A related phenomenon occurs during the production of components of the type III secretion apparatus in *Shigella flexneri*, which are encoded by a virulence plasmid (Penno et al., 2005; Penno et al., 2006; Penno and Parsot, 2006). Here, synthesis of mxiE, a transcription activator, depends upon transcript slippage in a U_9 tract that lies in the overlap between two open reading frames, resulting in synthesis of the full-length protein (this is in contrast to the *T. thermophilus* situation noted above in which slippage is required for production of short and long forms of the dnaX protein, both of which are functional). A similar situation occurs in the *S. flexneri spa13* gene, in which slippage within an A_{10} tract is required for synthesis of a full-length protein. Slippery tracts (A_9) also occur in the *mxiA* and *spa33* genes, but in these cases slippage disrupts the ORF and results in *decreased* production of functional products.

Interestingly, slippage in the *mxiE* gene affects both transcription and translation of the downstream *mxiD* gene, whose $5'$ end overlaps the distal portion of *mxiE*. In this case, it was suggested that premature termination of translation in the unshifted *mxiE* RNA results in polar effects and a decrease in transcription downstream of the slippery site due to coupling of transcription and translation (Newton et al., 1965; Penno and Parsot, 2006).

In an analysis of multiple bacterial genomes, Baranov et al. found that the largest number of genes with the potential to use transcriptional slippage for expression of open reading frames was found in IS elements, suggesting that this may be common mechanism in the expression of these genetic components (Baranov et al., 2005). A similar strategy may be used to fuse ORFs that constitute the *Staphylococcus aureus mapW* gene, which encodes a class II major histocompatibility-like protein that is involved in modulation of the host T-cell immune response (Baranov et al., 2005; Kuroda et al., 2001; Lee et al., 2002) and may promote antigenic variation and pathogenicity in *Helicobacter pylori* (Alm et al., 1999; Tomb et al., 1997). These results suggest that the functional use of slippery sequences may be important for genetic elements of limited size or where variability in the protein product may provide a selective advantage.

Potentially slippery sequences are particularly abundant in the bacterial endosymbionts of insects which have a small genome (160–790 kb) and a low G + C content (20–29%) (Tamas et al., 2008). Up to 50% of the genes in these organisms may contain tracts of A_{10} or greater, and many of these are pseudogenes that contain frameshifts within the poly(A) tracts. While it had earlier been suggested that the transcription systems of these organisms might be less prone to slippage

(Baranov et al., 2005), a recent study in which cloned cDNA copies of transcripts from these cells were examined reported that transcript slippage frequencies could be as high as 70% and that transcript slippage is required to rescue the function of some genes (Tamas et al., 2008). This seemingly inefficient strategy may be peculiar to these species, which have undergone strong selection for a small genome and benefit from a small population size.

The second situation in which transcript slippage may be beneficial to the cell is if it compensates for a frameshift mutation in the same or nearby region of an essential gene. An example of this in eukaryotic cells was first observed in a family with hypobetalipoproteinemia that resulted from a deletion of a single C residue within an A_6CA_3 tract. While this mutation is expected to result in a truncated protein, frameshifting in the uninterrupted A_9 tract due to transcript slippage restored the reading frame in ~10% of the transcripts, allowing the synthesis of a full-length functional protein (Linton et al., 1992, 1997). Similar observations have been made in a family with mild to moderately severe hemophilia A due to a deletion of a single T residue in an A_8TA_2 tract (Young et al., 1997) and in a canine gene involved in cyclic neutropenia (due to an insertion of an additional A residue in an A_9 tract (Benson et al., 2004)).

Transcript and/or replication slippage has also been implicated in human familial colorectal cancer due to a T to A substitution in an A_3TA_4 sequence (resulting in an A_8 tract) (Laken et al., 1997; Raabe et al., 1998). Based upon the detection of mutations in this region in nearly half of the tumors studied, Laken et al. concluded that the disease effects were most likely due to replication slippage in somatic tissues. However, as pointed out by Raabe et al., these aberrant proteins could also have resulted from transcript slippage (Raabe et al., 1998). Interestingly, truncated proteins that would arise from transcript slippage were detected in a coupled in vitro transcription/translation system that utilized phage RNAP and template DNA having the A_8 tract, but not in patient tissue, suggesting that the formation of these proteins may be the result of recoding in the in vitro system (Laken et al., 1997). As noted below, T7 RNAP is prone to transcript slippage in A_8 tracts, and thus the use of such coupled systems to explore recoding events should be undertaken with caution.

Slippage may not be limited to A tracts >8 nt in mammalian cells, as Ba et al. have reported slippage within A_6 tracts (Ba et al., 2000). Using a functional assay in which RT-PCR products were cloned into yeast that carried a p53-responsive reporter, they were able to detect alterations in transcripts of the *p53* gene in livers of rats with a high incidence of hepatitis and hepatoma. The majority of changes (>50%) were due to the insertion of an extra A residue in each of three A_6 tracts. Interestingly, slippage was observed preferentially in one of the three A_6 tracts, which may indicate either a context-dependent bias in slippage or preferential degradation of some slipped or poorly translated mRNAs as a result of RNA surveillance systems (Isken and Maquat, 2007). The frequency of slippage was significantly higher in diseased or aged rats, or upon exposure of hepatic cell lines to alcohol, suggesting that disease or tissue damage may result in a decrease in transcriptional fidelity. A survey of rat genes indicated that A_6 or U_6 tracts occur in a number of other genes, and the authors reported that they have also observed insertion of a single A residue in A_6 runs in the human APC tumor suppressor gene (cf., Linton et al., 1997; Raabe et al., 1998).

19.3 Slippage in Viral Systems

Transcription slippage has been reported in a number of viral systems, where it is often used to direct the synthesis of more than one protein from the same coding region. A paradigm for this phenomenon involves synthesis of the P protein in the nonsegmented negative-strand RNA (NNV) paramyxoviruses, in which additional nontemplated G residues are incorporated downstream of an A_nG_n tract (U_nC_n in the RNA template). Unlike the conventional mechanism invoked to explain slippage by DNA-dependent RNAPs, which involves slippage of the transcript without movement of the RNAP (see Fig. 19.1), it is proposed that the paramyxovirus RNA-dependent RNAP backtracks along the RNA template along with the associated transcript (a realignment that is facilitated by the formation of nondestabilizing rG:rU bps between the newly synthesized RNA and the template) followed by subsequent extension (Kolakofsky et al., 2005); notably, realignment by this mechanism may occur over intervals > 1 nt. Transcript slippage has also been observed to play a role in the synthesis of the glycoproteins of Ebola virus (also an NNV), which involves the addition of an A residue in an A_7 tract (Sanchez et al., 1996; Volchkov et al., 1995, 2001) and in the expression of hepatitis C virus (a positive-sense, single strand RNA virus) core protein (Ratinier et al., 2008). Lastly, slippage has been shown to play a role in pausing and the addition of poly(A) segments at both the 5′ and 3′ ends of mRNAs synthesized by the DNA-dependent RNAP of vaccinia virus (Deng and Shuman, 1997).

19.4 Slippage in Nonhomopolymeric Tracts

In addition to slippage in homopolymeric tracts, it has been reported that transcript slippage may occur in simple, direct repeats, such as $(GA)_n$ and $(CAG)_n$, and this phenomenon, known as "molecular misreading," has been implicated in a number of age-related disorders such as Alzheimer's disease, Down's syndrome, and diabetes (Fabre et al., 2002; van den Hurk et al., 2001; van Leeuwen et al., 2000, 2006). These diseases are characterized by the accumulation of aggregates of misfolded proteins that are not efficiently cleared by the proteasome system. However, the frequency of slipped transcripts in Alzheimer's and Down's patients is quite low and does not appear to differ significantly from that in unaffected patients (Gerez et al., 2005). This led Wills and Atkins (Wills and Atkins, 2006) to propose that frameshifting during translation of wild-type mRNAs may be responsible for these effects, rather than transcription slippage. However, either model would have to account for the increased accumulation of misfolded proteins late in disease, and it is therefore necessary to distinguish between the rate of accumulation of the aberrant protein and its rate of synthesis. The accumulation of misfolded wild-type proteins may be enhanced by the presence of altered forms of the protein that arise by frameshifting. In this regard, de Pril et al. (de Pril R. et al., 2006) have noted that failures in protein quality control mediated by the ubiquitin–proteasome system appear to be involved in a number of neurological disorders and that a modified form of ubiquitin (UBB+1, which itself arises by slippage during synthesis of the UBB mRNA) may interfere with proteasome function if it accumulates to a critical level. We suggest that the synthesis of aberrant proteins that arise by slippage (or frameshifting) and escape surveillance by the proteasome system may nucleate misfolding and aggregation of their normal counterparts. Even low levels of synthesis of the aberrant proteins would accelerate the accumulation of the deleterious forms, resulting in earlier onset of clinical symptoms.

19.5 Transcript Slippage During Initiation

Numerous reports have indicated that transcript slippage occurs during the early stages of initiation, and although this phenomenon is not likely to affect recoding (the nature of the protein product encoded by these transcripts), the phenomenon deserves mention at this point as knowledge of its features may inform our understanding of slippage during elongation.

The affinity of the RNAP for its promoter represents an energy barrier that must be surmounted before the initiation complex clears the promoter and enters the elongation phase. As a result, the polymerase is held at the promoter, while transcript is extended downstream until sufficient strain builds up to allow promoter release. This accounts for the release of abortive products during initiation, but may also contribute to the production of slippage products, as the affinity of the RNAP for the promoter inhibits translocation, allowing transcript slippage to compete with extension.

Transcript slippage has been observed in many bacterial promoters when there are homopolymeric tracts at or near the start site (Guo and Roberts, 1990; Harley et al., 1990; Jacques and Susskind, 1990; Liu et al., 1994; Parker, 1986; Xiong and Reznikoff, 1993; Jin, 1996). In addition, slippage involving increments of 2 or 3 nt has also been observed during initiation at promoters having short direct repeats in the first 3–4 nt of the transcribed region (Severinov and Goldfarb, 1994; Borukhov et al., 1993; Pal and Luse, 2002), suggesting that displacement and reannealing of the transcript over intervals >1 nt may be possible. Such a mechanism may be peculiar to initiation, as it has been shown that short abortive products that have been released from the IC may rebind and serve as primers for subsequent initiation events. However, Luse et al. have demonstrated that slippage by 2 nt in CUCUCU tracts by human pol II occurs even as the transcription complex extends downstream to +23. The efficiency of slippage in the CU tracts drops off sharply at intervals that correspond to transitions that occur during the formation of a fully processive EC;

nevertheless, these observations suggest that slippage by >1 nt may also be possible during elongation. Unlike slippage during elongation, which usually involves poly(U) and poly(A) tracts, slippage during initiation has also been observed within poly(G) tracts ((Imburgio et al., 2000; Martin et al., 1988; Meng et al., 2004); but see also (Burch et al., 1997)).

Aside from its mechanistic implications, it is important to note that slippage during initiation may play a significant regulatory role in vivo. For example, transcription initiation at several pyrimidine biosynthetic and salvage operons in *E. coli* is controlled by reiterative transcription (stuttering) that is sensitive to the intracellular concentration of UTP. Furthermore, attenuation control of *pyrG* expression in *Bacillus subtilis* is mediated by CTP-sensitive reiterative transcription that involves the repetitive addition of G residues to the 5′ end of the *pyrG* transcript (for review, see (Turnbough, Jr. and Switzer, 2008)).

19.6 Transcript Slippage During Elongation–In Vitro Studies

In view of the evidence for transcript slippage as a significant issue in vivo, there have been surprisingly few studies of the mechanism of slippage in vitro. Most models of transcript slippage have assumed that the RNA product moves backward along the template (away from the active site and toward the 5′ end of the RNA) without movement of the RNAP and that subsequent addition of the incoming NTP results in products that are larger than encoded by the template (idling; see Fig. 19.1). However, this model does not account for all of the observations made when a polymerase encounters a slippery tract.

For example, earlier studies of T7 RNAP (Macdonald et al., 1993) revealed that transcription of an extended poly(U) tract (U_{40}) resulted in synthesis of a heterogeneous population of products that were both larger and smaller than expected and that when the concentration of UTP was reduced, the sizes of the products became less dispersed and approached a size that corresponded to the incorporation of only 8–12 UMP residues (Fig. 19.2). This observation indicated that the polymerase (together with the associated transcript) may slide forward on the template without incorporating substrate and that resumption of polymerization after sliding through the U tract resulted in smaller products (forward sliding, see Fig. 19.3). At low UTP concentrations, forward sliding is competitive with polymerization, and the smaller products predominate. At moderate UTP concentrations, where transcript extension is competitive with slippage, both forward and backward slidings could occur (as well as the idling mode noted above). Both idling and backward slidings can result in synthesis of products that are larger than expected and so cannot be distinguished on the basis of the size of the transcripts that are produced.

The observation that the slippage products reached a limit size that corresponded to incorporation of 8–12 nt at low concentrations of UTP suggested that this might correspond to the length of the RNA:DNA hybrid in the T7 RNAP

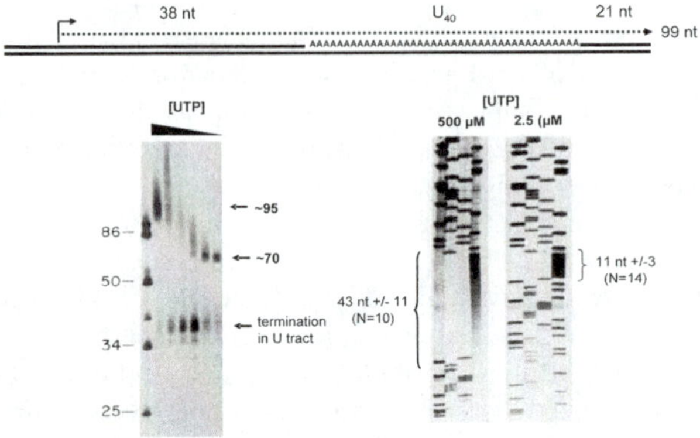

Fig. 19.2 Sliding of RNAP on extended homopolymeric tracts. A template that encodes a U_{40} tract (*top*) was transcribed by T7 RNAP in the presence of varying concentrations of UTP (400–2 μM) and the products were resolved by gel electrophoresis (*left*). At moderate concentrations of UTP the runoff products were heterogeneous and were both larger and smaller than predicted by the template (95 nt). As the UTP concentration was reduced the products became smaller and more homogeneous and appeared to approach a limit size of ~70 nt. Some products of a size that would correspond to termination of transcription within the U_{40} tract were also observed. Transcripts synthesized at 500 and 2.5 uM UTP were characterized by cDNA cloning and sequencing (*right*). The RNA products had the sequence predicted by the template on either side of the U_{40} tract, but varied within this region: at 500 uM UTP the average U length was 43 nt, +/− 11nt (10 clones analyzed); at 2.5 uM UTP the average U length was 11 nt, +/−3 nt (14 clones analyzed. Figure adapted from Macdonald et al. (1993)

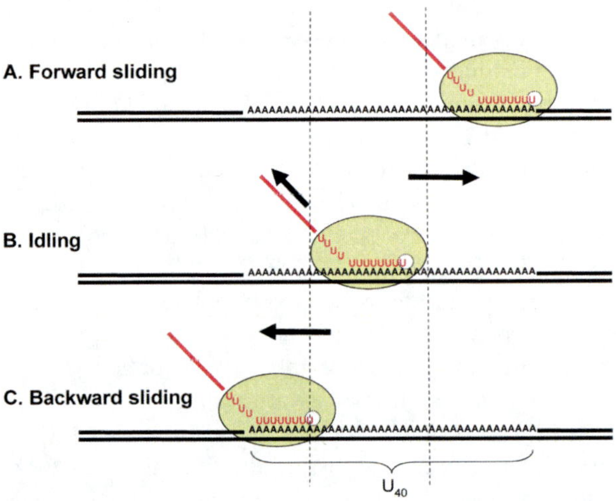

Fig. 19.3 Potential modes of transcript slippage. Slippage of the transcript relative to the DNA template on extended homopolymeric tracts may result from three different mechanisms. During *idling* (polymerization without translocation, middle panel; and see Fig. 19.1) the transcript slips and is subsequently extended without movement of the RNAP along the template; resumption of normal elongation results in products that are longer than expected. During *sliding* (top and bottom panels) translocation of the elongation complex (along with the transcript) occurs without polymerization. Resumption of normal elongation results in products that are shorter than expected (*forward sliding*) or longer than expected (*backward sliding*)

EC (Macdonald et al., 1993), and indeed, subsequent structural studies confirmed that the hybrid in the T7 EC is 8–9 bp. These observations are consistent with the notion that efficient slippage occurs when the hybrid consists of a homopolymeric tract along its length. Subsequent studies have shown that minimal length of poly(A) or poly(U) tracts required for efficient slippage by T7 RNAP is 8 nt (Molodtsov et al., submitted). On such templates the frequency of slippage is surprisingly high; under conditions where transcript extension beyond the U tract is proceeding at half maximal rates, slippage products account for over 80% of the transcripts. Interestingly, when the U tract was limited to 8 nt, which would constrain the complex from forward or backward sliding (which requires additional U-encoding regions downstream and upstream; see Fig. 19.3), nearly all of the products were larger than expected, consistent with an idling mode of slippage in this case (ibid).

19.7 Structural and Mechanistic Considerations of Translocation

Normally, each cycle of nucleotide incorporation is coupled to translocation of the active site downstream by 1 nt. What features of the RNAP or the nucleic acid components of the EC are responsible for maintaining the accuracy of this register and the force needed to advance the enzyme? It is known that the RNA:DNA hybrid is an important contributor to lateral stability of the EC and that shorter or weaker hybrids result in instability (Macdonald et al., 1993; Pal and Luse, 2003; Sidorenkov et al., 1998). What implications, if any, may be gleaned from the observation that translocation is uncoupled from polymerization during transcript slippage? For example, during idling it is proposed that polymerization occurs without translocation, while during sliding translocation is thought to occur without polymerization (see Fig. 19.3).

Structural analyses of both single subunit and multisubunit RNAPs have revealed close interactions between the RNAP and the nucleic acid scaffold on which it operates. Some of these interactions are responsible for processes such as unwinding of the duplex DNA downstream, displacing the nascent transcript from the upstream end of the hybrid, and reannealing of the duplex DNA at the trailing edge of the transcription bubble. Other elements are involved in the nucleotide addition cycle itself, and movement of some of these elements may punctuate and be coupled to the translocation event. There are currently two schools of thought as to what provides the force necessary to move the enzyme during each cycle of translocation (see (Bar-Nahum et al., 2005; Landick, 2004; Sousa, 2005; Yin et al., 1995) and refs therein). In the *power stroke model*, release of pyrophosphate following each cycle of nucleotide incorporation results in movement of one or more elements in the RNAP that are thought to push against a component of the nucleic acid scaffold. In the *Brownian ratchet model*, RNAP movement is driven by thermal diffusion, and equilibrium between the pre- and post-translocated sites of the active site is modulated by binding of the incoming NTP, much like a pawl in a ratchet, to lock the enzyme in the proper register. These considerations do not rule out the possibility that there may be more than one element involved in lateral stability and movement of the RNAP. For example, the fit of the RNA:DNA binding cavity to the hybrid may involve interdigitations that limit the directionality and movement of the hybrid to 1 bp increments while other elements provide the driving force.

At a minimum, the observation that translocation is uncoupled from polymerization during sliding of the transcription complex indicates that formation of the phosphodiester bond and the release of pyrophosphate are not required to move the RNAP along the template. This is consistent with what is observed during "backtracking," in which stalled complexes move backward while reestablishing the hybrid in the upstream direction and displacing the 3′ end of the transcript (Komissarova and Kashlev, 1997; Nudler et al., 1997). By themselves, however, these observations do not contribute much to our understanding of how the forces required for translocation are transmitted, as slippage of the transcript may occur in a passive manner, independently of the motive force for translocation (Guajardo et al., 1998). On the other hand, if the translocation force is applied, at least in part, against the transcript, then we might expect that a polymerase positioned within a slippery tract (e.g., a U_{16} tract) vs. a nonslippery tract (N_{16}) might be less able to proceed past a physical roadblock, as the translocation force would be applied against a slippage-prone vs. a slippage-resistant RNA. As shown in Fig. 19.4 this is the outcome that is observed. Here, increasing concentrations of a mutant form of the restriction enzyme EcoRI that binds tightly to the DNA but does not cleave were bound to a template downstream of a U_{16} or an N_{16} tract and the ability of T7 RNAP to proceed through the block during multiple rounds of transcription was determined. At lower concentrations the presence of the block had a much greater effect on progression through the U_{16} tract than through the N_{16} tract and resulted in a dramatic increase in the production of slippage products.

While the decreased ability to traverse the roadblock suggests that part of the force required for translocation may be applied against the transcript (a power stroke), this may also reflect the possibility that weak rU:dA or rA:dT bps result in a weak "pawl," allowing the Brownian ratchet to slip. Regardless of the mechanism, the observation that a downstream roadblock enhances slippage may have implications for regulatory mechanisms. While such physical barriers may include proteins or histones bound to the template, the local configuration of the DNA duplex may also affect the response of the RNAP to slippery tracts. This may account for earlier reports that slippage in vivo may vary at different locations within the same gene (Ba et al., 2000). Such modulating influences might affect pausing and termination, as well as the balance of proteins that would be produced as a result of shifts in the reading frame of the transcripts. Moreover, as slippage and polymerization appear to be competitive under physiological conditions (see (Wagner et al., 1990; Meng et al., 2004)) changes in NTP levels under various conditions may also modulate the frequency of transcript slippage.

As noted above, the observation that translocation is uncoupled from polymerization during transcript slippage suggests that one or more of the components involved in control of polymerase movement must be interacting with the RNA component of the RNA:DNA hybrid. A potential clue to the structural elements

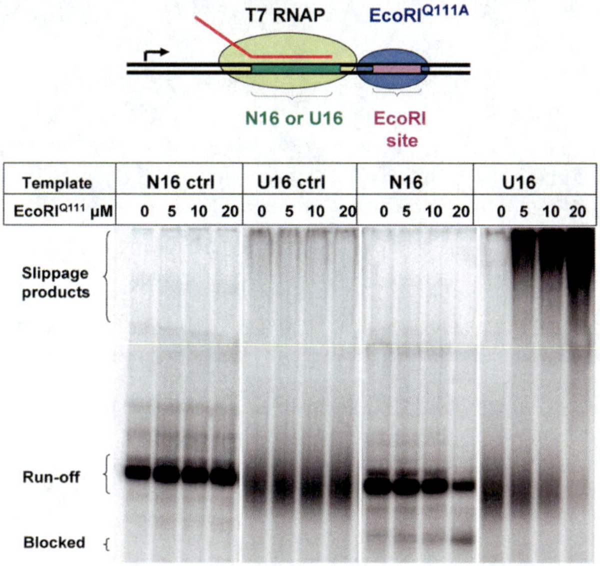

Fig. 19.4 A bound protein roadblock inhibits translocation and enhances slippage in poly(U) tracts. Templates that encode U_{16} or nonhomopolymeric (N_{16}) tracts just upstream of an EcoRI-binding site (U_{16} RI; N_{16} RI) or control (ctrl) templates that lacked the binding site were transcribed for 10 min in the presence of increasing concentrations of the EcoRI-mutant Q111A, which binds tightly to the recognitions site but does not cleave DNA, and the products were resolved by gel electrophoresis. Transcription from the N_{16} RI template resulted in homogeneous runoff products of the expected size, and the amount of this product decreased in the presence of high concentrations of the roadblock. In contrast, transcription from the U_{16} RI template gave rise to a heterogeneous set of runoff products due to slippage in the U tract (see Fig. 19.3); the synthesis of these products was markedly more sensitive to the presence of the roadblock, and the decrease in their abundance was accompanied by an increase in the appearance of larger slippage products that migrate near the top of the gel. The latter products arise from slippage before the roadblock, without further extension, or by enhanced slippage before the roadblock followed by subsequent extension

that may be involved comes from analysis of mutants of HIV reverse transcriptase (RT) that show increased slippage during replication, which map to a region of the enzyme that interacts closely with the primer:template (Hamburgh et al., 2006) (see Fig. 19.5). The significance of this observation with regard to RNAP function depends upon the fact that RT is a member of the structurally related pol I class of single subunit nucleotide polymerases that also includes T7 RNAP. Superposition of the T7 RNAP EC structure on the HIV RT structure reveals elements in T7 RNAP that are positioned in a similar manner to those implicated in slippage by RT and may play a similar role. One of these elements is part of a DX_2GR motif that is conserved among other members of the pol I family. In T7 RNAP this element has been proposed to be part of a primer:template grip, in which R425 interacts with the 3′ end of the RNA primer, and it has been shown that mutations in this motif result in altered transcript slippage during initiation (Imburgio et al., 2002). A similar role may be performed by the "clamp" element in multisubunit RNAPs (Landick, 2001).

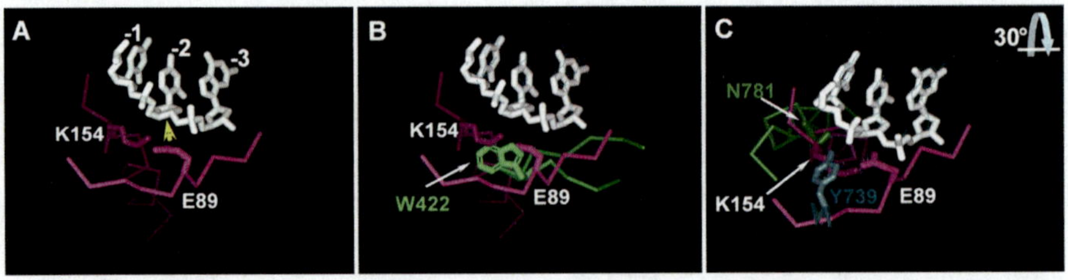

Fig. 19.5 Structural determinants of slippage in HIV RT and T7 RNAP.(**A**) In the HIV RT structure (1RTD), E89 (*magenta*) interacts with the phosphate backbone of the template strand (TS, *shown in white*) at position −2 (*yellow arrow*); the position of the side chain of E89 is stabilized by a salt bridge with K154 (Hamburgh et al., 2006).(**B**) The structure of T7 RNAP elongation complex (1MSW) was aligned with the RT structure by superimposing the RNA:DNA hybrid of the T7 complex on the DNA duplex of the RT complex. A portion of the TS of the T7 complex is *shown in light blue*. W422 of T7 RNAP (a part of the DX_2GR motif; *light green*) takes the position equivalent to that of the E89-K154 salt bridge in the RT complex.(**C**) Two other residues in T7 RNAP, Y739 at the N-terminal base of specificity loop (*shown in teal*) and N781 in α-helix adjacent to the C-terminal base of specificity loop (*dark green*) are superimposable on K154 of RT

19.8 Molecular Mechanisms of Transcript Slippage

In discussions of transcript slippage, it has generally been thought that slippage involves denaturing of the RNA:DNA hybrid, shifting of the transcript along its entire length, and reannealing in a new register. However, it is not clear whether this is the actual mechanism by which transcript slippage occurs, or whether there are other pathways that are less thermodynamically costly. Three potential pathways for transcript slippage are shown in Fig. 19.6.

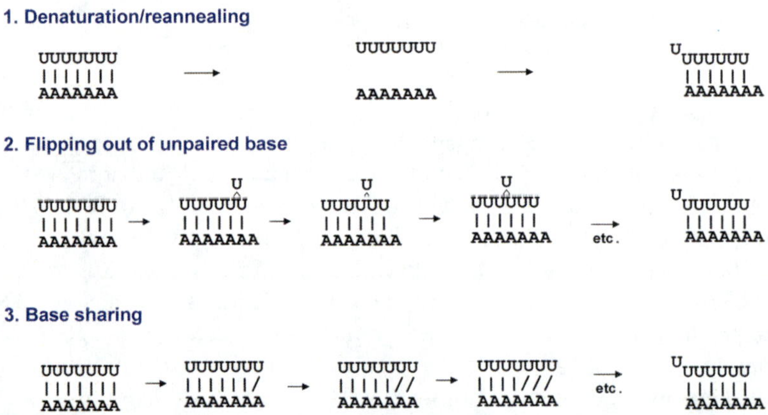

Fig. 19.6 Potential mechanisms of transcript slippage. Three potential mechanisms that would facilitate movement of the transcript relative to the template in the RNA:DNA hybrid are illustrated, see text for details

The first pathway (Fig. 19.6, scheme 1) involves melting and reannealing of the hybrid as described above. The stability of rU:dA base pair hybrids is extraordinarily low, and rU_8:dA_8 duplexes are not expected to be stable under physiological conditions (predicted free energy of formation, $\Delta G°_{37}= +1.7$ kcal/mol) (Sugimoto et al., 1995). Nevertheless, surprisingly little termination is observed within U tracts, even under limiting UTP concentrations ((Macdonald et al., 1993). It is therefore apparent that protein:nucleic acid contacts within the hybrid-binding cavity of the RNAP stabilize the homopolymeric duplex, allowing transcription to proceed without disassociation. In general, the fit of the hybrid-binding cavity to the RNA:DNA hybrid in currently known EC structures is rather tight and would not appear to tolerate wholesale denaturation and reannealing of the hybrid, as would be required by this model. Whether alternate configurations of the RNA:DNA hybrid other than the A-form duplex or an alternate conformation of the hybrid-binding cavity observed in crystal structures would be possible is an open question. Other conformations of the RNAP or hybrid that might allow slippage by this mechanism may exist.

It should be noted that while the stability of rA_8;dT_8 hybrids is higher than that of rU_8:dA_8 hybrids ($\Delta G°_{37}= -3.9$ kcal/mol) efficient slippage is observed in A_8 tracts. In contrast, little slippage is observed in $(AU)_4$ tracts, which have a lower stability ($\Delta G°_{37}= -2.3$ kcal/mol) than A_8 tracts. This may be due to fact that slippage and subsequent extension of nonhomopolymeric $(AU)_4$ tracts requires realignment by an increment of 2 nt vs. 1 nt for the homopolymeric tract, which may introduce a greater thermodynamic barrier.

In the second pathway (Fig. 19.6, scheme 2), displacement and flipping out of one or more bases in either the RNA (or template) would allow backward (or forward) movement of the 3′ end of the RNA relative to the template by a corresponding interval; subsequent propagation of the misalignment along the length of the hybrid by a "domino-like effect" would complete the shift in register. Whether such mismatches would be tolerated within the hybrid RNA:DNA hybrid-binding cavity of the RNAP is not clear. Structural and kinetic studies of pol I DNAPs indicate that there is considerable flexibility in protein structure and in the organization of nucleic acids that allow these DNAPs to accommodate deviations from normal duplex structure (Garcia-Diaz et al., 2006; Johnson and Beese, 2004; Ling et al., 2001; Tippin et al., 2004; Zang et al., 2005). Recent studies of fidelity during extension of primer:template assemblies by single subunit and multisubunit RNAPs indicated that RNAPs may tolerate a "flipped out" template base in the substrate-binding site, which lies downstream of the RNA:DNA hybrid cavity (Kashkina et al., 2006; Pomerantz et al., 2006). However, in contrast to DNAPs, RNAPs preferentially make substitution errors rather than deletion errors (which would require the accommodation of a flipped out base in the hybrid-binding cavity) (ibid). It therefore appears that such perturbations in helical structure in the hybrid-binding cavity are not as well tolerated by RNAPs.

The third pathway (Fig. 19.6, scheme 3) involves a mechanism of base sharing, in which bases may become involved in alternate pairing opportunities that would promote shifting of the hybrid. The basis for this proposal comes from the observation that homoduplexes of dA:dT exhibit an extraordinarily high propeller twist that may be stabilized by the formation of cross strand hydrogen bonds involving the adenine N-6 amine of a base on one strand and the thymine O-4 of the succeeding base on the opposite strand (Fig. 19.7A) (Yoon et al., 1988). We speculate that reorientation of the bases in the twisted configuration, either in a stepwise manner (domino effect) or a coordinated manner, could result in realignment of the two strands with minimal energy cost (Fig. 19.7B). This is particularly so if the ends of one of the strands were not constrained by normal pairing in either the upstream or downstream direction (as would be the case in the RNA:DNA hybrid, where the upstream end of the transcript is displaced and no longer involved in base pairing with the template and the downstream end is the 3′ terminus of the transcript). A similar mechanism has been proposed for slippage during replication by DNAP and HIV RT (Timsit, 1999; Hamburgh et al., 2006).

We note that the alternate configuration involving cross strand hydrogen bonds has only been demonstrated for DNA duplexes, and it is not known whether extended homopolymeric rU:dA or rA:dT hybrids would undergo a similar change in conformation (it has been reported that sequence-specific conformation changes of this type are attenuated in A-form vs. B-form DNA duplexes (Timsit, 1999)). We also note that the polarity of strand realignment that would be supported by this process would facilitate forward but not backward slippage of the transcript, as it would extend the 3′ end of the RNA primer toward, rather than away from, the active site. However, as noted above, the RNA:DNA hybrid-binding cavity may stabilize alternate configurations of the duplex, and these might allow a similar mechanism to support backward sliding.

With regard to alternate conformations of the RNA:DNA hybrid, structural analysis of complexes of HIV RT with a polypurine tract (Sarafianos et al., 2001) demonstrates unzipping and slippage of the RNA:DNA hybrid with a polarity that would be consistent with backward slippage of the transcript in RNAP (Fig. 19.8). The hybrid in polypurine tracts of HIV RT complexes is insensitive to the intrinsic RNaseH activity of RT, allowing the protected RNA to later serve as a primer for DNA synthesis. Structural analysis of this region in HIV RT complexes reveals a noncanonical form of the hybrid that involves both slipped and mismatched bases. This alternate structure is not observed in an unbound nucleic acid complex, but is stabilized or induced in the RT complex (reminiscent of the observation that RNAP stabilizes an otherwise unstable rU:dA hybrid).

In addition to facilitating slippage, the ability of the RNAP to induce or stabilize the formation of alternate structures in the hybrid might serve as the basis for regulatory signals. For example, T7 RNAP and pol II each recognize a similar sequence-specific pause/arrest signal that does not appear to involve any secondary structure in the RNA (Hawryluk et al., 2004; He et al., 1998). One possibility is that an alternate RNA:DNA hybrid structure formed at this sequence is recognized by the RNAP as part of the pausing/termination signal. Such signals involving noncanonical hybrid structures may be important in a variety of regulatory events involving pausing and termination.

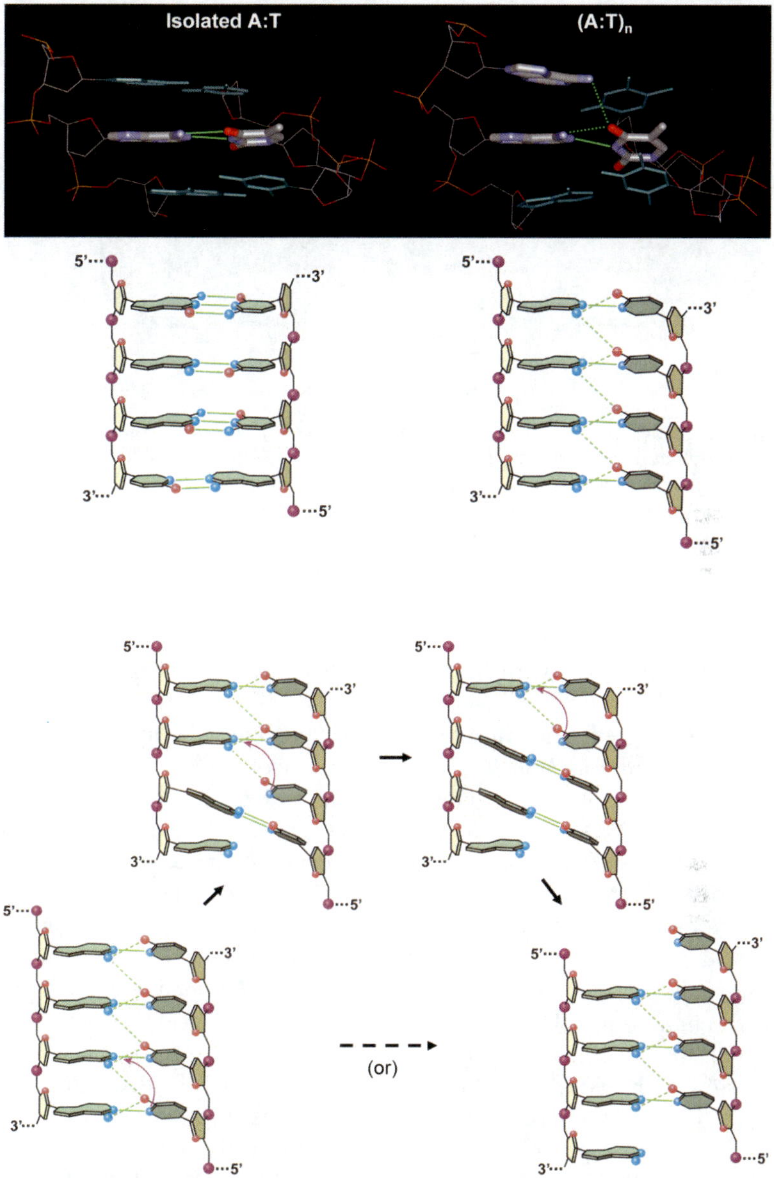

Fig. 19.7 Cross strand hydrogen bonds may facilitate transcript slippage by base sharing. Panel (**A**) Structural data indicate that A:T base pairs in homopolymeric A_n:T_n tracts exhibit a high propeller twist that may be stabilized by cross strand hydrogen bonds between the adenine N-6 amine of a base pair on one strand and the thymine O-4 of the succeeding base pair on the opposite strand. This occurs only on two or more successive adenines (Yoon et al., 1988). Panel (**B**) Reorientation of the bases in the twisted configuration, either in a stepwise manner (domino effect) or a coordinated manner could result in realignment of the two strands with minimal energy cost

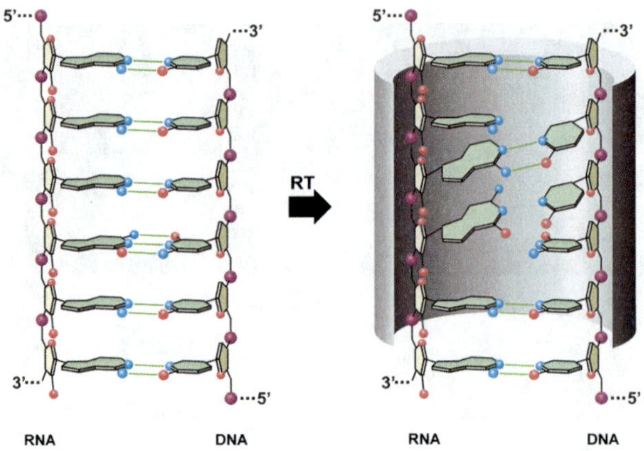

Fig. 19.8 Unzipping and slippage of the RNA:DNA hybrid in the polypurine tract of an HIV–RT complex. Structural analysis of HIV–RT complexes in association with the "polypurine tract" reveals a noncanonical form of the RNA:DNA hybrid that involves both slipped and mismatched bases (*right panel*, pdb 1HYS) (Sarafianos et al., 2001). This alternate hybrid structure is not observed in an unbound nucleic acid complex (pdb 1G4Q, (Kopka et al., 2003)) but is stabilized in the RT complex

19.9 Transcript Slippage During Termination

There is increasing evidence that transcript slippage may be involved in the process of termination (Macdonald et al., 1993; Larson et al., 2008; Toulokhonov and Landick, 2003). Intrinsic termination signals utilized by bacterial multisubunit RNAPs encode an RNA that can fold into a stable G:C-rich stem-loop structure followed by a U-rich region immediately downstream. The observation that the single subunit T7-like RNAPs also terminate at such signals suggests a common mechanism of termination that involves thermodynamic and structural features of the nucleic acid components, rather than the structure of the RNAPs. Recent single molecule studies with *E coli* RNAP indicate that at termination signals in which there is an uninterrupted U tract downstream from the stem-loop, termination may involve an RNA shearing mechanism in which formation of the stem-loop results in steric clash with the exit pore of the RNAP, causing shearing (slippage) of the transcript in the RNA: DNA hybrid and inactivation of the complex (Larson et al., 2008). We have performed experiments on stem-loop type terminators with T7 RNAP and have found that a poly(A) signal placed downstream of the stem-loop also results in termination, demonstrating that slippery sequences other than poly(U) may function in the termination process (Molodtsov et al., submitted).

Termination by the eukaryotic RNAPs (pol I, II, and III) is less well understood, but in all cases appears to involve an A:U-rich sequence or poly(A) or poly(U) tracts. While this may reflect the inherent instability of RNA:DNA hybrids, it is possible that transcript slippage, or the formation and recognition of noncanonical hybrids, may be involved. These effects may be modulated by secondary structure in the nascent transcript or the template or by binding of proteins or chromatin structure in the vicinity of the signal (as noted above). Transcript slippage has been specifically invoked in the case of termination by yeast pol I, which involves a U-rich element upstream of a binding site for the termination factor Reb1p (Reeder and Lang, 1997). When this region is uninterrupted (e.g., a U_9 tract) pol I engages in reiterative slippage in the presence of Reb1p, resulting in an extended pause but little termination. When there is an interruption in the U-run, slippage results in a mismatch in the hybrid, failure to extend the transcript, and termination, suggesting that backward sliding of the transcript and the active site may be an initial step in termination.

19.10 Concluding Remarks

In the end, evolutionary considerations leave us with a question, and possibly a clue. If slippage by RNAPs is deleterious, there would appear to be two ways to circumvent this problem. The first is to select against the occurrence of slippery tracts within coding regions, which is what many organisms seem to have done (Baranov et al., 2005). However, this would seem to be an inefficient strategy and would limit genomic flexibility during evolution. The other, and more direct strategy, would be to evolve RNAPs that are less prone to slip. The observation that the yeast mitochondrial RNAP does not slip as well as the related T7 RNAP (see above) indicates that such RNAP variants can exist.

We are therefore left with the conclusion that the continued presence in cells of RNAPs that can slip confers a positive advantage. While transcript slippage (or stuttering) may be useful for regulatory events that are known to occur during initiation, other means to regulate transcript initiation are available, Similarly, although slippage appears to play a role in termination at certain signals, other means to terminate are possible. The interesting possibility remains that there are other phenomena that occur during gene expression that rely upon slippage and its modulation. For example, the use of alternate reading frames as a result of slippage may be modulated under different conditions. The presence of blocking proteins or chromatin, or of alternate structures (sequences) in the DNA may render some tracts more or less prone to slippage. While only extended runs of A's or U's have thus far been implicated in efficient slippage, there have been reports that slippage may occur in simple direct repeats as well. Further experiments will be required to examine possible slippery sequences in vivo and in vitro under various conditions.

Importantly, as noted above, the mechanism of transcript slippage may depend upon the ability of the RNAP to accommodate noncanonical conformations of the RNA:DNA hybrid. This plasticity may be required for the recognition of other signals that, while not directly concerned with slippage, involve alternate hybrid structures. The advantage to the cell to maintain an RNAP that is capable of such transitions, even though it would allow transcript slippage, may explain the continued presence of slippery RNAPs during evolution. There also exists the possibility that novel transcription factors enhance the ability of the RNAP to tolerate or respond to such alternate hybrids.

In conclusion, the phenomenon of transcript slippage is likely to be far more significant than previously appreciated and may underlie a number of human diseases or mechanisms of cellular dysfunction.

Acknowledgments These studies were supported by grants from the National Institutes of Health (GM38147) and from the Foundation of UMDNJ to WTM. We are grateful to Chuck Turnbough, Don Luse, Sergei Borukhov, Dimitriy Markov, Steven Emanuel, and Maria Savkina for helpful comments, and to Mr. Raymond Castagna for technical support. We thank Craig Martin for pointing out to us the special properties of A_n:T_n homoduplexes that might provide a basis for transcript slippage, and Irina Artsimovitch and Evgeny Nudler for the gift of $EcoRI^{Q111A}$.

References

Alm RA, Ling LS, Moir DT, King BL, Brown ED, Doig PC, Smith DR, Noonan B, Guild BC, deJonge BL, Carmel G, Tummino PJ, Caruso A, Uria-Nickelsen M, Mills DM, Ives C, Gibson R, Merberg D, Mills SD, Jiang Q, Taylor DE, Vovis GF, Trust TJ (1999) Genomic-sequence comparison of two unrelated isolates of the human gastric pathogen Helicobacter pylori. Nature 397:176–180

Ba Y, Tonoki H, Tada M, Nakata D, Hamada J, Moriuchi T (2000) Transcriptional slippage of p53 gene enhanced by cellular damage in rat liver: monitoring the slippage by a yeast functional assay. Mutat Res 447:209–220

Bar-Nahum G, Epshtein V, Ruckenstein AE, Rafikov R, Mustaev A, Nudler E (2005) A ratchet mechanism of transcription elongation and its control. Cell 120:183–193

Baranov PV, Hammer AW, Zhou J, Gesteland RF, Atkins JF (2005) Transcriptional slippage in bacteria: distribution in sequenced genomes and utilization in IS element gene expression. Genome Biol 6:R25

Benson KF, Person RE, Li FQ, Williams K, Horwitz M (2004) Paradoxical homozygous expression from heterozygotes and heterozygous expression from homozygotes as a consequence of transcriptional infidelity through a polyadenine tract in the AP3B1 gene responsible for canine cyclic neutropenia. Nucleic Acids Res 32:6327–6333

Borukhov S, Sagitov V, Josaitis CA, Gourse RL, Goldfarb A (1993) Two modes of transcription initiation in vitro at the rrnB P1 promoter of *Escherichia coli*. J Biol Chem 268: 23477–23482

Burch CL, Danaher RJ, Stein DC (1997) Antigenic variation in Neisseria gonorrhoeae: production of multiple lipooligosaccharides. J Bacteriol 179:982–986

Chamberlin M, Berg P (1962) Deoxyribonucleic acid-directed synthesis of ribonucleic acid by an enzyme from *Escherichia coli*. Proc Natl Acad Sci USA 48:81–94

de Pril R, Fischer DF, van Leeuwen FW (2006) Conformational diseases: an umbrella for various neurological disorders with an impaired ubiquitin-proteasome system. Neurobiol Aging 27:515–523

Deng L, Shuman S (1997) Elongation properties of vaccinia virus RNA polymerase: pausing, slippage, 3' end addition, and termination site choice. Biochemistry 36:15892–15899

Fabre E, Dujon B, Richard GF (2002) Transcription and nuclear transport of CAG/CTG trinucleotide repeats in yeast. Nucleic Acids Res 30:3540–3547

Garcia-Diaz M, Bebenek K, Krahn JM, Pedersen LC, Kunkel TA (2006) Structural analysis of strand misalignment during DNA synthesis by a human DNA polymerase. Cell 124:331–342

Gerez L, de HA, Hol EM, Fischer DF, van Leeuwen FW, van SH, Benne R (2005) Molecular misreading: the frequency of dinucleotide deletions in neuronal mRNAs for beta-amyloid precursor protein and ubiquitin B. Neurobiol Aging 26:145–155

Guajardo R, Gopal V, Lopez P, Sousa R (1998) NTP concentration effects on initial transcription by T7 RNAP indicate that translocation occurs through passive sliding and reveal that divergent promoters have distinct NTP concentration requirements for productive initiation. J Mol Biol 281:777–792

Guo HC, Roberts JW (1990) Heterogeneous initiation due to slippage at the bacteriophage 82 late gene promoter in vitro. Biochemistry 29:10702–10709

Hamburgh ME, Curr KA, Monaghan M, Rao VR, Tripathi S, Preston BD, Sarafianos S, Arnold E, Darden T, Prasad VR (2006) Structural determinants of slippage-mediated mutations by human immunodeficiency virus type 1 reverse transcriptase. J Biol Chem 281:7421–7428

Harley CB, Lawrie J, Boyer HW, Hedgpeth J (1990) Reiterative copying by *E. coli* RNA polymerase during transcription initiation of mutant pBR322 tet promoters. Nucleic Acids Res 18:547–552

Hawryluk PJ, Ujvari A, Luse DS (2004) Characterization of a novel RNA polymerase II arrest site which lacks a weak 3' RNA-DNA hybrid. Nucleic Acids Res 32:1904–1916

He B, Kukarin A, Temiakov D, Chin-Bow ST, Lyakhov DL, Rong M, Durbin RK, McAllister WT (1998) Characterization of an unusual, sequence-specific termination signal for T7 RNA polymerase. J Biol Chem 273:18802–18811

Imburgio D, Anikin M, McAllister WT (2002) Effects of substitutions in a conserved DX(2)GR sequence motif, found in many DNA-dependent nucleotide polymerases, on transcription by T7 RNA polymerase. J Mol Biol 319:37–51

Imburgio D, Rong M, Ma K, McAllister WT (2000) Studies of promoter recognition and start site selection by T7 RNA polymerase using a comprehensive collection of promoter variants. Biochemistry 39:10419–10430

Isken O, Maquat LE (2007) Quality control of eukaryotic mRNA: safeguarding cells from abnormal mRNA function. Genes Dev 21:1833–1856

Jacques JP, Kolakofsky D (1991) Pseudo-templated transcription in prokaryotic and eukaryotic organisms. Genes Dev 5:707–713

Jacques JP, Susskind MM (1990) Pseudo-templated transcription by *Escherichia coli* RNA polymerase at a mutant promoter. Genes Dev 4:1801–1810

Jin DJ (1996) A mutant RNA polymerase reveals a kinetic mechanisms for the switch between nonproductive stuttering synthesis and productive initiation during promoter clearance. J Biol Chem 271:11659–11667

Johnson SJ, Beese LS (2004) Structures of mismatch replication errors observed in a DNA polymerase. Cell 116:803–816

Kashkina E, Anikin M, Brueckner F, Pomerantz RT, McAllister WT, Cramer P, Temiakov D (2006) Template misalignment in multisubunit RNA polymerases and transcription fidelity. Mol Cell 24:257–266

Kolakofsky D, Roux L, Garcin D, Ruigrok RW (2005) Paramyxovirus mRNA editing, the "rule of six" and error catastrophe: a hypothesis. J Gen Virol 86:1869–1877

Komissarova N, Kashlev M (1997) RNA polymerase switches between inactivated and activated states By translocating back and forth along the DNA and the RNA. J Biol Chem 272: 15329–15338

Kopka ML, Lavelle L, Han GW, Ng HL, Dickerson RE (2003) An unusual sugar conformation in the structure of an RNA/DNA decamer of the polypurine tract may affect recognition by RNase H. J Mol Biol 334:653–665

Kuroda M, Ohta T, Uchiyama I, Baba T, Yuzawa H, Kobayashi I, Cui L, Oguchi A, Aoki K, Nagai Y, Lian J, Ito T, Kanamori M, Matsumaru H, Maruyama A, Murakami H, Hosoyama A, Mizutani-Ui Y, Takahashi NK, Sawano T, Inoue R, Kaito C, Sekimizu K, Hirakawa H, Kuhara S, Goto S, Yabuzaki J, Kanehisa M, Yamashita A, Oshima K, Furuya K, Yoshino C, Shiba T, Hattori M, Ogasawara N, Hayashi H, Hiramatsu K (2001) Whole genome sequencing of meticillin-resistant Staphylococcus aureus. Lancet 357:1225–1240

Laken SJ, Petersen GM, Gruber SB, Oddoux C, Ostrer H, Giardiello FM, Hamilton SR, Hampel H, Markowitz A, Klimstra D, Jhanwar S, Winawer S, Offit K, Luce MC, Kinzler KW, Vogelstein

B (1997) Familial colorectal cancer in Ashkenazim due to a hypermutable tract in APC. Nat Genet 17:79–83

Landick R (2004) Active-site dynamics in RNA polymerases. Cell 116:351–353

Landick R (2001) RNA Polymerase Clamps Down. Cell 105:567–570

Larsen B, Wills NM, Nelson C, Atkins JF, Gesteland RF (2000) Nonlinearity in genetic decoding: homologous DNA replicase genes use alternatives of transcriptional slippage or translational frameshifting. Proc Natl Acad Sci USA 97:1683–1688

Larson MH, Greenleaf WJ, Landick R, Block SM (2008) Applied force reveals mechanistic and energetic details of transcription termination. Cell 132:971–982

Lee LY, Miyamoto YJ, McIntyre BW, Hook M, McCrea KW, McDevitt D, Brown EL (2002) The Staphylococcus aureus Map protein is an immunomodulator that interferes with T cell-mediated responses. J Clin Invest 110:1461–1471

Ling H, Boudsocq F, Woodgate R, Yang W (2001) Crystal structure of a Y-family DNA polymerase in action: a mechanism for error-prone and lesion-bypass replication. Cell 107:91–102

Linton MF, Pierotti V, Young SG (1992) Reading-frame restoration with an apolipoprotein B gene frameshift mutation. Proc Natl Acad Sci USA 89:11431–11435

Linton MF, Raabe M, Pierotti V, Young SG (1997) Reading-frame restoration by transcriptional slippage at long stretches of adenine residues in mammalian cells. J Biol Chem 272: 14127–14132

Liu C, Heath LS, Turnbough CL Jr. (1994) Regulation of pyrBI operon expression in *Escherichia coli* by UTP-sensitive reiterative RNA synthesis during transcriptional initiation. Genes Dev 8:2904–2912

Macdonald LE, Zhou Y, McAllister WT (1993) Termination and slippage by bacteriophage T7 RNA polymerase. J Mol Biol 232:1030–1047

Martin CT, Muller DK, Coleman JE (1988) Processivity in early stages of transcription by T7 RNA polymerase. Biochemistry 27:3966–3974

Meng Q, Turnbough CL Jr, Switzer RL (2004) Attenuation control of pyrG expression in Bacillus subtilis is mediated by CTP-sensitive reiterative transcription. Proc Natl Acad Sci U S A 101:10943–10948

Newton WA, Beckwith JR, Zipser D, Brenner S (1965) Nonsense mutants and polarity in the lac operon of *Escherichia coli*. J Mol Biol 14:290–296

Nudler E, Mustaev A, Lukhtanov E, Goldfarb A (1997) The RNA-DNA hybrid maintains the register of transcription by preventing backtracking of RNA polymerase. Cell 89:33–41

Pal M, Luse DS (2002) Strong natural pausing by RNA polymerase II within 10 bases of transcription start may result in repeated slippage and reextension of the nascent RNA. Mol Cell Biol 22:30–40

Pal M, Luse DS (2003) The initiation-elongation transition: lateral mobility of RNA in RNA polymerase II complexes is greatly reduced at +8/+9 and absent by +23. Proc Natl Acad Sci USA 100:5700–5705

Parker RC (1986) Synthesis of in vitro ColE1 transcripts with 5'-terminal ribonucleotides that exhibit noncomplementarity with the DNA template. Biochemistry 25:6593–6598

Penno C, Hachani A, Biskri L, Sansonetti P, Allaoui A, Parsot C (2006) Transcriptional slippage controls production of type III secretion apparatus components in Shigella flexneri. Mol Microbiol 62:1460–1468

Penno C, Parsot C (2006) Transcriptional slippage in mxiE controls transcription and translation of the downstream mxiD gene, which encodes a component of the Shigella flexneri type III secretion apparatus. J Bacteriol 188:1196–1198

Penno C, Sansonetti P, Parsot C (2005) Frameshifting by transcriptional slippage is involved in production of MxiE, the transcription activator regulated by the activity of the type III secretion apparatus in Shigella flexneri. Mol Microbiol 56:204–214

Pomerantz RT, Temiakov D, Anikin M, Vassylyev DG, McAllister WT (2006) A mechanism of nucleotide misincorporation during transcription due to template-strand misalignment. Mol Cell 24:245–255

Raabe M, Linton MF, Young SG (1998) Long runs of adenines and human mutations. Am J Med Genet 76:101–102

Ratinier M, Boulant S, Combet C, Targett-Adams P, McLauchlan J, Lavergne JP (2008) Transcriptional slippage prompts recoding in alternate reading frames in the hepatitis C virus (HCV) core sequence from strain HCV-1. J Gen Virol 89:1569–1578

Reeder RH, Lang WH (1997) Terminating transcription in eukaryotes: lessons learned from RNA polymerase I. Trends Biochem Sci 22:473–477

Sanchez A, Trappier SG, Mahy BW, Peters CJ, Nichol ST (1996) The virion glycoproteins of Ebola viruses are encoded in two reading frames and are expressed through transcriptional editing. Proc Natl Acad Sci USA 93:3602–3607

Sarafianos SG, Das K, Tantillo C, Clark AD, Jr., Ding J, Whitcomb JM, Boyer PL, Hughes SH, Arnold E (2001) Crystal structure of HIV-1 reverse transcriptase in complex with a polypurine tract RNA:DNA. EMBO J 20:1449–1461

Severinov K, Goldfarb A (1994) Topology of the product binding site in RNA polymerase revealed by transcript slippage at the phage lambda PL promoter. J Biol Chem 269:31701–31705

Sidorenkov I, Komissarova N, Kashlev M (1998) Crucial role of the RNA:DNA hybrid in the processivity of transcription. Mol Cell 2:55–64

Sousa R (2005) Machinations of a Maxwellian demon. Cell 120:155–156

Sugimoto N, Nakano S, Katoh M, Matsumura A, Nakamuta H, Ohmichi T, Yoneyama M, Sasaki M (1995) Thermodynamic parameters to predict stability of RNA/DNA hybrid duplexes. Biochemistry 34:11211–11216

Tamas I, Wernegreen JJ, Nystedt B, Kauppinen SN, Darby AC, Gomez-Valero L, Lundin D, Poole AM, Andersson SG (2008) Endosymbiont gene functions impaired and rescued by polymerase infidelity at poly(A) tracts. Proc Natl Acad Sci USA 105:14934–14939

Timsit Y (1999) DNA structure and polymerase fidelity. J Mol Biol 293:835–853

Tippin B, Kobayashi S, Bertram JG, Goodman MF (2004) To slip or skip, visualizing frameshift mutation dynamics for error-prone DNA polymerases. J Biol Chem 279:45360–45368

Tomb JF, White O, Kerlavage AR, Clayton RA, Sutton GG, Fleischmann RD, Ketchum KA, Klenk HP, Gill S, Dougherty BA, Nelson K, Quackenbush J, Zhou L, Kirkness EF, Peterson S, Loftus B, Richardson D, Dodson R, Khalak HG, Glodek A, McKenney K, Fitzegerald LM, Lee N, Adams MD, Hickey EK, Berg DE, Gocayne JD, Utterback TR, Peterson JD, Kelley JM, Cotton MD, Weidman JM, Fujii C, Bowman C, Watthey L, Wallin E, Hayes WS, Borodovsky M, Karp PD, Smith HO, Fraser CM, Venter JC (1997) The complete genome sequence of the gastric pathogen Helicobacter pylori. Nature 388:539–547

Toulokhonov I, Landick R (2003) The flap domain is required for pause RNA hairpin inhibition of catalysis by RNA polymerase and can modulate intrinsic termination. Mol Cell 12: 1125–1136

Turnbough CL Jr, Switzer RL (2008) Regulation of pyrimidine biosynthetic gene expression in bacteria: repression without repressors. Microbiol Mol Biol Rev 72:266–300

van den Hurk WH, Willems HJ, Bloemen M, Martens GJ (2001) Novel frameshift mutations near short simple repeats. J Biol Chem 276:11496–11498

van Leeuwen FW, Fischer DF, Kamel D, Sluijs JA, Sonnemans MA, Benne R, Swaab DF, Salehi A, Hol EM (2000) Molecular misreading: a new type of transcript mutation expressed during aging. Neurobiol Aging 21:879–891

van Leeuwen FW, Kros JM, Kamphorst W, van SC, de Vos RA (2006) Molecular misreading: the occurrence of frameshift proteins in different diseases. Biochem Soc Trans 34: 738–742

Volchkov VE, Becker S, Volchkova VA, Ternovoj VA, Kotov AN, Netesov SV, Klenk HD (1995) GP mRNA of Ebola virus is edited by the Ebola virus polymerase and by T7 and vaccinia virus polymerases. Virology 214:421–430

Volchkov VE, Volchkova VA, Muhlberger E, Kolesnikova LV, Weik M, Dolnik O, Klenk HD (2001) Recovery of infectious Ebola virus from complementary DNA: RNA editing of the GP gene and viral cytotoxicity. Science 291:1965–1969

Wagner LA, Weiss RB, Driscoll R, Dunn DS, Gesteland RF (1990) Transcriptional slippage occurs during elongation at runs of adenine or thymine in *Escherichia coli*. Nucleic Acids Res 18:3529–3535

Wills NM, Atkins JF (2006) The potential role of ribosomal frameshifting in generating aberrant proteins implicated in neurodegenerative diseases. RNA 12:1149–1153

Xiong XF, Reznikoff WS (1993) Transcriptional slippage during the transcription initiation process at a mutant lac promoter in vivo. J Mol Biol 231:569–580

Yin H, Wang MD, Svoboda K, Landick R, Block S, Gelles J (1995) Transcription against an applied force. Science 270:1653–1657

Yoon C, Prive GG, Goodsell DS, Dickerson RE (1988) Structure of an alternating-B DNA helix and its relationship to A-tract DNA. Proc Natl Acad Sci USA 85:6332–6336

Young M, Inaba H, Hoyer LW, Higuchi M, Kazazian HH Jr, Antonarakis SE (1997) Partial correction of a severe molecular defect in hemophilia A, because of errors during expression of the factor VIII gene. Am J Hum Genet 60:565–573

Zang H, Goodenough AK, Choi JY, Irimia A, Loukachevitch LV, Kozekov ID, Angel KC, Rizzo CJ, Egli M, Guengerich FP (2005) DNA adduct bypass polymerization by Sulfolobus solfataricus DNA polymerase Dpo4: analysis and crystal structures of multiple base pair substitution and frameshift products with the adduct 1,N2-ethenoguanine. J Biol Chem 280:29750–29764

Part V
Appendix

Chapter 20
Computational Resources for Studying Recoding

Andrew E. Firth, Michaël Bekaert, and Pavel V. Baranov

Abstract The rapid growth in the quantity of available sequence data has made necessary the development of efficient computational tools for its analysis. Substantial progress has been made in the development of tools for the identification and prediction of genes that are expressed via standard decoding. However, since recoded genes embrace only a minority of all genes and since their prediction requires different approaches, they are frequently neglected and as a result are often mis-annotated in the public databases or even left undetected during the annotation process. This chapter aims to describe available computer tools designed for the identification and analysis of recoded genes and public databases that collect information related to recoding. In addition, we also discuss how standard tools for sequence analysis can be used for these purposes.

Contents

P.V. Baranov (✉)
Biochemistry Department, University College Cork, Ireland
e-mail: brave.oval.pan@gmail.com

J.F. Atkins, R.F. Gesteland (eds.), *Recoding: Expansion of Decoding Rules Enriches Gene Expression*, Nucleic Acids and Molecular Biology 24,
DOI 10.1007/978-0-387-89382-2_20,

20.1 Recoding in the Genomic Era

The worldwide efforts in the pursuit to decipher the human genome have produced a number of useful acquisitions, including the development of cost-efficient technologies for high-throughput nucleic acid sequencing and extensive computational techniques for the analysis of sequence data. The development of these techniques has resulted in an explosion of sequence information, illustrated by an almost exponential growth in the number of sequences stored in GenBank (Fig. 20.1). The near universality of the genetic code and the rules of standard decoding have allowed the development of sophisticated and efficient computational algorithms for the identification of protein coding sequences, and annotation of the corresponding genes, in newly sequenced genomes. While many scientists involved in the computational analysis of nucleic acid sequences are enjoying the prosperous bonanza brought about by these developments, those involved in computational identification and

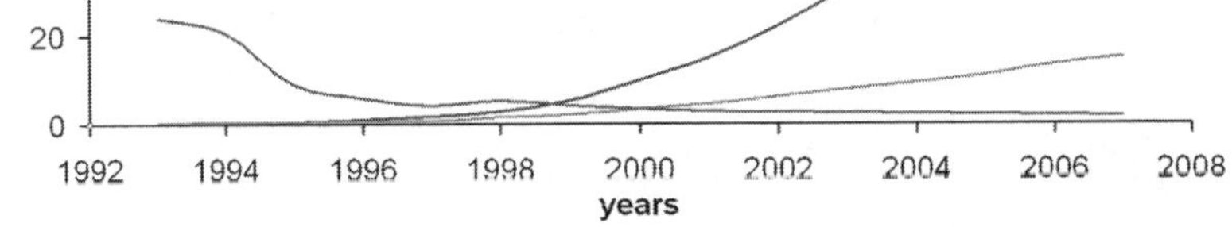

Fig. 20.1 Relative growth of recoding events annotated in GenBank. The *blue curve* represents the number of sequences in GenBank at the end of each year. The data were taken from the GenBank release 165.0 notes (15 April 2008, available at ftp://ftp.ncbi.nih.gov/genbank/gbrel.txt). The *green curve* indicates the number of sequences whose descriptions contain at least one of the following keywords: 'programmed frameshifting', 'translational frameshifting', 'programmed frameshift', 'translational frameshift', 'ribosomal slippage', 'ribosome slippage', 'ribosomal frameshift', 'ribosomal frameshifting'. It should be noted that, due to inconsistency in terminology, it is possible that some instances of programmed ribosomal frameshifting are described in GenBank without any of the above keywords. Hence, these data represent a tendency, rather than the absolute number of annotated instances of programmed frameshifting. The *red curve* indicates the relative proportion of sequences with programmed frameshifting among all sequences in GenBank

prediction of 'recoded' genes, perhaps, will not strongly object to a comparison of their activity to a placer gold mining technique employed by ancient natives living near the Caucus shore of the Black Sea, who were using sheep fleeces to collect golden flecks from sediments deposited in mountain streams (Strabo et al. 1854).

The rate of discovery of 'recoded' genes does not keep up with the pace of nucleic acid sequencing. Figure 20.1 illustrates the growth in the number of sequences in GenBank whose descriptions contain one or more keywords associated with a particular type of recoding, programmed ribosomal frameshifting. While the growth in the number of sequences in GenBank is roughly exponential, the increase in annotated recoding events is better approximated by a linear trend. As a result, the relative proportion of known genes that use recoding is decreasing each year. It seems counterintuitive that recoding events – always considered as rare – were relatively *more* abundant a decade ago, when researchers were not armed with effective tools for high-throughput genomics and proteomics research. The progressively increasing comparative scarcity of recoding events is obviously illusory and is indicative of a lack of progress in annotating homologs of known recoded genes, and in finding novel recoded genes. One reason for this is that it is simply much harder to identify recoded genes than canonically decoded genes. This illusory shortage of recoded genes can create difficulties in securing funding to study them. Meanwhile, the interest of mainstream researchers is also diminishing, with the view that there is little urgency in studying rare events. The consequence of this state of affairs is an over-simplified perception of the global picture of gene expression.

Certain progress is taking place due to the efforts of recoding enthusiasts, so that a number of tools for finding recoded genes have been, or are being, developed. A challenge for the future is to incorporate such tools into the mainstream genome annotation pipelines. The aim of this chapter is to give a description of existing computational resources primarily dedicated to studying recoding and also to describe more general computational tools that can be used for such purposes.

20.2 Databases of Recoding Events

A few databases have been created within the last decade in which genes that utilise recoding have been collected and annotated. They differ in a number of ways, such as the type of recoding events, methods for their identification and the ways in which these sequences are described (see Table 20.1, where these features are summarised for each database).

20.2.1 Recode Database

Until recently, most of the genes that were known to use recoding were found serendipitously, rather than as a result of systematic investigation. Such an unsystematic origin of recoded genes, in combination with their sequence and functional diversity, resulted in equally sundry descriptions of recoding events and redundant terminology. This situation is reflected in the annotations of recoded genes in major

Table 20.1 Comparison of public databases containing information on genes that use recoding during their expression

Database	References	Type of sequence	Number of genes	Major supporting evidence
Recode recode.genetics.utah.edu	Baranov et al. (2001, 2003)	All recoding events	516	Published literature, manual curation
Recode-2 recode.ucc.ie	Bekaert et al. (2010)	All recoding events	1292	Published literature, computational prediction, semi-manual curation
FSDB wilab.inha.ac.kr/fsdb	Moon et al. (2007)	Programmed ribosomal frameshifting	253	Extracted from other databases, computational prediction, manual curation
PRFDB cbmgintra.umd.edu/prfdb/	Jacobs et al. (2007; Belew et al. (2008)	−1 programmed ribosomal frameshifting comprising a heptamer slippery site and downstream RNA secondary structure	Over 4000	Computational prediction, no manual curation, putative candidates
SelenoDB www.selenodb.org	Castellano et al. (2008)	Eukaryotic selenoproteins and genes involved in selenocysteine incorporation or biosynthesis	81	Computational prediction, published literature, manual curation
ISfinder www-is.biotoul.fr	Siguier et al. (2006)	IS elements, including those that use frameshifting	NA	Computational prediction, manual curation

sequence databases, where numerous terminological descriptions denote very similar events. The level of detail at which these events are described also fluctuates dramatically. To overcome this problem, a group of scientists, many of whom are authors of this book, have endeavoured to bring the knowledge of recoding events under a common umbrella. This idea inspired the establishment of the first sequence database dedicated to recoding, the Recode database, which is currently located at http://recode.genetics.utah.edu (Baranov et al. 2001, 2003).

The database consists of sequences of genes that use recoding, annotated manually based on published literature. The database is designed to be used primarily by human beings. Annotations are embedded into html code to differentially highlight certain sequence elements involved in the stimulation of recoding events, according to a universal annotation scheme which is embedded into the Recode logo (Fig. 20.2). Each entry in the database is linked to PubMed abstracts of published papers used to derive information related to a particular entry. The

Detailed Information

back to summary : 1
Gene: oaz1
Type: +1 frameshifting
Kingdom: eukaryota
Latin: Homo Sapiens
Common: human, man
Accession: D78361.1
Cis-elements: UGA stop signal,3' RNA pseudoknot (kissing loops),5' other
Trans-factor: polyamines
Function of Recoding: Efficiency of ODC antizyme biosynthesis is modulated by polyamine levels; sensor for an autoregulatory circuit
Product: ODC antizyme:1. Inhibition and targeting ODC for degradation; 2.Regulation of polyamine transport
Evidence: Experimental
References: Tewari,D.S., Qian,Y., Thornton,R.D., Pieringer,J., Taub,R.,Mochan,E. and Tewari,M.Biochim. Biophys. Acta 1209 (2), 293-295 (1994)
Yang,D., Takii,T., Hayashi,H., Itoh,S., Hayashi,M. and Onozaki,K.Biochem. Mol. Biol. Int. 38 (5), 957-964 (1996)
Ivanov IP, Matsufuji S, Murakami Y, Gesteland RF, Atkins JF. : EMBO J 2000 Apr 17;19(8):1907-17
Comments: not-well characterized upstream stimulator

>oaz1_p_euk_hsap

```
   1 CGAAGCGGAG TGAGAAGGAC GGTCCCGTTC GTCACGAACG TCCGTCCCTC
  51 GTCCGTCGCG GCTCGGACCC GGTCCGGGAG AGCAGTGGCG GTGGTGCGGA
 101 GAGACGCAGC GGAGGTTTTC CTGGTTTCGG ACCCCAGCGG CCGGATGGTG
 151 AAATCCTCCC TGCAGCGGAT CCTCAATAGC CACTGCTTCG CCAGAGAGAA
 201 GGAAGGGGAT AAACCCAGCG CCACCATCCA CGCCAGCCGC ACCATGCCGC
 251 TCCTAAGCCT GCACAGCCGC GGCGGCAGCA GCAGTGAGAG TTCCAGGGTC
 301 TCCCTCCACT GCTGTAGTAA CCCGGGTCCG GGGCCTCGGT GGTGCTCCTG
 351 ATGCCCCTCA CCCACCCCTG AAGATCCCAG GTGGGCGAGG GAATAGTCAG
 401 AGGGATCACA ATCTTTCAGC TAACTTATTC TACTCCGATG ATCGGCTGAA
 451 TGTAACAGAG GAACTAACGT CCAACGACAA GACGAGGATT CTCAACGTCC
 501 AGTCCAGGCT CACAGACGCC AAACGCATTA ACTGGCGAAC AGTGCTGAGT
 551 GGCGGCAGCC TCTACATCGA GATCCCGGGC GGCGCGCTGC CCGAGGGGAG
 601 CAAGGACAGC TTTGCAGTTC TCCTGGAGTT CGCTGAGGAG CAGCTGCGAG
 651 CCGACCATGT CTTCATTTGC TTCCACAAGA ACCGCGAGGA CAGAGCCGCC
 701 TTGCTCCGAA CCTTCAGCTT TTTGGGCTTT GAGATTGTGA GACCGGGGCA
 751 TCCCCTTGTC CCCAAGAGAC CCGACGCTTG CTTCATGGCC TACACGTTCG
 801 AGAGAGAGTC TTCGGGAGAG GAGGAGGAGT AGGGCCGCCT CGGGGCTGGG
 851 CATCCGGCCC CTGGGGCCAC CCCTTGTCAG CCGGGTGGGT AGGAACCGTA
 901 GACTCGCTCA TCTCGCCTGG GTTTGTCCGC ATGTTGTAAT CGTGCAAATA
 951 AACGCTCACT CCGAATTAGC GGTGTATTTC TTGAAGTTTA ATATTGTGTT
1001 TGTGATACTG AAGTATTTGC TTTAATTCTA AATAAAAATT TATATTTTAC
1051 AATTTTATTC CGG
```

Fig. 20.2 Logo of the Recode database and a typical entry (human oaz1 gene). This figure shows a typical entry in the Recode database and how the Recode logo can be used to decipher annotation of the sequence. The logo illustrates different types of sequences involved in recoding events, coloured in the same manner in the sequence below. In this example, the shift site is underlined, the stop codon is in *red*, and the stimulatory pseudoknot structure is shown with the first stem in *green* and the second stem in *violet*

web interface allows searching of the database using keywords or browsing it based on three main categories: organism/taxon, name of the gene and the type of recoding. Due to the rapid growth of sequence information, manual curation of new entries became extremely laborious and impractical, and so the Recode database has not been updated on a regular basis since 2004. During preparation of this chapter a new version of the Recode database (Recode 2, available at http://recode.ucc.ie) has been developed (Bekaert et al. 2009). Recode 2 integrates existing and to-be-developed tools for finding new cases of recoding. Major sequence repositories have been scanned for new recoding events that were used to populate the database. While the current version of the Recode database does not cover discoveries made since 2004, it still remains a useful resource for people working in the field and is the oldest and least specialised database dedicated to Recoding.

20.2.2 Frameshift Database (FSDB)

The Frameshift database (FSDB) is a database that specialises in a particular type of recoding, programmed ribosomal frameshifting (Moon et al. 2007). It was created in 2006 by researchers in South Korea and is available at http://wilab.inha.ac.kr/fsdb/. A large proportion of the data deposited into FSDB was obtained from the Recode database described above and also from the database of RNA pseudoknots, PseudoBase (van Batenburg et al. 2001). In addition, the authors populated FSDB with certain viral genes using −1 frameshifting and bacterial release factor 2 (RF2) genes using +1 frameshifting. These sequences were identified by the FSFinder program (Moon et al. 2004), with the help of the ARFA program (Bekaert et al. 2006) in the case of RF2 genes. Both programs are described in more detail later in this chapter.

While FSDB preserves most of the features available in the Recode database, it also has a significantly enhanced graphical interface (Fig. 20.3) through the use of the FSFinder output format, thus providing a graphical representation of reading phases, the mutual organisation of open reading frames and the locations of potential slippery sequences and stimulatory signals. A graphical representation of potential stimulatory secondary RNA structures, as generated by the PseudoViewer program (Han et al. 2002; Han and Byun 2003; Byun and Han 2006), is also provided. In certain cases the information on such structures is taken from experimental studies, while in other cases it is based on predictions by the pknotsRG program (Reeder and Giegerich 2004). While this certainly helps with the visualisation of potential stimulatory RNA structures, in certain cases – such as in Fig. 20.3 – an incorrectly predicted structure is given instead of the experimentally demonstrated one (Matsufuji et al. 1995).

Fig. 20.3 A typical FSDB entry (human oaz1 gene). A typical FSDB entry is shown for the same gene as in Fig. 20.2. The *panel at the top* gives a short description of the entry with references to relevant literature and the nucleic acid sequence in GenBank. *Middle-upper*, a plot of ORFs is given with *blue vertical lines* representing stop codons and *red lines* representing start codons. The overlapping pair of ORFs that comprise the antizyme coding sequence are highlighted in *yellow*. *Middle-lower*, a more detailed description of the frameshift cassette is given, with the frameshift site highlighted in *blue* and the stimulatory RNA structure in green; the stimulatory structure is also shown as visualised by PseudoViewer. The *panel at the bottom* gives a description of the model used to find this frameshift cassette

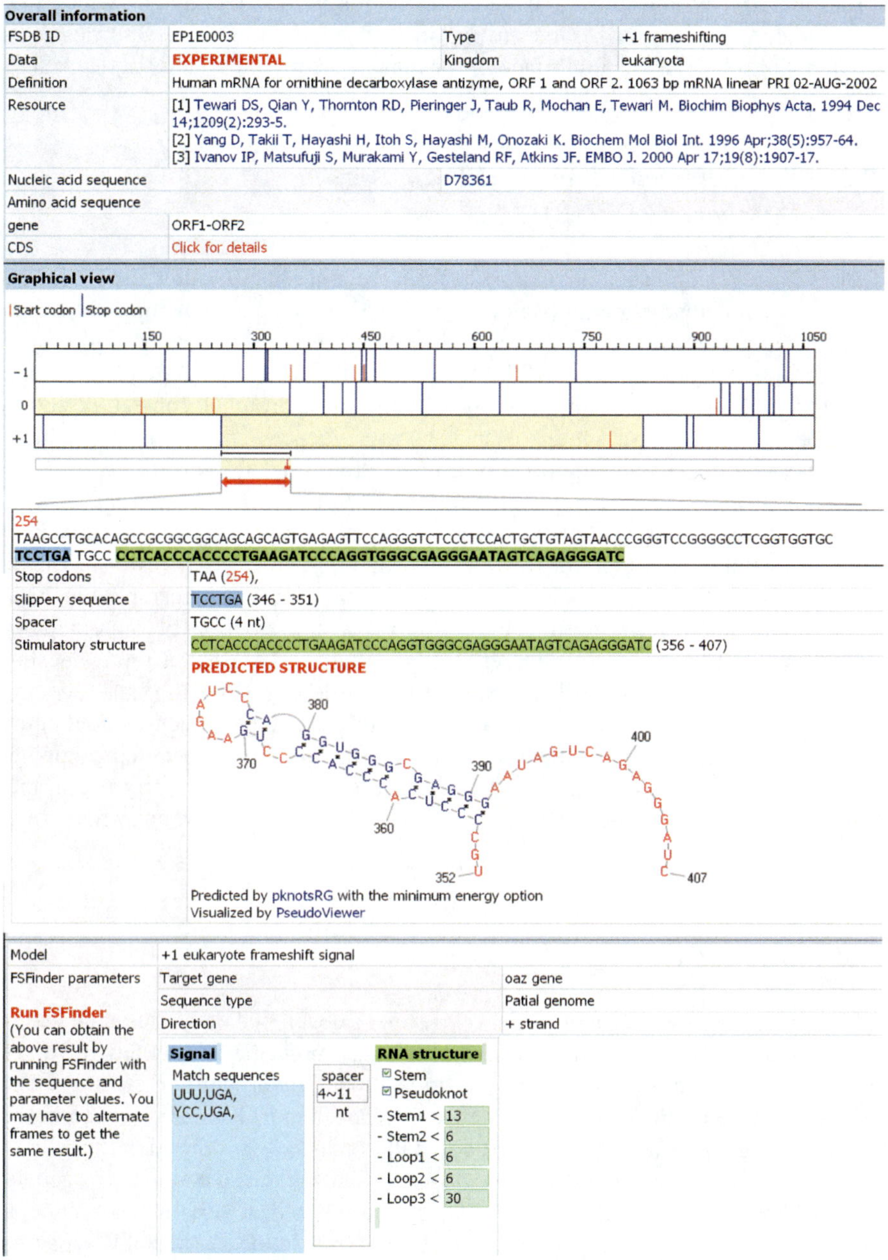

Fig. 20.3 (continued)

In summary, FSDB is a valuable resource for the exploration of genes that use programmed ribosomal frameshifting in their expression; the FSDB design is appealing and the graphical interface is convenient. More importantly, a few FSDB entries are currently not available in the Recode database.

20.2.3 Programmed Ribosomal Frameshifting Database (PRFDB)

PRFDB (http://cbmgintra.umd.edu/prfdb/) is a recent public resource developed by the group of Jonathan Dinman (Belew et al. 2008). It is a catalogue of mammalian and yeast genes that contain putative −1 frameshifting cassettes comprising a slippery heptamer sequence accompanied by a potential stimulatory 3′ RNA structure, as identified with an algorithm developed by the same group (Jacobs et al. 2007). It has been shown that a subset of such yeast sequences chosen for experimental verification do indeed support efficient −1 ribosomal frameshifting, with the frameshifting efficiency being higher than 50% in certain cases (Jacobs et al. 2007). Despite this supporting evidence, it needs to be taken into account that ribosomal frameshifting has not been verified experimentally for the great majority of the sequences in PRFDB. There is also no evidence either for widespread phylogenetic conservation of the frameshifting cassettes or for conservation of protein products synthesised via frameshifting. While this putative character of the frameshift candidates catalogued in PRFDB is a drawback in comparison to other recoding databases, it is also its advantage. PRFDB is a rich source of sequences containing putative −1 frameshift cassettes which are worthy of further detailed computational and experimental investigations. In this regard, it is particularly useful that the entire database is available in a machine readable format, convenient for further processing.

20.2.4 SelenoDB

SelenoDB (http://www.selenodb.org) is a recently developed database dedicated to information relating to selenoproteins, as well as molecular machinery involved in selenocysteine biosynthesis and co-translational incorporation of selenocysteine into proteins in eukaryotic organisms (Castellano et al. 2008). SelenoDB took advantage of recently developed computational tools for the identification of selenoprotein-encoding genes. Selenocysteine has no dedicated codon in the genetic code of any known organism; instead it is incorporated at certain UGA codons, whose standard role is to signal for termination of translation. A special RNA structure in the 3′ UTRs of eukaryotic mRNAs, termed the SECIS element, is responsible for the redefinition of UGA codons as Sec codons (see Chapters 1 and 2). Certain specific sequence and secondary structure characteristics of SECIS elements allowed the development of accurate patterns for the prediction of SECIS elements in nucleic acid sequences (Kryukov et al. 1999, 2003; Lescure et al.

1999). SECIS predictions, in turn, have been used to instruct modified gene prediction algorithms to ignore UGA codons as terminators when searching for new selenoprotein genes (Castellano et al. 2001). In addition, evolutionary conservation of selenoproteins, and the existence of homologs with cysteines in positions corresponding to the selenocysteines in selenoproteins, allowed the identification of putative novel selenoproteins using comparative sequence analysis (Castellano et al. 2004). Combination of these independent approaches provided a robust and reliable method for the prediction of selenoproteins (Kryukov et al. 2003). SelenoDB is not limited to selenoproteins; it also contains annotations of genes encoding components of selenocysteine incorporation machinery and biosynthesis. SelenoDB has a convenient graphical interface for searching and browsing and may be used to generate outputs useful for both manual human expert exploration and for flexible computational processing. It should also be noted that SelenoDB allows third party submissions of novel selenoprotein annotations which, in combination with meticulously detailed documentation, creates a solid platform for future collaborative development of this superb resource dedicated to the biology of the 'twenty-first' amino acid.

20.2.5 ISfinder

ISfinder (www-is.biotoul.fr) is not dedicated primarily to recoding events. Instead, it is a database of bacterial Insertion Sequences – providing information on sequences, annotation, names and nomenclature of IS elements (Siguier et al. 2006). The reason, we have decided to mention this database in this chapter is that many IS elements – those of the IS3 family in particular – use programmed ribosomal frameshifting in their expression (Baranov et al. 2006). Therefore, for a number of IS elements that contain overlapping open reading frames, a putative frameshift sequence is annotated in ISfinder. A number of such sequences have unusual characteristics and cannot be found in other databases. Therefore ISfinder can be a useful resource for finding novel prokaryotic shift-prone sequences for further detailed computational and experimental studies.

20.3 Approaches and Methods for Finding Recoded Genes

A number of computer programs specifically designed for identifying new instances of recoding will be described in the next section. Typically, these programs use some combination of three broad approaches: (1) search for genes that bear homology to known genes that use recoding, as in ARFA (Bekaert et al. 2006) and OAF (Bekaert et al. 2008); (2) search for specific signals within the nucleotide sequence that resemble signals known to stimulate various types of recoding, e.g. a X XXY YYZ heptanucleotide followed by a 3′ RNA secondary structure for −1 frameshifting (Hammell et al. 1999; Bekaert et al. 2003; Byun et al. 2007; Jacobs et al. 2007;

Theis et al. 2008) and (3) search for sequences that appear to be coding but that do not have an obvious canonical translation mechanism (e.g. ORFs lacking an AUG codon or 3′ ORFs on bicistronic mRNAs). In this section, however, some of the types of individual software tools that may be combined to make such algorithms will be discussed. We will be interested in tools for homology searching, tools for locating generalised patterns in nucleotide sequences, tools for RNA secondary structure prediction and tools for identifying 'unusual' coding sequences.

Searches for new cases of recoding have often concentrated on regions where two long and apparently coding ORFs overlap or, in the case of stop codon readthrough, abut (Namy et al. 2003; Moon et al. 2004; Wills et al. 2006). Increasingly, however, novel examples of recoding involve one ORF that is relatively short (Lin et al. 2007; Chung et al. 2008; Firth et al. 2008). Longer ORFs are more easily recognisable as coding sequences and, in instances where canonical translation seems problematic, the potential for recoding has often long since been investigated. In general, recoding events involve two ORFs, and either one has the potential to be short. For example, in the bacterial *prfB* gene it is the initial ORF (ORF1) that is short, and recoding (here +1 frameshifting) plays a role in regulating translation of the long frameshift ORF (ORF2) to produce functional release factor 2 (see Chapter 14). Conversely, in the bacterial *dnaX* gene, ORF2 is very short (in *Escherichia coli* it comprises just a single codon) and −1 frameshifting results in generation of a shorter protein product encoded by just the first two-thirds of ORF1. It is believed that the frameshifting mechanism ensures a fixed equimolar ratio of the two subunits of DNA polymerase III – tau (full-length product of standard translation) and gamma (truncated product from ribosomal frameshifting) – reviewed in Baranov et al. (2002a). For such cases, two distinct strategies are particularly useful: (1) to search for phylogenetically conserved known recoding signals in order to find recoding sites where one ORF is too short to identify with gene-finding software (e.g. Firth et al. 2008) and (2) to search for short coding sequences with sensitive gene-finding software, in order to find sites where the recoding signals do not conform to known patterns (e.g. Chung et al. 2008). In both, the key to robust computational detection (and hence well-directed experimental follow-up) is phylogenetic conservation. Some cases of recoding are conserved across vast evolutionary distances, e.g. antizyme +1 frameshifting is present from yeast to vertebrates, though the nature of the stimulatory signals involved may show much more variation (Ivanov and Atkins 2007). Thus comparative approaches are particularly powerful when sequences from a large number of closely related organisms are available – providing a high total divergence (i.e. summed over a phylogenetic tree) but with moderate pairwise divergences (i.e. no individual pair of sequences is too divergent). With the sequence data now available in GenBank and other databases, there are many opportunities for useful comparisons, e.g. *Drosophila* species, mammals, yeast species, higher plants, prokaryote clades and individual virus species.

Nonetheless, it is apparent that not all cases of recoding are conserved over great evolutionary distances. Indeed some cases may be species specific and undetectable by comparative methods. Computationally, such cases may be identified if both ORFs are long or if the recoding signals conform to already well-characterised patterns. As an example, non-comparative computational searches for −1 frameshifting signals in yeast, and other organisms, have revealed hundreds of potential frameshift sites (deposited into PRFDB; see previous section).

20.3.1 Homology Searching

Tools for sequence similarity analysis form an important component of many programs for identifying genes that use recoding, in particular for the systematic identification of potential homologs of known recoded genes. There is a large body of computer programs that can be used for the identification of potential homologs; here we mention only three of them, based mainly on their popularity. Techniques used for the analysis of pairwise sequence similarity can also be used for the identification of similar sequences (potential homologs) in large sequence data sets. However, since speed of computation becomes crucially important when searching large data sets, tools that use heuristic approaches, such as BLAST and FASTA, prevail over tools for constructing optimal alignments using dynamic programming algorithms. A more sensitive and biologically meaningful alternative to sequence similarity searches is provided by methods that use position-specific sequence profiles or probabilistic models created via the analysis of a set of similar sequences, an example being HMMER.

BLAST (Altschul et al. 1990, 1997) is by far the most popular tool for searching for sequence similarities in large data sets. It performs pairwise comparisons of sequences – seeking regions of local similarity – using a heuristic approach that is over 500 times faster than the Smith–Waterman algorithm (Smith and Waterman 1981). BLAST can perform hundreds or even thousands of sequence comparisons in a matter of minutes. The speed and relatively high accuracy of BLAST are among the key technical innovations of the BLAST programs. The original BLAST algorithm can be conceptually divided into three stages. In the first stage, BLAST searches for exact matches of small fixed-length words. Exact matches to these words are known as seeds. In the second stage, BLAST tries to extend the match in both directions, starting at the seed. If a high-scoring ungapped alignment is found, the database sequence passes on to the third stage. In the third stage, BLAST performs a gapped alignment between the query sequence and the database sequence using a variation of the Smith–Waterman algorithm. Statistically significant alignments are then displayed to the user. Due to the popularity of BLAST and its significant role in contemporary biology, an entire book with the same title (Korf 2003) has been dedicated to this outstanding computational tool.

FASTA (Lipman and Pearson 1985) is another heuristic method for local sequence alignment. It is slower than BLAST but still faster than the Smith–Waterman algorithm. The FASTA algorithm first searches for short sequences (called ktups – abbreviation for k tuples, or ordered sequences of k residues) that occur in both the query sequence and the sequence database. Then, the algorithm scores the ungapped alignments that contain the most identical ktups. These ungapped alignments are tested for their ability to be merged into a gapped alignment without reducing the score below a threshold. For those merged alignments that score over the threshold, an optimal local alignment of that region is then computed using a Smith–Waterman type of algorithm. FASTA ktups are shorter than BLAST words. Smaller ktup sizes increases sensitivity at the expense of speed. The current FASTA package contains programs for protein:protein, DNA:DNA and protein:translated DNA (with frameshifts). Recent versions of the package include special translated search algorithms that correctly handle frameshift errors when comparing nucleotide to protein sequence data.

Similarity between two sequences can occur by chance and does not necessarily indicate evolutionary relationship. Furthermore, different positions within a sequence evolve at different speeds due to their differential importance for the biological function. To take this into account, a position-scoring system needs to be used, where similarities between two sequences are scored differently depending on their position. Information on position-specific constraints cannot be derived from a single sequence but instead requires alignment of multiple sequences. PSI-BLAST (Altschul et al. 1997) uses an elegant solution to this problem. It starts with a single sequence to find the best hits, and these are used to build a position-specific scoring system that is used for further iterative searches with increased sensitivity during each subsequent iteration.

A popular alternative is the use of profile-HMMs (Krogh et al. 1994; Eddy 1998), where multiple alignments are described in terms of Hidden Markov Models. The alignment is represented in terms of states (such as match, mismatch, gap, deletion, end or beginning of the sequence) and transitions between the states, which are assigned parametric probability values estimated from an initial alignment. Then a sequence data set can be searched for sequences that match the model. HMMER 2 (http://hmmer.janelia.org/) is a popular computational tool that can be used for the generation of profile-HMM models from sequence alignments as well as for the analysis of large sequence data sets. HMMER is far more sensitive than standard BLAST or FASTA, but significantly slower. For the latter reasons, specialised programs for finding recoding genes, ARFA and OAF, described later in this chapter (both use HMMER as the key internal module), also use a relaxed FASTA search to reduce the sequence data set by excluding the most likely true negatives prior to the HMMER step. According to the HMMER project web page, a third version of the program is currently under development. It is claimed to have a speed comparable to that of BLAST.

20.3.2 Pattern Searching

In contrast to searches for homologs of known recoded genes, an initial search for *novel* recoding candidates may involve a search for particular patterns in nucleotide sequences. An example is the X XXY YYZ slippery heptanucleotide pattern characteristic of many eukaryotic -1 frameshift sites (here XXX represents any three identical nucleotides, YYY represents AAA or UUU and Z represents A, U or C)

(Brierley and Pennell 2001). Another example is the Tobacco mosaic virus-like stop codon readthrough site characterised by UAG CAR YYA (here Rs represent purines and Ys represents pyrimidines) (Skuzeski et al. 1991). There are many programs that can be used to search for particular nucleotide patterns in primary sequence and, in fact, such programs are not difficult to write from scratch for particular cases. By way of example, one commonly used and versatile program is PatScan, a.k.a. scan_for_matches (Dsouza et al. 1997). PatScan allows the user to define any nucleotide pattern using standard IUPAC nucleotide codes and also allows the user to specify repeats, spacer regions, reverse complements (e.g. for RNA hairpin structures) – including user-specified pairing rules (e.g. whether or not to allow G:U base-pairs), user-specified maximum numbers of mismatches, insertions and deletions in pattern-matches, and alternative patterns (i.e. 'or' notation). Patterns can easily include pseudoknots. It should be noted though that PatScan is not the ideal tool for identifying RNA secondary structures since, despite the flexibility, the overall 'global' structure has to be specified a priori and, furthermore, no optimisation takes place – the structure returned (if any) is the first that matches the input pattern. A common method is to select all sites in an input sequence database that match a very general primary pattern, and then use RNA prediction software (see below) to search for sites with appropriate near-by RNA secondary structures. PatScan works on single input sequences rather than alignments, but it is generally a trivial matter to post-process results to select only pattern-matches that align to equivalent pattern-matches in all (or nearly all) sequences of a sequence alignment. Another useful pattern-finding program with similar, but extended, functionality is RNAmotif (Macke et al. 2001).

RNA secondary structures are an integral component of many recoding sites. Searching for further occurrences of very specific structures known to stimulate recoding events may reveal new instances of recoding. Ab initio RNA structure prediction programs are discussed below, but there are also a number of programs that have been designed for locating occurrences of specific user-defined 'query' structures. In effect, such programs are similar to BLAST, FASTA and HMMER, but score homology to the query in both primary sequence and secondary structure. Several such programs are compared in Freyhult et al. (2007). One popular program – Infernal (Nawrocki and Eddy 2007) – uses covariance models (profile stochastic context-free grammars), which may be thought of as an extension of profile-HMMs to include secondary structure (Eddy and Durbin 1994). As with HMMER, Infernal uses a query profile built from a multiple alignment. Another program, RSEARCH (Klein and Eddy 2003), uses a covariance model with a single-sequence RNA structure query and 'RIBOSUM' matrices to score homology in both unpaired and paired regions. Such programs can be much slower than BLAST, FASTA and HMMER, but have higher sensitivity for finding distant RNA secondary structure homologs. Note, however, that neither Infernal nor RSEARCH can handle pseudoknots. Locomotif (Reeder et al. 2007a) is an alternative program that uses a thermodynamic approach and has a convenient graphical interface in which the user may define a motif comprising stems, loops and any required length restrictions and primary sequence motifs.

20.3.3 RNA Structure Prediction

There are now a large number of methods and programs available for RNA secondary structure prediction. Many of these have been reviewed and compared previously (Gardner and Giegerich 2004; Gruber et al. 2008a). Structure prediction programs may be classified in several ways as to (1) whether or not they can predict pseudoknots; (2) whether they fold single sequences or fold a set of homologous sequences; and (3) of those that fold multiple homologous sequences, whether they work on pre-aligned sequences, or fold and align simultaneously, or perform a structural alignment (i.e. independent of primary sequence). A wide range of methods are used – the most popular being minimum free energy prediction (Zuker and Stiegler 1981), including suboptimal solutions, and base-pairing probabilities derived using the partition function (McCaskill 1990). Other methods include mutual information and compensatory mutations, stochastic context-free grammars, folding kinetics, graph theory, and genetic algorithms.

Popular single-sequence methods include Mfold (Zuker 2003) and RNAfold (Hofacker 2003). One problem with single-sequence methods is that, for longer sequences, there can be many near-optimal solutions. Predictions based on such methods may suffer from inaccuracies in the thermodynamic parameters, nucleotide chemical modifications, failure to take account of co-transcriptional folding, intermolecular interactions, etc. Further, the mere potential for existence of a predicted structure, even if relatively stable, gives little indication as to whether or not it is biologically relevant (Rivas and Eddy 2000). Methods that fold alignments give much more powerful predictions, as even a moderate amount (e.g. 10%) of random nucleotide divergence has been shown to disrupt most predicted structures (Schuster et al. 1994). Such programs include RNAalifold (Hofacker et al. 2002), which makes use of thermodynamics and compensatory mutations, and Pfold (Knudsen and Hein 2003), which uses a phylogenetic stochastic context-free grammar (phylo-SCFG). RNA secondary structures that stimulate recoding frequently overlap coding sequences, and thus sequence evolution is constrained not only by the requirement to maintain a functional RNA secondary structure but also by the requirement to maintain a functional peptide sequence (sometimes in multiple reading frames). RNA-DECODER (Pedersen et al. 2004) is a phylo-SCFG-based program that explicitly models coding constraints as well as RNA secondary structure. RNAz [thermodynamic stability and structure conservation; (Washietl et al. 2005b)] and EvoFold [phylo-SCFG; (Pedersen et al. 2006)] have been developed and used for finding local conserved structures in genome-wide screens. Applied to alignments of the human genome with other vertebrates, both programs identified tens of thousands of potential conserved RNA secondary structures (Washietl et al. 2005a; Pedersen et al. 2006). The predictions are accessible on the UCSC Genome Browser (Karolchik et al. 2008). Such structures, where they overlap mRNAs and, in particular, coding sequences, may prove to be a useful source for the identification of recoding candidates. Indeed several of the predictions were mapped to known to SECIS amd SRE elements (see chapter 2) and antizyme frameshift sites.

One drawback with folding pre-aligned sequences is that, given sufficient sequence divergence, conserved secondary structures may shift relative to the primary sequence alignment. Furthermore, more divergent sequences (e.g. <65% pairwise identity) can be difficult to align accurately, and inaccuracies in the alignment can destroy the secondary structure signal (Washietl and Hofacker 2004). Ideally, therefore, one should try to use many sequences but with pairwise identities above ~70%. When sequences are too divergent to accurately align on primary sequence, different approaches should be used. An exact solution for simultaneously folding and aligning a group of sequences is provided by the Sankoff algorithm (Sankoff 1985); however, it is prohibitively expensive in terms of CPU resources. Thus heuristic approaches or additional constraints need to be applied. Examples include FOLDALIGN (Havgaard et al. 2005), Dynalign (Mathews and Turner 2002), CARNAC (Touzet and Perriquet 2004) and LocARNA (Will et al. 2007; Gruber et al. 2008b).

The ViennaRNA package (Hofacker 2003) is a popular suite of RNA secondary structure prediction and related software. The programs can be used via a webserver (Gruber et al. 2008b), but can also be downloaded and easily integrated into recoding site prediction pipelines. The programs include (see above) RNAfold, RNAalifold and RNAz, which run, more or less, in $O(n^3)$ time and $O(n^2)$ memory space (where n is the sequence length). For long alignments (including genomic data) such programs must be run in sliding windows or on discrete segments. However, a major shortcoming of these and many other RNA folding programs is their inability to predict pseudoknots, which are crucial for many types of recoding (see Chapter 7), although potentially one or more non-pseudoknotted stems may still be identified. Prediction of pseudoknots is generally much more CPU intensive so can be impractical on a genomic scale. If, however, potential recoding sites have already been identified through simple pattern scans (see above; possibly restricted to just phylogenetically conserved pattern-matches), then pseudoknot prediction for local sequence becomes more practical (Jacobs et al. 2007). One of the earliest pseudoknot prediction programs, PKNOTS (Rivas and Eddy 1999), runs in $O(n^6)$ time and $O(n^4)$ space and so is only practical for sequences not much longer than a hundred nucleotides. Longer sequences may be analysed using a sliding window, though any pseudoknots that involve longer range base pairing will be missed. There are faster pseudoknot-predicting programs that use heuristic methods to obtain near-optimal solutions and/or are only able to identify more restricted classes of pseudoknots. PknotsRG (Reeder et al. 2007b), for example, runs in $O(n^4)$ time and $O(n^2)$ space. PknotsRG also has an efficient sliding window mode, can predict suboptimal structures, and has been modified for locating -1 frameshift candidates (Theis et al. 2008). PKNOTS and pknotsRG work on single sequences. One interesting and relatively fast pseudoknot-prediction algorithm is ILM [Iterated Loop Matching (Ruan et al. 2004)], which can work on single sequences or pre-aligned sequences, using thermodynamics or comparative information or both. Initially it folds a sequence alignment without allowing pseudoknots. The most stable stem is identified, saved and its component nucleotides are removed from further consideration. Then the remaining alignment is refolded, and the process is repeated until no further compatible stems are found. The final structure, comprising the sum of all the saved stems, may contain pseudoknots but it is not guaranteed to be optimal. ILM runs in $O(\sim n^3)$ time and $O(n^2)$ space.

20.3.4 Coding Potential

A combination of the above approaches can be used for finding instances of recoding that have similarities with previously described cases. In contrast, an approach that does not rely on previous knowledge is the identification of protein coding sequences that are unlikely to be translated in a standard manner. Such sequences may be in-frame extensions of known coding sequences (stop codon readthrough candidates), out-of-frame extensions or internal overlapping coding sequences (frameshift candidates), or may even be disjoint from but proximal to known coding sequences, as in T4 phage gene 60 bypassing (Herr et al. 2000). They are often not annotated in mRNAs, either due to their short length, lack of initiator codon or, in the case of eukaryotes, a common belief that mRNAs are monocistronic.

Short coding sequences pose a particular problem, since the probability of an ORF existing in a random sequence negatively correlates with its length, and also since short ORFs contain insufficient information to reliably assess coding potential. Comparative sequence analysis offers a way forward, for example through the detection of substitution patterns characteristic of protein coding sequences (Lin et al. 2008), such as synonymous substitutions being favoured over non-synonymous substitutions or $K_a/K_s < 1$ (when evolving under purifying selection). These and other methods have been applied on a genomic scale in *Drosophila* (using alignments of up to 12 species), resulting in the identification of four putative frameshift sites, up to 149 readthrough sites, and 135 new ORFs in polycistronic mRNAs, with some detected ORFs as short as 15 codons (Lin et al. 2007). Characteristics such as K_a/K_s are more difficult to interpret when the same sequence encodes multiple proteins in different reading frames (Nekrutenko et al. 2005). Thus, in order to identify short overlapping coding sequences, it is more appropriate to explicitly model the evolutionary constraints imposed by dual coding, as in the gene-finding program MLOGD (Firth and Brown 2005; Firth and Brown 2006). MLOGD has been used to discover short overlapping coding sequences in virus genomes – some of which apparently involve novel cases of recoding (Chung et al. 2008).

Despite various other campaigns to identify 'difficult' coding sequences in yeast (Kellis et al. 2003), vertebrates (Chung et al. 2007), prokaryotes (Harrison et al. 2003) and so on, there are still many taxa with available sequence data that have not yet been analysed in detail. Furthermore, the continually increasing density of phylogenetic sampling will allow the continued discovery of ever shorter coding sequences and coding sequences with shallower phylogenetic distribution, even in the more well-studied taxa such as mammals (e.g. ENCODE Project Consortium 2007). These 'difficult' coding sequences fall into a variety of classes besides recoding candidates. Although hopefully weeded out through phylogenetic comparison, some apparent readthrough or frameshift candidates may be pseudogenes. Apparently bicistronic mRNAs may be translated to produce two separate proteins via leaky scanning, reinitiation, internal ribosome entry or ribosome shunting rather than to produce a fusion protein via frameshifting, readthrough or bypassing. Perhaps most perniciously, though, such phenomena, including many overlapping coding sequences, are very often a result of alternative splice isoforms (Lin et al. 2007; Chung et al. 2007). Another potential source of false recoding candidates is RNA editing. For example, in humans, specific C-to-U editing is known to be used for the generation of stop codons at the mRNA level (UAA) from corresponding sense codons (CAA) in the DNA template (Chen et al. 1987). A-to-I editing is very abundant in humans (Athanasiadis et al. 2004; Kim et al. 2004; Levanon et al. 2004; Li et al. 2009) and, when this occurs at a stop codon, it can mimic readthrough since inosine is read as guanosine by the decoding machinery.

20.4 Computer Programs Specifically Designed for Finding Recoding Events

Here we describe a few computational tools that were developed quite recently specifically to identify recoded genes.

20.4.1 FSFinder

FSFinder (Moon et al. 2004) and its descendant FSFinder 2.0 (Byun, Moon and Han 2007) are programs specifically designed for the identification and visualisation of cases of programmed ribosomal frameshifting. The current version of FSFinder is available at http://wilab.inha.ac.kr/fsfinder2/.

FSFinder uses pattern searching: regions of two overlapping ORFs are searched for patterns characteristic of particular types of frameshift signals. The current version of FSFinder allows flexible user-designed pattern searching, as well as searching for pre-designed patterns – including a −1 frameshift cassette (comprising a X XXY YYZ heptanucleotide followed by a stem loop or RNA pseudoknot structure), an RF2 frameshift cassette (a Shine–Dalgarno-like sequence upstream of a CUU UGA C motif), and an antizyme frameshift cassette (UUU UGA or YCC UGA followed by an RNA structure), as illustrated in Fig. 20.4A. FSFinder produces a graphical output where the positions of these patterns and the ORFs involved are highlighted in a plot that outlines all possible ORFs in the three reading frames, as well as descriptions of full or partial pattern-matches (Fig. 20.4B).

FSFinder has a number of limitations. On the one hand, it generates false negatives in those cases where the real frameshift site does not match the specified pattern. This is not a major problem when the frameshifting cassette is strictly conserved as in the case of RF2 genes (Baranov et al. 2002b), but it can be

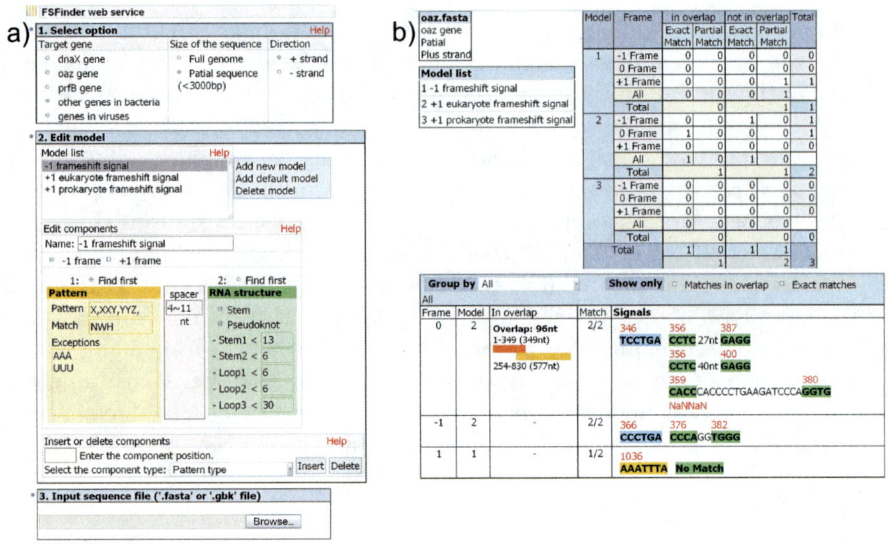

Fig. 20.4 FSFinder interface and output. (**a**) The interface to FSFinder – allowing one to upload a sequence file and to choose model parameters for the analysis. (**b**) The output of FSFinder. A plot of ORFs is shown at the top; features in the plot are described in the same way as in Fig. 20.3. A description of putative signals involved in the frameshifting event is given below. Here, it can be seen that four alternative structures can be predicted downstream of the shift site

a substantial problem in those cases where there is significant diversity in the frameshifting cassette, as in antizyme genes from distant organisms. On the other hand, a source of false positives derives from the fact that there are almost no universal frameshift-prone patterns; thus the same pattern cannot always be used to search for frameshifting candidates in all organisms (Ivanov and Atkins 2007). In addition, it needs to be noted that FSFinder only works under Microsoft Windows (it requires the Microsoft .Net framework). However, the web interface of FSFinder2 can be used online and therefore is platform independent. Another limitation is that, while FSFinder provides a friendly user-oriented interface, it is inconvenient for high-throughput applications and there is no easy way to embed it into more complex computational pipelines. Despite these limitations, FSFinder is a unique handy computational tool that can be used for the exploration of pattern occurrences in regions of ORF overlaps and for the prediction of potential instances of programmed ribosomal frameshifting.

20.4.2 ARFA

Automated Release Factor Annotation (ARFA) is a software tool for the identification of genes encoding class I bacterial release factors (Bekaert et al. 2006). ARFA analyses a given sequence for the presence of sequences encoding one of the three known class-I release factors, RF1, RF2 and RFH. RF2 is encoded in two overlapping ORFs and requires programmed ribosomal frameshifting to take place during translation (see Chapter 19). ARFA can detect the entire RF2 coding sequence and discriminate between RF2s encoded by a single ORF and RF2s encoded by overlapping ORFs. Precise detection of the frameshifting cassette is trivial since it is always located at the end of the first ORF (the ORF1 stop codon is an essential regulatory component and can be used as a reference point).

ARFA uses two external modules, FASTA and HMMER (see previous section). FASTA is applied only for large sequence data sets (>20 kb) and is used as a rapid front-end filter to reduce the number of unrelated sequences. In the subsequent step, a slower but more sensitive HMMER search is applied, using models created for different release factor paralogs and separate models for each of the two RF2 ORFs. HMMER is also used to produce matching scores for the frameshifting cassette, using an HMM built from the nucleotide sequences of previously known RF2 frameshift sites. Low-scoring frameshift sites are in fact of particular interest, since they may incorporate unusual features, such as a non-UGA stop codon, a non-CUU slippery codon, or a skewed Shine–Dalgarno sequence (see Chapter 14).

ARFA can be used via a web interface (http://recode.ucc.ie/arfa or mirror site http://recode.genetics.utah.edu/arfa; limited capacity) or the source code may be downloaded and used locally for large-scale analyses or incorporation into genome analysis pipelines. As input, ARFA takes either sequences in FASTA format or GenBank accession numbers. It outputs annotation of the release factor coding sequence, including annotation of the frameshift site in RF2, in either GenBank or XML formats (see last section). In terms of performance, ARFA correctly discriminates between release factor paralogs and detects RF2 frameshifting sites with virtually 100% specificity and sensitivity. For comparison, analysis of bacterial genome annotations available in 2006 demonstrated that only about 20% of RF2 genes using ribosomal frameshifting were correctly annotated. This emphasises the importance of integration of such simple but highly accurate tools as ARFA and OAF (described below) into the standard pipelines for genome annotation and analysis.

20.4.3 OAF

Ornithine Decarboxylase Antizyme Finder (OAF) is a specialised software tool for the identification of genes encoding antizyme and for annotating the site of programmed ribosomal frameshifting therein (Bekaert et al. 2008). The function of antizyme and the role of programmed ribosomal frameshifting in regulation of its expression are described in Chapter 13.

Conceptually OAF is similar to ARFA (described above). It also uses a combination of FASTA and HMMER searches, where the HMM models are derived from the two ORFs composing the antizyme coding sequence. OAF uses a set of several HMM models that are also used to discriminate between antizyme paralogs and between major phylogenetic groups. The latter feature is particularly useful in the analysis of data derived from EST projects for particular species, since it allows the identification of sequences potentially deriving from contaminant organisms.

OAF also uses nearly the same input/output scheme as ARFA, and performs with nearly the same accuracy. Similarly, OAF allows online (limited capacity) and local (full version) analyses. The web interface and source code for OAF are available at http://recode.ucc.ie/oaf and http://recode.genetics.utah.edu/oaf.

20.4.4 SECISearch

The existence of a special RNA signal, i.e. SECIS (see Chapter 1 and 2), that specifies recoding of UGA codons as selenocysteine in selenoprotein mRNAs, and its specific sequence and structural characteristics in eukaryotes, allows the identification of potential selenoprotein-encoding genes via searches for RNA structures similar to known SECIS elements within the 3′ UTRs of eukaryotic mRNAs. SECISearch has been developed for just this purpose (Kryukov, Kryukov and Gladyshev 1999; Kryukov et al. 2003). The SECISearch 2.0 web interface is available at http://genome.unl.edu/SECISearch.html.

SECISearch is based on a combination of two modules. The first module uses PatScan to search for sequence patterns matching known specific sequence characteristics of SECIS elements, such as the conserved quadruplet within the SECIS structure and the potential for complementary interactions to form the characteristic RNA secondary structure for SECIS. The second module uses the ViennaRNA package to build the entire RNA structure and calculate its free energy. SECISearch allows the user to specify a number of parameters during the search. The sequence pattern corresponding to the conserved quadruplet and the secondary RNA structure consensus used for the pattern search can be specified in several ways (modes) allowing very restrictive searches of the most conserved consensus SECIS structures or more relaxed searches allowing a larger number of deviations from the consensus structure.

Clearly the choice of a mode provides a trade-off between a lower number of false positives but a higher number of false negatives (strict modes) and an increased number of false positives but decreased number of false negatives (loose modes). In addition, the program allows four optional filters that eliminate potential structures with parameters that are known to be incompatible with the SECIS, for example the Y-filter eliminates Y-shaped structures and the S-filter eliminates structures with stems shorter than 8 base pairs. It is also possible for the user to specify the free energy threshold.

The existence of a structure similar to SECIS does not guarantee that the subject mRNA encodes a selenoprotein, since the structure may also appear randomly in sequences. However, a combination of SECISearch with an analysis of homologous coding sequences provides an efficient and reliable method for the identification of selenoprotein-encoding genes.

20.4.5 FreqAnalysis

FreqAnalysis (available at http://gesteland.genetics.utah.edu/freqAnalysis/) is a program for analysing the distribution of *k*-length patterns (or *k*-mers) in nucleotide sequences in a reading frame-dependent manner for the purpose of identifying aberrantly rare patterns (Shah et al. 2002). Few relatively short sequences are known to be able to promote efficient ribosomal frameshifting in particular species. Perhaps the best example is a heptamer frameshifting site in Ty1 transposase from *Saccharomyces cerevisiae*. The sequence is a combination of two codons in two alternative frames C.UU A.GG C, where spaces separate codons in the initial frame and dots in the +1 frame after frameshifting. The efficiency of frameshifting (in *S. cerevisiae*) at this sequence is about 40% (Belcourt and Farabaugh 1990). The authors of FreqAnalysis have reasoned that sequences like this might be deleterious if they occur in a gene that requires standard decoding, since so highly efficient frameshifting would result in the synthesis of a large amount of aberrant truncated proteins [but cf. (Jacobs et al. 2007)]. Therefore, such sequences should be avoided in protein coding regions of genomes. They further reasoned that if there are sequences that have similar frameshift-prone properties then they should also, as a general rule, be avoided. Therefore identification of aberrantly underrepresented sequences in the genome may lead to the identification of patterns that are prone to ribosomal frameshifting. This was the idea behind the design of the program FreqAnalysis. To identify aberrantly underrepresented patterns in coding sequences, FreqAnalysis takes sequences either in GenBank or FASTA format and compares the abundance of different heptamers with expected values (taking into account codon usage bias).

FreqAnalysis was tested on the genome of *S. cerevisiae* and a number of aberrantly underrepresented heptamers were identified, several of which had been previously shown to induce ribosomal frameshifting. Experimental analysis of ribosomal frameshifting on newly identified heptamers demonstrated that they are indeed capable of promoting +1 ribosomal frameshifting at an efficiency greatly exceeding background levels of frameshifting errors (Shah et al. 2002).

FreqAnalysis alone cannot be used for the prediction of recoding events, as experimental investigation of detected candidate patterns is essential. Analyses of other frameshift-prone patterns, such as those collected in PRFDB (Jacobs et al. 2007) and also A AA.A AA.G (Gurvich et al. 2003) (which triggers −1 frameshifting in *E. coli*) have demonstrated that, even though some negative selection against such sequences is evident, overall they are still moderately abundant. Perhaps the level of negative selection and consequent rarity of shift-prone sequences depends on the efficiency of frameshifting, in which case FreqAnalysis would be suitable only for the detection of highly efficient frameshift-prone sequences. On the other hand, since the rarity of particular sequences relates to a number of other biases besides codon bias, it is possible that certain rare sequences may turn out not to be shift-prone at all.

Nonetheless FreqAnalysis is a powerful tool for the identification of potentially shift-prone patterns. It allows the analysis of codon patterns of different types, not necessarily zero-phase heptamers as described above. The analysis of genomes by FreqAnalysis has been performed only for *S. cerevisiae*, where it produced interesting results. Further application of FreqAnalysis to other genomes could yield new taxon-specific shift-prone patterns, and follow-up analyses may reveal novel recoded genes.


```
- <arfaxml version="1.1">
  - <analysis>
    - <program>
        <prog-name>ARFA.pm</prog-name>
        <prog-version>1.4</prog-version>
      </program>
    - <date>
        <day>29</day>
        <month>10</month>
        <year>2009</year>
      </date>
    - <parameter>
        <evalue>1e-20</evalue>
        <table>11</table>
      </parameter>
    </analysis>
  - <sequence id="my_sequence.seq">
    - <input>
        <seq type="dna" length="1099">ATGTTTGAAATTAATCCGGTAAATAATCGCATTCAGGACCT...</seq>
      </input>
    - <output id="my_sequence">
      - <gene id="my_sequence.1">
          <coord>1..1099</coord>
          <name>prfB</name>
          <seq type="dna" length="1099">ATGTTTGAAATTAATCCGGTAAATAATCGCATTCAGGACC...</seq>
        </gene>
      - <cds id="my_sequence.2">
          <coord>join(1..75,77..1099)</coord>
          <name>prfB</name>
          <note>RF2 ORF0: 1.6e-09 ORF1: 2.2e-278</note>
          <product>Bacterial release factor 2</product>
          <seq type="prt" length="365">MFEINPVNNRIQDLTERSDVLRGYLDYDAKKERLEE...</seq>
          <model hmm="RF2">2.2e-278</model>
          <model hmm="RF2_ORF0">1.6e-09</model>
        </cds>
      - <frameshift id="my_sequence.3" evalue="0.067">
          <sd spacer="3">TAGGGGG</sd>
          <seq type="rna" length="17">TAGGGGGTATCTTTGAC</seq>
          <psite>CTT</psite>
          <asite>TGA</asite>
          <downstream>C</downstream>
        </frameshift>
      </output>
    </sequence>
  </arfaxml>
```


Fig. 20.5 Example of RecodeML output. This listing is an example of a recoding event produced by the ARFA program. For each event, a name or a definition is provided along with a unique number to identify the event. This information is contained within the header element; also included in the header element are keywords. The format is easily used as input to further software tools, but is also human readable

20.5 XML Format to Describe Recoding Events

It has been argued in the first section of this chapter that a lack of adequate progress in the identification and annotation of recoded genes is in part due to poor integration of existing methods into standard annotation pipelines. To help overcome this problem, we propose a RecodeML (Recode Markup Language) XML format that we hope will become standard for the description of recoding events, if adopted by others. XML-based formats have a number of advantages compared to other existing standards, since they allow a flexible annotation of a large number of sequence elements (stimulatory signals for instance) including those that are yet to be discovered. For the comprehensive description of advantages of XML-based languages and their current widespread use in bioinformatics resources see Romano (2008).

The listing in Fig. 20.5 is an example of a recoding event described in the proposed RecodeML format, as produced by the ARFA program. For each event, a name or a definition is provided, along with a unique number to identify the event. This information is contained within the header element; also included in the header element are keywords. Note how each keyword is contained within its own <keyword> tag. This makes data exchange and keyword-based searching much more efficient. The full Document Type Definition file is available at http://recode.ucc.ie/RecodeML.dtd.

Acknowledgments We are grateful to Drs. Sergi Castellano and Kyungsook Han for careful reading of the manuscript and useful comments. This work was supported by funds from Science Foundation Ireland.

References

Altschul SF, Gish W, Miller W, Myers EW, Lipman DJ (1990) Basic local alignment search tool. J Mol Biol 215:403–410

Altschul SF, Madden TL, Schaffer AA, Zhang J, Zhang Z, Miller W, Lipman D J (1997) Gapped BLAST and PSI-BLAST: a new generation of protein database search programs. Nucl Acids Res 25:3389–3402

Athanasiadis A, Rich A, Maas S (2004) Widespread A-to-I RNA editing of Alu-containing mRNAs in the human transcriptome. PLoS Biol 2:e391

Baranov PV, Fayet O, Hendrix RW, Atkins JF (2006) Recoding in bacteriophages and bacterial IS elements. Trends Genet 22:174–181

Baranov PV, Gesteland RF, Atkins JF (2002a) Recoding: translational bifurcations in gene expression. Gene 286:187–201

Baranov PV, Gesteland RF, Atkins JF (2002b) Release factor 2 frameshifting sites in different bacteria. EMBO Rep 3:373–377

Baranov PV, Gurvich OL, Fayet O, Prere MF, Miller WA, Gesteland RF, Atkins JF, Giddings MC (2001) RECODE: a database of frameshifting, bypassing and codon redefinition utilized for gene expression. Nucl Acids Res 29:264–267

Baranov PV, Gurvich OL, Hammer AW, Gesteland RF, Atkins JF (2003) Recode 2003. Nucl Acids Res 31:87–89

Bekaert M, Atkins JF, Baranov PV (2006) ARFA: a program for annotating bacterial release factor genes, including prediction of programmed ribosomal frameshifting. Bioinformatics 22: 2463–2465

Bekaert M, Bidou L, Denise A, Duchateau-Nguyen G, Forest JP, Froidevaux C, Hatin I, Rousset JP, Termier M (2003) Towards a computational model for -1 eukaryotic frameshifting sites. Bioinformatics 19:327–335

Bekaert M, Firth AE, Zhang Y, Gladyshev VN, Atkins JF, Baranov PV (2009) Recode-2: new design, new search tools, and many more genes. Nucl Acids Res e-pul ahead of print

Bekaert M, Ivanov IP, Atkins JF, Baranov PV (2008) Ornithine decarboxylase antizyme finder (OAF): fast and reliable detection of antizymes with frameshifts in mRNAs. BMC Bioinformatics 9:178

Belcourt MF, Farabaugh PJ (1990) Ribosomal frameshifting in the yeast retrotransposon Ty: tRNAs induce slippage on a 7 nucleotide minimal site. Cell 62:339–352

Belew AT, Hepler NL, Jacobs JL, Dinman JD (2008) PRFdb: a database of computationally predicted eukaryotic programmed −1 ribosomal frameshift signals. BMC Genomics 9:339

Brierley I, Pennell S (2001) Structure and function of the stimulatory RNAs involved in programmed eukaryotic-1 ribosomal frameshifting. Cold Spr Harb Symp Quant Biol 66: 233–248

Byun Y, Han K (2006) PseudoViewer: web application and web service for visualizing RNA pseudoknots and secondary structures. Nucl Acids Res 34:W416–W422

Byun Y, Moon S, Han K (2007) A general computational model for predicting ribosomal frameshifts in genome sequences. Comput Biol Med 37:1796–1801

Castellano S, Gladyshev VN, Guigo R, Berry MJ (2008) SelenoDB 1.0 : a database of selenoprotein genes, proteins and SECIS elements. Nucl Acids Res 36:D332–338

Castellano S, Morozova N, Morey M, Berry MJ, Serras F, Corominas M, Guigo R (2001) In silico identification of novel selenoproteins in the *Drosophila melanogaster* genome. EMBO Rep 2:697–702

Castellano S, Novoselov SV, Kryukov GV, Lescure A, Blanco E, Krol A, Gladyshev VN, Guigo R (2004) Reconsidering the evolution of eukaryotic selenoproteins: a novel nonmammalian family with scattered phylogenetic distribution. EMBO Rep 5:71–77

Chen SH, Habib G, Yang CY, Gu ZW, Lee BR, Weng SA, Silberman SR, Cai SJ, Deslypere JP, Rosseneu M et al. (1987) Apolipoprotein B-48 is the product of a messenger RNA with an organ-specific in-frame stop codon. Science 238:363–366

Chung BY, Miller WA, Atkins JF, Firth AE (2008) An overlapping essential gene in the Potyviridae. Proc Nat Acad Sci USA 105:5897–5902

Chung WY, Wadhawan S, Szklarczyk R, Pond SK, Nekrutenko A (2007) A first look at ARFome: dual-coding genes in mammalian genomes. PLoS Comput Biol 3:e91

Dsouza M, Larsen N, Overbeek R (1997) Searching for patterns in genomic data. Trends Genet 13:497–498

Eddy SR (1998) Profile hidden Markov models. Bioinformatics 14:755–763

Eddy SR, Durbin R (1994) RNA sequence analysis using covariance models. Nucl Acids Res 22:2079–2088

Firth AE, Brown CM (2005) Detecting overlapping coding sequences with pairwise alignments. Bioinformatics 21:282–292

Firth AE, Brown CM (2006) Detecting overlapping coding sequences in virus genomes. BMC Bioinformatics 7:75

Firth AE, Chung BY, Fleeton MN, Atkins JF (2008) Discovery of frameshifting in Alphavirus 6K resolves a 20-year enigma. Virol J 5:108

Freyhult EK, Bollback JP, Gardner PP (2007) Exploring genomic dark matter: a critical assessment of the performance of homology search methods on noncoding RNA. Genome Res 17: 117–125

Gardner PP, Giegerich R (2004) A comprehensive comparison of comparative RNA structure prediction approaches. BMC Bioinformatics 5:140

Gruber AR, Bernhart SH, Hofacker IL, Washietl S (2008a) Strategies for measuring evolutionary conservation of RNA secondary structures. BMC Bioinformatics 9:122

Gruber AR, Lorenz R, Bernhart SH, Neubock R, Hofacker IL (2008b) The vienna RNA websuite. Nucl Acids Res 36:W70–74

Gurvich OL, Baranov PV, Zhou J, Hammer AW, Gesteland RF, Atkins JF (2003) Sequences that direct significant levels of frameshifting are frequent in coding regions of *Escherichia coli*. EMBO J 22:5941–5950

Hammell AB, Taylor RC, Peltz SW, Dinman JD (1999) Identification of putative programmed -1 ribosomal frameshift signals in large DNA databases. Genome Res 9:417–427

Han K, Byun Y (2003) PSEUDOVIEWER2: Visualization of RNA pseudoknots of any type. Nucl Acids Res 31:3432–3440

Han K, Lee Y, Kim W (2002) PseudoViewer: automatic visualization of RNA pseudoknots. Bioinformatics 18(Suppl 1):S321–S328

Harrison PM, Carriero N, Liu Y, Gerstein M (2003) A "polyORFomic" analysis of prokaryote genomes using disabled-homology filtering reveals conserved but undiscovered short ORFs. J Mol Biol 333:885–892

Havgaard JH, Lyngso RB, Gorodkin J (2005) The FOLDALIGN web server for pairwise structural RNA alignment and mutual motif search. Nucl Acids Res 33:W650–653

Herr AJ, Atkins JF, Gesteland RF (2000) Coupling of open reading frames by translational bypassing. Annu Rev Biochem 69:343–372

Hofacker IL (2003) Vienna RNA secondary structure server. Nucl Acids Res 31:3429–3431

Hofacker IL, Fekete M, Stadler PF (2002) Secondary structure prediction for aligned RNA sequences. J Mol Biol 319:1059–1066

Ivanov IP, Atkins JF (2007) Ribosomal frameshifting in decoding antizyme mRNAs from yeast and protists to humans: close to 300 cases reveal remarkable diversity despite underlying conservation. Nucl Acids Res 35:1842–1858

Jacobs JL, Belew AT, Rakauskaite R, Dinman JD (2007) Identification of functional, endogenous programmed -1 ribosomal frameshift signals in the genome of *Saccharomyces cerevisiae*. Nucl Acids Res 35:165–174

Karolchik D, Kuhn RM, Baertsch R, Barber GP, Clawson H, Diekhans M, Giardine B, Harte RA, Hinrichs AS, Hsu F, Kober KM, Miller W, Pedersen JS, Pohl A, Raney BJ, Rhead B, Rosenbloom KR, Smith KE, Stanke M, Thakkapallayil A, Trumbower H, Wang T, Zweig AS, Haussler D, Kent WJ (2008) The UCSC Genome Browser Database: 2008 update. Nucl Acids Res 36:D773–779

Kellis M, Patterson N, Endrizzi M, Birren B, Lander ES (2003) Sequencing and comparison of yeast species to identify genes and regulatory elements. Nature 423:241–254

Kim DD, Kim TT, Walsh T, Kobayashi Y, Matise TC, Buyske S, Gabriel A (2004) Widespread RNA editing of embedded alu elements in the human transcriptome. Genome Res 14: 1719–1725

Klein RJ, Eddy SR (2003) RSEARCH: finding homologs of single structured RNA sequences. BMC Bioinformatics 4:44

Knudsen B, Hein J (2003) Pfold: RNA secondary structure prediction using stochastic context-free grammars. Nucl Acids Res 31:3423–3428

Korf I, Yandell M, Bedell J (2003) BLAST: O'Reilly and Associates Inc

Krogh A, Brown M, Mian IS, Sjolander K, Haussler D (1994) Hidden Markov models in computational biology. Applications to protein modeling. J Mol Biol 235: 1501–1531

Kryukov GV, Castellano S, Novoselov SV, Lobanov AV, Zehtab O, Guigo R, Gladyshev VN (2003) Characterization of mammalian selenoproteomes. Science 300:1439–1443

Kryukov GV, Kryukov VM, Gladyshev VN (1999) New mammalian selenocysteine-containing proteins identified with an algorithm that searches for selenocysteine insertion sequence elements. J Biol Chem 274:33888–33897

Lescure A, Gautheret D, Carbon P, Krol A (1999) Novel selenoproteins identified in silico and in vivo by using a conserved RNA structural motif. J Biol Chem 274: 38147–38154

Levanon EY, Eisenberg E, Yelin R, Nemzer S, Hallegger M, Shemesh R, Fligelman ZY, Shoshan A, Pollock SR, Sztybel D, Olshansky M, Rechavi G, Jantsch MF (2004) Systematic identification of abundant A-to-I editing sites in the human transcriptome. Nature Biotech 22: 1001–1005

Li JB, Levanon EY, Yoon JK, Aach J, Xie B, Leproust E, Zhang K, Gao Y, Church GM (2009) Genome-wide identification of human RNA editing sites by parallel DNA capturing and sequencing. Science 324:1210–1213

Lin MF, Carlson JW, Crosby MA, Matthews BB., Yu C, Park S, Wan KH, Schroeder AJ, Gramates LS, St Pierre SE, Roark M, Wiley KL Jr, Kulathinal RJ, Zhang P, Myrick KV, Antone JV, Celniker SE, Gelbart WM, Kellis M (2007) Revisiting the protein-coding gene catalog of *Drosophila melanogaster* using 12 fly genomes. Genome Res 17:1823–1836

Lin MF, Deoras AN, Rasmussen MD, Kellis M (2008) Performance and scalability of discriminative metrics for comparative gene identification in 12 Drosophila genomes. PLoS Computat Biol 4:e1000067

Lipman DJ, Pearson WR (1985) Rapid and sensitive protein similarity searches. Science 227:1435–1441

Macke TJ, Ecker DJ, Gutell RR, Gautheret D, Case DA, Sampath R (2001) RNAMotif, an RNA secondary structure definition and search algorithm. Nucleic Acids Res 29:4724–4735

Mathews DH, Turner DH (2002) Dynalign: an algorithm for finding the secondary structure common to two RNA sequences. J Mol Biol 317:191–203

Matsufuji S, Matsufuji T, Miyazaki Y, Murakami Y, Atkins JF, Gesteland RF, Hayashi S (1995) Autoregulatory frameshifting in decoding mammalian ornithine decarboxylase antizyme. Cell 80:51–60

McCaskill JS (1990) The equilibrium partition function and base pair binding probabilities for RNA secondary structure. Biopolymers 29:1105–1119

Moon S, Byun Yand Han K (2007) FSDB: a frameshift signal database. Computat Biol Chem 31:298–302

Moon S, Byun Y, Kim HJ, Jeong S, Han K (2004) Predicting genes expressed via -1 and +1 frameshifts. Nucl Acids Res 32:4884–4892

Namy O, Duchateau-Nguyen G, Hatin I, Hermann-Le Denmat S, Termier M, Rousset JP (2003) Identification of stop codon readthrough genes in *Saccharomyces cerevisiae*. Nucl Acids Res 31:2289–2296

Nawrocki EP, Eddy SR (2007) Query-dependent banding (QDB) for faster RNA similarity searches. PLoS Computat Biol 3:e56

Nekrutenko A, Wadhawan S, Goetting-Minesky P, Makova KD (2005) Oscillating evolution of a mammalian locus with overlapping reading frames: an XLalphas/ALEX relay. PLoS Genetics 1:e18

Pedersen JS, Bejerano G, Siepel A, Rosenbloom K, Lindblad-Toh K, Lander ES, Kent J, Miller W, Haussler D (2006) Identification and classification of conserved RNA secondary structures in the human genome. PLoS Computat Biol 2:e33

Pedersen JS, Meyer IM, Forsberg R, Simmonds P, Hein J (2004) A comparative method for finding and folding RNA secondary structures within protein-coding regions. Nucl Acids Res 32: 4925–4936

Reeder J, Giegerich R (2004) Design, implementation and evaluation of a practical pseudoknot folding algorithm based on thermodynamics. BMC Bioinformatics 5:104

Reeder J, Reeder J, Giegerich R (2007a) Locomotif: from graphical motif description to RNA motif search. Bioinformatics 23:i392–400

Reeder J, Steffen P, Giegerich R (2007b) pknotsRG: RNA pseudoknot folding including near-optimal structures and sliding windows. Nucl Acids Res 35:W320–324

Rivas E, Eddy SR (1999) A dynamic programming algorithm for RNA structure prediction including pseudoknots. J Mol Biol 285:2053–2068

Rivas E, Eddy SR (2000) Secondary structure alone is generally not statistically significant for the detection of noncoding RNAs. Bioinformatics 16:583–605

Romano P (2008) Automation of in-silico data analysis processes through workflow management systems. Briefings Bioinformat 9:57–68

Ruan J, Stormo GD, Zhang W (2004) ILM: a web server for predicting RNA secondary structures with pseudoknots. Nucl Acids Res 32:W146–149

Sankoff D (1985) Simultaneous solution of the RNA folding, alignment and protosequence problems. Siam J Appl Math 45:810–825

Schuster P, Fontana W, Stadler PF, Hofacker IL (1994) From sequences to shapes and back: a case study in RNA secondary structures. Proc Royal Soc London B 255:279–284

Shah AA., Giddings MC, Parvaz JB, Gesteland RF, Atkins JF, Ivanov IP (2002) Computational identification of putative programmed translational frameshift sites. Bioinformatics 18: 1046–1053

Siguier P, Perochon J, Lestrade L, Mahillon J, Chandler M (2006) ISfinder: the reference centre for bacterial insertion sequences. Nucl Acids Res 34:D32–36

Skuzeski JM, Nichols LM, Gesteland RF, Atkins JF (1991) The signal for a leaky UAG stop codon in several plant viruses includes the two downstream codons. J Mol Biol 218:365–373

Smith TF, Waterman MS (1981) Identification of common molecular subsequences. J Mol Biol 147:195–197

Strabo, Hamilton HC, Falconer W (1854) The geography of Strabo. H. G. Bohn, London

Theis C, Reeder J, Giegerich R (2008) KnotInFrame: prediction of -1 ribosomal frameshift events. Nucl Acids Res 36:6013–6020

Touzet H, Perriquet O (2004) CARNAC: folding families of related RNAs. Nucl Acids Res 32:W142–W145

van Batenburg FH, Gultyaev AP, Pleij CW (2001) PseudoBase: structural information on RNA pseudoknots. Nucl Acids Res 29:194–195

Washietl S, Hofacker IL (2004) Consensus folding of aligned sequences as a new measure for the detection of functional RNAs by comparative genomics. J Mol Biol 342:19–30

Washietl S, Hofacker IL, Lukasser M, Huttenhofer A, Stadler PF (2005a) Mapping of conserved RNA secondary structures predicts thousands of functional noncoding RNAs in the human genome. Nature Biotech 23:1383–1390

Washietl S, Hofacker IL, Stadler PF (2005b) Fast and reliable prediction of noncoding RNAs. Proc Natl Acad Sci USA 102:2454–2459

Will S, Reiche K, Hofacker IL, Stadler PF, Backofen R (2007) Inferring noncoding RNA families and classes by means of genome-scale structure-based clustering. PLoS Computat Biol 3:e65

Wills NM, Moore B, Hammer A, Gesteland RF, Atkins JF (2006) A functional -1 ribosomal frameshift signal in the human paraneoplastic Ma3 gene. J Biol Chem 281:7082–7088

Zuker M (2003) Mfold web server for nucleic acid folding and hybridization prediction. Nucl Acids Res 31:3406–3415

Zuker M, Stiegler P (1981) Optimal computer folding of large RNA sequences using thermodynamics and auxiliary information. Nucl Acids Res 9:133–148

Index

J.F. Atkins, R.F. Gesteland (ed.), *Recoding: Expansion of Decoding Rules Enriches Gene Expression*, Nucleic Acids and Molecular Biology 24,
DOI 10.1007/978-0-387-89382-2, © Springer Science+Business Media, LLC 2010